Structure and Dynamics of Asymmetric Interactions

Naohito Chino

Structure and Dynamics of Asymmetric Interactions

Springer

Naohito Chino
Emeritus Professor
Aichi Gakuin University
Nagoya, Aichi, Japan

ISBN 978-981-97-8268-0 ISBN 978-981-97-8269-7 (eBook)
https://doi.org/10.1007/978-981-97-8269-7

This Springer imprint is published by the registered company Springer Nature Singapore Pte Ltd.
The registered company address is: 152 Beach Road, #21-01/04 Gateway East, Singapore 189721, Singapore

This book is dedicated with love and affection
to my wife, Taeko,
and to our sons and daughter
Hitoshi, Kei, and Akane.

Preface

Asymmetric interaction is frequently observed when two or more people or things communicate with or react to each other. For example, one-sided love or hate among members of informal groups is typical examples. A group of hens and cocks show a special asymmetric interaction called pecking order. The number of browsing times as well as the spread of computer viruses from one site to another on the Internet also yields asymmetric interactions. Biosynthetic pathways of proteins in mammals have one-sided paths and cycles. Dopaminegic neurons have complicated asymmetric interactions between various parts of the brain. Hereafter, we shall call people, things, and groups, 'members', but call them 'nodes' in the context of graph theory.

This book integrates the fundamentals of asymmetric multidimensional scaling, spectral graph theory, graph embedding theory, and various dynamical system theories in order to deal with the static and dynamic aspects of asymmetric interactions. In this way, the book provides a comprehensive introduction to theories and methods for analyzing phenomena observed universally in social, behavioral, economical, geographical, biological, neural, chemical reactions, and other networks. The topics addressed here include the notions of asymmetric similarity matrices, graph spectra, dimension reduction, and difference and differential equations to describe the dynamics of networks, the bifurcation of vector fields, Mandelbrot sets, fractals and chaos, and Hilbert spaces. Illustrated by carefully chosen examples and supported by extensive simulation studies, this book is highly recommended to readers who seek to discover static asymmetric structures among members or nodes. It also appeals to those who want to understand the kinds of dynamics that are theoretically possible in their research domain.

This book is composed of three parts. Part I is concerned with the *structure* of asymmetric interaction, Part II its *dynamics*, and Part III includes supplementary chapters to Parts I and II. After an introduction (Chap. 1) which provides a bird's-eye view of asymmetric interactions among elements, we shall introduce the methods for analyzing the structure of asymmetric interactions based on the author's several papers on asymmetric MDS (concluding in 2012) in Chap. 2 through 7 of Part I. In Part II, we shall introduce the two methods for analyzing the dynamics of asymmetric interactions in Chaps 8 and 9, the former being a bifurcation model of changes in group structure called *Dynamical* System *Scaling* (abbreviated as DYNASCAL) written by the author and one of his colleagues and the latter being an elementary theory of *dynamical weighted digraph* (abbreviated as DWD). DWD is a theory of *complex dynamics* on weighted digraphs, which the author has been developing since 2000.

Since the major objective of this book is to provide applied researchers whose fields are not necessarily mathematics with the theories and methods for analyzing asymmetric interactions, we exclude proofs of all theorems which appear in this book, except for a few cases. Instead, we refer to all the corresponding literature in which proofs appeared.

I am indebted to many researchers in writing this book. First and foremost is the late Kenichi Shiraiwa, who was a source of inspiration and advice. Second is Masanori Nakagawa, who has been one of my colleagues, and also provided useful comments to this book. Third is Yoshio Takane, who provided various comments on my papers for a long time and extended his helpful hand to me frequently. Akinori Okada encouraged me to write this book first and I wish to express my thanks to him. I am also indebted to anonymous reviewers for their valuable suggestions and comments on an earlier version of this book. Gregory Lewis Rohe read the manuscript carefully and polished its English. Nevertheless, I am personally responsible for all the errors in this book.

Yutaka Hirachi, senior editor of Springer Nature Japan, encouraged me to write this book. I extend my deep appreciation to him.

Nagoya, Japan
July 2024

Naohito Chino

Contents

Chapter 1
Bird's-Eye View of Asymmetric Interactions

1.1 Introduction

Asymmetric interactions are frequently observed when two or more people or things communicate with or react to each other. For example, one-sided love and hate among members of any informal group are typical examples [16]. Trade imbalance is a result of asymmetric interactions among nations [16, 18, 19]. The confusion matrix among Morse codes shows asymmetric interactions (e.g., Rothkopf [87]; Shepard [93]). The car switching matrix among car segments shows asymmetric interactions (e.g., Harshman et al. [41]; Okada and Imaizumi [81]). A group of hens and cocks show a special asymmetric interaction called pecking order among the groups in ethology (e.g., Masure and Allee [70]). The spread of computer viruses from one site to another on the Internet also yields asymmetric interactions [4]. Biosynthetic pathways of proteins in mammals have one-sided paths and cycles (e.g., Imai and Guarente [52]). A new model of brain regions involved in learning and memory assumes complicated asymmetric interactions between various parts of the brain (e.g., Meeter et al. [73]). The payoff matrices of "rock-paper-scissors" in the context of the theory of the evolutionally stable strategy in biology represent hypothesized asymmetric interactions among strategies (Chino [18]; Sato et al. [91]). The hyperlink matrix in Google's PageRank algorithms represents asymmetric interactions among web pages [3]. The weight matrix also represents assumed asymmetric interactions among hidden layers in recurrent neural networks (e.g., Goodfellow et al. [34]). Hereafter we shall call people, things, groups, 'members', but call them 'nodes' or "vertices' in the context of graph theory.

Before proceeding, it will be natural and appropriate to consider two aspects of the asymmetric interactions exemplified above. One is its *structure*, and the other its *dynamics*. By the structure, we mean the static aspect of asymmetric interactions among members or nodes. A typical example of the static aspect is represented by an observed or hypothetical *directed graph (digraph)* shown by a diagram. The

N. Chino, *Structure and Dynamics of Asymmetric Interactions*,
https://doi.org/10.1007/978-981-97-8269-7_1

static aspect is also represented by an observed or hypothetical *asymmetric similarity matrix* (hereafter abbreviated as *ASM*) which is associated with a *digraph*, each of whose elements denotes a strength of interaction from one member to another member. A third example is a *latent configuration* of members or nodes in a certain metric space, which is estimated from the corresponding ASM or a digraph using various methods including asymmetric multidimensional scaling (abbreviated as asymmetric MDS), spectral graph theory, and so on. In Sect. 1.2 we shall show a body of examples which represent the static aspect of asymmetric interactions using ASM and digraph, and discuss the relation between them.

Theories and methods for analyzing the *structure of asymmetric interactions* have been developed in many different branches of science. Major ones might be *asymmetric multidimensional scaling* in psychometrics, *spectral graph theory* (abbreviated as SGT) in mathematics, and *graph embedding* (abbreviated as GE) in statistics and machine learning. However, these theories and methods developed in different branches of science have not necessarily referred to each other. As a result, each of them has not succeeded to take over some of the important knowledge developed in the other branches. In Sect. 1.3.1, we shall show typical examples and results of applications of the structure of asymmetric interactions using these theories and methods.

As for the *dynamics of asymmetric interactions*, it will be appropriate first to notice that there exists an interesting equivalence relation among *digraph*, *(recurrent) difference equation*, and *differential equation*, which has been pointed out, for example, by Beer [6]. Stated another way, any digraph can implicitly assume changes in locomotion of nodes over time. Therefore, difference equations as well as differential equations enable us to describe the changes in the digraph over time.

In any case, theories and methods for analyzing the dynamics of asymmetric interactions among members using the difference equation or the differential equation have been developed in several different branches of science, as with those for analyzing the structure of asymmetric interactions. One of them is concerned with the dynamics of *biological networks*. Another is the dynamics of *evolutionally stable strategies in biology and game theory* (e.g., Imai and Guarente [52]; Maynard and Price [71]). A third is the dynamics of neural networks in neural science (e.g., Aizenberg and Aizenberg [2]; Carta et al. [13]; Goodfellow et al. [34]; Hirose [44]; Hopfield [48]). A fourth is the dynamics of *chemical reactions networks in chemistry.* A fifth is the dynamics of *social networks in social and behavioral sciences.* We shall discuss them in Sect. 1.3.2, using typical examples and applications to these dynamics.

1.2 Asymmetric Similarity Matrix and the Corresponding Digraph

In the previous section, we have referred to several examples of asymmetric interactions among elements. These are considered as observations of asymmetric interactions sampled at a particular instant of time. If we consider a *general digraph*, ASM can be viewed as a *weight matrix* associated with the digraph. By a general digraph we mean the weighted digraph with multiple edges and loops. In the following, we shall show various observed ASMs and the corresponding digraphs. We show MATLAB codes for some of these digraphs in the appendix of this chapter.

Example 1.1 (sociometric data) Table 1.1 shows part of the sociomatrix among 10 male students in a senior-high school [16]. This data is measured by a 7-point rating scale. Therefore, the scale is considered to be measured at an interval scale level. The scores range from 1 to 7 for the most negative and positive attraction, respectively. According to this scale, for example, rater 1 somewhat likes rater 2, but rater 2 dislikes rater 1 quite a lot. In contrast, rater 2 likes rater 3 quite a lot, and rater 3 only somewhat likes rater 2.

Figure 1.1 is the digraph associated with this sociomatrix. This figure depicts symmetric and asymmetric relationships among three students vividly. Weight on the loop of each vertex of the graph designates the self-similarity of each student.

Table 1.1 Part of the sociomatrix among students in a senior high, which was adapted from Chino [16]

Rater/ratee	1	2	3
1	4	5	5
2	1	4	7
3	4	5	7

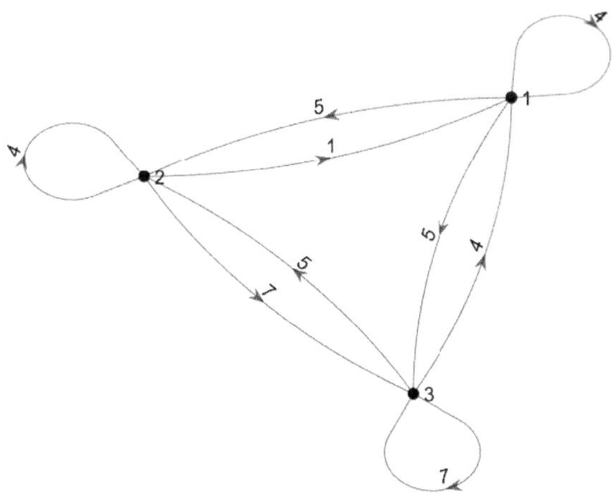

Fig. 1.1 Digraph associated with the sociomatrix shown in Table 1.1

Example 1.2 (international trade data) Table 1.2 shows the international trade data among four nations, and Fig. 1.2 is the corresponding digraph to this table [18]. It is apparent from Fig. 1.2 that the amount of trade imbalance is the greatest between USA and China among four nations in 2015.

Example 1.3 (confusion matrix) Table 1.3 shows part of the confusion matrix among Morse codes in Rothkopf [87] which is shown in Shepard [93]. 598 subjects, who were unfamiliar with the Morse code, were simply asked to judge whether two successively presented signals were the same or different. In this table the row symbols were sent first while the column symbols were sent second. The off-diagonal element in any given cell specifies the percent of the total opportunities that the corresponding pair of signals were judged to be the same, while the diagonal element indicates the percent of the correct response to each symbol.

Figure 1.3 shows the digraph associated with this table. It should be noted that node 2 is located almost at the center of the digraph. It is apparent from Fig. 1.3 that even if the number of nodes is six the resultant digraph becomes complicated.

Example 1.4 (car switching matrix) Table 1.4 shows part of the 1979 car switching matrix among 16 car segments which was reproduced from Okada and Imaizumi [81]. Table 1.4 is a rescaled matrix of the original matrix, Table 4 of Harshman et al. [41], in order to remove the size differences in the entries. Okada and Imaizumi

Table 1.2 The international trade data among Japan, America, China, and Russia in 2015

From/to	1. Japan	2. USA	3. China	4. Russia
1. Japan	43,480	1,382	1,200	55
2. USA	736	189,592	1,161	71
3. China	1,764	4,832	119,684	348
4. Russia	173	164	333	13,755

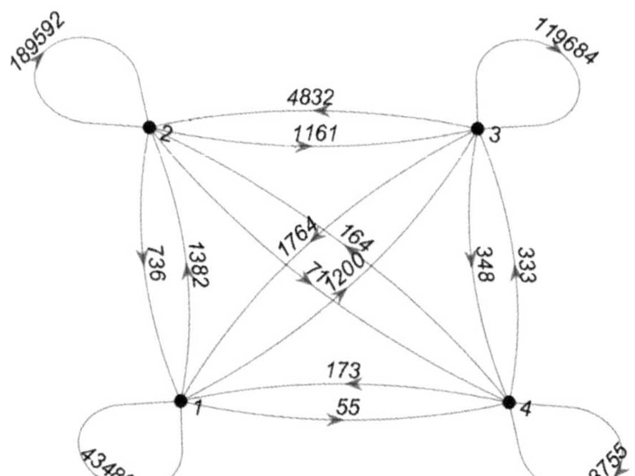

Fig. 1.2 Digraph associated with the trade data in Table 1.2

Table 1.3 Part of the confusion matrix among Morse codes, which was adapted from Rothkoph [87]

From/to	1 (.-)	2 (-…)	3 (-.---)	4 (--..)	5 (----.)	6 (-----)
1 (.-)	92	4	7	3	2	3
2 (-…)	5	84	30	42	4	4
3 (-.---)	9	23	86	23	23	16
4 (--..)	3	46	42	87	15	15
5 (----.)	3	14	21	24	91	78
6 (-----)	9	3	15	20	81	94

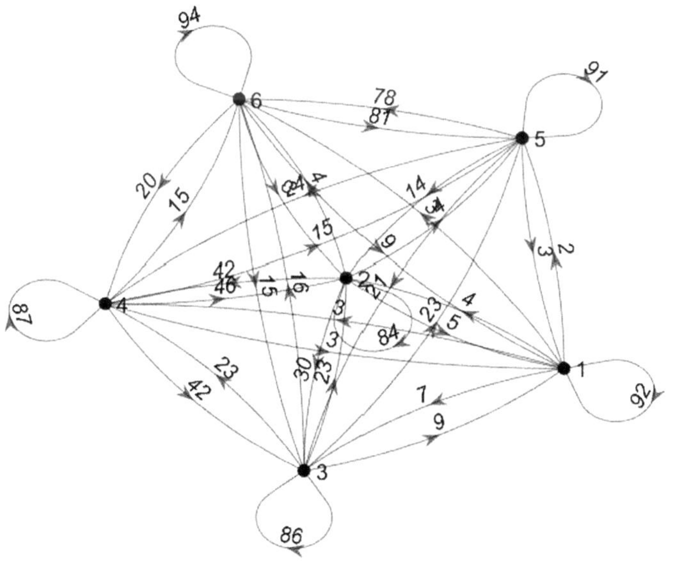

Fig. 1.3 Digraph associated with the confusion matrix in Table 1.3

removed the differences by multiplying each row and column of the original matrix by rescaling coefficients. Figure 1.4 is the corresponding digraph to this table. In Table 1.4, SUBD, SUBC, COMI, STDL, and LUXI, denote subcompact/domestic, subcompact/captive imports, import compact, low price standard, and luxury imports, respectively.

Example 1.5 (farmyard data of poultry) Table 1.5 shows part of the pecking orders of domestic chickens in matrix notation, which was adapted from Table I of Masure and Allee [70], and Fig. 1.5 is the corresponding digraph to this table. In the original table the status of each individual in a flock of 13 brown leghorn pullets is shown on the basis of the peck-order. These pullets were observed for 60 days in order to determine their peck-order.

Table 1.4 Part of the car switching matrix among 16 car segments which was adapted from Okada and Imaizumi [81]

From/to	SUBD	SUBC	COMI	STDL	LUXI
SUBD	10,291	4634	1404	2042	797
SUBC	10,140	24,462	4049	3133	1637
COMI	6733	8734	23,064	1270	4646
STDL	13,355	6394	3307	35,429	2045
LUXI	659	1769	6438	1077	278,391

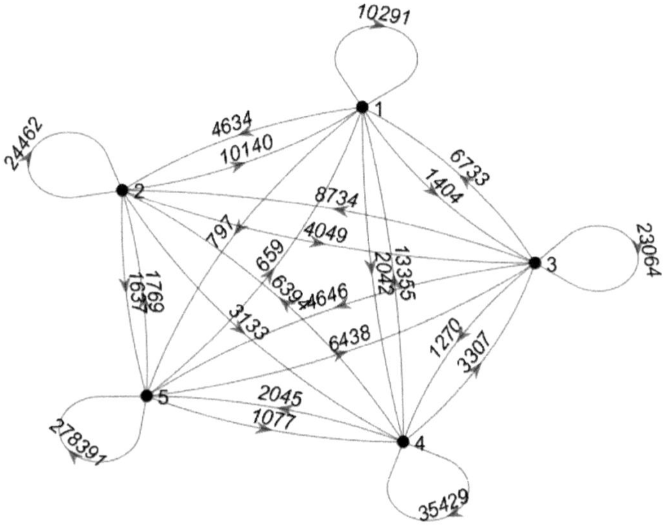

Fig. 1.4 Digraph associated with the car switching matrix shown in Table 1.4

Table 1.5 Part of the pecking orders of domestic chickens, which was adapted from Table I of Masure and Allee [70]

From/to	1	2	3	4	5
1		1	1	1	1
2			1	1	1
3				1	
4					1
5			1		

(Individuals 1, 2, 3, 4, and 5 in Table 1.5 correspond respectively to RW, RR, YY, M, and Y in Table I of Masure and Allee ([70], p. 309).)

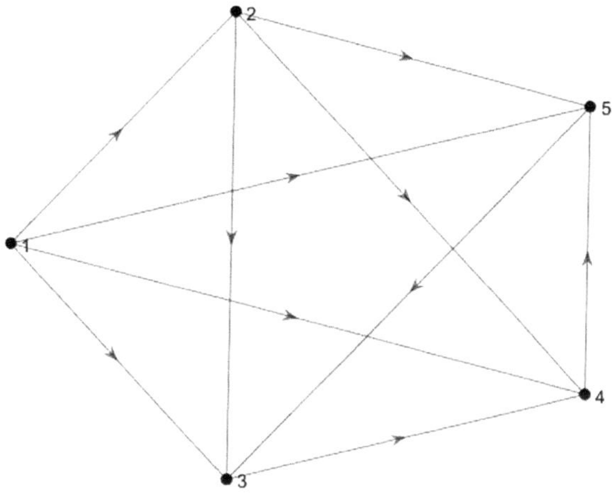

Fig. 1.5 Digraph associated with the ASM of the pecking orders shown in Table 1.5

It is apparent from Fig. 1.5 that individual 1 (RW) is the *despot* of the flock concerning the pecking order. Moreover, there exists a *directed 3-cycle* (e.g., Brouwer and Haemers [10], p. 8) among individuals 3, 4, 5.

Remark Rushen [89] observed the *development of pecking orders* in mixed sex groups of domestic chickens. The birds were observed 3 days a week from hatching until the birds had reached 25 weeks of age. The most reasonable conclusion of this study was that dominance relationships were established after only a short series of encounters, possibly just one. According to Rushen, Schjelderup-Ebbe [92] first described the pecking orders of domestic chickens.

Example 1.6 (biosynthetic data) Table 1.6 shows part of the adjacency matrix constructed from the biosynthetic pathway (digraph) of 11 proteins in mammals in Imai and Guarente [52]. In this table, abbreviated symbols, NAD, NIC, NMN, and NR, denote nicotinamide adenine dinucleotide, nicotinamide, nicotinamide mononucleotide, and nicotinamide riboside, respectively. The original 11×11 adjacency matrix was constructed by Chino ([18, 19]) from Fig. 2b of Imai and Guarente (op. cit.) which shows the NAD^+ biosynthetic pathway in mammals.

Figure 1.6 shows the digraph associated with this adjacency matrix. This graph is a central part of the NAD^+ biosynthetic pathway in Fig. 2b which includes two directed cycles.

As for the adjacency matrix and other matrices related to graph and digraph, we shall briefly define them later in this chapter. For further details, see Sect. 11.4 of Chap. 11.

Example 1.7 (neural network) Table 1.7 shows the adjacency matrix which was constructed from a schematic pathway of brain regions that play a role in incremental learning by Meeter et al. [73], and Fig. 1.7 is the digraph associated with this table.

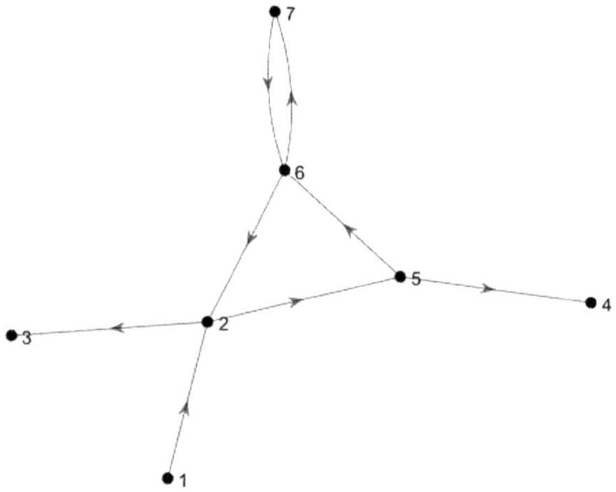

Fig. 1.6 Digraph associated with the adjacency matrix shown in Table 1.6

Table 1.6 Part of the adjacency matrix constructed from the biosynthetic pathway of proteins in mammals

From/to	1	2	3	4	5	6	7
1. Deamido-NAD	0	1	0	0	0	0	0
2. NAD	0	0	1	0	1	0	0
3. O-acetyl-ADP-ribose	0	0	0	0	0	0	0
4. 1-methyl-nicotinamide	0	0	0	0	0	0	0
5. NIC	0	0	0	1	0	1	0
6. NMN	0	1	0	0	0	0	1
7. NR	0	0	0	0	0	1	0

Table 1.7 The adjacency matrix which was constructed from a schematic pathway of the brain regions that play a role in incremental learning by Meeter et al. [73]

From/to	1	2	3	4	5	6	7
Hippocampus		1	1			1	1
Basal ganglia							
Amygdala							
Neocortex		1	1		1	1	
Parahippocampus	1	1	1			1	
Cerebellum							
Septum	1						

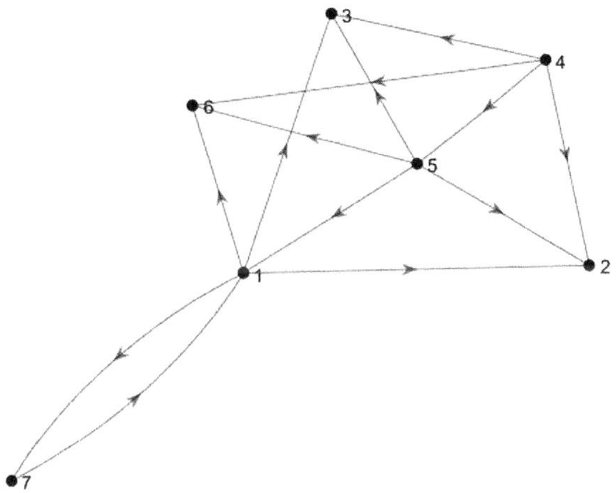

Fig. 1.7 Digraph associated with the schematic pathways shown in Table 1.7

Table 1.8 A pay-off matrix of "rock-paper-scissors", which was adapted from Sato et al. [91]

From/to	1 rock	2 paper	3 scissors
1 rock	0.25	– 1.00	1.00
2 paper	1.00	0.25	– 1.00
3 scissors	– 1.00	1.00	0.25

Example 1.8 (pay-off matrix) Table 1.8 shows a pay-off matrix of a generalized rock-paper-scissors game in Sato et al. [91], where the diagonal elements ε's in Eq. (1.3) of their paper are equal to 0.25 in this table, and Fig. 1.8 is the digraph associated with this table.

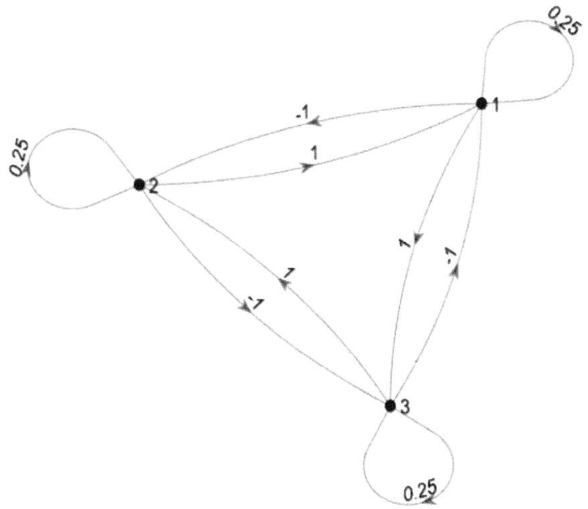

Fig. 1.8 Digraph associated with a payoff matrix of a rock-paper-scissors game in Sato et al. [91]

Example 1.9 (average pay-off matrix) Table 1.9 shows an average pay-off matrix in simulated intraspecific contests for five different strategies which appeared in Maynard Smith and Price [71]. The row denotes the strategies of the contestant who receives the pay-off, while the column denotes that of the opponent. For the details of these strategies, see Maynard Smith and Price (op. cit.). Table 1.9 is the result of two thousand contests of each type in a computer simulation, and the number in a given row and column is the pay-off gained by the row strategy when the opponent uses the column strategy. For example, in contests between "Mouse" and "Proper-Retaliator", the average pay-offs are 17.2 to "Mouse" and 56.7 to "Proper-Retaliator". Figure 1.9 is a digraph associated with the pay-off matrix in Table 1.9.

Example 1.10 (web page/adjacency matrix) Table 1.10 shows an example of the *adjacency matrix* of eight web pages with links. As for the definition of the adjacency matrix, see Sect. 11.4. This matrix was constructed from a digraph of the web pages shown in Austin [3]. Figure 1.10 is the digraph associated with this table.

Example 1.11 (web page/hyperlink matrix) In contrast, the *hyperlink matrix* $\boldsymbol{H}$ corresponding to the adjacency matrix is shown in Table 1.11. This matrix is the transposed matrix of the original hyperlink matrix shown in Austin (op. cit., p. 4).

Table 1.9 An average pay-off matrix in simulated intraspecific contests for five different strateges, adapted from Maynard Smith and Price [71]

From/to	1. Mouse	2. Hawk	3. Bully	4. Retaliator	5. Proper-Retaliator
1. Mouse	29.0	19.5	19.5	29.0	17.2
2. Hawk	80.0	– 19.5	74.6	– 18.1	– 18.9
3. Bully	80.0	4.9	41.5	11.9	11.2
4. Retaliator	29.0	– 22.3	57.1	29.0	23.1
5. Rroper-Retaliator	56.7	– 20.1	59.4	26.9	21.9

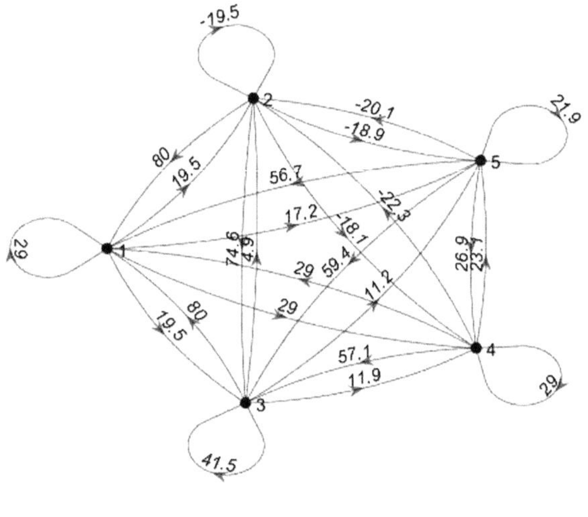

Fig. 1.9 Digraph associated with an average payoff matrix in simulated intraspecific contests for five different strategies

Table 1.10 An adjacency matrix of 8 web pages with asymmetric links

From/to	1	2	3	4	5	6	7	8
1	0	1	1	0	0	0	0	0
2	0	0	0	1	0	0	0	0
3	0	1	0	0	1	0	0	0
4	0	1	0	0	1	1	0	0
5	0	0	0	0	0	1	1	1
6	0	0	0	0	0	0	0	1
7	1	0	0	0	1	0	0	1
8	0	0	0	0	0	1	1	0

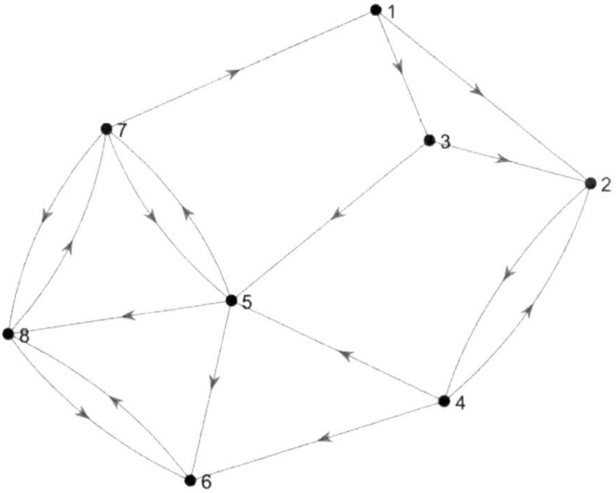

Fig. 1.10 Digraph associated with the adjacency matrix shown in Table 1.9

Table 1.11 Transposed hyperlink matrix of 8 web pages with asymmetric links, which was adapted from Austin (op. cit.)

From/to	1	2	3	4	5	6	7	8
1	0	1/2	1/2	0	0	0	0	0
2	0	0	0	1	0	0	0	0
3	0	1/2	0	0	1/2	0	0	0
4	0	1/3	0	0	1/3	1/3	0	0
5	0	0	0	0	0	1/3	1/3	1/3
6	0	0	0	0	0	0	0	1
7	1/3	0	0	0	1/3	0	0	1/3
8	0	0	0	0	0	1/2	1/2	0

The hyperlink matrix represents asymmetric links among web pages, and is determined as follows, according to Austin: Suppose that page P_j has l_j links. If one of those links is to page P_i, then P_j will pass on $1/l_j$ of its importance to P_i.

The importance ranking of P_i is then the sum of all the contributions made by pages linking to it. That is, if we denote the set of pages linking to P_i by B_i, then

$$I(P_i) = \sum_{P_j \in B_i} \frac{I(P_j)}{l_j}. \tag{1.1}$$

The hyperlink matrix, $H_{ij} = [H_{ij}]$ in which the entry in the ith row and jth columns is defined as

$$H_{ij} = \begin{cases} 1/l_j & \text{if } P_j \in B_i \\ \text{otherwise} & \end{cases}, \tag{1.2}$$

Note that the entries of the hyperlink matrix are all nonnegative, and the sum of the entries in a column is one unless the page corresponding to that column has no links.

Fig. 1.11 shows the digraph associated with the transposed hyperlink matrix shown in Table 1.11.

Example 1.12 (chemical complexes/adjacency matrix) A rather different kind of asymmetric interaction from those cited above is known in chemistry. Table 1.12 is an adjacency matrix between the four complexes, $X_1 + 2X_2, X_3, 2X_1 + X_2$, and

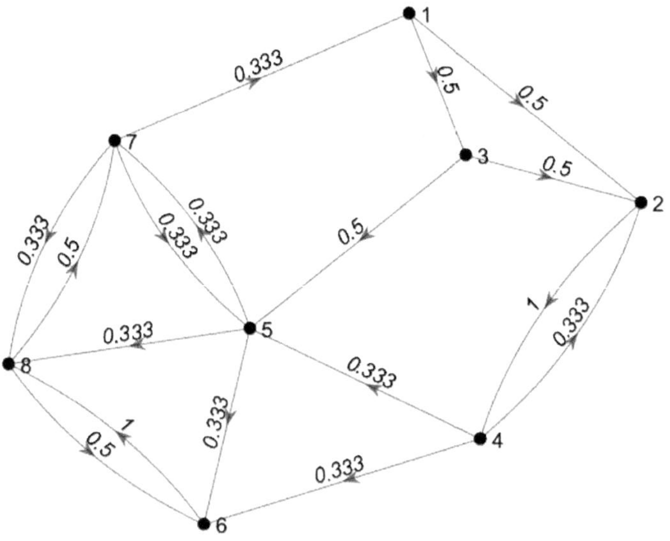

Fig. 1.11 Digraph associated with the transposed hyperlink matrix shown in Table 1.11

X_4, composed of the four chemical substrates, X_1, X_2, X_3 and X_4 Rao et al. [86] the corresponding digraph which is shown in Fig. 1.12 indicates a chemical reaction network, in which the four chemical complexes are composed of at most two chemical substrates out of the four substrates. As a result, *vertices* in this network denote these *chemical complexes*. Moreover, numbers, 1, 2, 3, 4 and 5 attached to five *edges* of the digraph denote, respectively, the *chemical reactions*, $v_1, v_2, v_3, v_4 and v_5$, to these complexes.

As shown in Table 1.12 as well as Fig. 1.12, both of them clearly depict the relation between complex and reaction.

However, they do not explicitly tell us the relation between substrate and reaction. To do this, we need to define the *stoichiometric matrix*, ***S***, which has traditionally played a fundamental role in analyzing the chemical reaction network. As will be discussed in detail in Chap. 11, in general the stoichiometric matrix can be represented by the product of the two kinds of matrices, ***Z*** and $\boldsymbol{B}_R$, where ***Z*** is the *complex*

Table 1.12 An adjacency matrix between the four complexes composed of the four chemical substrates, $X_1 + 2X_2, X_3, 2X_1 + X_2$, and X_4, which was constructed from the digraph shown in Fig. 2 of Rao et al. (op. cit.)

Complexes	1	2	3	4
1.$X_1 + 2X_2$	0	1	0	0
2.X_3	1	0	1	0
3.$2X_1 + X_2$	0	0	0	1
4.X_4	0	1	0	0

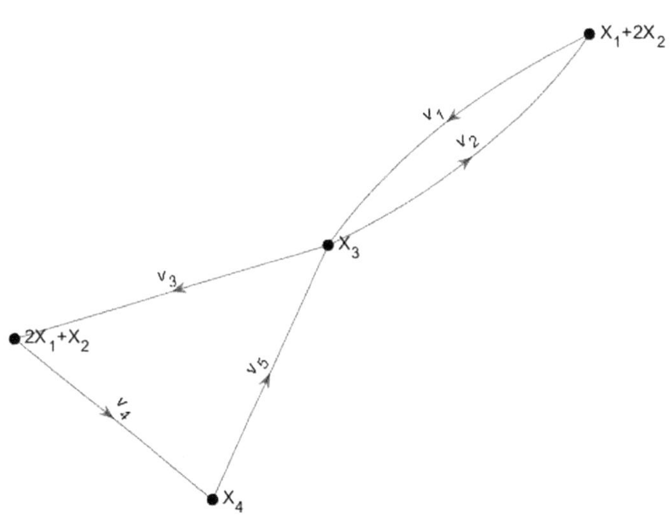

Fig. 1.12 Digraph associated with an adjacency matrix between the four complexes composed of the four chemical substrates, which was adapted from Fig. 2. of Rao et al. (op. cit.)

stoichiometric matrix which represents the relation between substrate and complex, while $\boldsymbol{B}_R$ is the *incidence matrix* of the complexes which specifically describes the relation between complex and reaction in the chemical reaction network (e.g., Rao et al. [86]). In the latter matrix, chemical reactions correspond to edges of a digraph. Here, it should be noticed that the sign of the incidence matrix defined by Rao et al., op. cit., is opposite from that defined by Brouwer and Hoemers [10]. Therefore, if we denote their incidence matrix by $\boldsymbol{B}$, then $\boldsymbol{S}$ must be equal to $-\boldsymbol{ZB}$.

As will be discussed in Sect. 1.3.2, stoichiometric matrix plays a crucial role in describing the dynamics of chemical reaction network.

Example 1.13 (allorecognition system in a reef-building coral) Table 1.13 shows an adjacency matrix of a hierarchy among 8 *Stylophora pistillata* colonies of a reef-building coral, which was constructed from Fig. 3 of Chadwick-Furman and Rinkevich [15]. It is known that corals have a variety of color forms that exhibit the same morphology, growth rate and sexual reproductive pattern. In this *two-year field study*, two major morphs, i.e., yellow morph and purple morph were chosen. In Table 1.13, colonies B and D are yellow, while the other colonies are purple. This table shows a nontransitive hierarchy among the 8 colonies at 2 years after initiation of contact.

Figure 1.13 is a digraph associated with the adjacency matrix of a hierarchy among 8 *Stylophora pistillata* colonies of a reef-building coral. This figure is essentially the same as that shown in Fig. 3 of Chadwick-Furman and Rinkevich (op. cit.). Arrows indicate the direction of *overgrowth*. As discussed in their paper, seven colonies, A, C, D, E, F, G, and M, exhibit a linear hierarchy of overgrowth interactions (where colony G > A > E > M = C > F > D). However, yellow colony B emerges as the one which changes this hierarchy to a nontransitive (or circular) hierarchy. Moreover, this colony overgrows the dominant G and two other purple colonies, E and M, but is overgrown by both the subordinate yellow D and the dominant A.

Example 1.14 (electrical circuit) Table 1.14 shows weight matrices corresponding to two kinds of electrical circuit with three terminals, α, β, and γ. Here, in the electrical circuit, nodes of a graph are usually called *terminals*, while edges are

Table 1.13 An adjacency matrix of a hierarchy among 8 *Stylophora pistillata* colonies of a reef-building coral

From/to	A	B	C	D	E	F	G	M
A		1	1	1	1	1		
B					1	1	1	1
C				1		1		1
D		1						
E				1		1		1
F		1		1				
G	1		1	1	1	1		
M			1			1		

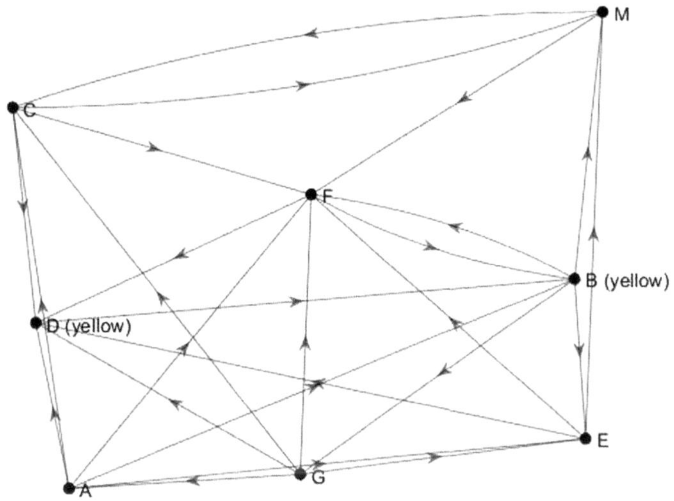

Fig. 1.13 The digraph associated with the adjacency matrix of a hierarchy among 8 *Stylophora pistillata* colonies of a reef-building coral

usually called *branches*. In this example, we consider an electrical circuit with a *resistor* marked by *R*, an *inductor* marked by *L*, and a *capacitor* marked by *C*. Therefore, i_R, i_L and $\mathrm{i_C}$ in Table 1.14 denotes the *current* through the resister, inductor, and capacitor, respectively. These currents are generally measured by real numbers. However, there is a strong restriction on these currents according to *Kirchhoff's current law* (KCL). As a result, the equation

$$\mathrm{i_R} = \mathrm{i_L} = -\mathrm{i_C} \tag{1.3}$$

holds for these currents. On the other hand, v_R, v_L, and v_C in Table 1.15 denote the *voltage* across each branch, $V(\beta) - V(\alpha)$, $V(\alpha) - V(\gamma)$, and V $(\beta) - V(\gamma)$, respectively. Here, for example, $V(\alpha)$ is the voltage at the terminal α. There is also a strong restriction on these voltages according to *Kirchhoff's voltage law* (KVL). As a result, the equation

$$v_R + v_L - v_C = 0 \tag{1.4}$$

Table 1.14 Weight matrix corresponding to the current circuit with three terminals

From/to	α	β	γ
α	0	0	$\mathrm{i_L}$
β	$\mathrm{i_R}$	0	$\mathrm{i_C}$
γ	0	0	0

Table 1.15 Weight matrix corresponding to the voltage circuit with three terminals

From/to	α	β	γ
α	0	0	v_L
β	v_R	0	v_C
γ	0	0	0

holds for these voltages.

Figure 1.14a, b show the digraph associated with the weight matrices shown in Tables 1.14 and 1.15, respectively.

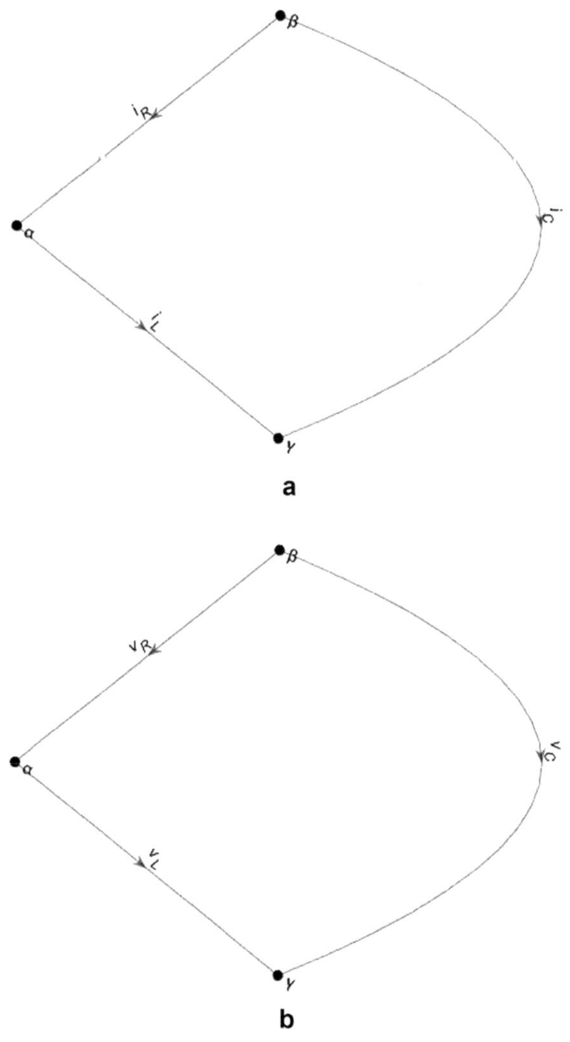

Fig. 1.14 **a** The digraph associated with the weight matrix of an electrical circuit, which is shown in Table 1.14. **b.** The digraph associated with the weight matrix of an electrical circuit, which is shown in Table 1.15

Remark In addition to the restrictions on currents and voltages, that is, (1.3) and (1.4), there is the third restriction on the register characteristic. It is concerned with the relationship on i_R and v_R, and is written as.

$$v_R = f(i_R). \tag{1.5}$$

If R denotes a *linear resistor*, then f is linear and (1.5) is a statement of *Ohm's law*. If R denotes a *nonlinear resistor*, (1.5) is called a *generalized Ohm's law* (e.g.. Hirsch and Smale [45], p. 213).

As will be discussed in Sect. 1.3.2, the current and the voltage introduced above play a crucial role in describing the dynamics of electrical circuit.

Example 1.15 (food web) Table 1.16 shows a weight matrix corresponding to a *food web* which is depicted in Fig. 1c of McCann et al. [72]. In Table 1.16, R denotes a resource density, C_1 denotes that of the first consumer species, C_2 denotes that of the second consumer species, and P denotes that of the top predator. On the other hand, Ω_{ij} are preference parameters.

Figure 1.15 shows a digraph associated with a weight matrix of a food web adapted from Fig. 1c of McCann et al. [72]. In Fig. 1c, preference parameters, $\Omega_{(C_1R)}$ and $\Omega_{(C_2R)}$ are not depicted. So, we also dropped them in Fig. 1.15.

Table 1.16 A weight matrix of a food web which is depicted in Fig. 1c of McCann et al. [72]

From/to	R	C_1	C_2	P
R	0	$\Omega_{(C_1R)}$	$\Omega_{(C_2R)}$	0
C_1	0	0	0	$\Omega_{(PC_1)}$
C_2	0	0	0	$1-\Omega_{(PC_1)}$
P	0	0	0	0

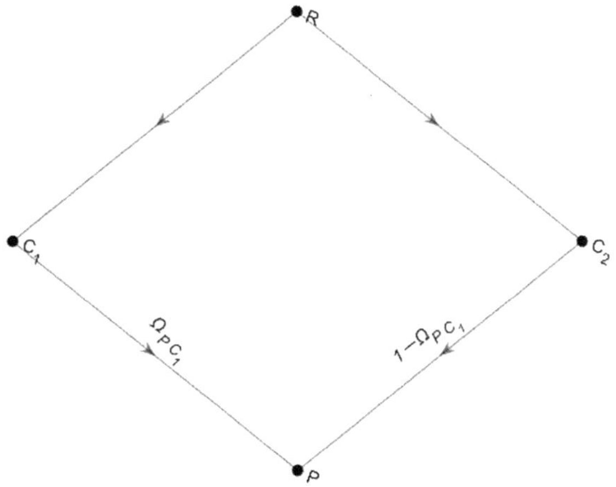

Fig. 1.15 A digraph associated with a weight matrix of a food web adapted from Fig. 1c of McCann et al. [72]

Table 1.17 A transition matrix of order 3

From/to	1	2	3
1	0.5	0.25	0
2	0.5	0.5	0.5
3	0	0.25	0.5

McCann et al. (op. cit.) analyze a set of *nonlinear* differential equations whose *state space* is composed of four densities of species, P, C_1, C_2, and R.

Example 1.16 (Markov chains) In Markov chains, some experiment is conducted with a set of k outcomes, or states $\{s_1, s_2, \cdots, s_k\}$. The experiment is repeated such that the conditional probability $p_{ij} = p(s_i|s_j)$ of the state s_i, $1 \leq i \leq k$, occurring on the $(n+1)$ th repetition depends only on the state s_j occurring on the nth repetition of the experiment. Thus, we have

$$\sum_{i=1}^{k} p_{ij} = 1, \quad 1 \leq j \leq k. \tag{1.6}$$

In Markov chains, this matrix $\boldsymbol{S} = \left(p_{ij}\right)$ of order k is called a *transition matrix.*

Table 1.17 shows an example of a transition matrix of order 3, which appears in Elaydi ([28], p. 144). Let us consider this matrix as a weight matrix of a digraph.
Figure 1.16 shows a digraph associated with the weight matrix in Table 1.17.

Remark Let $p_i(n)$ denote the probability that the state s_i will occur on the nth repetition of the experiment, $1 \leq j \leq k$. Then,

$$\boldsymbol{p}(n+1) = \boldsymbol{S}\boldsymbol{p}(n), \quad n = 1, 2, 3, \tag{1.7}$$

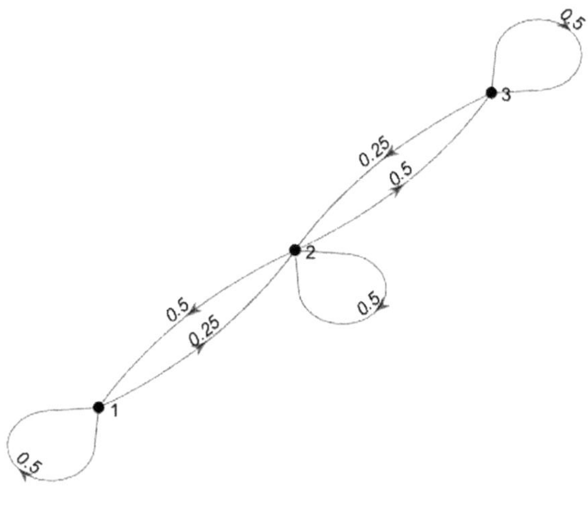

Fig. 1.16 A digraph associated with the weight matrix in Table 1.17

where $\boldsymbol{p}(n) = (p_1(n), p_2(n), \cdots, p_k(n))$ is the probability vector. It should be noticed that in Markov chains the transition matrix is a constant matrix.

Example 1.17 (directed 5-cycle as a special case of a circulant matrix) A directed 3-cycle which appeared in Example 1.6 is a special case of a *circulant matrix* (e.g., Gray [36]) or sometimes called *cyclic matrix* (e.g., Berlin and Kac [8]) which is shown in Table 1.18.

Table 1.19 shows an adjacency matrix corresponding to a *directed 5-cycle*, which is similar in form as the directed 3-cycle discussed above. It is also a special case of a circulant matrix in which $c_0 = 0, c_1 = 1, c_2 = c_3 = c_4 = 0$.

Figure 1.17 shows the digraph associated with an adjacency matrix of a directed 5-cycle shown in Table 1.19.

In general, directed n-cycle has the nth roots of unity (e.g., Brouwer and Haemers [10], p. 8). The nth roots of unity play a fundamental role in some of the neural network models.

Example 1.18 (circulant matrix of order 3) Table 1.20 shows an example of a circulant matrix of order 3, where $c_0 = 2, c_1 = 3, c_2 = 1$ in Table 1.18.

Figure 1.18 is the digraph associated with the circulant matrix of order 3 in Table 1.20.

The eigenvalues of a circulant matrix of order n can be expressed as a linear function of the nth roots of unity (e.g., Berlin and Kac [8], p. 828; Gray [36], p. 32), as with the n-cycle.

Table 1.18 A general form of the circulant or cyclic matrix

From/to	1	2	3	4	...	n
1	c_0	c_1	c_2	...	...	$c_{(n-1)}$
2	c_{n-1}	c_0	c_1	c_2		·
3		c_{n-1}	c_0	c_1		·
·						c_2
·						c_1
n	c_1	c_2	c_3		c_{n-1}	c_0

Table 1.19 The adjacency matrix corresponding to a directed 5-cycle

From/to	1	2	3	4	5
1		1			
2			1		
3				1	
4					1
5	1				

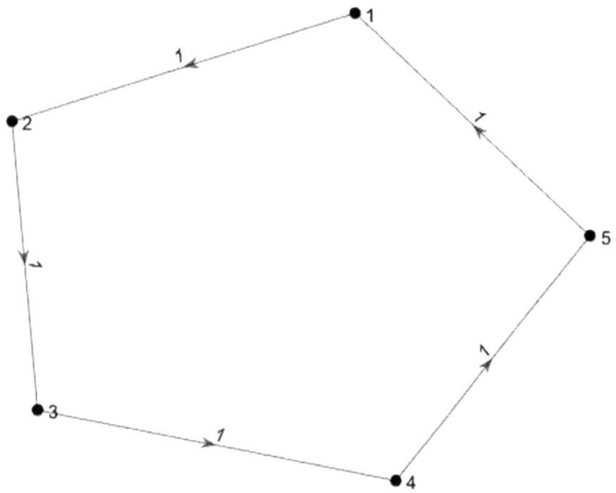

Fig. 1.17 Digraph associated with an adjacency matrix of a directed 5-cycle shown in Table 1.19

Table 1.20 a circulant matrix of order 3

From/to	1	2	3
1	2	1	3
2	3	2	1
3	1	3	2

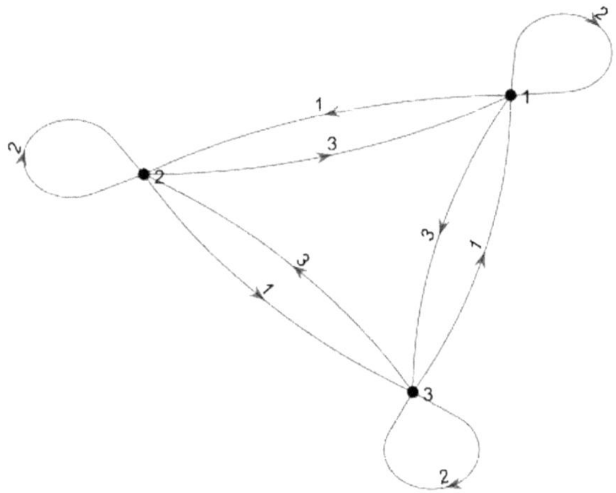

Fig. 1.18 Digraph associated with a circulant matrix of order 3

Example 1.19 (circulant matrix of order 4) Table 1.21 shows an example of a circulant matrix of order 4, where $c_0 = 2, c_1 = 4, c_2 = 1, c_3 = 3$ in Table 1.18. As shown in Fig. 1.19, circulant matrix of order 4 has pairs of directed edges between all pairs of vertices unlike the directed 4-cycle.

Remark Circulant matrix was named after Catalan [14] (e.g., Carlson, [12]). Berlin and Kac (op. cit.) introduce the circulant matrix shown in Table 1.18, which he calls

Table 1.21 A circulant matrix of order 4

From/to	1	2	3	4
1	2	4	1	3
2	3	2	4	1
3	1	3	2	4
4	4	1	3	2

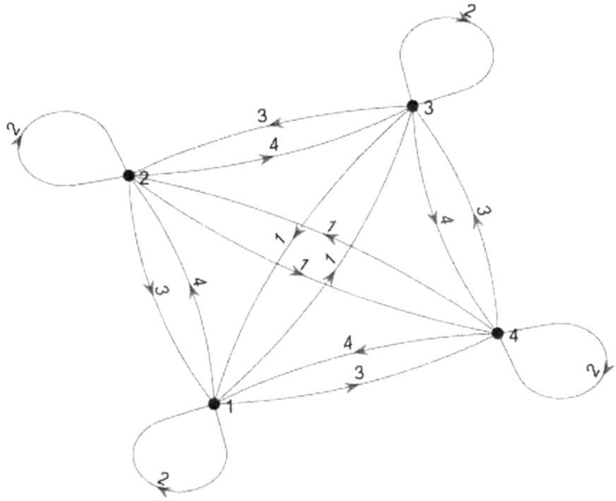

Fig. 1.19 Digraph associated with a circulant matrix of order 4

the cyclic matrix, in defining the Gaussian model which is one of the generalized versions of the Ising model of a ferromagnet. They solve the eigenvalue problem of the original circulant matrix first, and then solve that of a symmetric circulant matrix.

Example 1.20 (Toeplitz matrix) Table 1.22 shows a *Toeplitz matrix* of order n. The Toeplitz matrix has a property that each descending diagonal from left to right is constant. The circulant matrix in Table 1.18 is a special case of Toeplitz matrices (e.g., Gray, op. cit., p. 1; Lancaster and Tismenetsky [62], p. 68).

Figure 1.20 is the digraph associated with a Toeplitz matrix of order 3 in Table 1.23.

Remark Toeplitz matrices have applications to stochastic time series (e.g., Gray, op. cit.). For example, covariance matrices of weakly stationary processes are Toeplitz.

Example 1.21 (magic square of order 3) Table 1.24 shows a magic square of order 3 expressed in matrix form. It is the ancient Chinese Lo Shu magic square of order 3, which is composed of the first nine consecutive integers and has a magic line constant of 15 for rows, columns, and diagonals (e.g., Loly et al. [66]). In general, a classical magic square of order n has the line sum (magic constant)

$$C_n = n(n^2 + 1)/2. \tag{1.8}$$

Table 1.22 A Toeplitz matrix of order n

From/to	1	2	3	...	n
1	t_0	t_{-1}	t_{-2}		$t_{-(n-1)}$
2	t_1	t_0	t_{-1}		
3	t_2	t_1	t_0		$\vdots$
$\vdots$				$\ddots$	
n	t_{n-1}		...		t_0

(Here, the (i, j) element of the Toeplitz matrix T is t_{i-j}.)

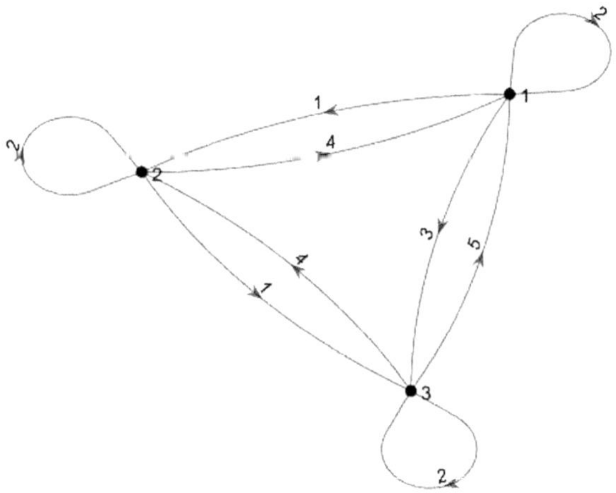

Fig. 1.20 The digraph associated with a Toeplitz matrix of order 3 in Table 1.22

Table 1.23 A Toeplitz matrix of order 3

From/to	1	2	3
1	2	1	3
2	4	2	1
3	5	4	2

Table 1.24 The ancient Chinese Lo Shu magic square of order 3, which is expressed in matrix form

From/to	A	B	C
A	4	9	2
B	3	5	7
C	8	1	6

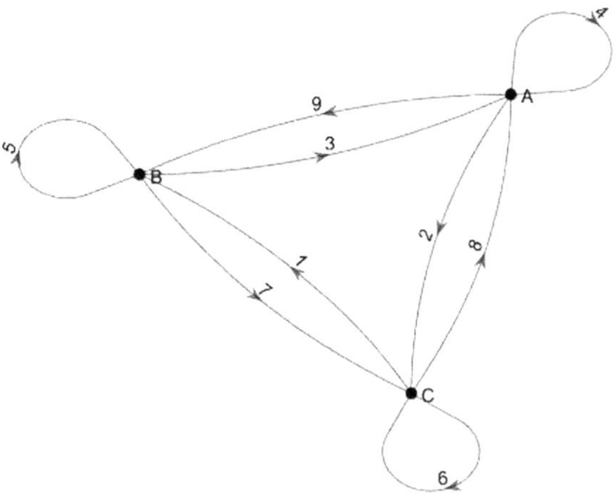

Fig. 1.21 The digraph associated with the Lo Shu magic square

A magic square can also be considered as a digraph. Figure 1.21 is the digraph corresponding to the Lo Shu square. At a glance, it looks like the digraph associated with the circulant matrix of order 3 shown in Fig. 1.18. However, it is apparent from their definitions that the magic square is more restrictive than the circulant matrix.

Example 1.22 (magic square of order 4) Table 1.25 is a magic square of order 4. It is the magic square which appears in the famous Dürer's engraving, *Melencolia I* (e.g., Dürer [26]; Sohm [96]). The magic constant of this square is 34. Figure 1.22 shows the digraph corresponding to the magic square in Table 1.25.

Remark Examples 1.21 and 1.22 are called *classical* magic squares in that elements of these matrices are composed of consecutive integers starting with 1. However, as will be discussed in more detail in Chapter 11, magic squares are defined also in the domains of real numbers and complex numbers nowadays (e.g., Fox [30]; Ward III [109]).

Magic squares have been developed in various parts of the world since a special magic square (i.e., the Lo Shu square) was initiated in ancient China (e.g., Ibrahim and Salman [51]). Recently, magic squares have been applied, for example, to the phenomena in quantum physics (e.g., De Las Cuevas et al. [24]; Iyer [56]; O'Neill [82]).

Table 1.25 The magic square of order 4 which appears in the famous Dürer's engraving, *Melencolia I*

From/to	A	B	C	D
A	16	3	2	13
B	5	10	11	8
C	9	6	7	12
D	4	15	14	1

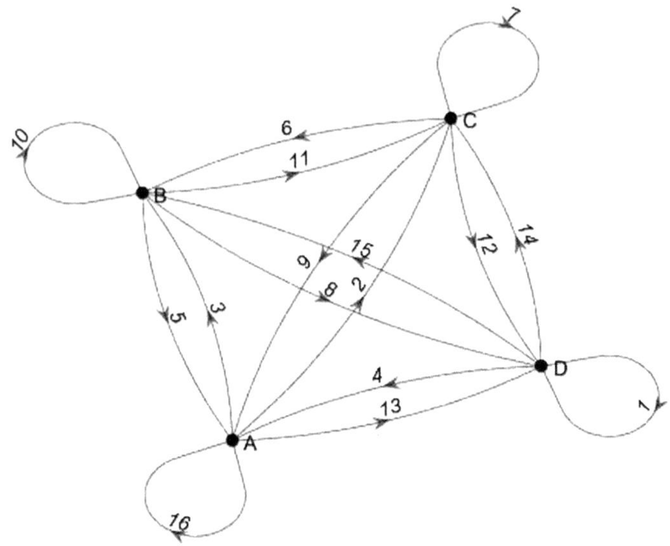

Fig. 1.22 The digraph corresponding to the magic square in Table 1.22

Example 1.23 Hermitian adjacency matrix) There exists a family of graphs in which some of the edges are undirected and others are directed. Such a graph is called a *mixed graph*. Mixed graphs are usually assumed to include neither multiple edges nor loops. Table 1.26 is the ordinary adjacency matrix corresponding to a digraph with multiple edges. We will be able to depict the graph as in Fig. 1.23 if we permit multiple edges in such a graph.

In contrast, Table 1.27 is a *Hermitian adjacency matrix*, which was introduced recently by Liu and Li [65]. As can be seen in this table, it includes imaginary number *i*, and it is easy to see that it is a *Hermitian matrix*. As for the definition of Hermitian matrix, see Chapter 11. In view of a graph, mixed graph includes an *undirected edge set* be $E_0(D)$ and a *directed edge set* be $E_1(D)$ in a *mixed graph D*.

If we consider the matrix defined in Table 1.27 as a weight matrix corresponding to the digraph with multiple edges, then the digraph can be written as in Fig. 1.24. The mixed graph associated with the Hermitian matrix shown in Table 1.27 is drawn

Table 1.26 An ordinary adjacency matrix in which some of the edges are undirected and others are directed

From/to	1	2	3	4
1	0	0	0	1
2	1	0	1	0
3	1	1	0	0
4	1	0	1	0

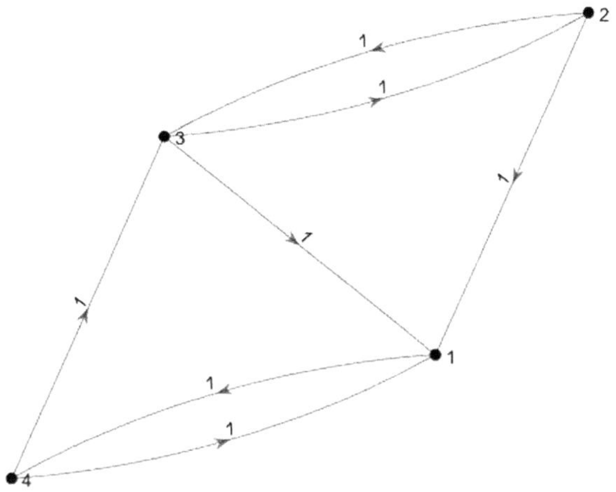

Fig. 1.23 An unusual expression of a mixed graph using multiple edges

Table 1.27 A Hermitian adjacency matrix of order 3, which appears in Fig. 2.1 of Liu and Li (op. cit., p. 184)

From/to	1	2	3	4
1	0	−i	−i	1
2	I	0	1	0
3	i	1	0	−i
4	1	0	I	0

in the right side of Fig. 2.1 of Liu and Li (op. cit., p. 184) in a different way. Since they do not permit any multiple edges in their mixed graph, multiple edges

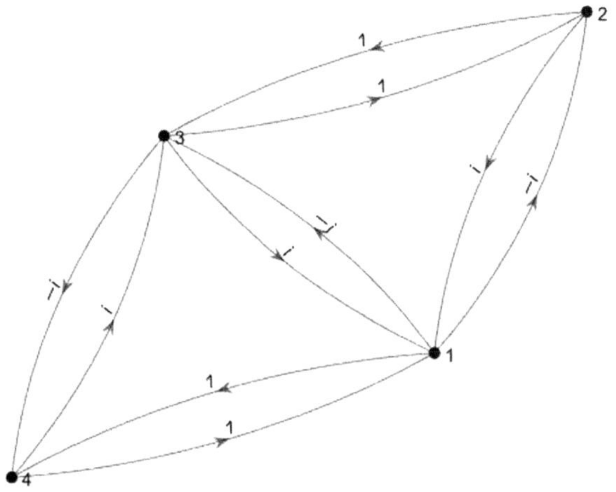

Fig. 1.24 Digraph associated with a Hermitian matrix in Table 1.24 which is considered as a weight matrix

with the same weight '1' in Fig. 1.24 are replaced by an undirected edge in their graph, and those with the different weights, i.e., 'i' and '$-i$', in Fig. 1.24, is replaced by the directed edge with the weight 'i'. Of course, they show no weights in their graph.

Remark Since a complex adjacency matrix called Hermitian adjacency matrix was introduced by Liu and Li (op. cit.), there has been a body of research on Hermitian adjacency matrices (e.g., Greaves et al. [37]; Moher [75]; Yuan et al. [112]). Of course, such complex adjacency matrices are meaningful from the view point of mathematics. However, such complex matrices can neither be observed in vivo nor *in daily life*, and thus these matrices are *hypothetical* or *theoretical.*

In contrast, Chino and Shiraiwa [21] proposed an asymmetric MDS model called Hermitian form model (abbreviated as HFM). As will be discussed in detail in Chap. 4, in HFM any *real* ASM assumed to be measured at a ratio scale level is transformed into a general *Hermitian matrix* first, and then objects are embedded in a *Hilbert space* or in an *indefinite metric space* depending on the eigenvalues of the Hermitian matrix. In HFM, we distinguish clearly between a *real* ASM and (*complex*) Hermitian matrix associated with the ASM. That is, a real ASM is the one *observed* in vivo *or in daily life,* while the transformed Hermitian matrix is *hypothetical or theoretical.* The following example shows a real ASM and the Hermitian matrix corresponding to the ASM.

Example 1.24 (Hermitian matrix associated with a real ASM) Table 1.28 shows an international trade data among Japan, America, China, and Russia in 2015, and appeared elsewhere [18, 19]. This type of data is usually preprocessed using a log-transformation before we analyze it. Table 1.29 is the log-transformed trade data among four nations.

Table 1.28 An international trade data among four nations as a real ASM

From/to	1. Japan	2. USA	3. China	4. Russia
1. Japan	43,480	1382	1200	55
2. USA	736	189,592	1161	71
3. China	1764	4832	119,684	348
4. Russia	173	164	333	13,755

Table 1.29 The log-transformed trade data among four nations

From/to	1. Japan	2. USA	3. China	4. Russia
1. Japan	10.6801	7.2313	7.0901	4.0073
2. USA	6.6012	12.1526	7.0570	4.2627
3. China	7.4753	8.4830	11.6926	5.8522
4. Russia	5.1533	5.0999	5.8081	9.5292

Table 1.30 The Hermitian matrix associated with the log transformed ASM in Table 1.29

From/to	1. Japan	2. USA	3. China	4. Russia
1. Japan	10.6801	6.9163 + 0.3150i	7.2827 − 0.1926i	4.5803 − 0.5730i
2. USA	6.9163 − 0.3150i	12.1526	7.7700 − 0.7130i	4.6813 − 0.4186i
3. China	7.2827 − 0.1926i	7.7700 + 0.7130i	11.6926	5.8302 + 0.0220i
4. Russia	4.5803 + 0.5730i	4.6813 + 0.4186i	5.8302 − 0.0220i	9.5292

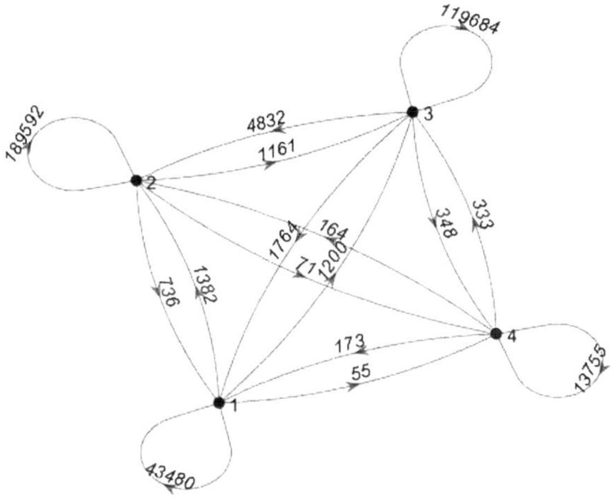

Fig. 1.25 The digraph associated with the observed trade data of four nations

Table 1.30 shows the Hermitian matrix computed from the transformed ASM in Table 1.29. using (4.1) in Chap. 4. It is apparent from (4.1) that there is a one-to-one correspondence between a real ASM and the Hermitian matrix associated with the ASM.

Given the three tables above, we can depict three digraphs associated with these tables. However, we only show two digraphs associated with two tables, Tables 1.28 and 1.30, in Figs. 1.25 and 1.26, respectively, for simplicity. It should be stressed here, again, that Table 1.28 is a real ASM observed in daily life, while Table 1.30 is a complex ASM which cannot be observed in daily life. Likewise, Fig. 1.25 is a digraph associated with an *observed* ASM, while Fig. 1.26 is a *hypothetical* digraph associated with the observed ASM and cannot be observed in daily life. Our Hermitian matrix may be contrasted with that defined by Liu and Li (op. cit.).

1.3 Two Aspects of Asymmetric Interactions

In Sect. 1.2 we have shown various ASMs and the corresponding digraphs, which imply asymmetric interactions among objects or nodes. These data are, in nine cases out of ten, observed in vivo, in field surveys, or in daily life. As was introduced in

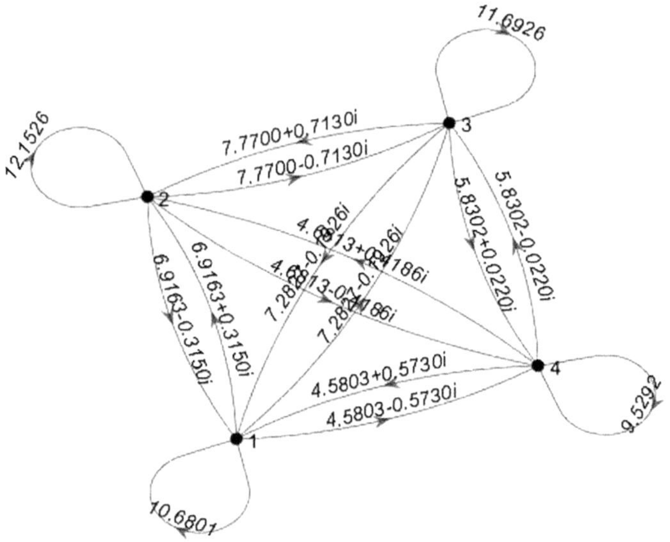

Fig. 1.26 The digraph associated with the Hermitian matrix computed from the transformed ASM in Table 1.26

the introductory section, asymmetric interaction has two aspects, that is, its structure and its dynamics.

As regards the structure, we may distinguish between *surface structure* and *latent structure*. An observed ASM or a digraph among objects and nodes, respectively, may be considered as a surface structure, while a configuration of objects or nodes obtained by an *asymmetric MDS* including a symmetric MDS, *spectral graph theory*, or singular value decomposition, etc., may be considered as a latent structure.

On the other hand, we are frequently interested in the dynamic aspect of asymmetric interactions, i.e., the dynamics of asymmetric interactions among objects or nodes. For example, the sociomatrix shown in Table 1.1 is merely a snapshot of the sociometric structure observed at a specific instant, and it will change over time dynamically through complicated asymmetric interactions among members. The pecking order matrix shown in Table 1.5 is also a snapshot of the pecking orders of some domestic chickens observed at a specific instant. As pointed out in the remark of Example 1.5, Rushen [89] observed the development of pecking orders in mixed sex groups of domestic chickens for 25 weeks, and found interesting results. Table 1.12 shows the relationships among four chemical complexes and five chemical reactions. These relationships directly indicate changes in chemical complexes over time through asymmetric interactions among complexes. The adjacency matrix of a hierarchy among 8 Stylophora pistillate colonies of a reef-building coral of Table 1.13 is a snapshot of a hierarchy of colonies, observed in two-year field study. Even in a simple electrical circuit shown in Table 1.14 and 1.15, current and voltage change dynamically over time. In a food web shown in Example 1.15, densities of a resource, consumers, and a predator change dynamically as time proceeds, according to the complicated interactions among them. Such dynamics have traditionally been modeled by some *differential equations* or *difference equations*.

In Sect. 1.3.1, we shall introduce some of the traditional methods for the analysis of the structure of asymmetric interactions which were listed in Sect. 1.2 and illustrate some of the results of applications of these methods to some empirical data. For further details of asymmetric MDS, see subsequent chapters in Part I of this book.

In Sect. 1.3.2, we shall bring in some of the traditional methods for the analysis of the dynamics of asymmetric interactions which were listed in Sect. 1.2 as well as those listed in the beginning of Sect. 1.1 and illustrate some of the results of applications of these methods to empirical data. For further details of some of these methods, see subsequent chapters in Part II of this book.

1.3.1 Traditional Analyses of Structures of Asymmetric Interactions

In this section we shall introduce some of the *asymmetric MDSs* (hereafter we shall abbreviate *asymmetric MDS* as *AMDS*) in psychometrics, *spectral graph theory* (hereafter abbreviated as *SGT*), and *graph embedding* (hereafter abbreviated as *GE*) in statistics and machine learning, using some of the examples listed in Sect. 1.2 as well as those in the literature. As will be described in detail in Chap. 11, AMDS is a method for embedding elements in a certain dimensional space such that the estimated asymmetric interpoint distances between elements fit as closely as possible to the observed asymmetric similarities between elements. Asymmetric distance function differs from model to model for AMDS. For further detail, see Chap. 7.

Asymmetric MDS

In asymmetric MDS, we must choose an appropriate model in analyzing an observed ASM or hypothetical ASM, from the view point of *scale level* of the ASM. For example, if it is considered as measured at an *interval level*, it will be an appropriate choice, for example, to choose Chino's ASYMSCAL. Example 1.1 in Sect. 1.2 is one such ASM. Original data of this example is shown in Tables 2.3 and 2.4 of Chap. 2, and the results of application of these data are shown in Figs. 2.9 and 2.10 in Chap. 2. Chino's ASYMSCAL embeds configurations of members in two-dimensional or three-dimensional *Euclidean space*, and the skewness between members is simultaneously represented in this space by an oriented area of the parallelogram spanned by a pair of position vectors corresponding to two members. As for the scale levels of measurement in psychometrics, see Sect. 10.3 in Chap. 10. Although Chino's ASYMSCAL enables us to estimate an *additive constant* which compensates for the weakness of interval scale level of measurement, its dimension of the configuration is restricted to *three*.

By contrast, the scale level of the international trade data shown in Example 1.2 is considered as *ratio*. Examples 1.3, 1.4, 1.5, 1.10, 1.11, 1.14, 1.15, 1.16, 1.17, 1.21,

1.22, etc. are considered as typical examples with ratio scale level of measurement. Here, it should be noted that Examples 1.14, 1.15, 1.16, 1.17, 1.19, 1.20, 1.21, 1.22 are hypothetical data and that Examples 1.14, 1.15, 1.16, 1.17, 1.19, 1.21, and 1.22 might possibly be observed.

In any case, HFM proposed by Chino and Shiraiwa [21] provides us with an appropriate model for the data measured at a ratio scale level. In HFM, members are embedded in a (complex) *Hilbert space* or an *indefinite metric space* depending on the nature of the observed (real) ASM. In fact, analysis of the trade data by HFM concludes that this data has a Hilbert space structure as a whole (see, Chino [18, 19]).

Figure 1.27 shows the configuration of members on a complex plane associated with the second largest eigenvalue of the log-transformed trade data, which is shown in Fig. 5.6 of Chap. 5. As discussed there, the amount of export from C (China) to A (USA) is greater than that from A to C. Similar trade imbalances are indicated from C to J, and from J to A. It should be noted that the *orientation of the configuration* is crucial in interpreting it. In HFM, the orientation is *clockwise* as in Fig. 1.27 if the eigenvalue corresponding to the configuration is positive, while it is *counterclockwise* if the eigenvalue is negative. For more detail, see, for example, Chino [18] in Chap. 5.

The scale level of the original car switching matrix in Example 1.4 can be considered as ratio, since it is a *count data*. Here, count data is sometimes called an *absolute scale* data (Suppes and Zinnes [98]). In any case, some researchers analyze the car switching data by considering it as ordinal data. Okada and Imaizumi (op. cit.) analyze this data by their nonmetric AMDS and obtain a configuration of members in a Euclidean space. In their augment distance model, however, a circle with different radius is attached to each point of the configuration and the skewness of the distance between members is interpreted using the difference in the radii. For more detail, see Fig. 4.1 in their paper.

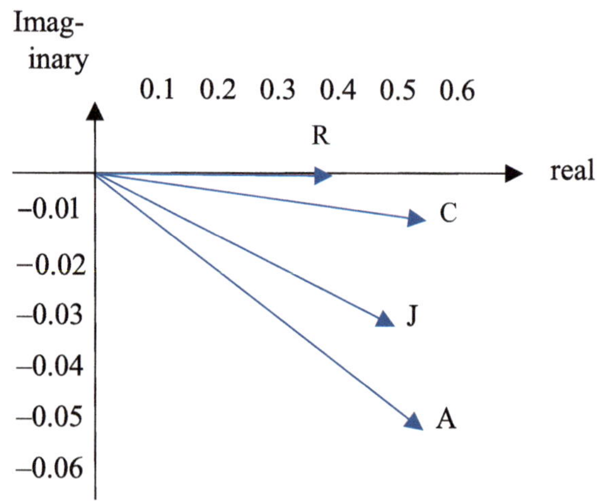

Fig. 1.27 The enlarged configuration of four nations. Here, the positive direction of this figure is clockwise because the eigenvalue is positive in this case

Spectral Graph Theory

In spectral graph theory, the notion of *spectrum* plays a central role in discussing properties of a graph. Although the spectrum has a lot of origins mainly in physics, it means an eigenvalue of a matrix in mathematics. Particularly in spectral graph theory, the relation between graph properties and the spectrum of the *adjacency matrix* or *Laplacian matrix* is studied as, for example, pointed out by Brouwer and Haemers [10]. Here, they call a Laplacian matrix a Laplace matrix.

Before proceeding, it will be appropriate to define briefly the four matrices associated with an undirected graph without *multiple edges and loops*: The adjacency matrix of $\boldsymbol{A}_u$, the *degree matrix* $\boldsymbol{D}_u$, the Laplacian matrix $\boldsymbol{L}_u$, and the *signless Laplacian matrix* $\boldsymbol{Q}_u$ (for greater detail, see Sect. 11.4). In short, $\boldsymbol{A}_u = \{a_{ij}\}$ is the 0–1 matrix whose (i,j) element is *one* if there is an edge from the vertex i to vertex j and *zero* otherwise. $\boldsymbol{D}_u$ is a diagonal matrix whose (i,i) element is equal to the degree of the ith vertex of a graph. $\boldsymbol{L}_u = \boldsymbol{D}_u - \boldsymbol{A}_u$. $\boldsymbol{Q}_u = \boldsymbol{D}_u + \boldsymbol{A}_u$. All of these matrices are *symmetric*. In contrast, all of Tables 1.6, 1.7, 1.10, 1.12, 1.13, 1.19 and 1.26 shown in Sect. 1.2 are *asymmetric*, since the corresponding graphs are *digraphs* (e.g., directed graphs).

In traditional spectral graph theory, a major interest had been in studying the relation between graph properties and the spectrum of the adjacency matrix or Laplacian matrix associated with *undirected graphs*. One of the reasons for this may be that the spectrum is closely related to the notion of *centrality* of a graph. Here, a centrality means the most important vertices within a graph, and there has been a variety of centrality measures in the literature (e.g., Bavelas [5]; Bonacich [11]; Freeman [31]; Katz [58], Negre et al. [77]). Well-known measures are *degree centrality* and *eigenvector centrality*. The degree centrality utilizes the degree matrix $\boldsymbol{D}_u$, and the eigenvector centrality is usually computed using the eigenvector corresponding to the largest eigenvalue of the adjacency matrix $\boldsymbol{A}_u$ or the Laplacian matrix $\boldsymbol{L}_u$.

Table 1.31 shows an adjacency matrix associated with the undirected graph depicted in Fig. 1.28.

Since an adjacency matrix associated with an undirected graph is symmetric as pointed out above, its spectra are all *real*. In fact, the spectra of the above matrix are 2.1701, 0.3111, −1.000, and −1.4832. Of course, the largest eigenvalue, which is called the *spectral radius* in matrix theory (e.g., Lancaster and Tismenetsky [62], p. 359), is 2.1701 in this case. The eigenvector corresponding to the largest eigenvalue

Table 1.31 An adjacency matrix associated with the undirected graph without multiple edges and loops

From/to	v_1	v_2	v_3	v_4
v_1	0	1	1	1
v_2	1	0	0	0
v_3	1	0	0	1
v_4	1	0	1	0

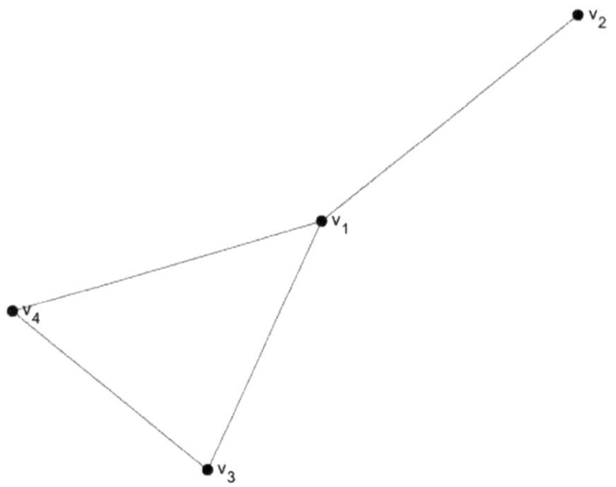

Fig. 1.28 An undirected graph without *multiple edges and loops* corresponding to the adjacency matrix in Table 1.31

is (0.6116, 0.2818, 0.5227, 0.5227). As a result, we can conclude that vertex (or node) v_1 (in Fig. 1.28) has the largest eigenvector centrality measure.

At this point, it should be noticed that our primary concern in this section is to consider the structure of asymmetric interactions among objects or nodes. To do so, it is appropriate and necessary to analyze an ASM and the corresponding digraph (not a graph as in the traditional spectral theory), considering a *holistic structure of the configuration of nodes* spanned by the two eigenvectors or all the eigenvectors of some asymmetric similarity matrix.

Of course, there are a few families of ASMs and associated digraphs which have been analyzed in traditional spectral theory. For example, Examples 1.10 and 1.11 in Sect. 1.2 show such ASMs. Apparently, the adjacency matrix and the hyperlink matrix in these examples are asymmetric. However, these matrices have been analyzed by applying the famous *Perron-Frobenius theorem* (e.g., Brouwer and Haemers [10]; Lancaster and Tismenetsky [62]), which will be discussed in more detail in Sect. 11.4. According to the theorem, a real matrix, which is *nonnegative* and *irreducible*, has a *positive eigenvalue* and a *positive eigenvector* associated with the eigenvalue, even if it is asymmetric. In fact, the largest eigenvalue of the hyperlink matrix which is the transpose of the matrix shown in Table 1.11 is 1.0, and the corresponding eigenvector is (0.1400, 0.1576, 0.0700, 0.1576, 0.2276, 0.4727, 0.4201, 0.6886). If we renormalize this vector in such a way that the sum of its components is one, we have (0.0600, 0.0675, 0.0300, 0.0675 0.0975, 0.2025, 0.1800, 0.2950), which is equal to the stationary vector shown in Austine [3]. As a result, we can say that the most important page is Page 8 (i.e., node 8) in Fig. 1.11.

However, once we have an ASM, our interests may not be restricted within the scope of centrality measures. For example, which page is more frequently visited from which page? Such a question can be answered by applying an appropriate AMDS developed in psychometrics. What is the *holistic metric space* in which web pages are embedded using the eigenvectors of a hyperlink matrix? HFM enables us to answer the question if the data is measured at a ratio scale level. In fact, we applied

HFM to the transposed hyperlink matrix shown in Table 1.11, and the eigenvalues of the Hermitian matrix constructed from Table 1.11 were 1.1695, 0.9256, 0.3426, 0.2017, −0.1935, −0.5860, −0.8698, −0.9900. This means that the holistic metric structure of the configuration of 8 web pages is an *indefinite metric* space, since there exist eigenvalues with different signs. However, if we look at the configuration per each dimension, then each of these component configurations is considered as a one-dimensional *Hilbert space*.

Figures 1.29 shows the complex configuration associated with the largest eigenvalue, 1.1695. If we notice that the *positive direction* of the configuration in Fig. 1.29 is *clockwise* since the corresponding eigenvalue is positive, then it is apparent that Pages 2 and 3 are more frequently visited from Page 1 and Pages 6 and 8 are from Page 5.

Another family of ASMs and the associated digraphs which have recently been studied extensively in spectral graph theory are the Hermitian adjacency matrices of a mixed graph, an example of which is shown earlier in Table 1.27 in Sect. 1.2. As pointed out in the note of Example 1.23, Hermitian adjacency matrices are hypothetical in the sense that they cannot be observed in vivo or in daily life. Moreover, these matrices are special Hermitian matrices in that all the diagonal elements are

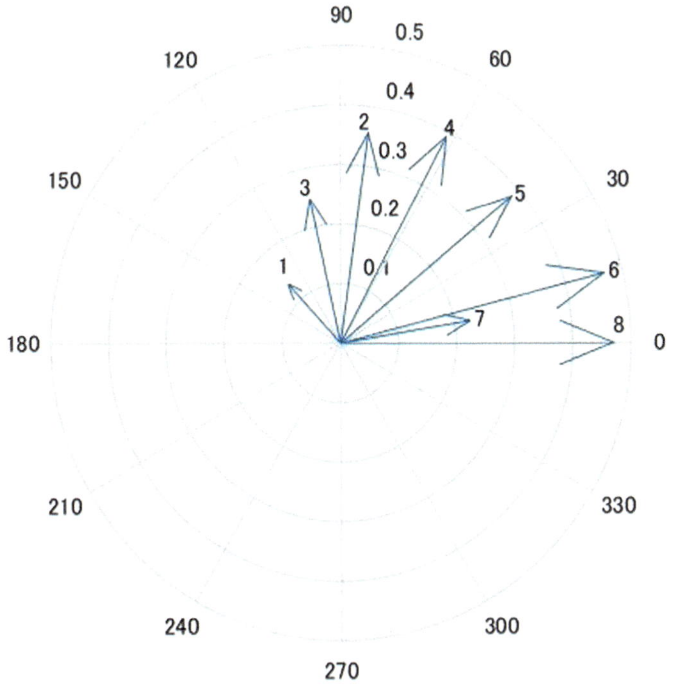

Fig. 1.29 Configuration of web pages on the complex plane associated with the largest eigenvalue, 1.1695. Since the eigenvalue is positive, the positive direction of the configuration in HFM is *clockwise*

zero in marked contrast to the fairly general Hermitian matrices utilized in HFM. For example, in the case of Table 1.27, the eigenvalues of the Hermitian adjacency matrix are $2, 1, -1, -2$. This means that the holistic metric structure of this Hermitian adjacency matrix has an *indefinite metric space*. As for the indefinite metric space, see, for example, Chino and Shiraiwa [21].

Graph Embedding

The scope of the theory of *graph embedding* is very wide, as will be discussed in more detail in Sect. 11.5. It ranges from *graph embedding in a surface* to various *dimensional reduction methods*. In a planarity algorithm for embedding a graph in a surface, one deals with the problem of whether a given graph can be embedded without any crossing edges in the plane (e.g., Hopcroft and Tarjan [47]). However, we will not discuss this algorithm because it is beyond the scope of this book.

By contrast, the following is a list of well-known examples of the dimensional reduction methods (e.g., Yan et al. [110]): Principal Component Analysis (PCA) [49], Linear Discriminant Analysis (LDA) [108], Locality Preserving Projections (LPP) [42, 43], Isometric Feature Mapping (ISOMAP) [101]; Tenenbaum et al. [102], Locally Linear Embedding (LLE) [88], and Laplacian Eigenmap [7]. Factor analysis (FA) [40] and traditional multidimensional scaling (MDS) (e.g., Ramsay [84, 85]; Takane [99]; Takane et al. [100]; Torgerson [104]) developed mainly in psychometrics can also be viewed as examples of the dimensional reduction methods. Each of these methods embeds objects in a certain low dimensional space, given some similarity or dissimilarity matrix among objects. It should be noted that all of these methods assume that these matrices are *symmetric*.

On the other hand, there exist a few graph embedding methods which assume some *asymmetric matrices*. One of them is the *singular value decomposition* (SVD). Usually, SVD assumes a *real* rectangular matrix, as we shall see in more detail in Sect. 11.2. Let $\boldsymbol{A}$ be an $m \times n$ of rank k. Then, there is an $m \times m$ orthogonal matrix $\boldsymbol{U}$ composed of the *left-singular vectors*, an $n \times n$ matrix $\boldsymbol{V}$ composed of the *right-singular vectors*, such that $\boldsymbol{U}^t\boldsymbol{A}\boldsymbol{V} = \boldsymbol{S}$ or $\boldsymbol{A} = \boldsymbol{U}\boldsymbol{S}\boldsymbol{V}^t$. Here the diagonal entries of $\boldsymbol{S}$ are called the *singular values* of $\boldsymbol{A}$ and can be arranged to be nonincreasing, all these entries are nonnegative and exactly k of them are strictly positive (e.g., Lawson and Hanson [63]). In the context of graph embeddings or dimension reduction methods, $\boldsymbol{A}$ may be an ASM. For example, Loly et al. [66] applies SVD to the Lo Shu magic square in Example 1.21 of Sect. 1.2, and examines its asymmetric structure. Eldén [29] introduces the *higher order* SVD called HOSVD [25], and applies it to hand written digits. Okada [80] analyzes an ASM among four societies applying the ordinary SVD to it, from the view point of an asymmetric MDS. Recently, Kroonenberg [59] illustrates various multiway extensions of SVD including HOSVD, inspired by De Lathauwer et al. (op. cit.).

Another method is an *asymmetric LPP* which was recently proposed by Iwai et al. [55]. They extended LPP in such a way that one can handle an asymmetric similarity matrix, by replacing the symmetric similarity matrix in the optimization function with an extended Hermitian matrix, the original one of which was proposed by Chino and Shiraiwa [21]. They applied the asymmetric LPP to the USPS handwritten digit database [50].

1.3.2 Traditional Analyses of Dynamics of Asymmetric Interactions

In this section we shall introduce some of the traditional analyses of dynamics of asymmetric interactions which use *differential equations* or *difference equations* in order to describe these dynamics. These include analyses of dynamics of social networks, biological networks, evolutional stable strategies in biology and game theory, electrical circuits, neural networks, and chemical reaction networks.

Example 1.1 in Sect. 1.2 is a snapshot of a *social network*, which is part of the sociomatrix among 20 male students in a senior high school [16]. It will be easy to imagine that the proximities between members of a formal group at an instant of time will change dynamically according to complex interactions among members in a group. A social psychologist, Newcomb [78] conducted a full-scale investigation, and obtained a set of longitudinal sociomatrices. He observed an acquaintance process of 17 previously unacquainted male students, who lived together in a fraternity-style house. The obtained self-reports were complete rank orderings of the other 16 by attraction during each of the 16 weeks of the experiment. As a preprocessing of these original ranking data, Chino and Nakagawa [20] applied a nonmetric MDS, SSA2 [39] to each of the sociomatrices, and obtained a set of longitudinal configurations of members. Then, they fitted a system of general two-dimensional *nonlinear differential equations* to this set of data, that is:

$$\begin{pmatrix} dx_1/dt \\ dx_2/dt \end{pmatrix} = \begin{pmatrix} f_1(x_1, x_2, b) \\ f_2(x_1, x_2, b) \end{pmatrix}, \tag{1.9}$$

where b is a certain parameter. Here, (1.9) is equivalent to (8.1) in Chap. 8. Functions f_1 and f_2 in (1.9) are approximated by cubic spline functions, given the set of longitudinal configurations. For more detail, see Chap. 8.

Finally, they found several qualitative patterns of the solution curves, applying the *qualitative theory of bifurcations of vector fields* (e.g., Guckenheimer and Holmes [38]) to the set of longitudinal configurations. These patterns are shown in Figs. 8. 5, 8.6 and 8.7 of Chino and Nakagawa (op. cit.) in Chap. 8. Recently, the author has been developing a *complex difference equation model* which enables us to predict changes in an ASM over time. Chap. 9 is an introduction to an elementary theory of *dynamical weighted digraph* which the author has been developing for several years.

Examples 1.5, 1.6, 1.7, 1.8, 1.13 and 1.15 are examples of *biological networks*. Of course, these are only a few examples of networks in biology. In this section, we shall deal with Example 1.15 [72] of food web which has a long history of research. In particular, the Lotka-Volterra equations are famous models for predator-prey interaction (i.e., Lotka [67, 68]; Volterra [105, 106]), which are composed of a system of *two-dimensional nonlinear differential equations*:

$$\begin{pmatrix} dx/dt \\ dy/dt \end{pmatrix} = \begin{pmatrix} rx - pxy \\ epxy - my \end{pmatrix}, \tag{1.10}$$

where x and y are the prey density and the predator density, respectively; r is the Malthusian rate of increase of the prey; m the mortality rate of the predator; p the predation rate; e the coefficient of conversion from prey to predator numbers or biomass [33]. On the other hand, MaCann et al. (op. cit.) consider a system of *4-dimensional nonlinear differential equations*, which is composed of *rational functions* of a resource density, the densities of the first and second consumer species, and a density of the top predator. They showed that the equations exhibited *chaotic dynamics* under certain conditions.

Examples 1.8 and 1.9 show the pay-off matrices in *evolutionarily stable strategies in biology and game theory*. In particular, Example 1.8 is a special case of the pay-off matrices of a generalized rock-paper-scissors game in Sato et al. [91], as introduced in Sect. 5.4 in this book. As will be described in (5.14) in this book, Sato et al. (op. cit.) investigated the problem of learning to play the game of rock-paper-scissors, using a system of *special nonlinear differential equations*, which we reproduce in the following (1.11):

$$\begin{bmatrix} \dot{x}_i \\ \dot{y}_j \end{bmatrix} = \begin{bmatrix} x_i\big((\boldsymbol{A}\mathbf{y})_i - \mathbf{x}^t\boldsymbol{A}\mathbf{y}\big) \\ y_j\big((\boldsymbol{B}\mathbf{x})_j - \mathbf{y}^t\boldsymbol{B}\mathbf{x}\big) \end{bmatrix}, i = 1, \cdots, n, j = 1, \cdots, m. \tag{1.11}$$

To be precise, Example 1.8 is one of the pay-off matrices, i.e., $\boldsymbol{A}$ for the first player, where $\varepsilon_x = 0.25$. In contrast, $\boldsymbol{B}$ is the pay-off matrix for the second player in this equation. Here, we reproduce (5.15) in this book, which defines $\boldsymbol{A}$ and $\boldsymbol{B}$, in the following (1.12):

$$\mathbf{A} = \begin{bmatrix} \varepsilon_x & -1 & 1 \\ 1 & \varepsilon_x & -1 \\ -1 & 1 & \varepsilon_x \end{bmatrix}, \mathbf{B} = \begin{bmatrix} \varepsilon_y & -1 & 1 \\ 1 & \varepsilon_y & -1 \\ -1 & 1 & \varepsilon_y \end{bmatrix} \tag{1.12}$$

Sato et al. consider a general case where $\varepsilon_x = -\varepsilon_y = \varepsilon$. In this case, $A_{ij} = -B_{ij}$, and (1.11) becomes a *Hamiltonian system* [46], as pointed out by Sato et al. They show that this system exhibits a chaotic orbit within a certain range of the ε.

Example 1.14 is an example of *electrical circuits*, which is called an *RLC circuit*. We can derive from it a differential equation that shows how the state of the circuit varies in time (e.g., Hirsch and Smale [45]). In addition to the KCL and KVL discussed in Example 1.14 shown in Sect. 1.2, we have to introduce the following three equations, as Hirsch and Smale (op. cit.) describe. One is the *generalized Ohm's law*,

$$f(i_R) = v_R, \tag{1.13}$$

which describes the functional relationship between the current i_R through the resister v_R. Another is the *Faraday's law*,

$$L\frac{di_L(t)}{dt} = v_L(t) \tag{1.14}$$

which describes the change in the current i_L through the inductor over time. The other is the condition imposed on the capacitor C,

$$C\frac{dv_C(t)}{dt} = i_C(t) \tag{1.15}$$

which describes the change in the voltage at the capacitor over time. Finally, these five equations can be simplified, considering the relations between them, as

$$\begin{pmatrix} dx/dt \\ dy/dt \end{pmatrix} = \begin{pmatrix} y - f(x) \\ -x \end{pmatrix}, \tag{1.16}$$

where $x = i_L$ and $y = v_C$. Hersch and Smale (op. cit., p. 214) discuss the derivation of (1.16) in some detail.

Example 1.7 is an example of *neural networks* (abbreviated as *NN*). As, for example, Goodfellow et al. [34] point out, the research history of neural networks (to be precise, *artificial neural networks*, *ANN*) is long and has various names and changing fortunes. ANN includes r*ecurrent neural network* (RNN), *cellular neural network* (CNN) [22], *convolutional neural network* (another CNN) (e.g., Fukushima [32]; LeCun et al. [64]) and so on. However, we shall introduce here only two of the RNNs, and show examples of equations which describe these networks. As one of the RNNs, Siegelmann et al. [95, Siegelmann [94] proposed the following model,

$$x_i(t+1) = \sigma\left(\sum_{j=1}^{N} a_{ij}x_j(t) + \sum_{j=1}^{M} b_{ij}u_j(t) + c_i\right), i = 1, \cdots, N \tag{1.17}$$

where N is the number of neurons, M is the number of external input signals, x_j are the activations of the neurons, u_j are the external inputs, and a_{ij}, b_{ij}, and c_j are the rational valued coefficients, also called constants or weights. On the other hand, the function σ is the simplest possible "sigmoid", that is,

$$\sigma(x) = \begin{cases} 0, & \text{if } x < 0 \\ x, & \text{if } 0 \le x \le 1 \\ 1, & \text{if } x > 1 \end{cases} \tag{1.18}$$

but various other sigmoidal functions may substitute it. They call this model the classical *analog recurrent neural network* (*ARNN*). Equation (1.17) can be written in matrix form as

$$\boldsymbol{x}(t+1) = \sigma(\boldsymbol{A}\boldsymbol{x}(t) + \boldsymbol{B}\boldsymbol{u}(t) + \boldsymbol{c}) \tag{1.19}$$

where $\boldsymbol{A}$ is an $N \times N$ weight matrix between the neurons, which specifies interactions between neurons, while $\boldsymbol{B}$ is an $\mathrm{N} \times M$ weight matrix between the input signals. As Siegelmann points out, (1.17) (or equivalently (1.19)) can be considered as a *control system* (e.g., Elaydi [28]) with a constant term $\boldsymbol{c}$ from an engineering view.

The other is a *complex-valued neural network* (*CVNN*) proposed by Müezzinoğlu et al. [76]. Their method is designed for a *complex-valued multistate Hopfield associative memory*. This network consists of n fully connected neurons, whose states at time instant k constitute the n-dimensional state vector $\boldsymbol{x}[k]$ of the network. Then the lth component of this vector at time $k+1$ is represented as

$$x_l[k+1] = \mathrm{csign}_K\left(e^{i(\pi/K)} \sum\nolimits_j w_{lj} x_j[k]\right), \tag{1.20}$$

where K is the resolution factor of the network, and determine the cardinality of the finite state space. The $n \times n$ weight matrix $\boldsymbol{W} = \left[w_{lj}\right]$ is a *Hermitian matrix* and its diagonal elements are all nonnegative (Jankowski et al. [57]). Letting ε_j be the nth root of unity, w_{lj} is defined as follows:

$$w_{lj} = \frac{1}{n} \sum_{k=1}^{n} \varepsilon_l^{(k)} \overline{\varepsilon}_j^{(k)}, \, l, j = 1, \cdots, n \tag{1.21}$$

This equation is a modified Hebb rule, as Jankowski et al. point out.

In updating the state of the lth neuron, all other states are kept unchanged. Equation (1.20) is repeated until the following *energy function* reaches a local minimum:

$$E(\boldsymbol{x}) = -\frac{1}{2} \boldsymbol{x}^* \boldsymbol{W} \boldsymbol{x}. \tag{1.22}$$

Here, $\boldsymbol{x}^*$ is the complex conjugate of $\boldsymbol{x}$. The *csign*$_K$function*csign*$_K$ in (1.20) is defined as follows:

$$\mathrm{csign}_K(u) = \begin{cases} e^0, & 0 \leq \arg(u) < 2\pi/K \\ e^{i[2\pi/K]}, & 2\pi/K \leq \arg(u) < 4\pi/K \\ e^{i[2\pi/K](K-1)}, & (K-1)2\pi/K \leq \arg(u) < 2\pi \end{cases}. \tag{1.23}$$

The complex-valued multistate Hopfield network proposed by Jankowski et al. (op. cit.) employs the complex neuron model whose origin dates back to Aizenberg et al. [1]. Hirose and his colleagues also utilize the complex neuron model (e.g., Hirose [44]).

Remark In matrix theory (1.21) is called a special Hermitian form defined on a complex vector space F^n, where $\boldsymbol{\varepsilon}_i = \left(\varepsilon_i^{(1)}, \varepsilon_i^{(2)}, \cdots, \varepsilon_i^{(n)},\right)$ and $\boldsymbol{\varepsilon}_i \in F^n$. As for a general Hermitian form, see, for example, Lancaster and Tismenetsky [62, p. 203], and a remark on Definition 10.38 in Chap. 10 of this book. On the other hand, the function defined by (1.22) is a general Hermitian form defined on a complex vector space F^n, where $\boldsymbol{x} \in F^n$. In HFM for asymmetric MDS developed by Chino and Shiraiwa [21], which is described in Chapter 4 of this book, *observed real ASM* is transformed into a Hermitian matrix first, and then it is modelled by a general Hermitian form.

Example 1.12 is an example of *chemical reaction networks*. As suggested in Sect. 1.2, stoichiometric matrix plays an important role in describing the dynamics of these networks. Table 1.32 shows it. Here, the (j, k) element of the $m \times r$ stoichiometric matrix $\boldsymbol{S}$ is obtained by subtracting the number of moles of jth substrate from that of the kth product, where m is the number of *chemical species* (or *substrates*) and r is the number of chemical *reactions* in the network.

As pointed out in Sect. 1.2, $\boldsymbol{S}$ can be computed as $\boldsymbol{Z}\boldsymbol{B}_R$ where $\boldsymbol{Z}$ is the $m \times c$ *complex stoichiometric matrix* and $\boldsymbol{B}_R$ is the $m \times c$ *incidence matrix* defined by Rao et al. [86], where c is the number of *chemical complexes* such as $X_1 + 2X_2$ which constitute the vertices in Fig. 1.12. In fact, $\boldsymbol{S}$ is computed as

$$\boldsymbol{S} = \boldsymbol{Z}\boldsymbol{B}_R = \begin{bmatrix} 1\,0\,2\,0 \\ 2\,0\,1\,0 \\ 0\,1\,0\,0 \\ 0\,0\,0\,1 \end{bmatrix} \begin{bmatrix} -1 & 1 & 0 & 0 & 0 \\ 1 & -1 & -1 & 0 & 1 \\ 0 & 0 & 1 & -1 & 0 \\ 0 & 0 & 0 & 1 & -1 \end{bmatrix} \tag{1.24}$$

In general, a chemical reaction network focuses on what happens to the *concentrations* of the various molecular species involved, which comprise a set of *reactants* and a set of *products*, as time proceeds. Let $\boldsymbol{x}$ be an m-dimensional *concentration vector* of *m molecular species* (or *chemical substrates*), and $\boldsymbol{v}(\boldsymbol{x})$ be an r-dimensional vector each element of which represents a *reaction rate* of the concentration of any substrate. Then, change in the concentrations of the species over time can be written as *a system of differential equations*:

$$\dot{\boldsymbol{x}} = \boldsymbol{S}\boldsymbol{v}(\boldsymbol{x}). \tag{1.25}$$

Table 1.32 A stoichiometric matrix of the chemical reaction network, whose digraph associated with the adjacency matrix is shown in Fig. 1.12 of Sect. 1.2

Substrates/reactions	v_1	v_2	v_3	v_4	v_5
X_1	– 1	1	2	– 2	0
X_2	– 2	2	1	– 1	0
X_3	1	– 1	– 1	0	1
X_4	0	0	0	1	-1

In chemical reaction networks, there is a basic law which prescribes the dynamics of the concentrations of the species, and it is called the *Law of Mass Action* [107]. This law leads to (1.25) as *polynomial* differential equations. In fact, as, for example, Rao et al. [86] describe,

$$v_j(\boldsymbol{x}) = k_j \exp\left\{\boldsymbol{Z}^t_{Y_j} \ln(\boldsymbol{x})\right\}, \mathrm{j} = 1, 2, \cdots, r \tag{1.26}$$

where $v_j(\boldsymbol{x})$ is the j-th component of $\boldsymbol{v}(\boldsymbol{x})$, Y_j denotes the substrate complex of the jth reaction, and $k_j \geq 0$ is the rate constant of the jth reaction. Here, it should be noticed that Y_j sometimes changes depending on reactions even if the pair of complexes concerning the reaction is the same. For example, in Fig. 1.12 in Sect. 1.2, Y_1 of the first reaction v_1 is $X_1 + 2X_2$ as a *reactant* (before the reaction) and is not X_3 as a *product* (after the reaction), while Y_2 of the second reaction v_2 is X_3 as a reactant and it not $X_1 + 2X_2$ as a product.

Let us now compute two reaction rates as examples. Noticing that $\boldsymbol{Z}^t_{Y_1}$ is the transpose of the first column of $\boldsymbol{Z}$ in (1.24), we have

$$v_1(\boldsymbol{x}) = k_1 \exp\left\{\begin{bmatrix} 1\ 2\ 0\ 0 \end{bmatrix}\begin{bmatrix} ln(x_1) \\ ln(x_2) \\ ln(x_3) \\ ln(x_4) \end{bmatrix}\right\} = k_1 x_1 x_2^2, \tag{1.27}$$

where $\boldsymbol{x} = \{x_i\}$. Similarly, notice that $\boldsymbol{Z}^t_{Y_5}$ is the transpose of the fourth column of $\boldsymbol{Z}$ in (1.24). The reason for this is that as can be seen in Fig. 1.12 the reactant (before the reaction v_5) is X_4. and not X_3 as a product in the reaction v_5. As a result, we have

$$v_5(\boldsymbol{x}) = k_5 \exp\left\{\begin{bmatrix} 0\ 0\ 0\ 1 \end{bmatrix}\begin{bmatrix} \ln(x_1) \\ \ln(x_2) \\ \ln(x_3) \\ \ln(x_4) \end{bmatrix}\right\} = k_5 x_4. \tag{1.28}$$

We can easily compute all the other three reaction rates in the same way.
Finally, using all the reaction rates, we have

$$\begin{aligned}\dot{\boldsymbol{x}} = \boldsymbol{S}\boldsymbol{v}(\boldsymbol{x}) &= \begin{bmatrix} -1 & 1 & 2 & -2 & 0 \\ -2 & 2 & 1 & -1 & 0 \\ 1 & -1 & -1 & 0 & 1 \\ 0 & 0 & 0 & 1 & -1 \end{bmatrix}\begin{bmatrix} k_1 x_1 x_2^2 \\ k_2 x_3 \\ k_3 x_3 \\ k_4 x_1^2 x_2 \\ k_5 x_4 \end{bmatrix} \\ &= \begin{bmatrix} -k_1 x_1 x_2^2 + k_2 x_3 + 2k_3 x_3 - 2k_4 x_1^2 x_2 \\ -2k_1 x_1 x_2^2 + 2k_2 x_3 + k_3 x_3 - k_4 x_1^2 x_2 \\ k_1 x_1 x_2^2 - k_2 x_3 - k_3 x_3 + k_5 x_4 \\ k_4 x_1^2 x_2 - k_5 x_4 \end{bmatrix},\end{aligned} \tag{1.29}$$

1.4 Discussion

In Sects. 1.1 and 1.2 we have given a brief overview of asymmetric interactions among objects or nodes which are observed in vivo *or in daily life* or *are hypothesized mathematically.* In general, there exist two aspects of asymmetric interaction to be exemplified. One is its *structure*, and the other its *dynamics*. As discussed in Sects. 1.1 and 1.3.1, the structure of the static aspect of asymmetric interaction is usually represented by a *directed graph* (i.e., *digraph*) or *ASM* (i.e., asymmetric similarity matrix). These two notions are closely related to each other, because in both of them objects or nodes can be embedded in a certain metric space. There is an immense literature on the static aspect of asymmetric interaction, which is distributed in many fields of science ranging from social and behavioral sciences to natural sciences, as overviewed in Sect. 1.2.

Works on graph and digraph have mainly been done in *graph theory* and *spectral graph theory*, while those on ASM have been done in *asymmetric MDS* in psychometrics. In traditional graph theory and spectral graph theory, major concern has been limited to graphs (not digraphs). Moreover, it seems that researchers engaged in these theories are not interested in the *scale levels of measurement* of the elements of the Laplacian matrix (which is considered as a weight matrix) associated with graph or digraph. On the other hand, researchers in psychometrics have long been interested in the scale levels of the elements of the data under study (e.g., Stevens [97]; Thurstone [103]), one of which is summarized in Sect. 10.3. As a result, researchers in psychometrics have developed various symmetric and asymmetric MDSs which respond to data with various scale levels of measurement. A famous symmetric MDS is ALSCAL developed by Takane et al. [100], which enables us to handle the data scaled at all the levels of measurement. As for asymmetric MDS, see, for example, Chino [17], which is reproduced in Chap. 7.

Another difference between spectral graph theory and asymmetric MDS is in the *assumption of the space* in which objects or nodes are embedded. In spectral graph theory, nodes of a graph or digraph are usually embedded in a *Euclidean space* by solving an eigenvalue–eigenvector decomposition of a Laplacian matrix. Such an embedding is used also in *graph embedding methods* in machine learning (e.g., Yan et al. [110]). We shall briefly discuss them in Sect. 11.5.

By contrast, for example, in symmetric MDS some researchers have also utilized a more general class of metric space other than the Euclidean space. One is a *special Minkowski space* in which Minkowski's r-metric distance is defined. Kruskal ([60]; [61], p. 116]) proposes to use "a class of distance function most often called the L_p or l_p metrics, but occasionally known as *Minkowski's r-metric distance*" defined as

$$d_{ij} = \left[\sum\nolimits_{l=1}^{t} \left|x_{it} - x_{jt}\right|^{r}\right]^{1/r}. \tag{1.30}$$

Kruskal points out that this distance includes *city-block distance* (or *Manhattan metric*), *Euclidean distance*, "L_∞*-metric*" for $r = 1, 2, \infty$, respectively. Equation (1.30) is a distance function which is deduced from a special case of the *general* l^p *norm* defined by

$$\|\mathbf{z}\| = \left(\sum_{i=1}^{\infty} |z_i|^p\right)^{1/p}, \tag{1.31}$$

where $\mathbf{z} = \{z_n\} \in l^p$ and z is a complex vector (e.g., Debnath and Mikusiński[23] p. 11). In matrix algebra, letting $\boldsymbol{x}$ be a typical vector defined on C^n (or R^n), the following norms are called the *Hölder norms* (e.g., Lancasterr and Tismenetsky [62], p. 351):

$$\|\boldsymbol{x}\| = \left(\sum_{i=1}^{n} |x_i|^p\right)^{1/p}, p \geq 1. \tag{1.32}$$

Another is a *Riemannian space*, which was applied to the binocular visual space by Indow and his colleagues (e.g., Indow et al. [53], Indow [54]).

In asymmetric MDS, various models have been proposed since Young [111] proposed a weighted Euclidian model called ASYMSCAL. As can be seen in Chap. 7 which basically reproduces Chino [17], these models are classified into a few groups. However, we classify here these models from a different view point. From a view point of *space structure*, these models can be classified into two. These are *augmented Euclidean space model* and *non-Euclidean space model* including *Hilbert space model, indefinite metric model, asymmetric Minkowski metric model.* According to this classification, most of the extant asymmetric MDS models can be said to belong to the augmented Euclidean space model. This model augments the Euclidean distance by adding something to it or changing something slightly in order to cope with the asymmetry of similarity matrix among objects. It should be noticed that a canonical analysis of skew-symmetry proposed by Gower [35] defined by (7.20) in Chap. 7 assumes a *symplectic structure* which has no Euclidean structure. HFM has a Hilbert space structure or indefinite metric structure, depending on the definiteness of a Hermitian matrix constructed uniquely from any ASM using (4.1) in Chap. 4. An asymmetric distance function proposed by Sato [90] gives an asymmetric Minkowski metric. Both of the Hilbert space model and the asymmetric Minkowski metric model are natural generalizations of the Euclidean space model, each metric of which has mathematically a *holistic metric structure*. Here, it should be noticed that the metric structure of the Hilbert space in HFM comes from the definition of the Hilbert space. That is, its metric structure is *induced* from a (complex) *inner product* which is a condition of the definition, as we shall see in more detail in Sect. 10.5 of Chap. 10.

As regards the dynamic aspect of asymmetric interaction, a variety of systems of difference as well as differential equations have been proposed until now, as discussed in Sect. 1.3.2. Some of these equations are based on the *established laws* which govern the phenomena in each of the discipline of science, while others are not. For example, even in a simple electrical circuit there exist several laws such as the *Ohm's law* and *Faraday's law* which govern the current and voltage in the circuit. As a result, the system of differential equations which describes the changes in the state of the circuit over time is determined uniquely like (1.16). Of course, the form of the function f in (1.13) depends on the characteristics of the resistor. Moreover, if there exist parameters which influence the behaviors of the system, this form must be augmented. For example, let such a parameter be the temperature of the resistor. Then, (1.16) is slightly modified and is written as

$$\begin{pmatrix} dx/dt \\ dy/dt \end{pmatrix} = \begin{pmatrix} y - f_\mu(x) \\ -x \end{pmatrix}. \tag{1.33}$$

Furthermore, let the function $f_\mu(x)$ be described by

$$f_\mu(x) = x^3 - \mu x \tag{1.34}$$

Then, this system exhibits interesting behavior such as a *limit cycle*, as discussed in Hirsch and Smale ([45], pp. 227–228). As for the fundamental behaviors of the systems of linear and nonlinear differential equations, see Chap. 12.

Another example is a chemical reaction network. As discussed in the previous section, this network is basically governed by the *law of mass action*, and the fundamental form of the system of differential equations which describes the change in the concentrations of the chemical species over time is written using (1.25). It is interesting to note that chemical reaction networks can have parameters such as temperature, as electrical circuits have, and the behaviors of the system change as parameter values change (e.g., Nicolis and Prigogine [79]).

However, in most of the networks mainly observed in social and behavioral sciences, it is rare that such strict laws are established. Therefore, in such a network it will be necessary and appropriate to consider general system models which possibly explain various phenomena.

Finally, we shall examine the space structure of the state space of these dynamical systems discussed above. Traditionally, in physical system position $\boldsymbol{x}$ and velocity $\boldsymbol{v}$ of a particle are considered as constituents of the state space of the system, each of which is contained in R^3 (e.g., Hirsch and Smale [45], p. 22). More generally, a set of nonlinear differential equations $\dot{\boldsymbol{x}} = f(\boldsymbol{x})$ with initial condition $\boldsymbol{x}(0) = \boldsymbol{x}_0$ is defined as a continuously differentiable $\left(C^1\right)$ map, $f : U \to R^n$, where $U \subset R^n$ be an open subset of *real Euclidean space* E^n (or of a *differentiable manifold* M) (Guckenheimer and Holmes [38], p. 3). A bit different definition of the state space of this differential equation is made by Hirsch and Smale (op. cit., p. 162). According to their definition, the real Euclidean space is replaced by a *normed vector space*. In quantum physics, a *stationary Schrödinger equation* is defined as a differential equation on a (*complex*) *Hilbert space* (e.g., Blank et al. [9]; Peebles [83]).

As regards difference equations, the state space of a general nonlinear difference equation, $x(n+1) = g(n, x(n))$, is defined as a map $g : Z^+ \times R \to R$, where Z^+ is the set of nonnegative integer and R is the set of real number (e.g., Elaydi [28], p. 2).

If x is an m-dimensional vector, the map is $g : Z^+ \times R^m \rightarrow R^m$. In control systems, a simple example of which is a nonhomogeneous system, $\boldsymbol{x}(n+1) = \boldsymbol{Ax}(n) + \boldsymbol{u}(n)$, $\boldsymbol{x}(n)$ is said to be composed of the *state variables* $x_1(n), x_2(n), \cdots, x_m(n)$ of the system, and a control $\boldsymbol{u}(n)$ is not included in the state variables (e.g., Elaydi, op. cit., p. 251). Usually, R^m might be identified as E^m. In *complex dynamical system*, its domain is a *complex plane* C (e.g., Milnor [74]), or an *n-dimensional complex space* C^n (e.g., Ebeling [27]).

In closing this chapter, we will raise a question: What is an appropriate space with which we unify both the structure and dynamics of asymmetric interactions among objects or nodes, given an observed ASM? A candidate for attaining this goal may be a *Hilbert space*. One reason for this is that any ASM measured at a ratio scale level can be transformed into a Hermitian matrix and that if this matrix is positive semidefinite (p.s.d.) we can embed objects or nodes in a finite-dimensional Hilbert space according to the Chino-Shiraiwa theorem [21]. In other words, *asymmetric similarity naturally induces a Hilbert space structure*. Even if the Hermitian matrix associated with an ASM is *indefinite*, one can examine one-dimensional Hilbert space structures corresponding to the positive eigenvalues of the Hermitian matrix. We shall go into details of this problem in Chaps. 4 and 5.

Another reason is that in many cases asymmetric interactions among objects or nodes might include *complicated nonlinear interactions* among them, and these interactions will change as time proceeds. In such cases, traditional linear and static models developed in psychometrics as well as statistics might not be able to capture these interactions effectively. As a result, it will be appropriate to introduce dynamical system theories in mathematics.

A third reason is that when we consider a dynamical system model for examining the complex structure and dynamics of asymmetric interactions among objects or nodes, we have candidate theories of complex dynamical systems as discussed above (e.g., Milnor [74]; Ebeling [27]). An elementary theory of *Dynamical (or Dynamic) Weighted Digraph* (abbreviated as DWD) which will be introduced in Chap. 9 is a goal of this book, which has not been completed yet.

Appendix

In this appendix, we shall show a few examples of the MATLAB codes which depict digraphs shown in Sect. 1.2. We assume the reader knows basic commands and functions of MATLAB. For a comprehensive guide to MATLAB, we recommend Martinez and Martinez [69]).

Example A.1 (codes for Fig. 1.1) The following MATLAB commands are written to draw Fig. 1.1 in Sect. 1.2. The first two commands are necessary to suppress the coordinate axes in the figure. The third command defines the sociomatrix shown in Table 1.1. For the arguments of the functions (e.g., set, digraph, and plot in this program), enter *help* command like "*help set*" on the command window of MATLAB.

```
set(gca,'Visible','off','NextPlot','replaceChildren').
ax = gca;
A = [4 5 5; 1 4 7; 4 5 7].
G = digraph(A);
plot(ax,G,'k','Layout','force','EdgeLabel',G.Edges.Weight).
print -depsc eps-files/sociomatrix-of-order3.eps.
```

Example A.2 (codes for Fig. 1.5) The following MATLAB commands are written to draw Fig. 1.5 in Sect. 1.2. The function G = digraph(s,t) in the fifth command specifies the edges of the digraph shown in Fig. 1.5 in pairs to represent the source and target nodes.

```
set(gca,'Visible','off','NextPlot','replaceChildren').
ax = gca;
s = [1 1 1 1 2 2 2 3 4 5];
t = [2 3 4 5 3 4 5 4 5 3];
G = digraph(s,t);
plot(ax,G,'k','Layout','force').
print -depsc eps-files/Springer-pecking-order.eps.
```

Example A.3 (codes for Fig. 1.12) The following MATLAB commands are written to draw Fig. 1.12 in Sect. 1.2. The third command specifies the adjacency matrix shown in Table 1.12. The fourth command specifies the node name of the digraph shown in Fig. 1.12, and the seventh command specifies the edge name. In specifying edge names, it should be noticed that we use underscores like 'V_1' which indicates V_1.

```
set(gca,'Visible','off','NextPlot','replaceChildren').
ax = gca;
A = [0 1 0 0; 1 0 1 0; 0 0 0 1; 0 1 0 0].
names = {'X_1 + 2X_2' 'X_3' '2X_1 + X_2' 'X_4'};
G = digraph(A,names);
h = plot(ax,G,'k','Layout','force','EdgeLabel',G.Edges.Weight).
h.EdgeLabel = {'v_1','v_2','v_3','v_4','v_5'};
print -depsc eps-files/chemical-reaction.eps.
```

References

1. Aizenberg, N. N., Ivaskiv, Yu. L., & Pospelov, D. A. (1971). A certain generalization of threshold functions (in Russian). *Doklady Akademii Nauk SSSR, 196*, 1287–1290.
2. Aizenberg, N. N. & Aizenberg, I. N. (1992). CNN based on multi-valued neuron as a model of associative memory for grey scale images. *CNNA'92 Proceedings of the second International Workshop on Cellular Neural Networks Applications* (pp. 36–41). Munich, Germany,
3. Austin, D. (2006). *How Google finds your needle in the Web's haystack*. Monthly essays on mathematical topics, The American Mathematical Society.

4. Balthrop, J., Forrest, S., Newman, M. E. J., & Williamson, M. M. (2004). Technological networks and the spread of computer viruses. *Science, 304*, 527–529.
5. Bavelas, A. (1948). A mathematical model for group structures. *Applied Anthropology, 7*, 16–30.
6. Beer, R. D. (1995). A dynamical systems perspective on agent-environment. *Artificial Intelligence, 72*, 173–215.
7. Belkin, M., & Niyogi, P. (2001). Laplacian eigenmaps and spectral techniques for embedding and clustering. *Advances in Neural Information Processing System, 14*, 585–591.
8. Berlin, T. H., & Kac, M. (1952). The spherical model of a ferromagnet. *Physical Review, 86*, 821–835.
9. Blank, J., Exner, P., & Havlíček, M. (1994). *Hilbert space operators in quantum physics*. AIP Press.
10. Brouwer, A. E., & Haemers, W. H. (2012). *Spectra of graphs*. Springer.
11. Bonacich, P. (1972). Factoring and weightin approaches to status scores and clique identification. *Journal of Mathematical Sociology, 2*, 113–120.
12. Carlson, D. H. (1982). Circulant matrices, by Philip, J. Davis, Wiley, New York, 1979 (book reviews). *Bulletin (New Series) of The American Mathematical Society, 7*, 421–422.
13. Carta, I., Chen, C. H., Schott, A. L., Dorizan, S., & Khodakhah, K. (2019). *Science, 363*, 248–257.
14. Catalan, E. (1846). Note sur le problème de Malfatti [Note on the Malfatti problem]. *Nouvelles annales de mathématiques: journal des candidats aux écoles polytechnique et normale, Serie 1, 5*, 60–64.
15. Chadwick-Furman, N., & Rinkevich, B. (1994). A complex allorecognition system in a reef-building coral: Delayed responses, reversals and nontransitive hierarchies. *Coral Reefs, 13*, 57–63.
16. Chino, N. (1978). A graphical technique for representing the asymmetric relationships between N objects. *Behaviormetrika, 5*, 23–40.
17. Chino, N. (2012). A brief survey of asymmetric MDS and some open problems. *Behaviormetrika, 39*, 127–165.
18. Chino, N. (2020a). How to use the Hermitian form model for asymmetric MDS. In T. Imaizumi, Nakayama, A., & Yokiyama, S. (Eds.), *Advanced Studies in Behaviormetrics and Data Science* (pp. 19–41). Springer Nature.
19. Chino, N. (2020). Hermitian symmetry on a Hilbert space – Its applications to some asymmetric phenomena. *Natural Science, 12*, 221–236.
20. Chino, N., & Nakagawa, M. (1990). A bifurcation model of change in group structure. *The Japanese Journal of Experimental Social Psychology, 29*, 25–38.
21. Chino, N., & Shiraiwa, K. (1993). Geometrical structures of some non-distance models for asymmetric MDS. *Behaviormetrika, 20*, 35–47.
22. Chua, L. O., & Yang, L. (1988). Cellular neural networks: Theory. *IEEE Transactions on Circuits and Systems, 35*, 1257–1272.
23. Debnath, L., & Mikusiński, P. (1990). *Introduction to Hilbert Spaces with Applications*. Academic Press.
24. De La Cuevas, C., Drescher, T., & Netzer, T. (2020). *Journal of Mathematical Physics, 61*, 111704–111715.
25. De Lathauwer, L., De Moor, B., & Vandewalle, J. (2000). A multilinear singular value decomposition. *SIAM Journal on Matrix Analysis and Applications, 21*, 1253–1278.
26. Dürer, A. (1514). Melencolia I. Engraving. National Gallery of Art. U.S.A.
27. Ebeling, W. (2007). *Functions of several complex variables and their singularities*. American Mathematical Society.
28. Elaydi, S. N. (1999). *An Introduction to Differential Equations*. Springer.
29. Eldén, L. (2007). *Matrix methods in data mining and pattern recognition*. Society for Industrial and Applied Mathematics.
30. Fox, C. (1956). Magic matrices. *Mathematical Gazette, 40*, 209–211.

31. Freeman, L. C. (1978/79). Centrality in social networks conceptual clarification. *Social Networks, 1*, 215–239.
32. Fukushima, K. (1975). Cognition: A self-organizing multilayered neural network. *Biological Cybernetics, 20*, 121–136.
33. Gatto, M. (2009). On Volterra and D'Ancona's footsteps: The temporal and spatial complexity of ecological interactions and networks. *Italian Journal of Zoology, 76*, 3–15.
34. Goodfellow, I., Bengio, Y., & Courville, A. (2016). *Deep Learning*. The MIT Press.
35. Gower, J. C. (1977). The analysis of asymmetry and orthogonality. In J. R. Barra, F. Brodeau, G. Romer, & B. van Cutsem (Eds.), *Recent Developments in Statistics* (pp. 109–123). North-Holland.
36. Gray, R. M. (2011). Taeplitz and circulant matrices: A review. In A. Barg (Ed.) *Foundations and trends in communications and information theory* (vol. 2, pp. 155–239). Norwell: now publishers.
37. Greaves, G., Mohar, B., & O, S. (2019). Interlacing families and the Hermitian spectral norm of digraphs. *Linear Algebra and its Applications, 564*, 201–208.
38. Guckenheimer, J., & Holmes, P. (1983). *Nonlinear oscillations, dynamical systems, and bifurcation of vector fields*. Springer-Verlag.
39. Guttman, L. (1968). A general nonmetric technique for finding the smallest coordinate space for a configuration of points. *Psychometrika, 33*, 469–506.
40. Harman, H. H. (1967). *Modern Factor Analysis* 2nd edn. Revised. The University of Chicago Press.
41. Harshman, R. A., Green, P. E., Wind, Y., & Lundy, M. E. (1982). A model for the analysis of asymmetric data in marketing research. *Marketing Science, 1*, 205–242.
42. He, X., & Niogi, P. (2003). *Proceedings of the 16th international conference on neural information processing systems*, pp. 153–160.
43. He, X., Yan, S., Hu, Y., Niyogi, P., & Zhang, H.-J. (2005). Face recognition using Laplacianfaces. *IEEE Transactions on Pattern Analysis and Machine Intelligence, 27*, 328–340.
44. Hirose, H. (2003). *Complex-valued neural networks, theories and applications*. World Scientific.
45. Hirsch, M. W., & Smale, S. (1974). *Differential equations, dynamical systems, and linear algebra*. Academic Press.
46. Hofbauer, J. (1996). Evolutionary dynamics for bimatrix games: a Hamiltonian system? *Journal of Mathematical Biology, 34*, 675–688.
47. Hopcroft, J., & Tarjan, R. (1974). Efficient planarity testing. *Journal of the Association for Computing Machinery, 21*, 540–568.
48. Hopfield, J. J. (1982). Neural networks and physical systems with emergent collective computational abilities. *Proceedings of the National Academy of Sciences of the United States of America, 79*, 2554–2558.
49. Hotelling, H. (1933). Analysis of a complex of statistical variables into principal components. *Journal of Educational Psychology, 24*(417–441), 498–520.
50. Hull, J. J. (1994). A database for handwritten text recognition research. *IEEE transactions on Pattern Analysis and Machine Intelligence, 16*, 550–554.
51. Ibrahim, A. A., & Salman, S. A. (2019). Some properties of magic squares of distinct squares and cubes. *Al-Mustansiriyah Journal of Science, 30*, 60–63.
52. Imai, S., & Guarente, L. (2014). NAD^+ and sirtuins in aging and disease. *Trends in Cell Biology, 24*, 464–471.
53. Indow, T., Inoue, E., & Matsushima, K. (1962). An experimental study of the Luneburg theory of binocular space perception (1). The 3- and 4-point experiments. *Japanese Psychological Research, 4*, 6–16.
54. Indow, T. (1968). Multidimensional mapping of visual space with real and simulated stars. *Perception and Psychophysics, 3*, 45–53.
55. Iwai, Y., Nishiyama, M., & Yoshimura, H. (2017). Asymmetric locality preserving projection and its application to k-nearest neighbor method. In: *Proceedings of the 15th IAPR International Conference on Machine Vision Applications* (pp. 42–45), Nagoya, Japan.

56. Iyer, R. (2021). Helmholtz Hamiltonian mechanics electromagnetic physics gagin generalizing mass-charge and charge-fields gage metrics to quantum relativity gage metrics. https://doi.org/10.20944/preprints202107.0685.v1
57. Jankowski, S., Lozowski, A., & Zurada, J. M. (1996). Complex-valued multistate neural associative memory. *IEEE Transactions on Neural Networks, 7*, 1491–1496.
58. Katz, L. (1953). A new status index derived from sociometric analysis. *Psychometrika, 18*, 39–43.
59. Kroonenberg, P. M. (2020). Multiway extensions of the SVD. In T. Imaizumi, A. Nakayama, & S. Yokoyama (Eds.), *Advanced studies in Behaviometrics and DATA SCIENCE, ESSAYS in Honor of Akinori Okada* (pp. 141–157). Springer.
60. Kruskal, J. B. (1964). Multidimensional scaling by optimizing goodness of fit to a nonmetric hypothesis. *Psychometrika, 29*, 1–27.
61. Kruskal, J. B. (1964). Multidimensional scalling: a numerical method. *Psychometrika, 29*, 115–129.
62. Lancaster, R., & Tismenetsky, M. (1985). *The theory of matrices with applications* (2nd ed.). Academic Press.
63. Lawson, C. L., & Hanson, R. J. (1974). *Solving least squares problem*. Prentice-Hall.
64. LeCun, Y., Boser, B., Denker, J. S., Henderson, D., Howard, R. E., Hubbard, W., & Jackel, L. D. (1989). Backpropagation applied to handwritten zip code recognition. *Neural Computation, 1*, 541–551.
65. Liu, J., & Li, X. (2015). Hermitian-adjacency matrices and Hermitian energies of mixed graphs. *Linear Algebra and its Applications, 466*, 182–207.
66. Loly, P., Cameron, I., Trump, W., & Schindel, D. (2009). Magic square spectra. *Linear Algebra and its Applications., 430*, 2659–2680.
67. Lotka, A. J. (1910). Contribution to the theory of periodic reactions. *Journal of Physical Chemistry, 14*, 271–274.
68. Lotka, A. J. (1924). *Elements of physical biology*. Williams & Wilkins.
69. Martinez, W. L., & Martinez, A. R. (2008). *Computational statistics handbook with MATLAB* (2nd ed.). Chapman & Hall/CRC.
70. Masure, R. H., & Allee, W. C. (1934). The social order in flocks of the common chicken and the pigeon. *The Auk, 51*, 306–327.
71. Maynard Smith, J., & Price, G. R. (1973). The logic of animal conflict. *Nature, 246*, 15–18.
72. McCann, K., Hastings, A., & Huxel, G. R. (1998). Weak trophic interactions and the balance of nature. *Nature, 395*, 794–798.
73. Meeter, M., Myers, C. E., & Gluck, M. A. (2005). Integrating incremental learning and episodic memory models of the hippocampal region. *Psychological Review, 112*, 560–585.
74. Milner, J. (1999). *Dynamics in one complex variable*. Vieweg.
75. Mohar, B. (2016). Hermitian adjacency spectrum and switching equivalence of mixed graph. *Linear Algebra and its Applications., 489*, 324–340.
76. Müezzinğlu, M. K., Güzeliş, C., & Zurada, J. M. (2003). A new design method for the complex-valued multistate Hopfield associative memory. *IEEE Transactions on Neural Networks, 14*, 891–899.
77. Negre, C. F. A., Morzan, U. N., Hendrickson, H. P., Pal, R., Lisi, G. P., Loria, J. P., Rivalta, I., Ho, J., & Batista, V. S. (2018). Eigenvector centrality for characterization of protein allosteric pathways. *Proceedings of the National Academy of Sciences of the United States of America, 115*, E12201–E12208.
78. Newcomb, T. M. (1961). *The Acquaintance Process*. Yew York: Holt. Rinehart and Winston.
79. Nicolis, G., & Prigogine, I. (1977). *Self-organization in Nonequilibrium Systems. From Dissipative Structures to order through Fluctuations*. John Wiley and Sons.
80. Okada, A. (2011). Analyzing social affinity for foreigners by spectral decomposition: asymmetric multidimensional scaling of EASS 2008 data. *Bulletin of the Japanese General Social Survey Research Center of Osaka University of Commerce, 8*, 119–128 (in Japanese).
81. Okada, A., & Imaizumi, T. (1987). Nonmetric multidimensional scaling of asymmetric proximities. *Behaviormetrika, 21*, 81–96.

82. O'Neill, C. C. (2021). Magic squares, the week force and the Higgs mechanism. https://doi.org/10.13140/RG.2.2.23910.96321/1
83. Peebles, P. J. E. (1992). *Quantum mechanics*. Princeton University Press.
84. Ramsay, J. O. (1977). Maximum likelihood estimation in multidimensional scaling. *Psychometrika, 42*, 241–266.
85. Ramsay, J. O. (1982). Some statistical approaches to multidimensional scaling data. *The Journal of the Royal Statistical Society Series A (General), 145*, 285–312.
86. Rao, S., van der Schaft, A., & Jayawardhana, B. (2013). A graph-theoretical approach for the analysis of model reduction of complex-balanced chemical reaction networks. *Journal of Mathematical Chemistry, 51*, 2401–2422.
87. Rothkopf, E. Z. (1957). A measure of stimulus similarity and errors in some paired associate-learning tasks. *Journal of Experimental Psychology, 53*, 94–101.
88. Roweis, S. T., & Saul, L. K. (2000). Nonlinear dimensionality reduction by locally linear embedding. *Science, 290*, 2323–2326.
89. Rushen, J. (1982). The peck orders of chickens: How do they develop and why are they linear? *Animal Behavior, 30*, 1129–1137.
90. Sato, Y. (1988). An analysis of sociometric data by MDS in Minkowski space. In K. Matssita (Ed.), *Statistical theory and data analysis II* (pp. 385–396). North-Holland.
91. Sato, Y., Akiyama, E., & Farmer, J. D. (2002). Chaos in learning a simple two-person game. *Proceedings of the National Academy of Science of the United States of America, 99*, 4748–4751.
92. Schjelderup-Ebbe, T. (1922). Beiträge zur Sozial-psychologie des Haushuhns [Observation on the social psychology of domestic fowls]. *Zeitschrift für Psychologie, 88*, 225–252.
93. Shepard, R. N. (1963). Analysis of proximities as a technique for the study of information processing in man. Human Factor, 33–48.
94. Siegelmann, H. T. (1995). Computation beyond the Turing limit. *Science, 268*, 545–548.
95. Siegelmann, H. T., & Sontag, E. D. (1995). On the computational power of neural nets. *Journal of Computer and System Sciences, 50*, 132–150.
96. Sohm, P. L. (1980). Dürer's "Melencolia I"; The Limits of Knowledge. *Studies in the History of Art, 9*, 13–32.
97. Stevens, S. S. (1946). On the theory of scales of measurement. *Science, 103*, 677–680.
98. Suppes, P., & Zinnes, J. L. (1963). Basic measurement theory. In R. D. Luce, R. R. Bush, & E. E. Galanter (Eds.), *Handbook of mathematical psychology* (Vol. 1, pp. 1–76). Wiley.
99. Takane, Y. (1981). Multidimensional successive categories scaling: a maximum likelihood method. *Psychometrika, 46*, 9–28.
100. Takane, Y., Young, F. W., & de Leeuw, J. (1977). Nonmetric individual differences multidimensional scaling: an alternative least squares method with optimal scaling features. *Psychometrika, 42*, 1–68.
101. Tenenbaum, J. B. (1998). Mapping a manifold of perceptual observations. In M. Jordan, M. Keams, & S. Solla (Eds.), *Advances in Neural Information Processing 10* (pp. 632–688). MIT press.
102. Tenenbaum, J. B., De Silva, V., & Langford, J. C. (2000). A global geometric framework for nonlinear dimensionality reduction. *Science, 290*, 2319–2323.
103. Thurstone, L. L. (1927). Psychophysical analysis. *The American Journal of Psychology, 38*, 368–389.
104. Torgerson, W. S. (1958). *Theory and Methods of Scaling*. Wiley.
105. Volterra, V. (1926). Variazioni e fluttazioni del numero d'individui in specie animali conviventi. *Memorie dell' Accademia dei Lincei, 6*, 31–113.
106. Volterra, V. (1926). Fluctuations in the abundance of a species considered mathematically. *Nature, 118*, 558–560.
107. Waage, P. & Guldberg, C. M. (1964). Studier over Affiniteten [Studies of affinities], *Transactions of the Scientific Society in Christiania*, pp. 35–45.
108. Wald, A. (1943). On a statistical problem arising in the classification of an individual into one of two groups. *Annals of Mathematical Statistics, 15*, 145–162.

109. Ward, J. E., III. (1980). Vector spaces of magic squares. *Mathematical Magazine, 53*, 108–111.
110. Yan, S., Xu, D., Zhang, B., Zhang, H-.J., Yang, Q., & Lin, S. (2007). Graph embedding and extensions: a general framework for dimensionality reduction. *IEEE transactions on pattern analysis and machine intelligence*, *29*, 40–51.
111. Young, F. W. (1975). An asymmetric Euclidean model for multi-process asymmetric data. *Paper presented at U.S.-Japan Seminar on MDS*, San Diego, U.S.A.
112. Yuan, Bo-J., Wang, Y., Gong, S.-C., & Qiao, Y. (2020). *Graphs and Combinatorics*, *36*, 1573–1584.

Part I
The Structure of Asymmetric Interactions

In part I of this book, which includes Chaps. 2 through 7, the structure of asymmetric interactions is discussed from the viewpoint of *asymmetric MDS* in psychometrics. These chapters are, for the most part, reproductions of papers written by the author, Chaps. 2, 3, 5, and 7 being written by the author alone and Chaps. 4 and 6 by the author and his colleagues. We corrected a few words and phrases found in the original papers and omitted some parts of the papers which were thought to be unnecessary. Moreover, we added short notes before the references of each of these papers.

In Chap. 4, which includes a key term in this book, that is, the *Hilbert space*, we added a simple MATLAB code for analyzing empirical data in the appendix. In Chap. 7, we reviewed the asymmetric MDS methods that have been developed up to 2012. The reader may notice from this review that major asymmetric MDS techniques assume certain *distance models*. By contrast, some of them assume certain *inner product space models*. Chapters 2 through 5 are concerned with the inner product space models. Particularly, Chap. 4 assumes the *Hermitian inner product space model*, which is an extension of the usual inner product space model to the *complex space*. The superiority of the inner product space models over distance models is that the *origin* of the space is uniquely determined. This property is very important and crucial in modeling the dynamics of asymmetric interactions among objects, as can be seen in Chap. 8 of Part II. Although the origin is set to the *centroid* of the configuration in most of the *symmetric MDS models* as well as *asymmetric MDS models* which assume distance models, such an origin is defined merely *statistically*, as will be introduced in Chap. 11 of Part III.

Finally, the superiority of the Hermitian inner product model of HFM in Chap. 4 over most of the asymmetric MDS models is that the holistic metric structure including the symmetric part and the skew-symmetric part is defined clearly using the *polar identity* defined by (4.40). By contrast, the holistic metric structure of most of the asymmetric MDS models which are usually called augmented distance models is *unclear* except for HFM and the two-dimensional model of Chino's ASYMSCAL. They merely assume that the symmetric part of the data has a Euclidean distance structure. As regards Chino's ASYMSCAL to be discussed in Chap. 2, there is no indication of the holistic metric structure. However, it is not difficult to prove that in the case of two dimensions, this model is equivalent to the special case of HFM, i.e., the one-dimensional HFM, except for the sign of the configuration of objects and the scale level assumption of this model. In any case, we can prove from the polar identity that in the Hermitian inner product model, various laws on the metric space such as *triangle inequality*, the *Parallelogram Law*, and the *Law of Cosines* hold. Such properties will be discussed in Chap. 10 in Part III.

Chapter 2
Chino's ASYMSCAL

2.1 Introduction

One of the areas of endeavor in recent multi-dimensional scaling research concerns the representation of the asymmetric relationships between objects (for example, Lingoes [12]; Young [15]; Hayashi [11]). But as Young [15] argued, there have been relatively few applications of this feature. Young [15] drew a distinction between single and multi-process data. That is, by Young, the difference is in the number of separate and independent processes which are involved in generating the data represented by a single matrix. If there is a separate process for each row (say) of the matrix, the data are called 'multi-process ones', whereas if there is one over-all process generating the data in all rows of the matrix, they are called 'single process ones'. Further, he suggested that if it is the case that multi-process asymmetric data are much more common than single-process asymmetric data, then it is clear why the successful MDS of asymmetric data is not very common. In the case of multi-process asymmetric data, his model would be one of the natural approaches, as it takes into consideration the differences between the several processes represented by a multi-process ones. But, in view of the fact that, for example, the amount of trade which belongs to multi-process data has been actually a serious problem between Japan and America, we think it fruitless to draw a strict distinction between single and multi-process data. Further, as will be shown later, we find somewhat difficulties in interpreting the configuration obtained by such a model as Young's, because we must also consider the weights in addition to the configuration. This difficulty increases in proportion to square of the number of the objects. This kind of arguments may be applicable to the model proposed by Lingoes [12], assuming that the two kinds of solutions are obtained, i.e. the column solution and the row solution. Now let's say that we are most interested in investigating the structure of the "skewness" between object s. Then it is sufficient for us to distinguish between 'comparable' and 'incomparable' data. That is, given an asymmetric data matrix, we may ask ourselves whether the pairs located at a symmetric part in the matrix are

N. Chino, *Structure and Dynamics of Asymmetric Interactions*,
https://doi.org/10.1007/978-981-97-8269-7_2

comparable or not in a practical sense. This distinction is much the same as Coombs' [5] or Takane et al.'s [13]. In any case, such an idea leads to a new model discussed in this paper.

2.2 The Data to Be Analyzed

Many data problems in the social and behavioral sciences involve some notation of the relationships between objects. Let the data be for N objects, say O_1, O_2, ..., O_N. We shall denote the degree of the relationship between O_j and O_k by R_{ik}. Here we assume the degree of the relationship R_{jk} to be defined at the interval levels of measurement in Stevens' terminology. Thus, we shall mainly refer to the degree R_{jk} which is assumed to be the conditional probability of some sort, of O_j, given O_k. For example, these may refer to the attraction each member of a group expresses towards the other members of the group, and the amount of trade each nation in the world has with other nations, etc. In any case, the data to be analyzed here must be 'comparable' in the sense mentioned in the previous section.

2.3 The Model

Generally, the degree of the relationship R_{jk} is defined by

$$R_{jk} = \boldsymbol{x}_j^t \boldsymbol{A} \boldsymbol{x}_k + c, \tag{2.1}$$

where parameters, $\boldsymbol{x}_j$ and $\boldsymbol{x}_k$ denote the unknown vectors of order p, corresponding to O_j and O_k respectively, and matrix $\boldsymbol{A}$ can be written as

$$\boldsymbol{A} = a\boldsymbol{I} + b\boldsymbol{I}^*, \tag{2.2}$$

where $\boldsymbol{I}^*$ is a skew-symmetric matrix of order p, and a, b, and c are the unknown parameters. Therefore, R_{ik} is a bilinear alternating form of the vectors $\boldsymbol{x}_j$, $\boldsymbol{x}_k$. Note that this is an asymmetric model since $R_{kj} = \boldsymbol{x}_k^t \boldsymbol{A} \boldsymbol{x}_j + c$, and since it is generally the case that $\boldsymbol{A}^t \neq \boldsymbol{A}$. Here we will restrict the problem within the two-dimensional case first. Then, matrix $\boldsymbol{I}^*$ is defined by

$$\boldsymbol{I}^* = \begin{pmatrix} 0 & 1 \\ -1 & 0 \end{pmatrix}. \tag{2.3}$$

Substitution of (2.3) into (2.2) yields

$$A = \begin{pmatrix} a & b \\ -b & a \end{pmatrix},$$

Which when substituted in (2.1) yields, in summation form as

$$R_{jk} = a\left(x_{j1}x_{k1} + x_{j2}x_{k2}\right) + b\left(x_{j1}x_{k2} - x_{j2}x_{k1}\right) + c. \tag{2.4}$$

From (2.4) it is easy to see that we suppose the scalar product and the cross product for the symmetric part and the asymmetric part, of the R_{jk}, respectively. Here we mean $\left(R_{jk} + R_{kj}\right)/2$ and $\left(R_{jk} - R_{kj}\right)/2$ by the symmetric part and the asymmetric part, respectively. By doing so, we intended to represent the complicated relationships between objects in "a" configuration as simply as possible. That is, $\left|x_{j1}x_{k2} - x_{j2}x_{k1}\right|$ which is the absolute value of the cross product, represents twice the area of the triangle with vertices P_0, P_j, P_k, where $\boldsymbol{x}_j = \overrightarrow{P_0P_j}$, $\boldsymbol{x}_k = \overrightarrow{P_0P_k}$.

Further, it is very convenient for us in interpreting the configuration obtained by this model that the sign of $x_{j1}x_{k2} - x_{j2}x_{k1}$ determines the orientation of the parallelogram.

Here let us consider the vectors $\boldsymbol{x}_\mathrm{j}$, $\boldsymbol{x}_\mathrm{k}$ in three-dimensional space for a while. We may employ a "right-handed" co-ordinate system without loss of generality. Then, Fig. 2.1 shows the direction of $\boldsymbol{x}_j \times \boldsymbol{x}_k = \overrightarrow{P_0R}$ which is such that the rotation about the line P_oR of the vector $\boldsymbol{x}_j = \overrightarrow{P_0P_j}$ into the vector $\boldsymbol{x}_k = \overrightarrow{P_0P_k}$ by an angle γ between 0 and π appears counterclockwise when viewed from R. By using this property of the cross product in three-dimensional space, we can determine the orientation of the parallelogram in two-dimensional space. Secondly, let us consider the three-dimensional case. Then, matrix $\boldsymbol{I}^*$ is defined by

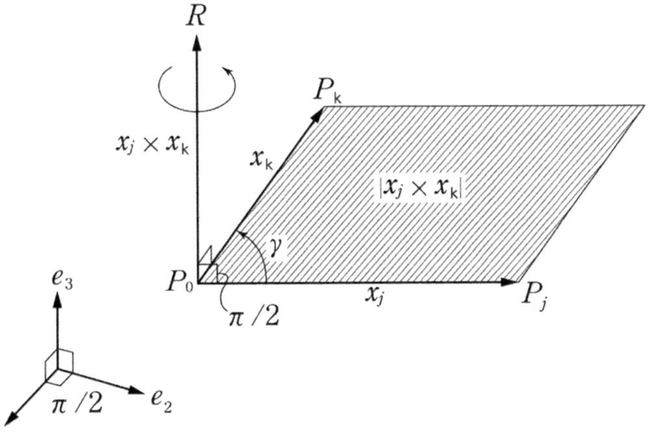

Fig. 2.1 Cross product $\boldsymbol{x}_j \times \boldsymbol{x}_k$ in right-handed coordinate system in three-dimensional space, where $\boldsymbol{e}_1 = (1, 0,0)$, $\boldsymbol{e}_2 = (0, 1,0)$, $\boldsymbol{e}_3 = (0, 0,1)$

$$\boldsymbol{l}^* = \begin{pmatrix} 0 & 1 & -1 \\ -1 & 0 & 1 \\ 1 & -1 & 0 \end{pmatrix} \tag{2.5}$$

Then, by the same argument employed to obtain (2.4)

$$\begin{aligned} R_{jk} &= a(x_{j1}x_{k1} + x_{j2}x_{k2} + x_{j3}x_{k3}) \\ &\quad + b\{(x_{j2}x_{k3} - x_{j3}x_{k2}) + (x_{j3}x_{k1} - x_{j1}x_{k3}) + (x_{j1}x_{k2} - x_{j2}x_{k1})\} + c. \end{aligned} \tag{2.6}$$

In this case, the second right-hand term of (2.6) without b can be interpreted geometrically as the volume of the parallelepiped spanned by the vectors $\boldsymbol{x}_{\mathrm{j}}$, $\boldsymbol{x}_{\mathrm{k}}$, and $\boldsymbol{l}$, where $\boldsymbol{l}^t$=(1, 1, 1). In determinant notation, it takes the form

$$\begin{aligned} det(x_j, x_k, l) &= (x_{j2}x_{k3} - x_{j3}x_{k2}) + (x_{j3}x_{k1} - x_{j1}x_{k3}) + (x_{j1}x_{k2} - x_{j2}x_{k1}) \\ &= (\boldsymbol{x}_j \times \boldsymbol{x}_k)^t \boldsymbol{l}, \end{aligned} \tag{2.7}$$

where vector $\boldsymbol{x}_j \times \boldsymbol{x}_k$ is the cross product in three-dimensional space.

As is well known, it is sometimes referred to as the triple vector product. Furthermore, generally, an ordered triple of three independent vectors $\boldsymbol{u}$, $\boldsymbol{v}$, $\boldsymbol{w}$ defines a certain orientation. Thus, in a right-handed co-ordinate system, we call the triple $\boldsymbol{x}_{\mathrm{j}}$, $\boldsymbol{x}_{\mathrm{k}}$, $\boldsymbol{l}$ oriented positively with respect to the system if it has the same orientation as the triple of vectors $\boldsymbol{e}_1, \boldsymbol{e}_2, \boldsymbol{e}_3$ shown in Fig. 2.1. This is one of the reasons why we adopted (2.5). It should be noted that the following equation holds by definition;

$$det(\boldsymbol{x}_j, \boldsymbol{x}_k, \boldsymbol{l}) = -det(\boldsymbol{x}_k, \boldsymbol{x}_j, \boldsymbol{l}).$$

Here we find some difficulties in interpreting the configuration in three-dimensional space, though the second right-hand term of (2.6) can be interpreted geometrically. But (2.7) helps us to interpret it in this case. That is, from (2.7), it is evident that the det $(\boldsymbol{x}_{\mathrm{j}}, \boldsymbol{x}_{\mathrm{k}}, \boldsymbol{l})$ consists of the sum of the projections onto each of the two-dimensional planes of the parallelogram spanned in space by vectors $\boldsymbol{x}_{\mathrm{j}}$, $\boldsymbol{x}_{\mathrm{k}}$. Let us denote the components of $\boldsymbol{x}_j \times \boldsymbol{x}_k$ by z_1, z_2, and z_3. For example, the expression

$$z_3 = x_{j1}x_{k2} - x_{j2}x_{k1}$$

is just the cross product of the two-dimensional vectors (x_{j1}, x_{j2}) and (x_{k1}, x_{k2}), as is shown in Fig. 2.2.

We will discuss some invariant properties of our model now. First, it is clearly invariant under the orthogonal rotation of the configuration. Secondly, it is invariant under the reflections of the configuration in a certain sense. That is, for example, it is sufficient for us to change the orientation of the pallalelogram under the reflections of the configuration in the two-dimensional case. Thirdly, and most notably, it is invariant under uniform stretching and shrinking of the configuration neglecting the constant c. In other words, it has the property called homogeneous.

Finally, it is possible to expand the cross products in higher dimensions in a sense. But, in view of the data analysis, we thought it fruitless to do so. Furthermore, as will

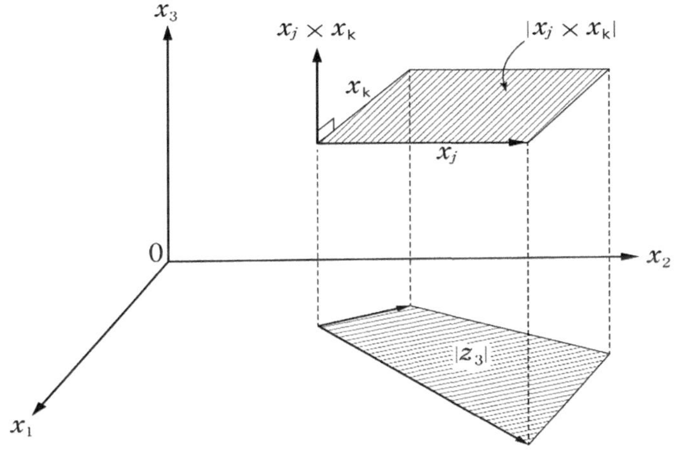

Fig. 2.2 Components of cross product $\boldsymbol{x}_j \times \boldsymbol{x}_k = (z_1, z_2, z_3)$ interpreted as projected areas

be shown in Sect. 2.10, our model is sufficiently enough to explain the asymmetric data even if the dimension is two, in most cases.

2.4 The Illustrative Data

Here, we shows an errorless data to illustrate our model. Table 2.1 shows the degrees of relationships between 6 points which are artificial data. These data were reconstructed from the assumed configuration in Fig. 2.3 using (2.4) of our model. The assumed parameters a, b, and c were 1, 1, and 0, respectively. Assuming that the 6 points represent 6 persons in a small group and the degree of relationships between them represents one's attraction score towards others, we find it easy to see the following relationships between them from the model, glancing at the configuration shown in Fig. 2.3. Here we may employ a right-handed coordinate system. Then the positive direction in this system is shown in Fig. 2.3. For convenience, we denote the center of the circle in Fig. 2.3 by P_0.

(1) For example, the skewness between persons 1 and 4 is the greatest of all, as the angle between the lines P_0P_1, and P_0P_4 is $\pi/2$. Further, we find it easy to see that person 4 likes person 1 very much, though person 1 doesn't like person 4 at all. It should be noted that the coordinate system is assumed to be right-handed.
(2) Persons 1 and 6 like each other, as the angle between the lines P_0P_1, and P_0P_6 is 0.
(3) For example, persons 1 and 3 hate each other, as the angle between the lines P_0P_1, and P_0P_3 is π.

Table 2.1 Reconstructed data from the assumed configuration shown in Fig. 2.3 (artificial data)

Rater/ratee	1	2	3	4	5	6
1	1	1	− 1	− 1	0	1/2
2	− 1	1	1	− 1	$-\sqrt{2}$	− 1/2
3	− 1	− 1	1	1	0	− 1/2
4	1	− 1	− 1	1	$\sqrt{2}$	1/2
5	$\sqrt{2}$	0	$-\sqrt{2}$	0	1	$\sqrt{2}/2$
6	1/2	1/2	− 1/2	− 1/2	0	1/4

Although diagonal elements were empty in the original table, we added them to this table

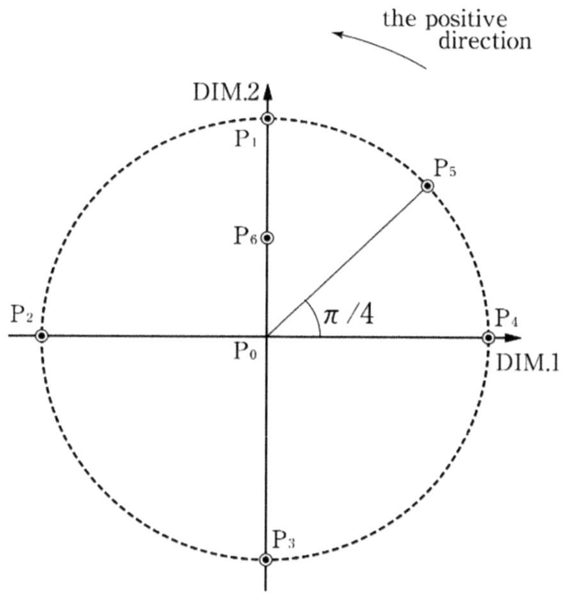

Fig. 2.3 The assumed configuration

(4) For example, person 4 likes person 5 very much, but person 5 neither likes nor dislikes him, as the angle between the lines P_0P_4 and P_0P_5 is $\pi/4$. Such a relationship might be called "unilateral love".

On the other hand, person 5 neither likes nor dislikes person 2, but person 2 hates him very much, as the angle between the lines POP5 and POP2 is $3\pi/4$. Such a relationship might be interesting in contrast to that of "unilateral love".

2.5 The Algorithm

For fallible data, we are given experimentally the N(N − 1) elements, R_{jk}. A reasonable procedure would seem to be to minimize the following function Q in a least square sense;

$$Q = \sum_{j \neq k} \sum_{k} \left(R_{jk} - \boldsymbol{x}_j^t \boldsymbol{A} \boldsymbol{x}_k - c\right)^2, \tag{2.8}$$

There are many strategies for attacking this problem via the gradient approach. We propose subdividing the parameters into two subsets, and then obtaining least squares estimates for one of the parameters subsets under the assumption that the remaining parameters are in fact known constants. The estimation is then alternately repeated for the first subset and then the other until all subsets have been so estimated. This process is iterated until convergence is obtained. This algorithm is a familiar alternating least squares (ALS) method.

A major reason for this is, the rapidity of this algorithm relative to DFP which is one of the variable metric methods and was found to be the most consistently successful, of the methods tried for unconstrained optimization by Box [2].

To minimize Q, we first take the derivative of Q with respect to $\boldsymbol{x}_h$, where subscript $h \neq j, k$. Then, we have

$$\partial Q / \partial \boldsymbol{x}_h = -2 \left\{ \sum_{j \neq h} \left(R_{jh} - \boldsymbol{x}_j^t \boldsymbol{A} \boldsymbol{x}_h - c\right) \boldsymbol{A}^t \boldsymbol{x}_j + \sum_{k \neq h} \left(R_{hk} - \boldsymbol{x}_h^t \boldsymbol{A} \boldsymbol{x}_k - c\right) \boldsymbol{A}^t \boldsymbol{x}_k \right\}. \tag{2.9}$$

Switching notation of the right member of (2.9) from k to j for later convenience, and rearranging its right member, and noting that $\boldsymbol{A} + \boldsymbol{A}^t = 2a\boldsymbol{I}$, we get

$$\partial Q / \partial \boldsymbol{x}_h = -2 \sum_{j \neq h} \left\{ \left(R_{jh} \boldsymbol{A}^t + R_{hj} \boldsymbol{A}\right) \boldsymbol{x}_j - \left(\boldsymbol{A}^t \boldsymbol{x}_j \boldsymbol{x}_j^t \boldsymbol{A} + \boldsymbol{A} \boldsymbol{x}_j \boldsymbol{x}_j^t \boldsymbol{A}^t\right) \boldsymbol{x}_k - 2ac\boldsymbol{x}_j \right\}. \tag{2.10}$$

Setting $\partial Q / \partial \boldsymbol{x}_h$ equal to zero, and rearranging, we finally have.

$$\boldsymbol{x}_h = \boldsymbol{C}_h^{-1} \sum_{j \neq h} \left\{ \left(R_{jh} \boldsymbol{A}^t + R_{hj} \boldsymbol{A}\right) - 2ac\boldsymbol{I} \right\} \boldsymbol{x}_{j,} \tag{2.11}$$

where $\boldsymbol{C}_h = \boldsymbol{A}^t \boldsymbol{Y} \boldsymbol{A} + \boldsymbol{A} \boldsymbol{Y} \boldsymbol{A}^t$ and $\boldsymbol{Y} = \sum_{j \neq h} \boldsymbol{x}_j \boldsymbol{x}_j^t$.

Equation (2.11) suggests the following iterative solution

$$\boldsymbol{x}_{h,t+1} = \boldsymbol{C}_{h,t}^{-1} \sum_{j \neq h} \left\{ \left(R_{jh} \boldsymbol{A}^t + R_{hj} \boldsymbol{A}\right) - 2a_t \boldsymbol{c}_t \boldsymbol{I} \right\} \boldsymbol{x}_{j,t.} \tag{2.12}$$

The values $\boldsymbol{x}_{j,t}$ obtained at trial t are used to obtain the next $\boldsymbol{x}_{h,t+1}$.

That (2.12) enjoys the usual properties of a gradient algorithm can be seen by rewriting it, with aid of (2.10), in the form

$$\boldsymbol{x}_{h,t+1} = \boldsymbol{x}_{h,t} - \frac{1}{2}\boldsymbol{C}_{h,t}^{-1}(\partial Q/\partial \boldsymbol{x}_h)_t.$$

This shows how the coordinates at trial $t + 1$ are adjusted from those of trial t by taking into account the derivatives as evaluated at trial t. For the details of getting this iterative equation, see Chino [3].

There are two different normalization questions that should be considered at this time. One refers to normalization of the configuration on the way of the iterative process to the final configuration. For computational purpose, it is desirable to keep the scale of the co-ordinates as invariant as possible throughout the calculations. Fortunately, as mentioned above, Eqs. (2.4) and (2.6) of our model have the desirable properties, especially the one called homogeneity. Accordingly, we have normalized the configuration by scaling each vector $\boldsymbol{x}_j$ so that the maximum $(|\boldsymbol{x}_1|, |\boldsymbol{x}_2|, \ldots |\boldsymbol{x}_N|)$ equals one at each trial t. The other has to do with normalization of the data. As we assumed the degrees R_{jk} to be defined at the interval levels of measurement, it is convenient to normalize them in such a way that the mean and the standard deviation of them equal zero and one, respectively.

Secondly, taking the derivatives of Q with respect to a, b, and c, respectively, and setting them equal to zero, we finally have

$$a = \sigma_{RP}/\sigma_{PP},\ \ b = \sigma_{RP}/\sigma_{PP},\ \ c = \overline{R}_{..} - a\overline{P}_{..}, \tag{2.13}$$

where

$$a = \sigma_{RP}/\sigma_{PP},\ \ b = \sigma_{RP}/\sigma_{PP},\ \ c = \overline{R}_{..} - a\overline{P}_{..},$$

$$\sigma_{PP} = \sum\sum_{j\neq k}\left(\boldsymbol{x}_j^t\boldsymbol{x}_k - \overline{P}_{..}\right)^2/\{N(N-1)\},$$

$$\sigma_{RP} = \sum\sum_{j\neq k}\left(R_{jk} - \overline{R}_{..}\right)\left(\boldsymbol{x}_j^t\boldsymbol{x}_k - \overline{P}_{..}\right)/\{N(N-1)\},$$

$$\sigma_{OO} = \sum\sum_{j\neq k}\left|\boldsymbol{x}_j \times \boldsymbol{x}_k\right|^2/\{N(N-1)\},$$

$$\sigma_{RO} = \frac{\sum\sum_{j\neq k} R_{jk}sign\left(\boldsymbol{x}_j \times \boldsymbol{x}_k\right)\left|\boldsymbol{x}_j \times \boldsymbol{x}_k\right|}{\{N(N-1)\}},$$

$$\overline{R}_{..} = \sum\sum_{j\neq k} R_{jk}/\{N(N-1)\},$$

$$\overline{P}_{..} = \sum\sum_{j\neq k} \boldsymbol{x}_j^t\boldsymbol{x}_k/\{N(N-1)\}.$$

and

$$\overline{P}_{..} = \sum\sum_{j \neq k} \boldsymbol{x}_j^t \boldsymbol{x}_k / \{N(N-1)\}.$$

We will present the flow of this algorithm later.

2.6 Convergence

It is desirable to obtain the second-order derivative Hessian matrix in order to help evaluate whether an extremum of the function Q has been found and to implement the more efficient Newton–Raphson method.

Thus, we need the derivatives of (2.10). Remembering $\sum_{j \neq h}\left(\boldsymbol{A}^t \boldsymbol{x}_j \boldsymbol{x}_j^t \boldsymbol{A} + \boldsymbol{A} \boldsymbol{x}_j \boldsymbol{x}_j^t \boldsymbol{A}^t\right)$ $= \boldsymbol{C}_h$, we obtain

$$\frac{\partial}{\partial \boldsymbol{x}_h}\left(\frac{\partial Q}{\partial \boldsymbol{x}_h}\right) = 2\boldsymbol{C}_h. \tag{2.14}$$

The determinant of the right-hand term of (2.14) in two-dimensional case can be written in summation notation as

$$\begin{aligned} \frac{1}{2}\det(\boldsymbol{C}_h) = (a^4 + b^4)&\left\{\sum_{j \neq h} x_{j1}^2 \sum_{j \neq h} x_{j2}^2\right\} - \left(\sum_{j \neq h} x_{j1} x_{j2}\right)^2 \\ &+ a^2 b^2 \left\{\sum_{j \neq h} x_{j1}^2 + \sum_{j \neq h} x_{j2}^2 + 2\left(\sum_{j \neq h} x_{j1} x_{j2}\right)^2\right\} \end{aligned} \tag{2.15}$$

The first right-hand term of (2.15) is greater than or equal to 0 on account of Schwartz's inequality.

Then it will be easily proved that the Hessian matrix is positive semidefinite. Accordingly, our procedure (2.12) is robust throughout the calculations except when either a or b becomes nearly equals to zero and that $x_{j1} \fallingdotseq \alpha x_{j2}$ $(j = 1, 2, \ldots, N$ and$j \neq h)$ in the two-dimensional case.

In the case of three-dimensions, we haven't proved whether the Hessian is positive definite or not. But consideration about this problem helps evaluate whether the matrix is positive definite or not. Rearranging the right member of the matrix $\boldsymbol{C}_h$ defined above, and remembering, $\boldsymbol{I}^{*t}\boldsymbol{Y}\boldsymbol{I}^* = \boldsymbol{I}^*\boldsymbol{Y}\boldsymbol{I}^{*t}$ we get,

$$\boldsymbol{C}_h = 2a^2\boldsymbol{Y} + 2b^2\boldsymbol{I}^*\boldsymbol{Y}\boldsymbol{I}^{*t}. \tag{2.16}$$

Though the matrix $\boldsymbol{Y}$ of the first right-hand term is generally positive semi-definite, the Hessian $\boldsymbol{C}_h$ would also be positive definite no matter what the matrix $\boldsymbol{I}^*\boldsymbol{Y}\boldsymbol{I}^{*t}$ is, if $\boldsymbol{Y}$ is positive definite and a^2 is sufficiently large relative to b^2. This observation may be basically the same as the one on which the Marquardt method rests (see, Bard [1]). Finally, while we have described the process (2.12) for convenience, computations may be accelerated and refined if we introduce a quasi-Newton algorithm in this process. Alternatively, the Hessian can be modified if necessary. Discussion of these ideas takes us away from our immediate objectives.

2.7 The Measure of Goodness of Fit

The total variance of R_{jk}, σ_{RR} would be partitioned into three parts, with aid of $\partial Q/\partial x_h = 0$, $(h = 1,2,\ldots,N)$ etc. Thus, we have

$$\sigma_{RR} = \sigma_{RP}^2/\sigma_{PP} + \sigma_{RO}^2/\sigma_{OO} + \sigma_\varepsilon^2.$$

Setting $\sigma_{RR} = 1$ without loss of generality, we get

$$1 = r_{RP}^2 + r_{RO}^2 + \sigma_\varepsilon^2, \tag{2.17}$$

where r_{RP} and r_{RO} represent the Pearson's product-moment coefficients of correlation between the R_{jk} and the scalar product, and the R_{jk} and the "areas" of the parallelogram spanned by vectors $\boldsymbol{x}_j$ and$\boldsymbol{x}_k$, respectively.

It is evident that (2.17) has a desirable property to evaluate goodness of fit of the data to our model. Thus, we can define F as goodness of fit to the model, where

$$F = r_{RP}^2 + r_{RO}^2.$$

Further, we can define r_{RO}^2 as a measure of "skewness" within N objects, too.

2.8 The Flow of Our Algorithm

As was discussed above, our algorithm involves two major phases. We will now present the flow of our algorithm in detail. The procedure consists of the following steps:

(1) Compute the configuration of r dimensions using Hayashi's eq-type quantification by symmetrizing the raw observations, where Hayashi's one [9] is essentially the same algorithm as was used in Guttman's SSA [8] as a tool for the initial approximation to the configuration.

(2) Set the initial configuration $x_{j,0}(j = 1,2, \ldots, N)$ using arbitrary two (three-) dimensional one selected from the co-ordinates of r dimensions obtained by eq-type quantification.
(3) Set $t = 1$.
(4) At the start of the t-th iteration, normalize the configuration and then compute a_t, b_t, and c_t using (2.13).
(5) Compute Q_t using (2.8), where Q_t represents the value of Q at trial t.
(6) If $t = 1$, proceed to step 11. Otherwise:
(7) Compute $D = (Q_{t-1} - Q_t)/\{N(N - 1)\}$.
(8) If $D < 0$, return to step 2.
(9) If $D > \varepsilon$, proceed to step 11, where ε is a small positive constant. Otherwise:
(10) If all the absolute values of the gradients are smaller thanδ, terminate and accept $x_{j,t}$ $(j = 1, 2, \ldots, N)$,a_t,b_t, and c_t where δ is a small positive constant. Otherwise:
(11) Compute $x_{j,t+1}$ for $j = 1,2, \ldots, N$ using (2.12).
(12) Replace t by $t + 1$.
(13) If $t > t$ max, return to step 2. Otherwise return to step 4.

2.9 Relations to Earlier Works

Almost all previously proposed methods for ASYMSCAL (multi-dimensional scaling of asymmetric data) can be classified into two groups, as was discussed in Sect. 2.1. On the contrary, the method for ASYMSCAL proposed in the present paper cannot be classified into those two groups. First and foremost, the present method can be distinguished from all previous ones in that it is designed to represent the asymmetric relationships and the orientations of the skewness between objects in a configuration. In particular, from a mathematical point of view, the present method assumes the model to be composed of the vector product. In this sense, our procedure might be regarded as a kind of tensor approximations. Secondly, the present method makes the assumption that the data to be analyzed must be defined at the interval level of measurement. Nevertheless, our method bears an obvious resemblance to MDS. First it becomes an inner product model if the parameter b becomes to zero in our model. Secondly, it becomes a special case of generalized Guttman-Lingoes' SSA-III, if we assume R_{jk} to be defined at the ordinal level of measurement and adopt a new algorithm appropriate to it. The reason why we call it 'a special case' is that our model might be restricted within two or three dimensions, even if generalized to a non-metric case.

2.10 Some Applications

We will now present some applications of our model to social and behavioral science problems.

First, we choose data which have been previously analyzed so that we can compare our results with those already published. Specifically, we employ data by Coombs et al. [6]. These data were reanalyzed by Young [15], and Hayashi [11]. We will compare our solution with those of Hayashi's ASYMSCAL-KL, and G-L's SSA-II. The data are shown in Table 2.2. We don't refer to Young's ASYMSCAL, because Hayashi's ASYMSCAL-KL model is equivalent to Young's except for the algorithm to solve it and that model enables us to compute the correlation ratio which is an explicit measure of goodness of fit to the model.

The two-dimensional configuration obtained by the analysis of Chino's ASYMSCAL is shown in Fig. 2.4. Pearson's product-moment coefficients, r_{RP} and r_{RO} were computed as a measure of goodness of fit to our model. These coefficients were $-$ 0.92 and 0.10 for r_{RP} andr_{RO}, respectively. An over-all measure of goodness of fit, i.e., F $(=r_{RP}^2 + r_{RO}^2)$, then becomes 0.86. In interpreting this configuration, it is important to keep in mind that in this case r_{RP} is negative. It is easy to see from this configuration that, JASP and JCP, and AJP and JExP are in a relationship which might be called "unilateral love", respectively. On the other hand, we find it easy to see that JApP, JEdP, and PKa "hate" one another, but JASP and JCPP "like" each other.

It is of interest to compare the configuration obtained by the analysis discussed here with that of Hayashi's. Figures 2.5 and 2.6 show Hayashi's results. There are some similarities between Figs. 2.4 and 2.6. But the former is different in philosophy from the latter. The correlation ratio, η^2 in Hayashi's solution was 0.83. At least in this case, it is evident that our solution is much efficient in form than that of Hayashi's. It should be noted, however, that these two measures of goodness of fit to the model are different in philosophy. Accordingly, it might be unable to compare them straightforwardly.

Table 2.2 The citation frequency among certain psychology journals (from Coombs [5])

From/to	1	2	3	4	5	6	7	8
1 AJP		43	12	26	23	9	22	11
2 JASP	41		35	54	21	36	39	42
3 JApP	20	32		33	14	3	49	4
4 JCPP	31	54	40		51	34	25	37
5 JCP	24	1	18	50		17	53	19
6 JEdP	13	27	7	30	15		44	6
7 JExP	2	56	47	45	52	46		38
8 PKa	16	29	8	28	10	5	48	

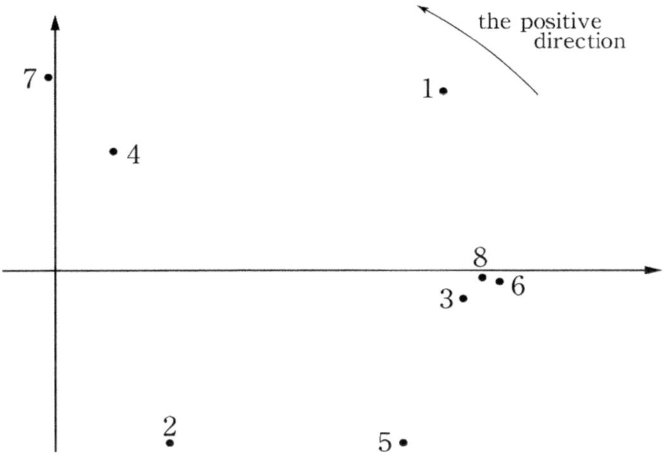

Fig. 2.4 The configuration obtained by the analysis of Chino's ASYMSCAL for the 8 journals

Fig. 2.5 The one-two plane of the weights obtained by the analysis of Hayashi's ASYMSCAL-KL for the 8 journals

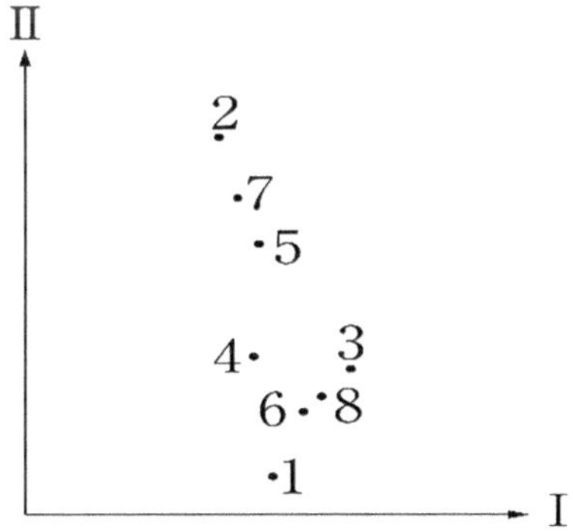

Fig. 2.6 The configuration obtained by the analysis of Hayashi's ASYMSCAL-KL for the 8 journals

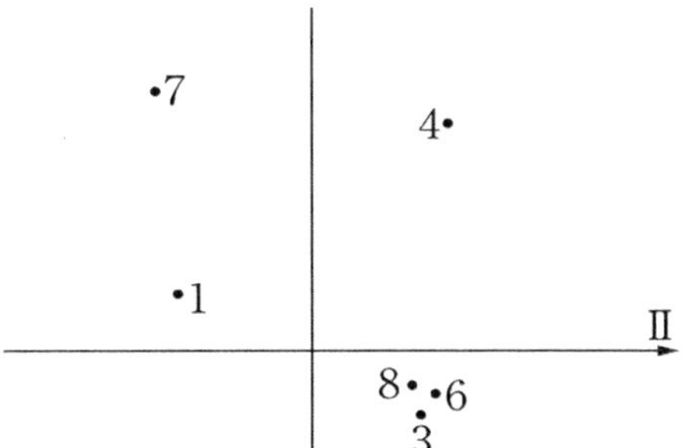

Further, while we must consider both weights and the configuration in interpreting the latter, only one configuration in interpreting the former is needed.

Next, we will compare the configuration obtained by the analysis discussed here with those of G-L's. For the purpose of comparison between them, we regarded Coombs data as dissimilarity one. Here we thought it permissible to regard the data as comparable in a practical sense. Accordingly, we computed both the column solution and the row one. These are shown in Figs. 2.7 and 2.8. The coefficients of alienation computed as a measure of goodness of fit were 0.06 and 0.04 for the column solution and the row one, respectively. Now we must consider two configurations in interpreting the result.

Here, it should be noted that the resemblance between Hayashi's configuration and G-L's configuration (especially, the column solution) is strikingly close, though Young has already pointed out such a case comparing his configuration with that of G-L's.

From the above discussion, we might conclude that the locations of journal 1 and that of journal 4 in the configuration obtained by the analysis of Chino's ASYMSCAL

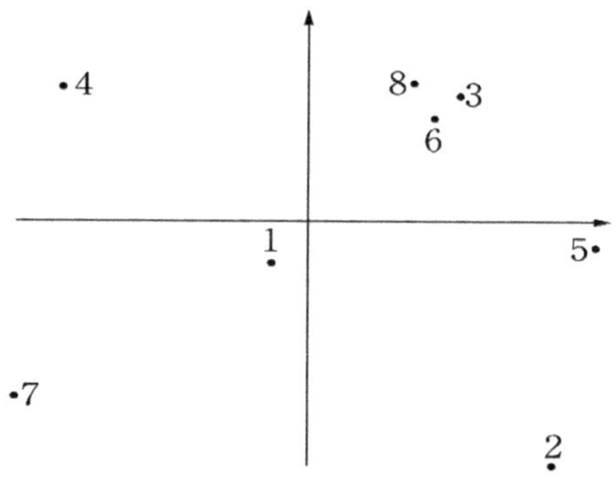

Fig. 2.7 The column solution obtained by the analysis of G-L's SSA-II

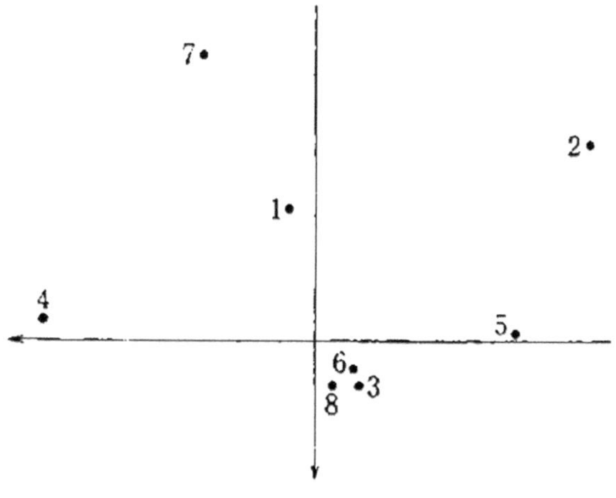

Fig. 2.8 The row solution obtained by the analysis of G-L's SSA-II

are somewhat different from Hayashi's one and G-L's one. This location seems to be rather important in the case of Chino's ASYMSCAL because the direction in the configuration is essential in this method. Paraphrase it to say, when viewed from the positive direction in Fig. 2.4, we might conclude that journal 4 is cited more frequently from the other journals than journal 1. This might be easily verified from Table 2.2.

Secondly, we will analyze two sets of sociometric data using Chino's ASYMSCAL, under the assumption that these data are comparable in a practical sense. The data shown in Tables 2.3 and 2.4 were collected from the same class at a senior high school on June and July, respectively. The class consisted of 47 girls. The girls were asked to rate the amount of liking each member of the class expresses towards own and the other members. We selected 20 girls from here in order to represent a typical pattern of the sociometric structure in this class. The results are shown in Figs. 2.9 and 2.10, which indicate the planes defined by dimensions two and three, one and two, respectively. The coefficients r_{RP} and r_{RO} were 0.66 and − 0.37, and 0.61 and − 0.53 for the data on June and July, respectively.

Table 2.3 The amount of liking each member of the class expresses towards own and the other members on June (The scores range from 1 to 7 for the most negative and positive attraction, respectively)

Subjects	1	2	3	4	5	6	7	8	9	10	11	12	13	14	15	16	17	18	19	20
1	4	3	4	3	5	5	6	4	7	7	4	7	5	4	4	6	6	6	6	7
2	2	3	4	6	7	6	5	5	4	5	4	4	4	4	4	5	4	4	3	5
3	4	4	3	5	5	4	4	3	4	5	3	4	4	4	3	4	4	4	3	5
4	4	7	6	3	7	7	4	6	4	5	4	5	5	7	6	7	4	4	4	3
5	1	7	6	7	4	7	6	6	6	5	1	5	4	6	6	6	5	4	6	1
6	4	5	4	6	5	7	4	4	4	4	5	4	4	6	5	4	4	4	5	4
7	4	5	4	4	3	3	6	6	4	6	4	5	4	5	5	5	5	3	5	3
8	2	4	4	4	5	4	5	2	4	4	4	4	4	4	4	4	3	4	5	3
9	6	5	5	5	5	6	5	5	4	6	5	5	5	5	6	5	5	7	5	5
10	4	4	4	3	4	4	4	4	4	4	4	4	4	4	4	4	5	4	4	4
11	3	4	4	5	4	3	4	4	4	4	1	3	4	4	5	4	4	2	4	4
12	4	4	4	4	5	4	4	4	4	5	3	6	4	4	4	4	4	4	4	4
13	5	3	4	3	4	5	4	4	4	5	4	3	4	4	4	4	4	4	3	4
14	4	4	4	4	6	7	4	4	4	4	4	4	4	3	5	4	4	4	6	4
15	3	4	3	4	5	5	4	4	4	4	4	4	4	4	3	4	4	4	4	3
16	6	6	4	6	6	4	6	4	4	7	1	6	2	5	6	5	7	5	2	4
17	5	5	4	4	5	5	4	3	4	6	4	4	4	4	4	5	6	4	4	3
18	4	4	4	4	4	4	4	4	4	4	3	4	5	4	5	4	4	4	4	4
19	6	6	3	4	7	6	4	6	4	6	4	4	4	6	5	6	4	4	2	4
20	5	3	4	4	5	5	5	3	4	5	4	4	4	4	3	5	4	3	5	5

Table 2.4 The amount of liking each member of the class expresses towards own and the other members on July

Subjects	1	2	3	4	5	6	7	8	9	10	11	12	13	14	15	16	17	18	19	20
1	4	6	6	4	7	6	7	4	7	7	6	7	6	6	6	7	7	7	7	7
2	3	4	4	4	4	4	4	4	4	4	4	4	4	4	4	4	4	4	4	3
3	4	4	4	6	6	4	5	1	4	6	4	5	4	5	3	4	5	4	3	4
4	4	6	6	4	6	6	4	4	4	5	4	5	5	6	6	6	1	5	6	3
5	1	7	6	7	1	6	5	6	6	6	2	6	6	6	6	6	6	5	7	1
6	4	6	4	6	1	7	4	4	4	4	5	4	4	6	5	5	4	4	4	4
7	4	4	4	4	1	3	5	5	4	6	4	5	4	4	4	4	5	4	5	4
8	3	4	4	4	5	5	5	2	4	4	4	1	4	4	4	4	3	4	4	4
9	7	6	6	6	6	7	6	6	4	7	6	6	6	6	6	7	7	7	6	6
10	4	3	4	1	4	4	4	4	4	4	3	4	4	4	4	4	4	4	4	4
11	4	4	4	1	4	1	4	4	4	4	2	4	4	6	4	4	4	4	6	6
12	4	4	4	1	4	4	5	4	4	4	4	5	4	4	4	4	4	4	4	4
13	4	4	4	4	5	5	4	1	5	5	4	4	3	5	5	5	4	4	4	4
14	4	5	4	5	5	7	4	5	4	5	4	4	4	3	5	4	4	5	7	4
15	4	4	3	4	4	4	4	4	4	4	4	4	4	4	3	4	5	4	4	3
16	6	6	4	6	6	6	6	3	6	7	4	6	6	6	6	6	7	5	6	6
17	5	5	4	1	4	5	4	4	5	5	4	4	4	4	5	5	5	4	4	4
18	4	4	4	5	4	4	4	4	4	4	4	4	5	4	4	4	4	4	4	4
19	6	6	4	5	6	6	4	6	4	5	4	4	4	6	5	5	4	4	2	4
20	6	3	4	5	6	5	5	4	4	5	6	4	4	4	3	4	4	4	5	5

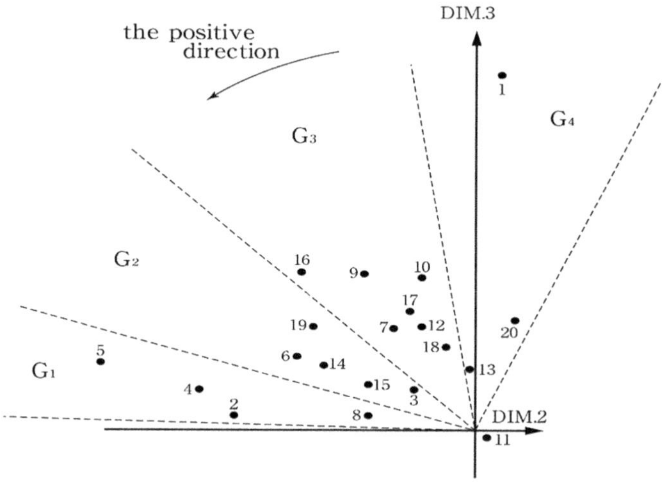

Fig. 2.9 The two–three plane obtained by the analysis of Chino's ASYMSCAL for the 20 senior high school students on June

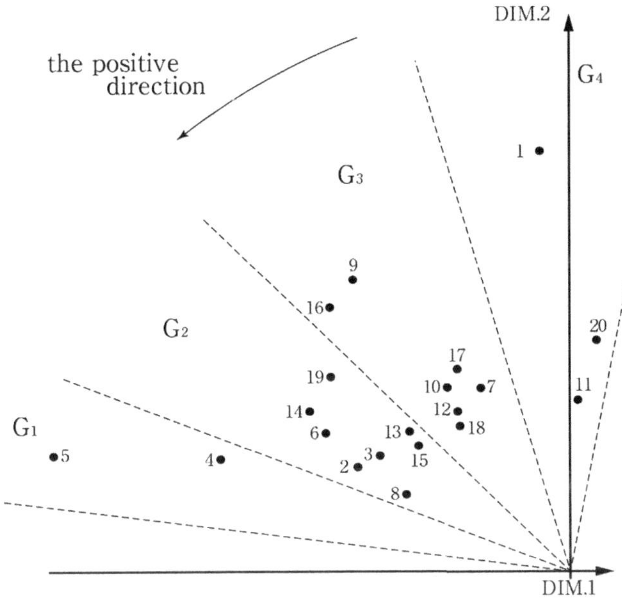

Fig. 2.10 The one-two plane obtained by the analysis of Chino's ASYMSCAL for the same students (as in Fig. 3.11) on July

We needed three-dimensional configurations in both cases, because two-dimensional ones didn't explain the observed amount of liking so much as was expected. It should be noted that in these cases we must return to the data matrix in determining the direction of the system because it is difficult for us to determine it from the sign of the r_{RO} in case of three-dimensions, whereas it is easy for us to do so in case of two dimensions. In these figures, an attempt has been made to divide the plane broadly into regions corresponding to distinct groups. Recall that our model has, not only the property of the scalar product, but also the property of the cross product, and that these are independent each other. These properties enable us to divide the plane into regions as if these spaces were Euclidian. Dotted lines, in these figures show these divisions. Let us denote these groups by G_1,G_2,G_3, and G_4, and interpret the result on June first.

At a glance, it is evident that student 11 doesn't belong to either group on June. He may be regarded as being disliked by all. Next, from the property of the cross product of our model, it is easily seen from Fig. 2.9 that the members of G_4 are relatively in a state called "unilateral love" towards the members of the other groups, whereas the members of G_1 are relatively in the reverse state. On the other hand, the members of G_2 are relatively liked by the other members of the class. On the contrary, the members of G_3. are relatively disliked by the other members of the class. Then, it is important to note that the degrees of R_{jk} and R_{kJ} corresponding to the objects located close at the origin are nearly equal to zero from our model, neglecting the parameter c. Thus, in this case, this means that students 3, 8, 11, and 13 relatively neither like

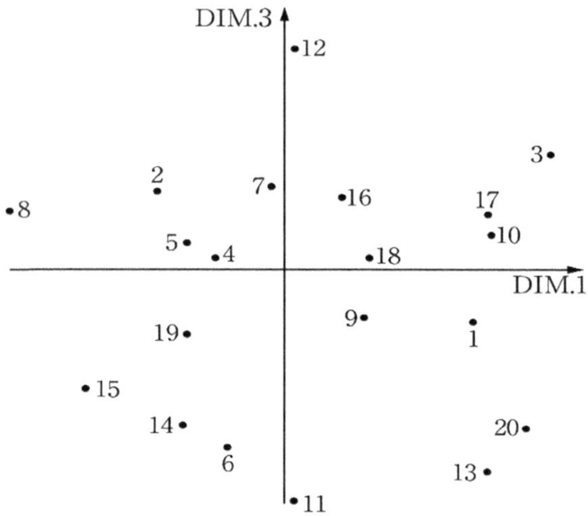

Fig. 2.11 The one-three plane of the column solution obtained by the analysis of SSA = II for the sociometric data on June

nor dislike the other members of the class except for student 11 and the reverse holds, too. In other words, students 3, 8, and 13 show indifference to the other members and the latter members do so to the members 3, 8, and 13.

Next, by comparing Fig. 2.9 with Fig. 2.10, we cannot help concluding that the sociometric structure of the class on July is very similar to that on June. Such a case would likely happen because a sociometric structure cannot be varied in so short a time.

On the contrary, the two column (row) solutions obtained by the analysis of SSA II are somewhat different with each other. Figures 2.11, 2.12, 2.13, and 2.14 show these results. The coefficients of alienation were 0.12, 0.11, 0.11, and 0.08 for Figs. 2.11, 2.12, 2.13, and 2.14, respectively. To our surprise, the groups G_1, G_2, G_3, and G_4 in Fig. 2.9 correspond broadly to the groups which belong to two-, three-, one-, and four-quadrant, in Fig. 2.11, respectively.

Thus, we see that a) Chino's ASYMSCAL not only reveals the same clustering as other algorithms in several cases, b) but also enables us to give information.

about the magnitude and the orientation of the skewness between objects. c) Chino's ASYMSCAL provides us with an overall measure of skewness between objects.

Finally, we will analyze the data of the amount of trade each nation in the world has with other nations, using Chino's ASYMSCAL. This set of data was obtained from *Statistical Yearbook 1974* by United Nations [14]. The data are shown in Table 2.5, and the result is shown in Fig. 2.15. The coefficients, r_{RP} and r_{RO} were 0.93 and 0.04, respectively. It is of interest to note that Japan is just in a state called "unilateral love". In other words, this may be the reason why she has been called "the economic animal".

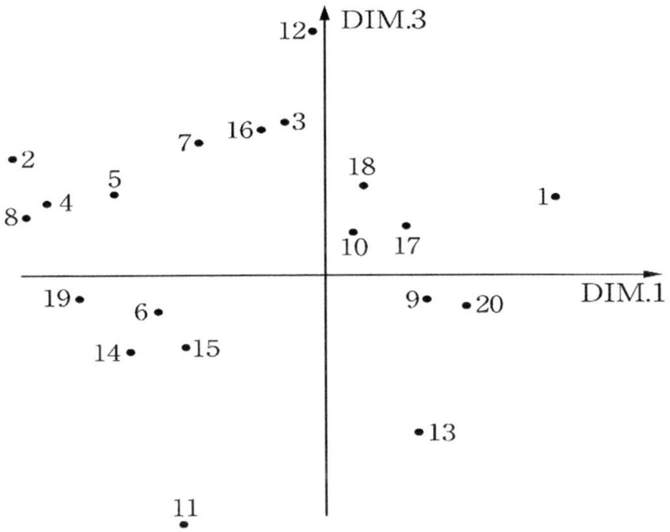

Fig. 2.12 The one-three plane of the row solution obtained by the analysis of SSA-II for the sociometric data on June

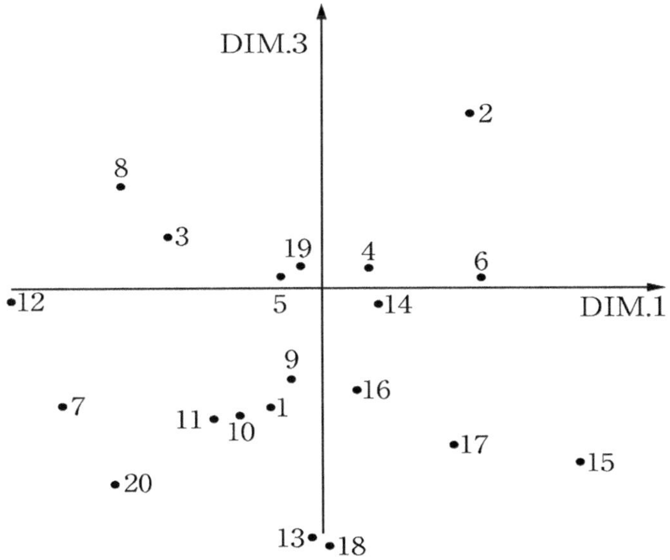

Fig. 2.13 The one-three plane of the column solution obtained by the analysis of SSA = II for the sociometric data on July

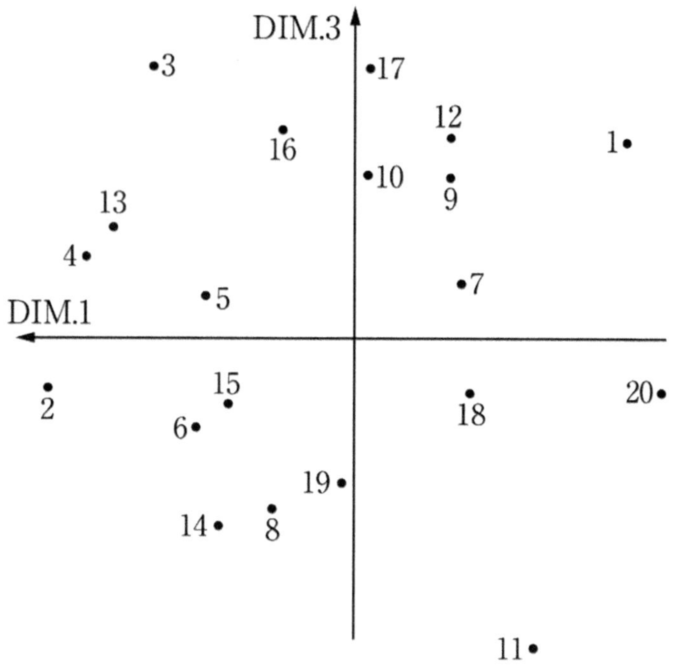

Fig. 2.14 The one-three plane of the raw solution obtained by the analysis of SSA-II for the sociometric data on July

Table 2.5 World exports by provenance and destination

From/to	EC	EA	US	JN	AN	AA	ME	OA	UR	EP
EC		25.67	15.67	2.84	2.42	9.62	5.59	5.07	2.66	6.49
EA	17.99		2.43	0.70	0.35	1.33	0.76	0.82	0.78	1.51
US	16.38	2.23		8.18	1.68	1.51	2.21	6.85	1.19	0.60
JN	4.40	1.32	9.55		1.46	2.48	1.61	8.93	0.49	0.32
AN	1.70	0.12	1.14	3.08		0.08	0.16	1.01	0.24	0.13
AA	11.60	0.88	1.92	0.98	0.04		0.20	0.35	0.63	0.67
ME	10.57	1.21	1.27	5.06	0.33	0.57		2.22	0.35	0.23
OA	5.33	0.46	6.94	6.96	0.79	0.63	0.59		0.62	0.36
UR	2.88	1.04	0.19	0.84	0.01	0.66	0.67	0.45		10.03
EP	5.38	1.49	0.33	0.14	0.55	0.63	0.64	0.30	10.56	

F.O.B. value in million U.S. dollars

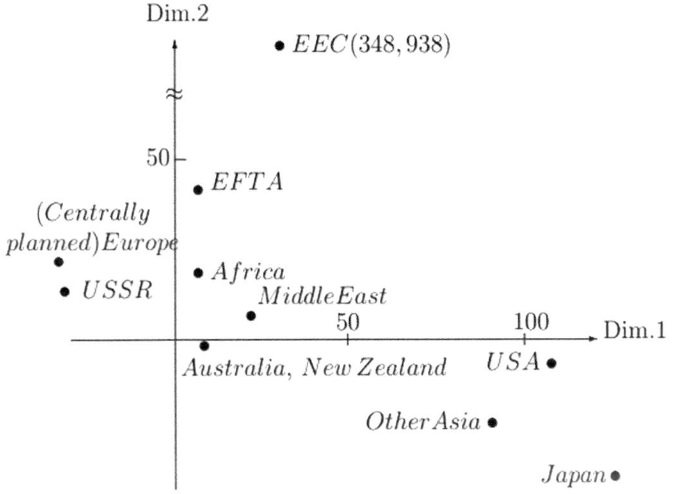

Fig. 2.15 The configuration obtained by the analysis of Chino's ASYMSCAL for the data of exports by provenance and destination (The positive direction of this configuration is *counterclockwise*). This configuration was adopted from Chino [4, Fig. 1]

2.11 Summary and Discussion

We proposed a new model for representing the asymmetric relationships between objects. The results presented here are very encouraging as to the validity and utility of a model accounting for the "skewness" between objects in terms of the "areas" of the parallelogram spanned by two vectors. There are a number of practical benefits of this method in interpreting the result. For example, while such a method as Young's ASYMSCAL requires the weights in addition to the configuration in interpreting the result, the method presented here not only reveals the clustering of objects but also enables us to give information about the magnitude and the orientation of the skewness between objects with only one configuration. Thus, the reader would confirm that the more the number of the point increases, the easier our model becomes in interpreting the result in contrast with G-L's or Young's. Another advantage is that our model has both the property of the scalar product and the property of the cross one and that they are independent with each other. For this reason, we can easily divide the obtained planes into regions corresponding to distinct groups as if these spaces were Euclidian. The other advantage is that our model provides us with an explicit over-all measure of skewness between objects, while the earlier works do not provide us with any explicit measure as far as we know. Thus, in these data such as in Table 2.3, the over-all magnitude of skewness between persons might be interpreted as the "latent energy" of skewness which is analogous to the "earthquake". If it increases, then this might be regarded as a sign of a "catastrophe" of the sociometric structure a group. We think such an inference would be possible from the psychological theories of consistency (for example, Heider's balance theory; Newcomb's AB-X model, etc.).

A principal disadvantage of the method is that it is limited to the case of three dimensions at most. Fortunately, however, our evidence to data indicates that our model accounts very well for data only in two dimensions, while further investigation is clearly indicated.

As described above our model may be classified as a "metric" one, though we assume the representation space to be "non-metric." Thus it requires the relatively strong assumption to the data. This is another disadvantage of our model. Explorations with real data have indicated, however, that this metric model yields sometimes solutions in practice which are virtually indistinguishable from those attained by the nonmetric ones, concerning the clustering of the objects.

Next, we cannot represent, for example, a person who hates all the other members of the group by any point in our space. In this case, the degrees of the relationships R_{jk} of O_j is constant for all k.

Further, we might not imagine the explicit distances between objects from our model. One might regard this as a disadvantage of the method. But, from our point of view, it would be fruitless to think that the metric notion should be applied to the asymmetric relationships, as we must be restricted by the "symmetric law" so far as we use a "metric" notion.

Finally, it should be noted that the data to be analyzed by our model is restricted to the one in which the degrees of the relationships R_{jk} and R_{kj} must be *comparable* with each other in a practical sense. That is, we didn't draw such a distinction as Coombs' [5] and Young's [15]. Though Young has already argued, we also think such a difference between two types of data is very subtle. Further, in view of the fact that, for example, the amount of trade which belongs to multi-process data in Young's terminology has actually been a serious problem between Japan and America, we think it fruitless to draw such a strict distinction between single and multi-process data as Young has done. This does not mean that such a distinction is not necessary. We think it might be necessary sometimes, but not always.

Acknowledgements The author wishes to express his thanks to the referees for their constructive comments. The author also wishes to acknowledge the valuable advice on both the theoretical aspects and on the format of this survey which has been given him by Yoshio Uchida, professor of Nagoya University. Helpful comments were provided by J. C. Lingoes, professor of The University of Michigan and also by Masao Ikuzawa, professor of Osaka City University. Acknowledgement is also made to Dr. Chikio Hayashi for his kindness in sending the author some results on the analysis of asymmetric relationships. Special acknowledgement is due to Masayuki Yamamoto, consultant of his school, for careful proofreading.

Notes

This chapter is the reproduction of Chino [3]. We corrected a few words and phrases of the errors found in the original paper, and omitted some parts of the paper which were thought to be unnecessary. Moreover, we redrew unclear lines and symbols in the figures found in the original paper.

It should be noted that the positive direction of the configuration shown in Fig. 2.3 of Chino's ASYMSCAL is different from that of HFM in Chap. 4. In Chino's ASYMSCAL the positive direction of the configuration in the two-dimensional case is *counterclockwise* as shown in Fig. 2.3, while in HFM it is *clockwise* in complex one-dimensional space which can be identified with the two-dimensional Euclidean space as explained in Chap. 5. The reason for this may be evident if we rewrite (5. 5), by setting l equals to 1 and transforming s_{jk} to R_{jk}, r_{jl} to x_{j1}, and c_{jl} to x_{j2}, as

$$R_{jk} = \lambda_1\left(x_{j1}x_{k1} + x_{j2}x_{k2}\right) + \lambda_1\left(x_{j2}x_{k1} - x_{j1}x_{k2}\right), \tag{2.18}$$

and rewriting this equation, by setting λ_1 equal to b, as

$$R_{jk} = \mathrm{b}\left(x_{j1}x_{k1} + x_{j2}x_{k2}\right) - \mathrm{b}\left(x_{j1}x_{k2} - x_{j2}x_{k1}\right). \tag{2.19}$$

It should be noted that the sign of the second-right hand term of (2.19) is minus, while that of (2.4) is plus.

References

1. Bard, Y. (1974). *Nonlinear parameter estimation*. Academic Press.
2. Box, M. J. (1966). A comparison of several current optimization methods, and the use of transformations in constrained problems. *The Computer Journal, 9*, 67–77.
3. Chino, N. (1978). A graphical technique for representing the asymmetric relationships between *N* objects. *Behaviormetrika, 5*, 23–40.
4. Chino, N. (2012). A brief survey of asymmetric MDS and some open problems. *Behaviormetrika, 39*, 127–165.
5. Coombs, C. H. (1964). *A theory of data*. Wiley.
6. Coombs, C. H., Dawes, R. M., & Tversky, A. (1970). *Mathematical psychology*. Prentice Hall.
7. Gower, J. C. (1977). The analysis of asymmetry and orthogonality. In J. R. Barra, F. Brodeau, G. Romer, & B. van Cutsem (Eds.), *Recent developments in statistics* (pp. 109–123). North Holland.
8. Guttman, L. (1968). A general nonmetric technique for finding the smallest coordinate space for a configuration of points. *Psychometrika, 33*, 469–506.
9. Hayashi, C. (1952). On the prediction of phenomena from qualitative data and the quantification of qualitative data from the mathematico-statistical point of view. *Annals of the Institute of Statistical Mathematics, 3*(2), 69–98.
10. Hayashi, C. (1974). Minimum dimension analysis. *Behaviormetrika, 1*, 1–24.
11. Hayashi, C. (1977). MDS ni okeru hitaishosei no mondai [The problem of asymmetry in MDS]. In *Proceedings of the 45-th annual conference of Japan Statistical Society at Fukuoka University*, Vol. 41.

12. Lingoes, J. C. (1973). *The Guttman-Lingoes nonmetric program series*. Mathesis Press.
13. Takane, Y., Young, F. W., & de Leeuw, J. (1977). Nonmetric individual differences multidimensional scaling: An alternative least squares method with optimal scaling features. *Psychometrika, 42*, 1–68.
14. United Nations. (1975). *Statistical year book 1974*. United Nations Publication.
15. Young, F. W. (1975). An asymmetric Euclidean model for multi-process asymmetric data. In *Paper presented at U.S.-Japan Seminar on MDS*, San Diego, U.S.A.

Chapter 3
GIPSCAL

3.1 Introduction

During the past twenty years, various multidimensional scaling (MDS) techniques which deal directly or indirectly with the asymmetric relationships among objects have been developed by many researchers.

The motivation of developing these asymmetric MDS's is, of course, to overcome a short-fall inherent in the traditional MDS techniques. That is, the traditional MDS's are not so powerful in cases when the similarity or dissimilarity data matrices are asymmetric by nature because they are based on the assumption that the similarity or dissimilarity data can be related to the interpoint distances in some metric space [10, 16, 34].

For this reason, some researchers have augmented the traditional MDS's by the assumption that the similarity or dissimilarity between objects is a function not only of interpoint distance in a metric space but also of the quantities related to the objects. For example, the squared distance is augmented by differential weights for dimensions in the weighted Euclidean model proposed first by Young [38, 39], Baker et al. [2] and Hayashi [13]. In Krumhansl's distance-density model, the interpoint distance is augmented by some measure of spatial density in the regions surrounding the points [15]. Weeks and Bentler [36, 37] and Saito [24, 25] have proposed modified distance models in which the interpoint distance is augmented by a few constants related to the points. Okada and Imaizumi [20, 21] have proposed a nonmetric version of a generalized modified distance model.

Some other researchers have taken different approaches which utilize metric spaces. For example, SSA2 (Smallest Space Analysis 2) by Guttman [10] and Lingoes [18] imposes row compatibility and column compatibility on data and yields two solutions in a metric space. Tobler's wind model explains asymmetries by the direction of wind assigned to mesh points on the configuration of objects [33]. Sato's model represents asymmetries using Randers' metric which is an asymmetric metric function [26, 27].

N. Chino, *Structure and Dynamics of Asymmetric Interactions*,
https://doi.org/10.1007/978-981-97-8269-7_3

Somewhat different or completely different approaches have been taken by some researchers. The feature matching model proposed by Tversky [35] represents the similarity or dissimilarity between two objects by a linear combination of the measures of the common and distinctive features of the two objects.

Chino [3, 5] has proposed a model involving a geometric generalization of scalar products. This model fits the scalar (inner) product and the magnitude of the cross product of solution vectors to the symmetric and skew-symmetric parts of the data, respectively. However, the model was defined only for two or three dimensions, first.

Canonical analysis of asymmetry proposed by Constantine and Gower [7, 9] decomposes dissimilarity data matrix into two parts, the symmetric and skew-symmetric parts, and then analyzes them separately. While the symmetric part is analyzed by some established distance model, the skew-symmetric part is analyzed by its canonical decomposition. Asymmetric relationships are interpreted in terms of areas of triangles and colinearities. Therefore, the canonical analysis has a strong resemblance to the model proposed by Chino.

DEDICOM proposed by Harshman [11] and Harshman et al. [12] can be said to be a more general inner product model than the model proposed by Chino. DEDICOM decomposes observed similarity data into some combination of few basic underlying relationships. Of concern here is the model formulated by Chino [3]. This model is written as:

$$S_{jk} = a\boldsymbol{x}_j^t\boldsymbol{x}_k + b\boldsymbol{x}_j^t\boldsymbol{I}^*\boldsymbol{x}_k + c + E_{jk}, \tag{3.1}$$

where the quantity S_{jk} is the observed similarity between objects O_j and O_k (similarity from O_j to O_k to be precise), while the matrix $\boldsymbol{I}^*$ is a special skew-symmetric matrix of form

$$\boldsymbol{I}^* = \begin{pmatrix} 0 & 1 \\ -1 & 1 \end{pmatrix}, \boldsymbol{I}^* = \begin{pmatrix} 0 & 1 & -1 \\ -1 & 0 & 1 \\ 1 & -1 & 0 \end{pmatrix}, \tag{3.2}$$

for two and three dimensions, respectively. Quantities, a, b, and c are unknown constants.

The quantity, $\boldsymbol{x}_j^t\boldsymbol{I}^*\boldsymbol{x}_k$ in the second right-hand term of (3.1) has a special meaning. In the two-dimensional case, it becomes

$$\boldsymbol{x}_j^t\boldsymbol{I}^*\boldsymbol{x}_k = x_{j1}x_{k2} - x_{j2}x_{k1}, \tag{3.3}$$

which is just the cross product of two vectors $\boldsymbol{x}_j$ and $\boldsymbol{x}_k$, the absolute value of which represents the area of the parallelogram spanned by the two vectors. It should be noted that the cross product in a two-dimensional space is a scalar.

The sign of the term $\boldsymbol{x}_j^t\boldsymbol{I}^*\boldsymbol{x}_k$ determines the orientation of the parallelogram. This orientation can be easily guessed by defining the positive direction in the one-two plane in the following way: let us choose the horizontal and vertical axes, respectively, as the first and second axes, and suppose that the sign of b in (3.1) is positive. Then we say that the positive direction in this plane is counterclockwise. If the sign of b is negative, we then say that the positive direction is clockwise. The orientation of the parallelogram is positive if the sense (counterclockwise or clockwise) in which the vertices formed by the tails of the two vectors follow each other is the same as that for the positive direction in the plane. The orientation is negative if this sense is different from that for the positive direction. The positive direction therefore indicates the orientation of asymmetry in the plane.

In the three-dimensional case, the quantity becomes

$$\boldsymbol{x}_j^t \boldsymbol{I}^* \boldsymbol{x}_k = \left(x_{j2}x_{k3} - x_{j2}x_{k1}\right) + \left(x_{j3}x_{k1} - x_{j1}x_{k3}\right) + \left(x_{j1}x_{k2} - x_{j2}x_{k1}\right). \tag{3.4}$$

In this case, the cross product or vector product of two vectors $\boldsymbol{x}_j$ and $\boldsymbol{x}_k$ is a vector and its first, second, and third components are, respectively, the first, second, and third right hand terms of (3.4) in a right-handed coordinate system. It is evident, from the forms of these terms in (3.4), that the absolute values of these terms represent areas of parallelograms in two-dimensional spaces. In fact, the first, second, and third terms correspond, respectively, to parallelograms projected onto the $x_2 - x_3, x_3 - x_1, x_1 - x_2$, planes of the parallelogram spanned in space by the vectors $\boldsymbol{x}_j, \boldsymbol{x}_k$. Figure 3.1 illustrates this. This means that the geometrical interpretation of the quantity in three dimensions can be reducible to that of the quantity in two dimensions.

As suggested in (3.4), the positive directions in these planes are all counterclockwise if we choose the horizontal and vertical axes, respectively, as the second and third, the third and first, and the first and second axes when the sign of b is positive. If we exchange the horizontal and vertical axes of, say, the third-first plane, the positive direction must also be changed. As we did in the two-dimensional case, if the sign of b in (3.1) is negative, all the positive directions in those planes should be altered accordingly.

One shortcoming of this model is that it was defined only for two or three dimensions. Chino [4, 5] has extended this model into higher dimensions. However, no discussion has been made concerning theoretical and practical implications of the extension. The current paper is devoted to this purpose.

In the next section, we shall introduce the extended model and discuss the theoretical implications of this model. We shall call the extended model GIPSCAL (a Generalized Inner Product multidimensional SCALing). In higher dimensions we can define neither the cross product nor the vector product. However, it will be shown that we can still define areas of parallelograms in higher dimensions and associate with the two vectors $\boldsymbol{x}_j, \boldsymbol{x}_k$ an area of a parallelogram, the square of which is a well-known quantity in numerical analysis.

In the third section, we shall refer to a psychological justification of the model. In the fourth section, we shall discuss the problem of double centering transformation. In the classical MDS by Torgerson [34], this transformation plays a fundamental role prior to the factoring of an inner product matrix. The double centering transformation of the observed similarity matrix is contrasted with that of the squared distance matrix in the classical MDS. In the fifth section, we shall show some examples of

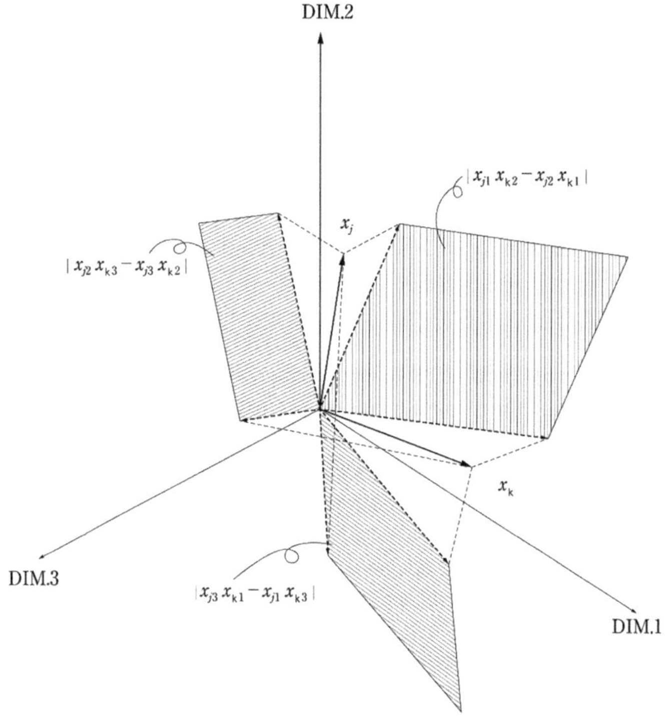

Fig. 3.1 Parallelograms projected onto three planes of the parallelogram spanned in space by the vectors $\boldsymbol{x}_j, \boldsymbol{x}_k$

application. In the discussion section, we shall show the necessity for developing a general maximum likelihood MDS which enables us to select the best model out of a variety of extant asymmetric MDS's.

3.2 The Model

Let the data be for N objects, say $O_1, O_2, \ldots, O_N$ which may be physical stimuli, classmates, companies, or nations and so forth. We shall denote the observed similarity from O_j to O_k by S_{jk}, assuming that the similarity is defined at the interval level of measurement in Stevens' terminology. In this sense, our model is a metric multidimensional scaling. The GIPSCAL model is written as:

$$S_{jk} = a\boldsymbol{x}_j^t\boldsymbol{x}_k + b\boldsymbol{x}_j^t\boldsymbol{I}_p^*\boldsymbol{x}_k + c + E_{jk}, \quad (3.5)$$

where a, b, and c are unknown constants, and E_{jk} is an error term, while vectors $\boldsymbol{x}_j, \boldsymbol{x}_k$ are coordinates of objects in a p-dimensional Euclidean space, which are estimated from similarity data, S_{jk}'s.

Note that this equation is, at a glance, the same as (3.1), which is a model equation of an ASYMSCAL proposed by Chino [3]. However, the difference between these two equations is that the vectors, $\boldsymbol{x}_j, \boldsymbol{x}_k$, are two-dimensional or three-dimensional in (3.1), while these are p-dimensional where p is any number which is greater than 2 in (3.5).

The matrix, $\boldsymbol{I}_p^*$ is a special skew-symmetric matrix of form,

$$\boldsymbol{I}_p^* = \left\{ sgn\begin{pmatrix} 12\cdots p \\ \cdots lm \cdots \end{pmatrix} \right\}, \tag{3.6}$$

where

$$sgn\begin{pmatrix} 12\cdots p \\ \cdots lm \cdots \end{pmatrix} = \begin{cases} 0, & \text{if the two indices } l, m \text{ are the same}, \\ 1, & \text{if } (\cdots lm \cdots) \text{is an even permutation of } (12\cdots p), \\ -1, & \text{if } (\cdots lm \cdots) \text{is an odd permutation of } (12\cdots p). \end{cases}$$

Of course, in the special case when p is less than four, $\boldsymbol{I}_2^*$ and $\boldsymbol{I}_3^*$ coincide with the corresponding matrices $\boldsymbol{I}_2^*$ of Chino's ASYMSCAL in two and three dimensions, respectively.

In this way, GIPSCAL is thought of as a natural extension of Chino's ASYMSCAL into higher dimensions, at least formally. However, at this point some questions may arise: is it possible in higher dimensions to consider the quantity $\boldsymbol{x}_j^t \boldsymbol{I}_p^* \boldsymbol{x}_k$ as the sum of the parallelograms projected onto two-dimensional planes? Can we associate with the two vectors $\boldsymbol{x}_j, \boldsymbol{x}_k$ a third vector uniquely? For, only in three dimensions can we define a vector product $\boldsymbol{x}_j \times \boldsymbol{x}_k$ that again is a vector. The answer to the first question is yes. Quantity, $\boldsymbol{x}_j^t \boldsymbol{I}_p^* \boldsymbol{x}_k$ in the extended model has a very desirable property that, on the l-m plane,

$$\boldsymbol{x}_j^t \boldsymbol{I}_p^* \boldsymbol{x}_k = \sum_{l \neq m}^{p} sgn\begin{pmatrix} 12\cdots p \\ \cdots lm \cdots \end{pmatrix} x_{jl} x_{km}, \tag{3.7}$$

becomes

$$sgn\begin{pmatrix} 12\cdots p \\ \cdots lm \cdots \end{pmatrix} \left(x_{jl} x_{km} - x_{jm} x_{kl} \right).$$

This means that the geometrical interpretation of the quantity in higher dimensions is reducible to that of the area of the parallelogram spanned by vectors $\boldsymbol{x}_j$ and $\boldsymbol{x}_k$ in two dimensions. Figure 3.1 is a special case when $p = 3$. In the case when the sign of b in (3.1) is positive, the positive direction in the l-m plane becomes counterclockwise if we choose the horizontal and vertical axes, respectively, as the l-th and m-th axes and if the permutation $(\cdots lm \cdots)$ is even.

The answer to the second question is no. As discussed in the previous section, in higher dimensions we cannot associate with the two vectors $\boldsymbol{x}_j, \boldsymbol{x}_k$ a third vector outside the plane spanned by the two vectors in a geometrical fashion. However, we can still associate with the two vectors the area of a parallelogram spanned by these vectors.

It is interesting to note that the square of the area is the Gramian or Gram determinant of the two vectors, which is well-known in numerical analysis and matrix algebra [8, 17].

The Gramian $g\left(\boldsymbol{x}_j, \boldsymbol{x}_k\right)$ is written as

$$\begin{aligned} g\left(\boldsymbol{x}_j, \boldsymbol{x}_k\right) &= \left(\boldsymbol{x}_j^t \boldsymbol{x}_j\right)\left(\boldsymbol{x}_k^t \boldsymbol{x}_k\right) - \left(\boldsymbol{x}_j^t \boldsymbol{x}_k\right)\left(\boldsymbol{x}_k^t \boldsymbol{x}_j\right), \\ &= \begin{vmatrix} \boldsymbol{x}_j^t \boldsymbol{x}_j & \boldsymbol{x}_j^t \boldsymbol{x}_k \\ \boldsymbol{x}_k^t \boldsymbol{x}_j & \boldsymbol{x}_k^t \boldsymbol{x}_k \end{vmatrix}. \end{aligned} \tag{3.8}$$

It can also be shown readily that the following relation holds between the Gramian of the two vectors $\boldsymbol{x}_j$, x_k and the right-hand term of (3.7):

$$g\left(\boldsymbol{x}_j, \boldsymbol{x}_k\right) = \sum_{\substack{l,m=1 \\ l<m}}^{p} \left(x_{jl}x_{km} - x_{jm}x_{kl}\right)^2, \tag{3.9}$$

Moreover, it is interesting to note that in general

$$g\left(\boldsymbol{x}_j, \boldsymbol{x}_k\right) \geq 0, \tag{3.10}$$

and that equality holds only if $\boldsymbol{x}_j$ and $\boldsymbol{x}_k$ are linearly dependent.

In any case, we can say that GIPSCAL fits the scalar product (inner product) and the area of the parallelogram spanned by two vectors to the symmetric and asymmetric part of observed similarity judgements, respectively.

3.3 Psychological Justification of the Model

One may think that the availability of GIPSCAL is only in the parsimonious geometrical representation of the structure in the data. However, GIPSCAL has another availability in that it provides a psychological justification of why asymmetry arises. For convenience, let us rewrite (3.5) as

$$S_{jk} = \boldsymbol{x}_j^t \tilde{\boldsymbol{x}}_k + c + E_{jk}, \tag{3.11}$$

or

$$S_{jkr} = \boldsymbol{x}_j^t \tilde{\boldsymbol{x}}_k + c + E_{jkr}, \tag{3.12}$$

where

$$\tilde{\boldsymbol{x}}_k = \left(a\boldsymbol{I} + b\boldsymbol{I}_p^*\right)\boldsymbol{x}_k = \boldsymbol{A}\boldsymbol{x}_k. \tag{3.13}$$

Here S_{jkr} denotes the similarity from O_j to O_k rated by judge J_r.

Both (3.11) and (3.12) tell us that perception of O_k to O_j is distorted such that

$$\tilde{\boldsymbol{x}}_k = \boldsymbol{A}\boldsymbol{x}_k, \tag{3.14}$$

where vector $\tilde{\boldsymbol{x}}_k$ is the perceived image of O_k, while $\boldsymbol{x}_k$ is the preimage of O_k. Figures 3.2 and 3.3 illustrate this. In this sense, the derived configuration by GIPSCAL is said to be a preimage configuration. In the field of person perception in social psychology, several general judgmental effects such as the halo effect and logical error have been known [28]. These effects may be combined to give rise to the perceived images discussed above.

It is of interest to note that the magnitude of the preimage of each object is related to the similarity to itself because, from (3.5)

$$S_{jj} = a\boldsymbol{x}_j^t \boldsymbol{x}_j + c + E_{jj}. \tag{3.15}$$

Fig. 3.2 Perceived image $\tilde{\boldsymbol{x}}_k$ of member O_k by member O

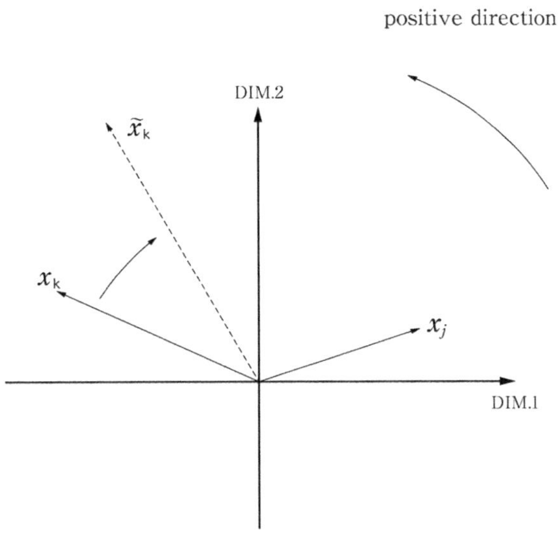

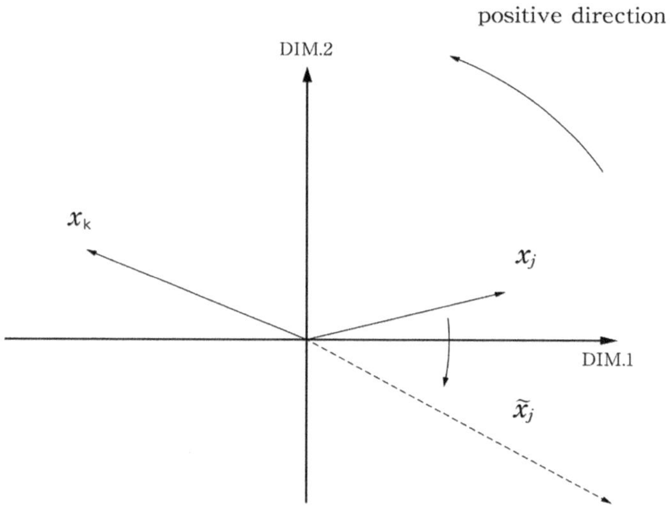

Fig. 3.3 Perceived image $\boldsymbol{x}_j$ of member O_j by member O_k

3.4 Double Centering Transformation as a Preprocessing of Original Data

Coordinates of the derived configuration by MDS techniques are arbitrary in the sense of being only defined up to a central dilation, a translation (origin shift), and a rotation. By contrast, coordinates obtained by GIPSCAL are not arbitrary for the origin shift. However, we can have a configuration whose centroid is the origin in the special case when the double centering transformation is permitted to an original full similarity data matrix (we shall call such a configuration origin centroid configuration). By the full similarity data matrix we mean the similarity data matrix $\boldsymbol{S}$ which has no missing observations. To show this, let us first rewrite (3.5) in matrix form as

$$\boldsymbol{S} = a\boldsymbol{X}\boldsymbol{X}^t + b\boldsymbol{X}\boldsymbol{I}_p^*\boldsymbol{X}^t + c\boldsymbol{J}_N + \boldsymbol{E}, \tag{3.16}$$

where $\boldsymbol{S}$ is an N by N similarity data matrix, $\boldsymbol{X}$ is the preimage configuration, $\boldsymbol{J}_N$ is an N by N matrix whose entities are all 1, and $\boldsymbol{E}$ is an N by N matrix of error terms. Second, we take note that the origin centroid configuration is obtained by premultiplying $\boldsymbol{X}$ by a well-known centering matrix [14].

$$\boldsymbol{G} = \boldsymbol{I} - \left(\boldsymbol{1}_N\boldsymbol{1}_N^t/N\right) \tag{3.17}$$

The matrix $\boldsymbol{G}$ has the following properties:

$$\boldsymbol{G}^t = \boldsymbol{G}, \boldsymbol{G}^2 = \boldsymbol{G}, \boldsymbol{G}\boldsymbol{1}_N = 0. \tag{3.18}$$

and is an orthogonal projection matrix. Here, $\mathbf{1}_N$ is an N-dimensional vector whose components all equal unity.

Then, by a double centering transformation of the matrix $\boldsymbol{S}$, (3.16) can be rewritten as

$$\boldsymbol{S}_c = \boldsymbol{GSG}^t = a\boldsymbol{X}_c\boldsymbol{X}_c^t + b\boldsymbol{X}_c\boldsymbol{I}_p^*\boldsymbol{X}_c^t + c\boldsymbol{GJ}_N G^t + \boldsymbol{GEG}^t, \tag{3.19}$$

where the matrix $\boldsymbol{X}_c$ is the origin centroid configuration. For the original full similarity data matrix, the third right-hand term of (3.19) is zero, since

$$\boldsymbol{GJ}_N\boldsymbol{G}^t = \boldsymbol{O}, \tag{3.20}$$

Here $\boldsymbol{O}$ is a matrix whose entities are all 0. It is interesting to note that the Young-Householder transformation is written as

$$\begin{aligned} \boldsymbol{P}_c &= \boldsymbol{X}_c\boldsymbol{X}_c^t = \boldsymbol{GPG}^t = -\boldsymbol{GTG}^t/2, \\ &= \left\{-\left(d_{jk}^2 - \overline{d}_{j\cdot}^2 - \overline{d}_{\cdot k}^2 + \overline{d}_{\cdot\cdot}^2\right)/2\right\}, \end{aligned} \tag{3.21}$$

and that

$$\boldsymbol{GSG}^t = \left\{s_{jk} - \overline{s}_{j\cdot}^2 - \overline{s}_{\cdot k}^2 + \overline{s}_{\cdot\cdot}^2\right\}, \tag{3.22}$$

where matrices $\boldsymbol{P}$ and $\boldsymbol{T}$ are an inner product matrix and a squared distance matrix, respectively. It is clear from (3.19) and (3.21) that double centering transformation of the similarity data matrix $\boldsymbol{S}$ leads to the following equation:

$$\boldsymbol{S}_c = -a\boldsymbol{GSG}^t/2 + b\boldsymbol{X}_c\boldsymbol{I}_p^*\boldsymbol{X}_c^t + c\boldsymbol{GJ}_N\boldsymbol{G}^t + \boldsymbol{GEG}^t. \tag{3.23}$$

Equation (3.23) describes the relation between similarity and distance in GIPSCAL. To be precise, it states that the doubly centered similarity is a linear combination of the squared distance between two vectors and the area of the parallelogram spanned by the two vectors except for constant and error terms in GIPSCAL.

To this point we have assumed the full similarity data matrix. However, in many cases we face data matrices with missing observations. In such cases, GIPSCAL provides near origin centroid configuration under the double centering transformation. This will be explained in the following section.

Some readers may doubt whether the double centering transformation of the original similarity matrix is appropriate. In fact, this transformation sometimes provides a better fit to the GIPSCAL model, but sometimes also a worse fit. Apparently, the results depend on the data.

Aside from the discussion of goodness of fit, there is no logical reason why we must choose the origin centroid configuration or near origin centroid one in the case of the inner product model. Thus, in the recent version of the GIPSCAL programme, this transformation is only one of the options in the preprocessing of original data.

3.5 The Algorithm

Suppose that we are given NT elements, of the S_{jk}. Let

$$e_{jk} = \begin{cases} 1, \textit{ if } S_{jk} \textit{ is given} \\ 0, \textit{ if } S_{jk} \textit{ is not given.} \end{cases} \tag{3.24}$$

Then

$$NT = \sum_{j-1}^{N} \sum_{k=1}^{N} e_{jk}. \tag{3.25}$$

A reasonable procedure would seem to be to minimize the following function Q in a least square sense:

$$Q = \sum_{j=1}^{N} \sum_{k=1}^{N} e_{jk} \left(S_{jk} - a\boldsymbol{x}_j^t \boldsymbol{x}_k - b\boldsymbol{x}_j^t \boldsymbol{I}_p^* \boldsymbol{x}_k - c \right)^2. \tag{3.26}$$

For attacking this problem, we have chosen a familiar alternating least squares method, which is the same algorithm taken as that in Chino's ASYMSCAL.

To minimize Q, we first take the derivative of Q of (3.26) with respect to $\boldsymbol{x}_h$, where $h \neq j, k$. Then, we have

$$\begin{aligned} \partial Q / \partial \boldsymbol{x}_h = -2 \Bigg\{ & \sum_{j-1}^{N} e_{jh} \left(S_{jh} - \boldsymbol{x}_j^t \boldsymbol{A} \boldsymbol{x}_h - c \right) \boldsymbol{A}^t \boldsymbol{x}_j \\ & + \sum e_{hk} \left(S_{hk} - \boldsymbol{x}_h^t \boldsymbol{A} \boldsymbol{x}_k - c \right) \boldsymbol{A} \boldsymbol{x}_k \Bigg\}, \end{aligned} \tag{3.27}$$

where $\boldsymbol{A} = a\boldsymbol{I} + b\boldsymbol{I}_p^*$, which was used in (3.13).

Switching notation of the second right-hand term of (3.27) from k to j for later convenience, and rearranging the right member of the resulting equation, and setting $\partial Q / \partial \boldsymbol{x}_h$ equal to zero and rearranging, we get

$$\boldsymbol{X}_h = \boldsymbol{C}_h^{-1} \left(\boldsymbol{A}^t \boldsymbol{u}_h + \boldsymbol{A} \boldsymbol{v}_h \right), \tag{3.28}$$

where

$$\boldsymbol{C}_h = \boldsymbol{A}^t \boldsymbol{Y}_h \boldsymbol{A} + \boldsymbol{A} \boldsymbol{Z}_h \boldsymbol{A}^t, \tag{3.29}$$

$$\boldsymbol{Y}_h = \boldsymbol{X}^t \boldsymbol{B}_h \boldsymbol{X}, \boldsymbol{Z}_h = \boldsymbol{X}^t \boldsymbol{D}_h \boldsymbol{X}, \tag{3.30}$$

$$\boldsymbol{B}_h = diag(e_{ih}), \boldsymbol{D}_h = diag(e_{hi}), \tag{3.31}$$

$$\boldsymbol{u}_h = \sum_{j=1}^{N} e_{jh}\left(S_{jh} - c\right)\boldsymbol{x}_j, \tag{3.32}$$

$$\boldsymbol{v}_h = \sum_{j=1}^{N} e_{hj}\left(S_{hj} - c\right)\boldsymbol{x}_j. \tag{3.33}$$

The form (3.28) suggests the iterative solution

$$\boldsymbol{x}_{h,t+1} = \boldsymbol{C}_{h,t}^{-1}\left(\boldsymbol{A}_t^t \boldsymbol{u}_{h,t} + \boldsymbol{A}_t \boldsymbol{v}_{h,t}\right), \tag{3.34}$$

The value $\boldsymbol{x}_{j,t}$ obtained at trial t is used to obtain the next $\boldsymbol{x}_{h,t+1}$. At each trial t, the configuration is normalized by scaling each vector $\boldsymbol{x}_j$ so that the maximum $(|\boldsymbol{x}_1|, |\boldsymbol{x}_2|, \ldots, |\boldsymbol{x}_N|)$ equals one. This is the same strategy as in Chino's ASYMSCAL. Secondly, taking the derivatives of Q of (3.26) with respect to c, setting it equal to zero, and solving for c, we find

$$c = \overline{S} - a\overline{P} - b\overline{O}. \tag{3.35}$$

where

$$\overline{S} = \sum_{j=1}^{N}\sum_{k=1}^{N} e_{jk} S_{jk}/NT, \tag{3.36}$$

$$\overline{P} = \sum_{j=1}^{N}\sum_{k=1}^{N} e_{jk}\boldsymbol{x}_j^t \boldsymbol{x}_k/NT, \tag{3.37}$$

and

$$\overline{O} = \sum_{j=1}^{N}\sum_{k=1}^{N} e_{jk}\boldsymbol{x}_j^t \boldsymbol{I}_p^* \boldsymbol{x}_k/NT. \tag{3.38}$$

Next, substituting for c in (3.26), we have

$$Q = \sum_{j=1}^{N}\sum_{k=1}^{N} e_{jk}\left\{\left(S_{jk} - \overline{S}\right) - a\left(\boldsymbol{x}_j^t \boldsymbol{x}_k - \overline{P}\right) - b\left(\boldsymbol{x}_j^t \boldsymbol{I}_p^* \boldsymbol{x}_k - \overline{O}\right)\right\}^2. \tag{3.39}$$

Taking the derivatives of Q with respect to a and b respectively, and setting $\partial Q/\partial a$ and $\partial Q/\partial b$ equal to zero respectively, we find

$$a = \left(\sigma_{sp}\sigma_o^2 - \sigma_{so}\sigma_{po}\right)/\left(\sigma_p^2\sigma_o^2 - \sigma_{po}^2\right), \tag{3.40}$$

$$b = \left(\sigma_{so}\sigma_p^2 - \sigma_{sp}\sigma_{po}\right)/\left(\sigma_p^2\sigma_o^2 - \sigma_{po}^2\right), \tag{3.41}$$

where

$$\sigma_{sp} = \sum_{j=1}^{N}\sum_{k=1}^{N} e_{jk}\left(S_{jk} - \overline{S}\right)\left(\boldsymbol{x}_j^t\boldsymbol{x}_k - \overline{P}\right)/NT, \tag{3.42}$$

$$\sigma_{so} = \sum_{j=1}^{N}\sum_{k=1}^{N} e_{jk}\left(S_{jk} - \overline{S}\right)\left(\boldsymbol{x}_j^t\boldsymbol{I}_p^*\boldsymbol{x}_k - \overline{O}\right)/NT, \tag{3.43}$$

$$\sigma_{po} = \sum_{j=1}^{N}\sum_{k=1}^{N} e_{jk}\left(\boldsymbol{x}_j^t\boldsymbol{x}_k - \overline{P}\right)\left(\boldsymbol{x}_j^t\boldsymbol{I}_p^*\boldsymbol{x}_k - \overline{O}\right)/NT, \tag{3.44}$$

$$\sigma_p^2 = \sum_{j=1}^{N}\sum_{k=1}^{N} e_{jk}\left(\boldsymbol{x}_j^t\boldsymbol{x}_k - \overline{P}\right)^2/NT, \tag{3.45}$$

$$\sigma_o^2 = \sum_{j=1}^{N}\sum_{k=1}^{N} e_{jk}\left(\boldsymbol{x}_j^t\boldsymbol{I}_p^*\boldsymbol{x}_k - \overline{O}\right)^2/NT, \tag{3.46}$$

We shall now discuss the problem of preprocessing the original similarity data. GIPSCAL has two options concerning this problem. One is the simple normalization of the original data, in which case s_{jk}'s are normalized in such a way that their mean and the standard deviation equal zero and one, respectively. The other is the double centering transformation. As pointed out in the previous section, in this case, the quantity c becomes zero if and only if the original data have no missing observations. Otherwise c is not equal to zero. For the right-hand term of (3.38) will not equal zero if we have asymmetric missing observations (i.e., either e_{jk} or e_{kj} is zero).

Next, we shall discuss the measure of goodness of fit. The total variance of S_{jk}, that is, σ_s^2 would be partitioned into two parts, with the aid of (3.39), (3.40), and (3.41). Thus, we have

$$\sigma_s^2 = \sigma_s^2\left(r_{sp}^2 + r_{so}^2 - 2r_{sp}r_{so}r_{po}\right)/\left(1 - r_{po}^2\right) + \sigma_e^2, \tag{3.47}$$

or

$$1 = \left(r_{sp}^2 + r_{so}^2 - 2r_{sp}r_{so}r_{po}\right)/\left(1 - r_{po}^2\right) + (\sigma_e/\sigma_s)^2, \tag{3.48}$$

where

$$r_{sp} = \sigma_{sp}/\left(\sigma_s\sigma_p\right), \tag{3.49}$$

$$r_{so} = \sigma_{so}/(\sigma_s\sigma_o), \tag{3.50}$$

$$r_{po} = \sigma_{po}/\left(\sigma_p\sigma_o\right), \tag{3.51}$$

and

$$\sigma_e^2 = Q_{min}/NT. \tag{3.52}$$

We can define the first right-hand term of (3.48) as the indicator of goodness of fit to the model. Thus, it is defined to be

$$F = \left(r_{sp}^2 + r_{so}^2 - 2r_{sp}r_{so}r_{po}\right)/\left(1 - r_{po}^2\right), \tag{3.53}$$

For the full similarity data matrix, we have

$$F = r_{sp}^2 + r_{so}^2, \tag{3.54}$$

which coincides with the indicator as in Chino's ASYMCAL.

Finally, we shall consider the problem of convergence. It is desirable to obtain the second-order derivative Hessian matrix in order to help evaluate whether an extremum of the function Q has been found and to implement the more efficient Newton–Raphson method. Remembering (3.29), we obtain

$$\partial/\partial \boldsymbol{x}_h(\partial Q/\partial \boldsymbol{x}_h) = 2\boldsymbol{C}_h, \tag{3.55}$$

From the definition of $\boldsymbol{B}_h$ and $\boldsymbol{D}_h$, it is clear that $\boldsymbol{B}_h^2 = \boldsymbol{B}_h$ and $\boldsymbol{D}_h^2 = \boldsymbol{D}_h$. Hence

$$\boldsymbol{Y}_h = \boldsymbol{X}^t\boldsymbol{B}_h\boldsymbol{X} = (\boldsymbol{B}_h\boldsymbol{X})^t(\boldsymbol{B}_h\boldsymbol{X}), \tag{3.56}$$

$$\boldsymbol{Z}_h = \boldsymbol{X}^t\boldsymbol{D}_h\boldsymbol{X} = (\boldsymbol{D}_h\boldsymbol{X})^t(\boldsymbol{D}_h\boldsymbol{X}). \tag{3.57}$$

Rewriting the Hessian matrix $\boldsymbol{C}_h$ by using (3.53) and (3.57), we have

$$\begin{aligned}\boldsymbol{C}_h &= \boldsymbol{A}^t\boldsymbol{Y}_h\boldsymbol{A} + \boldsymbol{A}\boldsymbol{Z}_h\boldsymbol{A}^t,\\ &= (\boldsymbol{B}_h\boldsymbol{X}\boldsymbol{A})^t(\boldsymbol{B}_h\boldsymbol{X}\boldsymbol{A}) + \left(\boldsymbol{D}_h\boldsymbol{X}\boldsymbol{A}^t\right)^t\left(\boldsymbol{D}_h\boldsymbol{X}\boldsymbol{A}^t\right),\\ &= \boldsymbol{P}_h^t\boldsymbol{P}_h + \boldsymbol{Q}_h^t\boldsymbol{Q}_h,\end{aligned} \tag{3.58}$$

where

$$P_h = B_h XA \text{ and } Q_h = D_h XA^t. \tag{3.59}$$

It is not difficult to verify that the Hessian C_h given by (3.58) is non-negative definite.

Moreover, it is apparent from (3.58) and (3.59) that the Hessian is positive definite if all the ranks of the matrices B_h, D_h, X, and A are p. Considering the case in which the Hessian is not positive definite, we recommend starting the iterative process of GIPSCAL from a variety of different initial configurations.

3.6 Examples of Application

We shall now show some examples of application. First, we will apply GIPSCAL to a set of errorless data which was shown as an illustration in Chino [3]. We add diagonal elements to the data to make a full similarity data matrix. In this case both options of preprocessing, that is, the simple normalization and the double centering, recovered the assumed configuration of the data almost perfectly in two dimensions except for a rotation and/or a translation (origin shift). However, these options differed slightly in the coefficients r_{SP} and r_{SO}. In the case of the simple normalization, r_{SP} and r_{SO} were 0.720 and 0.693, respectively. In the case of the double centering, these were 0.715 and 0.698. As expected, the constant term c in GIPSCAL was almost zero in the case of the double centering, while it was not zero in the case of the simple normalization. Figure 3.4 shows the derived configuration for the double centering.

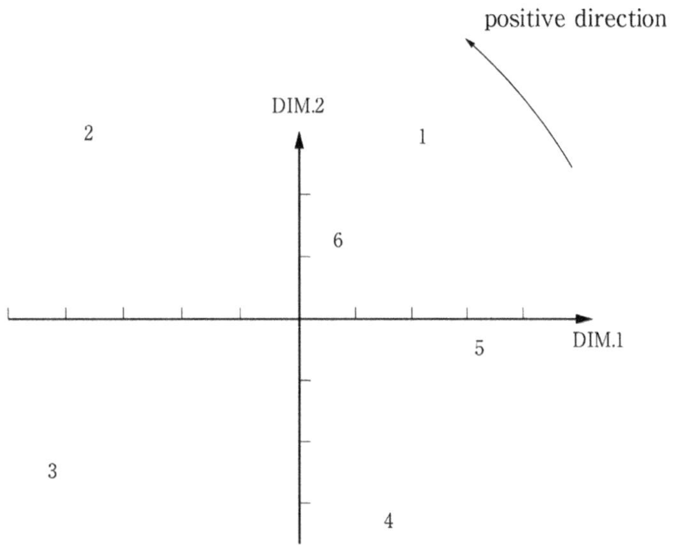

Fig. 3.4 Derived configuration for errorless data

Second, we will show the results of application to journal citation data, especially the data shown in Table 2 in Chino [3]. Figure 3.4 in that paper, of course, shows the derived configuration by Chino's ASYMSCAL for simply normalized data. The data were treated as similarity data. Values of F for doubly centered data were 0.835, 0.946, and 0.965 in two, three, and four dimensions, respectively.

We have chosen the three-dimensional configuration because the principal axis rotation of the four-dimensional configuration indicates that the latter configuration is almost degenerated. In other words, variation in the fourth dimension was very poor. Figures 3.5, 3.6 and 3.7 shows the planes defined by dimensions one and two, one and three, and two and three, respectively. We have not interpreted these principal axes because the number of journals is too small to assign them appropriate labels. In this case, r_{sp} and r_{so} were $-$ 0.907 and $-$ 0.350, respectively. As discussed in the introduction section, the sign of the constant b in our model, which is the same as that of r_{so}, determines the direction of asymmetry in each plane. Since the sign is negative, the positive directions in the planes defined by the above three are clockwise, counterclockwise, and clockwise, respectively. These.directions are shown in the planes. It should be noted that the sign of r_{sp} is negative. This means that the shorter the distance between journals the less frequent the citation. For, we treated the data as similarity data.

The one-two plane explains the major symmetric relations contained in the original data. For example, locations of JAP, JEdP, and Pka indicate that citations between these journals are very poor. The same is true for AJP and JCPP. On the other hand, JExP is almost on the line passing through the origin and is located on the opposite side of journals JAP, JEdP, and Pka. This means that citations between JExP and

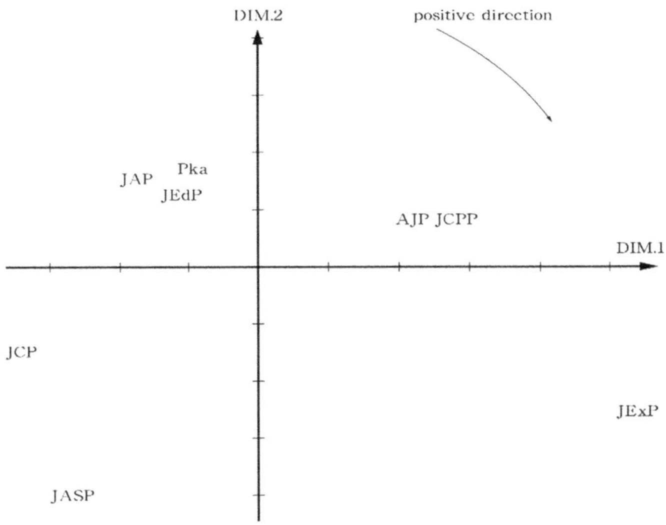

Fig. 3.5 The one-two plane for the journal citation data

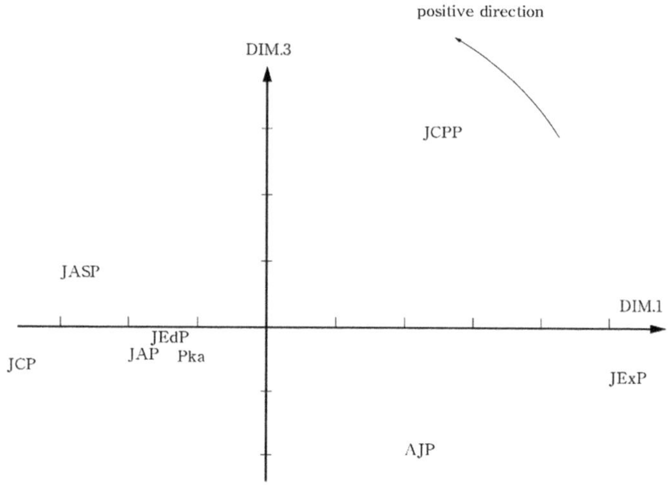

Fig. 3.6 The one-three plane for the journal citation data

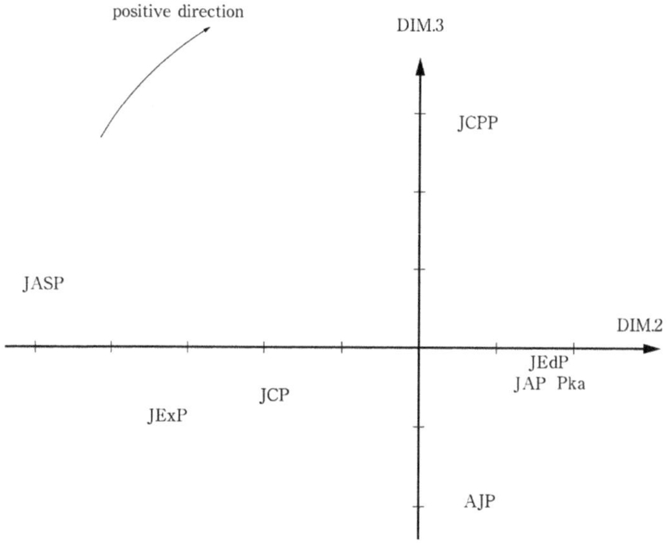

Fig. 3.7 The two–three plane for the journal citation data

each of these three journals are very frequent. Relation is the same between JCP and each of AJP and JCPP as well as between JASP and each of journals AJP and JCPP.

This plane also indicates the major asymmetric relations included in the data.

For example, the positive direction in this plane describes the citation surpluses from JExP to JASP. The similar relations hold from JASP to JCP, from JCP to Pka, and so on. The only major asymmetry which contradicts the data is the relation between

JCPP and JExP. This plane describes the citation surplus from JCPP to JExP, but it is not true.

The one-three plane draws the true citation surplus from JExP to JCPP. This can be readily confirmed if we notice that the positive direction in this plane is counterclockwise. This plane also emphasizes relatively poor citations between JASP, JAP, JCP, JEdP, and Pka and relatively frequent citations between JExP and each of the five journals. The two–three plane might depict the major symmetric relations and minor asymmetric relations included in the data. For example, the positive direction in this plane (i.e., clockwise) indicates the citation excesses from JCPP to each of JAP, JEdP, and Pka as well as from each of these three journals to AJP.

Finally, we will show the results of application to word association data gathered by Nakagawa [19]. Table 3.1 shows this. The subject was asked to rate the magnitude of association by a seven point rating scale. Thus, the scores range from 1 to 7 for the weakest association and the strongest association, respectively. We analyzed the data as similarity data. Since the goodness of fit was poor in lower dimensions, we obtained solutions up to six dimensions. Values of F for doubly centered data were 0.461, 0.573, 0.672, 0.759, and 0.779, in two, three, four, five and six dimensions, respectively. Considering the deceleration of the improvement of the goodness of fit as well as ease of interpretation, we tentatively adopted the five-dimensional configuration. The principal axis solution has indicated that this configuration is not degenerated. Figures 3.8, 3.9, 3.10, and 3.11 show the planes defined by dimensions one and two, one and three, one and four, and one and five, respectively. Although combinations of the five dimensions yield 10 planes, we selected the above four planes for simplicity as well as for space limitations. In this case, r_{sp} and were r_{so} 0.864 and $-$ 0.116, respectively.

The first dimension might be labelled natural-social because the first four (i.e., fish, plant, tree, nest) are related to natural resources and the last four (i.e., class, labor, fight, and knight) to social resources or events. The second dimension has an even less obvious interpretation: the first three (that is, iron, labor, and mother) are associated with stable impressions and the last three (i.e., fight, honor, and knight) with unstable impressions. In the same manner, we labelled the third, fourth, and fifth dimensions release-discrimination, soft-hard, and neutral-judge, respectively. Each plane defined by the combination of these dimensions describes various aspects of symmetric and asymmetric relations contained in the data. The positive directions in the above four planes are, respectively, counterclockwise, clockwise, counterclockwise, and clockwise, since the sign of r_{so} is positive.

The one-two plane describes relatively symmetric associations between words within subgroups. For example, honor, knight, and fight form a subgroup. Plant and tree constitute another subgroup. In contrast with the distance model, the words located on the line passing through the origin indicate that they also have symmetric relations. Thus, jack and iron, for example, constitute a third subgroup. This plane also draws some aspects of asymmetric relation. The major ones are association imbalances between yacht and labor as well as knight and iron. The positive direction in this plane indicates, for example, that although the intensity of association with yacht from labor is strong, the reverse is relatively weak.

Table 3.1 Word association data gathered by Nakagawa [19]

From/to	1	2	3	4	5	6	7	8	9	1	11	12	13	14	15	16	17	18	19	20
1		5	2	3	3	2	1	4	4	5	2	3	6	4	2	3	1	6	2	3
2	6		1	2	4	1	1	7	6	6	3	5	4	6	4	1	1	4	2	2
3	4	1		4	1	7	1	3	7	2	4	2	3	2	7	5	3	3	5	7
4	4	2	4		1	5	1	2	4	2	4	1	1	2	6	2	4	3	3	3
5	2	4	1	3		4	5	4	2	4	5	5	5	4	2	4	4	4	4	5
6	2	3	4	6	4		2	3	2	3	3	1	1	4	7	3	2	4	5	2
7	2	1	1	1	5	3		1	2	2	4	1	2	2	2	6	1	2	2	3
8	4	6	2	1	2	2	1		2	4	3	4	2	7	5	2	1	4	3	3
9	5	7	5	3	2	2	2	3		2	4	1	4	5	5	4	3	6	6	2
10	5	7	1	2	4	2	1	6	3		2	3	5	7	3	2	1	4	4	2
11	3	4	4	3	6	4	4	2	5	4		4	5	2	6	2	2	2	3	5
12	3	4	2	2	5	1	1	2	1	3	3		3	3	4	6	1	2	4	7
13	6	3	4	2	4	1	2	2	5	4	2	3		5	4	6	3	7	3	4
14	5	6	1	1	4	4	2	7	4	7	3	2	6		4	5	1	4	5	2
15	2	4	7	6	3	7	4	6	5	3	6	4	4T	6		4	4	3	5	7
16	4	2	6	2	5	6	6	2	4	6	2	5	6	7	5		1	4	4	6
17	4	1	2	4	4	4	1	1	2	1	3	1	3	2	6	3		3	4	2
18	5	6	4	3	3	2	2	4	6	5	5	1	7	4	3	4	5		5	4
19	3	2	5	3	4	6	1	4	6	5	4	3	4	6	6	5	3	6		3
20	4	2	7	3	4	2	4	3	5	1	4	7	7	2	7	6	1	4	3	

In the above table, words 1 through 20 denote label, fish, class, honor, relief, knight, jack, nest, yacht, plant, hope, mother, scale, tree, fight, iron, umpire, number, play, and labor, respectively

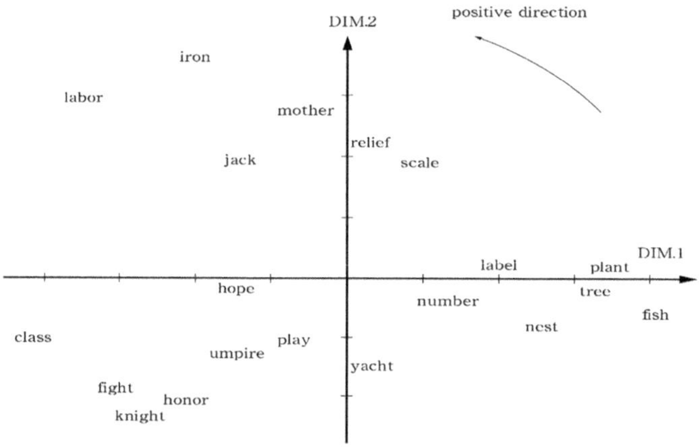

Fig. 3.8 The one-two plane for the word association data

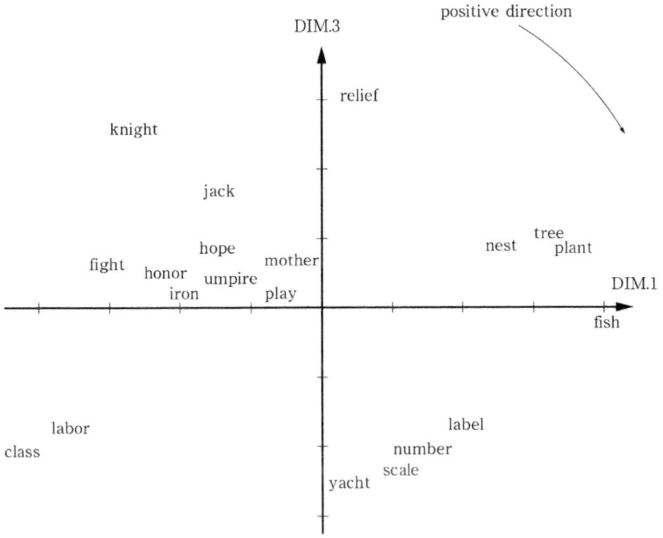

Fig. 3.9 The one-three plane for the word association data

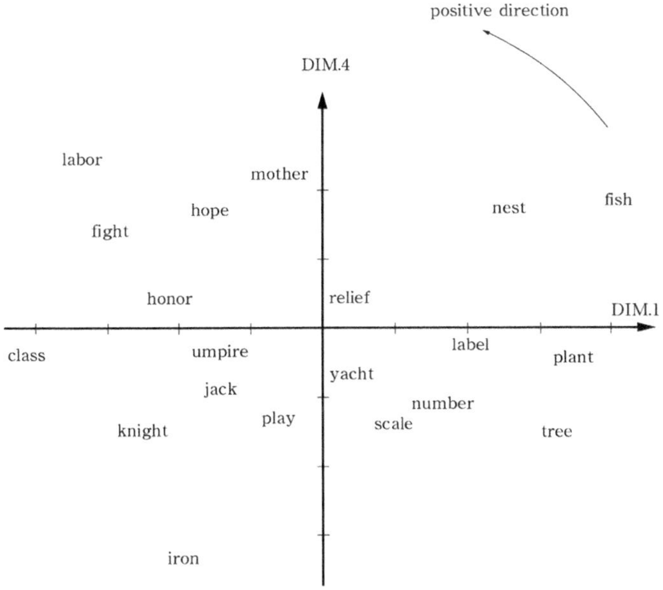

Fig. 3.10 The one-four plane for the word association data 7

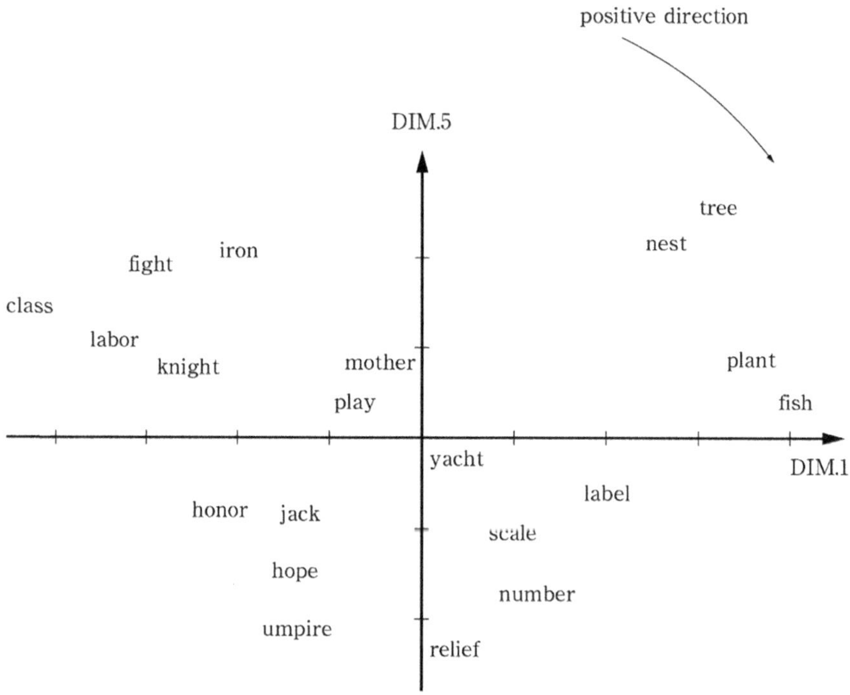

Fig. 3.11 The one-five plane for the word association data

There are several major symmetric and asymmetric relations depicted in this plane which contradict the data. For example, intensity of association between class and labor is the same in the original data, but this plane draws a relatively large imbalance of association since the area of the parallelogram spanned by location vectors of these two words in relatively big. Another contradiction is the asymmetric relation between plant and iron. According to the original data, the intensity of association with plant from iron is stronger than that with iron from plant. However, the positive direction suggests the reverse relation. In a similar manner, the relation indicated in this plane between words scale and labor contradicts the original data.

The one-three plane for the word association data, the one-three plane explains. Some of the relations which the one-two plane fails to describe. One is the symmetric relation between class and labor. Another is the asymmetric relation between relief and hope. This plane might mainly draw the asymmetric relation in the direction of the third dimension, that is, release-discrimination. The major three association imbalances in this direction are depicted between the pairs: class and knight, iron and knight, and hope and relief.

The one-four plane describes a major asymmetric relation between plant and iron, which neither the one-two plane nor the one-three plane can describe. That is, the positive direction in this plane describes the following relation: the intensity of association with plant from iron is strong, while the reverse weak.

The one-five plane describes major symmetric relations and minor asymmetric relations contained in the original data. As for symmetric relations, these include associations between words within subgroups: group 1 (fish and plant), group 2 (nest and tree), group 3 (class, fight, and labor), group 4 (scale and number). As for asymmetric associations, this plane describes many of the minor association imbalances between pairs nest and fight, tree and iron, label and tree, label and fish, umpire and number, honor and relief. The positive direction in this plane, which is clockwise, all coincides with the directions of association imbalances between these pairs included in the original data.

3.7 Discussion

In this paper we have introduced an extension of Chino's ASYMSCAL into higher dimensions and discussed theoretical and practical implications of the extension. GIPSCAL, which is the extended model, fits the inner product and the area of the parallelogram spanned by two vectors to the symmetric and skew-symmetric parts of original or preprocessed similarity judgements, respectively. The square of the area of the parallelogram is nothing other than the Gramian of these two vectors.

GIPSCAL not only permits parsimonious geometrical representations of both the asymmetric structure and symmetric structure of data but also permits a social psychological justification as to why asymmetry arises.

Although inner product models are not in the mainstream of MDS, we have shown that our generalized inner product model-GIPSCAL can be related to a distance model indirectly, if we assume a double centering transformation as a preprocessing of original similarity data. Here it should be noted that the famous Young-Housholder transformation, which plays a fundamental role in MDS techniques, is obtained through the double centering transformation of the inner product matrix produced from the configuration of objects whose origin is arbitrary. Figure 3.8, the one-five plane for the word association data.

One shortcoming of GIPSCAL is its employment of a metric MDS. Considering the nature of data in social and behavioral sciences, it will be better to develop a nonmetric asymmetric MDS as Okada and Imaizumi [21] have done.

Another shortcoming is that GIPSCAL is merely a representation model, as are most of the extant MDS techniques. In the near future some researchers may develop response models which can take specific response processes in which asymmetry arises. However, since the context in which asymmetry arises seems to be diverse in character, we believe that there will be some important roles which representation models play. For example, if we face a set of longitudinal relational data matrices and consider the derived configuration obtained at each time by an appropriate extant MDS as a snapshot of a group formation process, it seems to be better to construct a group dynamics model which accounts for changes in group structure over time rather than to construct a specific response model which accounts for the cause of asymmetry at each time. In such a case, representation models may catch the structure at each time naturally without preoccupation with the specific response process assumed.

One other short-fall is that GIPSCAL as well as all the extant asymmetric MDS's does not utilize information about judgmental errors. To utilize such information, the maximum likelihood principle should be applied, as has been done in symmetric MDS[22, 23, 29–31, 32]. This also enables us to compute AIC [1], and therefore enables us to compare extant asymmetric MDS's and thereby to choose the best model.

Acknowledgements The author wishes to express his thanks to anonymous referees for their constructive comments. Special thanks are due to W. B. Redekop for careful proofreading.

Notes

This chapter is the reproduction of Chino [6]. We redrew unclear lines and symbols in the figures found in the original paper.

GIPSCAL extends the two-dimensional model of Chino's ASYMSCAL described by (2.4) in Chap. 2 to the case of higher dimensions within the real number. By contrast, HFM discussed in Chap. 4 extends the model to that of higher dimensions within the complex number.

References

1. Akaike, H. (1974). A new look at the statistical model identification. *IEEE Transactions on Automatic Control, 19*, 716–723.
2. Baker, R. F., Young, F. W., & Takane, Y. (1977). An Asymmetric Euclidean model. In *Paper submitted for publication to Psychometrika.*
3. Chino, N. (1978). A graphical technique for representing the asymmetric relationships between N objects. *Behaviormetrika, 5*, 23–40.
4. Chino, N. (1979). Extension of Chino's ASYMSCAL into higher dimensions. In *Paper presented at the 7th annual meeting of the Behaviormetric Society of Japan*, Osaka, Japan (in Japanese).
5. Chino, N. (1980). A unified geometrical interpretation of the MDS techniques for the analysis of asymmetry and related techniques. In *Paper presented at the symposium on "Asymmetric multidimensional scaling" at the Spring Meeting of the Psychometric Society*, Iowa, U.S.A.
6. Chino, N. (1990). A generalized inner product model for the analysis of asymmetry. *Behaviormetrika, 27*, 25–46.
7. Constantine, A. G., & Gower, J. C. (1978). Graphical representation of asymmetric matrices. *Applied Statistics, 27*, 297–304.
8. Courant, R., & John, F. (1974). *Introduction to Calculus and analysis* (Vol. 2). Wiley.

9. Gower, J. C. (1977). The analysis of asymmetry and orthogonality. In J. R. Barra, F. Brodeau, G. Romer, & B. van Cutsem (Eds.), *Recent developments in statistics* (pp. 109–123). North Holland.
10. Guttman, L. (1968). A general nonmetric technique for finding the smallest coordinate space for a configura tion of points. *Psychometrika, 33*, 469–506.
11. Harshman, R. A. (1978). Models for analysis of asymmetrical relationships among N objects or stimuli. In *Paper presented at the First Joint Meeting of the Psychometric Society and The Society for Mathematical Psychology*. Hamilton, Canada.
12. Harshman, R. A., Green, P. E., Wind, Y., & Lundy, M. E. (1982). A model for the analysis of asymmetric data in marketing research. *Marketing Science, 1*, 205–242.
13. Hayashi, C. (1977). MDS ni okeru hataishosei no mondai [The problem of asymmetry in MDS]. In *Proceedings of the 45-th annual conference of Japan Statistical Society*, p. 41. Fukuoka University.
14. Horst, P. (1965). *Factor analysis of data matrices*. Holt, Rinehart and Winston Inc.
15. Krumhansl, C. L. (1978). Concerning the applicability of geometric models to similarity data: The interrelationship between similarity and spatial density. *Psychological Review, 85*, 445–463.
16. Kruskal, J. B. (1964). Multidimensional scaling by optimizing goodness of fit to a nonmetric hypothesis. *Psychometrika, 29*, 1–27.
17. Lancaster, P., & Tismenetsky, M. (1985). *The theory of matrices* (2nd ed.). Academic Press.
18. Lingoes, J. C. (1973). *The Guttman-Lingoes nonmetric program series*. Mathesis Press.
19. Nakagawa, M. (1986). Personal communication.
20. Okada, A., & Imaizumi, T. (1984). Geometric models for asymmetric similarity. In *Research Reports of School of Social Relations*, Rikkyo (St. Paul's) University.
21. Okada, A., & Imaizumi, T. (1987). Nonmetric multidimensional scaling of asymmetric proximities. *Behaviormetrika, 21*, 81–96.
22. Ramsay, J. O. (1977). Maximum likelihood estimation in multidimensional scaling. *Psychometrika, 42*, 241–266.
23. Ramsay, J. O. (1982). Some statistical approaches to multidimensional scaling data. *The Journal of the Royal Statistical Society Series A (General), 145*, 285–312.
24. Saito, T. (1983). A method of multidimensional scaling to obtain a sphere configuration. *Hokkaido Behavioral Science Report, Series-M, 4*.
25. Saito, T. (1986). Multidimensional scaling to explore complex aspects in dissimilarity judgement. *Behaviormetrika, 20*, 35–62.
26. Sato, Y. (1988). An analysis of sociometric data by MDS in Minkowski space. In K. Matsusita (Ed.), *Statistical theory and data analysis II* (pp. 385–396). North Holland.
27. Sato, Y. (1989). Minkowski keiryo o mochiita hitaisyo ruijisei no kyori hyogen ni tsuite. [On the distance representation of asymmetric dissimilarities by Minkowskimetric function]. *Computational Statistics, 2*, 35–45. (in Japanese).
28. Tagiuri, R. (1969). Person perception. In G. Lindzey & E. Aronson (Eds.), *The handbook of social psycholgy* (2nd ed., pp. 395–449, Vol. III). Addison-Wesley.
29. Takane, Y. (1978). A maximum likelihood method for nonmetric multidimensional scaling: 1. The case in which all empirical pairwise orderings are independent—Theory. *Japanese Psychological Research, 20*, 7–17.
30. Takane, Y. (1978). A maximum likelihood method for nonmetric multidimensional scaling: 2. The case in which all empirical pairwise orderings are independent—Evaluation. *Japanese Psychological Research, 20*, 105–114.
31. Takane, Y. (1981). Multidimensional successive categories scaling: A maximum likelihood method. *Psychometrika, 46*, 9–28.
32. Takane, Y., & Carroll, J. D. (1981). Nonmetric maximum likelihood multidimensional scaling from directional rankings of similarities. *Psychometrika, 46*, 389–405.
33. Tobler, W. (1976–77). Spatial interaction patterns. *Journal of Environmental Systems, 6*, 271–301.
34. Torgerson, W. S. (1958). *Theory and Methods of Scaling*. Wiley.

35. Tversky, A. (1977). Features of similarity. *Psychological Review, 84*, 327–352.
36. Weeks, D. G., & Bentler, P. M. (1980). *A restricted asymmetric multidimensional scaling model.* Paper presented at the Spring meeting of the Psychometric Society.
37. Weeks, D. G., & Bentler, P. M. (1982). Restricted multidimensional scaling models for asymmetric proximities. *Psychometrika, 47*, 201–208.
38. Young, F. W. (1975). An asymmetric Euclidean model for multi-process asymmetric data. In *Paper presented at U.S.-Japan Seminar on MDS*, San Diego, U.S.A.
39. Young, F. W. (1987). Multidimensional scaling: History, theory, and applications. Lawrense Erlbaum Associates.

Chapter 4
HFM

4.1 Introduction

In recent years much attention has been attracted to non-distance models for asymetric MDS. These include CASK (Canonical Analysis of SKew symmetry) proposed by Gower [13] and Constantine and Gower [9], DEDICOM (DEcomposition into DIrectional COMponents) by Harshman [15] and Harshman, Green, Wind, and Lundy [16], GIPSCAL (Generalized Inner Product multidimensional SCALing) by Chino [2–4], HCM (Hermitian Canonical Model) proposed originally by Escoufier and Grorud [12], and reformulated independently in a different way and amplified somewhat by Chino [5, 6].

Unfortunately, however, the geometrical structures of these models have been only partially understood. For example, Harshman [15] calls DEDICOM a non-spatial model. Chino [4] does not show the necessary and sufficient condition for GIPSCAL to be a Euclidean space model. Recently, Kiers and Takane [17] have proposed a generalized version of GIPSCAL and have shown the necessary and sufficient condition for DEDICOM and GIPSCAL to have an equivalent expression. But, their discussions are limited to pointing out that these two models can have such an expression. On the other hand, Chino [6] has pointed out that HCM can have some metric properties such as a finite-dimentional complex (f.d.c.) Hilbert space structure under a general condition, but his discussion still lacks a unified geometrical interpretation of these four models. Gower and Zielman [14] have discussed the non-Euclidean space properties of the symmetric part as well as the skew-symmetric part of a proximity matrix. But they neither discuss the concrete space structures of each of them nor refer to the structures as a whole.

In this paper we examine the geometrical structures of these models for asymmetric MDS in *error-free data measured at a ratio level* using several concepts which are little known to psychometricians, and provide a unified geometrical interpretation of their holistic structures. As a result we relate these models to each other. Results

N. Chino, *Structure and Dynamics of Asymmetric Interactions*,
https://doi.org/10.1007/978-981-97-8269-7_4

are summarized into the following five conclusions, which are proved in Sect. 4.2, where the terminology is fully explained:

(C1) Complex counterparts of DEDICOM, GIPSCAL, and HCM all assign a Hermitian form to the *Hermitian data matrix* $\boldsymbol{H}$ constructed uniquely from an original proximity matrix $\boldsymbol{S}$ such that

$$\boldsymbol{H} = \boldsymbol{S}_s + i\boldsymbol{S}_{sk}, \tag{4.1}$$

where $\boldsymbol{S}_s$ and $\boldsymbol{S}_{sk}$ are the symmetric part and the skew-symmetric part of the proximity data matrix, respectively. Incidentally, these models can be generally called the *Hermitian form models*.

(C2) The complex counterparts of DEDICOM, GIPSCAL, HCM, and other possible models for $\boldsymbol{H}$ are expressible in terms of finite-dimensional complex *Hilbert space* if $\boldsymbol{H}$ is *positive (negative) semi-definite*. Moreover, in this case, objects can be represented in the space.

(C3) The complex counterparts of DEDICOM, GIPSCAL, HCM, and other possible models of $\boldsymbol{H}$ are expressible in terms of an *indefinite metric* if $\boldsymbol{H}$ is indefinite.

(C4) The Hermitian form in DEDICOM, GIPSCAL, and HCM is composed of a *metric tensor* and an *exterior product* called the *2-form*, which correspond to the symmetric part and the skew-symmetric part of $\boldsymbol{S}$, respectively. Differences in form of these models depend upon the basis chosen for these two parts. On the one hand, GIPSCAL employs an orthogonal basis for both, while HCM chooses an orthonormal basis for the former and a *symplectic basis* for the latter. Thus, GIPSCAL and HCM have a standard Euclidean structure if $\boldsymbol{H}$ is positive (negative) semi-definite. On the other hand, DEDICOM does not choose a specific standard basis. CASK uses a symplectic basis, but does not assume any metric tensor.

(C5) We can also choose, for example, a symplectic basis for DEDICOM and GIPSCAL. In this case, the metric tensor is furnished with a weighted Euclidean space structure for DEDICOM and GIPSCAL. The generalized version of GIPSCAL as well as a DEDICOM version proposed recently by Kiers and Takane [17] are such cases. These conclusions are proved in the next section. In the third section we shall also prove that, under the general condition stated in (C2), the Hermitian form models have an interesting property, which traditional (squared) distance model does not. In the fourth section we shall prove that the general condition leads to an extension of the famous Young-Householder theorem (Young & Householder [21]) to complex Hilbert space.

4.2 Proofs of the Five Conclusions

Let us now prove seriatim the five conclusions stated in the introduction section. First, we define the complex counterparts of DEDICOM and GIPSCAL. These definitions can be made simply by putting the purely imaginary number i to the second right-hand terms of the following equations for DEDICOM and GIPSCAL:

$$S = \{s_{jk}\} = \boldsymbol{YAY}^t = \boldsymbol{YA}_s\boldsymbol{Y}^t + \boldsymbol{YA}_{sk}\boldsymbol{Y}^t, \tag{4.2}$$

and

$$S = a\boldsymbol{ZZ}^t + b\boldsymbol{ZL}_q\boldsymbol{Z}^t + c1_N1_N^t, \tag{4.3}$$

where $\boldsymbol{S}$ is a proximity matrix of order $(N \times N)$, $\boldsymbol{A}_s$ and $\boldsymbol{A}_{sk}$ are the symmetric part and the skew-symmetric part of the DEDICOM "core" matrix $\boldsymbol{A}$ of order $(p \times p)$ giving the directional relationships among the basic p types or dimensions, and $\boldsymbol{L}_q$, is the special skew-symmetric matrix of order $(q \times q)$ in the GIPSCAL model. Note that the constant c in Eq. (4.3) is zero, since we have assumed that the data is measured at a ratio level in this paper. In terms of (4.1), (4.2) becomes

$$\boldsymbol{H} = \{h_{jk}\} = \boldsymbol{YA}_s\boldsymbol{Y}^t + i\boldsymbol{YA}_{sk}\boldsymbol{Y}^t, \tag{4.4}$$

and (4.3) becomes,

$$\boldsymbol{H} = a\boldsymbol{ZZ}^t + i(b\boldsymbol{ZL}_q\boldsymbol{Z}^t). \tag{4.5}$$

From (4.4) and (4.5), we get

$$h_{jk} = \boldsymbol{y}_j^t\boldsymbol{A}_s\boldsymbol{y}_k + i\boldsymbol{y}_j^t\boldsymbol{A}_{sk}\boldsymbol{y}_k, \tag{4.6}$$

and

$$h_{jk} = a z_j^t z_k + i\left(b z_j^t \boldsymbol{L}_q z_k\right), \tag{4.7}$$

respectively.

As for HCM, there exists an orthonormal eigenbasis $\{\boldsymbol{u}_1, \boldsymbol{u}_2, \cdots, \boldsymbol{u}_N\}$ in C^N, which is composed of the eigenvectors of $\boldsymbol{H}$, since the starting matrix $\boldsymbol{H}$ of HCM is Hermitian (Lancaster & Tismenetsky, 1985). Thus we have

$$\boldsymbol{Hu}_i = \lambda_i\boldsymbol{u}_i, 1 \le i \le N, \tag{4.8}$$

Here, we have without loss of generality, $\lambda_i \neq 0 (1 \le i \le n)$, and $\lambda_{n+1} = \lambda_{n+2} = \cdots = \lambda_N = 0$. If we let

$$\boldsymbol{U} = \{\boldsymbol{u}_1, \boldsymbol{u}_2, \cdots, \boldsymbol{u}_n, \boldsymbol{u}_{n+1}, \boldsymbol{u}_{n+2}, \cdots, \boldsymbol{u}_N\} = (\boldsymbol{U}_1, \boldsymbol{U}_2),$$

then (4.8) can also be written as

$$\boldsymbol{HU} = (\boldsymbol{U}_1, \boldsymbol{U}_2)\begin{pmatrix} \boldsymbol{\Lambda} & \boldsymbol{O} \\ \boldsymbol{O} & \boldsymbol{O} \end{pmatrix} = \boldsymbol{U}\tilde{\boldsymbol{\Lambda}},$$

where $\boldsymbol{\Lambda} = diag(\lambda_1, \lambda_2, \cdots, \lambda_n)$. Then we have

$$\boldsymbol{H} = \boldsymbol{U}_1 \boldsymbol{\Lambda} \boldsymbol{U}_1^*, \tag{4.9}$$

since $\boldsymbol{U}$ is unitary. Note that (4.9) holds not only for HCM but also for DEDICOM as well as GIPSCAL, since the left-hand sides of (4.4) and (4.5) are both Hermitian. Moreover, if we put

$$\boldsymbol{U}_1 = \boldsymbol{U}_r + i\boldsymbol{U}_c, \boldsymbol{X} = (\boldsymbol{U}_r, \boldsymbol{U}_c), \tag{4.10}$$

Then we get

$$\boldsymbol{H} = \boldsymbol{X}\boldsymbol{\Omega}_s\boldsymbol{X}^t + i\boldsymbol{X}\boldsymbol{\Omega}_{sk}\boldsymbol{X}^t, \tag{4.11}$$

where

$$\boldsymbol{\Omega}_s = \begin{pmatrix} \boldsymbol{\Lambda} & \boldsymbol{O} \\ \boldsymbol{O} & \boldsymbol{\Lambda} \end{pmatrix}, \quad \boldsymbol{\Omega}_{sk} = \begin{pmatrix} \boldsymbol{O} & -\boldsymbol{\Lambda} \\ \boldsymbol{\Lambda} & \boldsymbol{O} \end{pmatrix}, \tag{4.12}$$

and $\boldsymbol{U}_r$ and $\boldsymbol{U}_c$ are the real part and the imaginary part of $\boldsymbol{U}_1$, respectively.

From (4.11) we have

$$h_{jk} = \boldsymbol{x}_j^t \boldsymbol{\Omega}_s \boldsymbol{x}_k + i\boldsymbol{x}_j^t \boldsymbol{\Omega}_{sk} \boldsymbol{x}_k, \tag{4.13}$$

Then, conclusion Cl is evident. For, putting $\varphi(\boldsymbol{\zeta}, \boldsymbol{\tau}) = \boldsymbol{\zeta}\boldsymbol{\Lambda}\boldsymbol{\tau}^*$, φ satisfies the following properties of the *Hermitian form* (Cristescu [10]; Lancaster and Tismenetsky [18]):

(i) $\varphi(\boldsymbol{\zeta}_1 + \boldsymbol{\zeta}_2, \boldsymbol{\tau}) = \varphi(\boldsymbol{\zeta}_1, \boldsymbol{\tau}) + \varphi(\boldsymbol{\zeta}_2, \boldsymbol{\tau})$,
(ii) $\varphi(\alpha\boldsymbol{\zeta}, \boldsymbol{\tau}) = \alpha\varphi(\boldsymbol{\zeta}, \boldsymbol{\tau})$,
(iii) $\varphi(\boldsymbol{\zeta}, \boldsymbol{\tau}) = \overline{\varphi(\boldsymbol{\tau}, \boldsymbol{\zeta})}$,

where $\boldsymbol{\tau}, \boldsymbol{\zeta}$ are row vectors in an n-dimensional complex vector space, and $\overline{\varphi(\boldsymbol{\tau}, \boldsymbol{\zeta})}$ is the complex conjugate of $\varphi(\boldsymbol{\tau}, \boldsymbol{\zeta})$. Moreover, from (4.9) we get $h_{jk} = \varphi(\boldsymbol{v}_j, \boldsymbol{v}_k)$, where $\boldsymbol{v}_j, \boldsymbol{v}_k$ are the row vectors corresponding to the j-th row and the k-th row of $\boldsymbol{U}_1$. Here, it should be emphasized that (4.9) holds not only for DEDICOM, GIPSCAL, and HCM but also for other possible models for $\boldsymbol{H}$.

In general, a Hermitian form is said to be positive-definite if $\varphi(\boldsymbol{\zeta}, \boldsymbol{\zeta}) \geq 0$ for any $\boldsymbol{\zeta}$, and is called a *(Hermitian) scalar product* if $\varphi(\boldsymbol{\zeta}, \boldsymbol{\zeta}) > 0$ for any $\boldsymbol{\zeta} \neq 0$. This means that we can define a *seminorm* on a vector space if, for example, $\varphi(\boldsymbol{\zeta}, \boldsymbol{\zeta}) \geq 0$ for any $\boldsymbol{\zeta}$ in such a way that

$$\|\boldsymbol{\zeta}\| = \sqrt{\varphi(\boldsymbol{\zeta}, \boldsymbol{\zeta})}, \tag{4.14}$$

In particular, if φ is positive for any $\boldsymbol{\zeta} \neq 0$, $\boldsymbol{\zeta}$ defines a *norm*. Since a normed vector space is a *pre-Hilbert space* if and only if the norm is generated by a scalar product, DEDCOM, GIPSCAL, HCM, and other possible models for $\boldsymbol{H}$ are expressible in terms of *f.d.c. Hilbert space*, since a finite-dimensional complex space is always *complete*. Of course, the positive semi-definiteness of $\boldsymbol{H}$ is a necessary and sufficient condition for these models to be expressible in terms of complex Hilbert space. The proof of conclusion *C2* is straightforward for HCM. It should be understood that the sign of $\boldsymbol{H}$ is altered when $\boldsymbol{H}$ is negative semi-definite. Such an $\boldsymbol{H}$ corresponds to some *dissimilarity data matrix*. The proof is trivial for DEDCOM, GIPSCAL, and other possible models for $\boldsymbol{H}$, since the decomposition defined by (4.1) is unique. Recently, Kiers and Takane (1992) have proved that a necessary and sufficient condition for these two models to have an equivalent expression is the positive-definiteness of $\boldsymbol{A}_s$ in (4.4). This is different from the condition for these two models to be expressible in terms of *f.d.c.* Hilbert space. Conclution *C3* is trivial, at this point, since $\varphi(\boldsymbol{\zeta}, \boldsymbol{\zeta})$ is not necessarily positive (or negative) if $\boldsymbol{H}$ is indefinite.

We shall now examine the features of the Hermitian form used in DEDICOM, GIPSCAL, and HCM. It is apparent, from (4.6), (4.7), and (4.13), that all of these models assume a symmetric bilinear form and a skew-symmetric bilinear form for the symmetric and the skew-symmetric parts of the proximity data, respectively. In the mathematical literatures these are also called the *metric tensor* and the *exterior product* or the *2-form*. The metric tensor is a $\begin{pmatrix} 0 \\ 2 \end{pmatrix}$ tensor at a point in a vector space, and thus a bilinear function of two vectors. Generally, a tensor of type $\begin{pmatrix} p \\ q \end{pmatrix}$ at a point in a vector space (more generally at a point of a manifold) is defined to be a real-valued multilinear function which takes as arguments p *one-forms* and q *vectors* (Schutz [19]). Here, a one-form is defined as a linear, real-valued function of vectors. Following Schutz, the metric tensor, G is defined by

$$G(\boldsymbol{\alpha}, \boldsymbol{\beta}) = G(\boldsymbol{\beta}, \boldsymbol{\alpha}) \equiv \boldsymbol{\alpha}^t \boldsymbol{\beta}, \tag{4.15}$$

where $\boldsymbol{\alpha}$ and $\boldsymbol{\beta}$ are two vectors in an l-dimensional vector space. Moreover, let us define matrix $\boldsymbol{M}$ of order $(l \times l)$ whose (j, k) entry is

$$m_{jk} = G\big(\boldsymbol{b}_j, \boldsymbol{b}_k\big) \equiv \boldsymbol{b}_j^t \boldsymbol{b}_k, \tag{4.16}$$

where $\boldsymbol{b}_j$ $(j = 1, 2, \cdots, l)$ constitute a basis of the vector space. The nonsingular symmetric matrix $\boldsymbol{M}$ can be diagonalized by using its eigenvalues and eigenvectors. If $\boldsymbol{M}$ is an identity matrix, the metric tensor is usually called the *Euclidean metric*, and the vector space is called the Euclidean space. More generally, the metric tensor is called a Euclidean metric if $\boldsymbol{M}$ is positive-definite or nagative-definite. Otherwise, it is called *indefinite*. A famous indefinite metric is a *Minkowski metric* in relativity theory in physics, where one eigenvalue is negative, and the others are all positive.

It is apparent that CASK has no metric tensor, since CASK can be written as

$$S_{sk} = \boldsymbol{X}\boldsymbol{\Gamma}\boldsymbol{Y}^t = \boldsymbol{X}\boldsymbol{\Gamma}\boldsymbol{K}\boldsymbol{X}^t, \tag{4.17}$$

where $\boldsymbol{\Gamma}$ is a diagonal matrix of order $(N \times N)$ of form

$$\boldsymbol{\Gamma} = diag(\gamma_1, \gamma_1, \gamma_2, \gamma_2, \cdots, (0)), \tag{4.18}$$

and $\boldsymbol{X}$ is an orthogonal matrix of order $(N \times N)$. Moreover, $\boldsymbol{K}$ is the matrix of order $(N \times N)$ which is defined by

$$\boldsymbol{K} = \begin{bmatrix} 0\ 1 & & & \\ -1\ 0 & & & \\ & \ddots & & \\ & & 0\ 1 & \\ & & -1\ 0 & \\ & & & (1) \end{bmatrix}. \tag{4.19}$$

The second component of the Hermitian form used in DEDICOM, GIPSCAL, and HCM, is a certain 2-form. In general, an exterior form of degree k, or a k-form, is a function ω, of k vectors in an n-dimensional real vector space R^n, which has the following properties:

(1) *k-linear*, i.e.,

$$\omega\left(\lambda_1\boldsymbol{\xi}_1' + \lambda_2\boldsymbol{\xi}_1'', \boldsymbol{\xi}_2, \cdots, \boldsymbol{\xi}_k\right) = \lambda_1\omega(\boldsymbol{\xi}_1', \boldsymbol{\xi}_2, \cdots, \boldsymbol{\xi}_k) + \lambda_2\omega\left(\boldsymbol{\xi}_1'', \boldsymbol{\xi}_2, \cdots, \boldsymbol{\xi}_k\right), \tag{4.20}$$

(2) *antisymmetric*, i.e.,

$$\omega(\boldsymbol{\xi}_{i_1}, \boldsymbol{\xi}_{i_2}, \ldots, \boldsymbol{\xi}_{i_k}) = (-1)^{\nu}\omega(\boldsymbol{\xi}_1, \boldsymbol{\xi}_2, \cdots, \boldsymbol{\xi}_k), \tag{4.21}$$

where

$$\nu = \begin{cases} 0, \textit{if the permutation}(i_1, i_2, \cdots, i_k) \textit{ is even}, \\ 1, \textit{if the permutation}(i_1, i_2, \cdots, i_k) \textit{ is odd}. \end{cases}$$

A well-known k-form is the oriented volume of the parallelepiped with edges $\boldsymbol{\xi}_1, \boldsymbol{\xi}_2, \cdots, \boldsymbol{\xi}_k$ in an oriented Euclidean space E^k. Likewise, a famous 2-form is the oriented area of the parallelogram spanned by two vectors $\boldsymbol{\xi}_1$ and $\boldsymbol{\xi}_2$ of the oriented Euclidean space E^2.

A 2-form can be written as a linear combination of the exterior products of two 1-forms. It should be noted that the value of the exterior product $\omega_1 \wedge \omega_2$ of two 1-forms ω_1 and ω_2 on the pair of vectors $\boldsymbol{\xi}_1, \boldsymbol{\xi}_2 \in R^n$ can be thought of as a projection of the oriented area spanned by the two vectors in an n-dimensional real vector space onto the two dimensional ω_1, ω_2-plane (Arnold [1]). Figure 4.1 illustrates this idea.

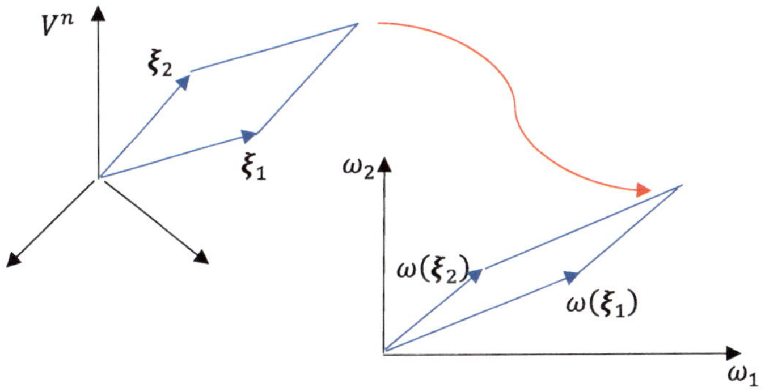

Fig. 4.1 $\omega_1 \wedge \omega_2$ as the image of the parallelogram

In this figure, $\omega(\boldsymbol{\xi}_1)$ and $\omega(\boldsymbol{\xi}_2)$ are the two-dimensional *images* of $\boldsymbol{\xi}_1$ and $\boldsymbol{\xi}_2$ in the n-dimensional vector space.

It is apparent, from (4.6), (4.7), (4.13), that $\boldsymbol{y}_j^t\boldsymbol{A}_{sk}\boldsymbol{y}_k$, $\boldsymbol{z}_j^t\boldsymbol{L}_q\boldsymbol{z}_k$, and $\boldsymbol{x}_j^t\boldsymbol{\Omega}_{sk}\boldsymbol{x}_k$ are all 2-forms and thus they can be a projection of the *oriented area* spanned by the two vectors in an n-dimensional real vector space onto a two-dimensional plane. Then, it will be natural to ask what are the differences of form among these three models. The answer is that *the variety of form depends upon what kind of framework we intend to choose to look at the projection of the oriented area*, i.e. the exterior product $\omega_1 \wedge \omega_2$. Mathematically, this is equivalent to saying what kind of basis we intend to choose. There are at least three strategies to do so. One is to choose an orthonormal basis on R^n and project the exterior product onto $\binom{n}{2}$ component planes. GIPSCAL employs this strategy. Another is to take a symplectic basis on R^{2n}, and CASK and HCM use this strategy. A third is not to take a specific standard basis, and this is the strategy of DEDICOM. The details of the first strategy can be explained as follows: In this case, we first consider q independent 1-forms $x_1, x_2, \cdots, x_q$, which are called the *basic forms*. The 2-forms $x_i \wedge x_j$ constructed from these basic forms have a very simple geometrical meaning (Arnold [1]): the value of $x_i \wedge x_j$ is the oriented area of the image of the parallelogram spanned by the two vectors $\boldsymbol{\xi}_1$ and $\boldsymbol{\xi}_2$ on the coordinate plane x_i, x_j under the projection parallel to the remaining coordinate functions.

It is interesting to note that (a) $\binom{q}{2}$ (=$q(q-1)/2$) 2-forms, $x_i \wedge x_j$, are linearly independent, and (b) every 2-form on the q-dimensional space with coordinate $x_1, x_2, \cdots, x_q$ can be uniquely represented in the form

$$\omega^2 = \sum_{i<j}^{q} a_{ij}(x_i \wedge x_j), \tag{4.22}$$

where

$$a_{ij} = \omega_b^2(\boldsymbol{e}_i, \boldsymbol{e}_j), \tag{4.23}$$

and $\boldsymbol{e}_i$ is the i-th basis vector (Arnold [1]). The ω_b^2 is the 2-form on the pair $\boldsymbol{e}_i, \boldsymbol{e}_j$. It is evident that the skew-symmetric part of GIPSCAL, *i.e.* $z_j^t \boldsymbol{L}_q z_k$, is nothing but the 2 form ω^2 in (4.22) in the special case when the q basis vectors $\boldsymbol{e}_1, \boldsymbol{e}_2, \ldots, \boldsymbol{e}_q$ are orthonormal.

The second strategy is to employ a symplectic basis. According to Arnold [1]: A *symplectic (linear) structure* on R^{2n} is defined as a nondegenerate bilinear skew symmetric 2-form given in R^{2n}. This form is called the *skew-scalar product* and is denoted by $[\boldsymbol{\eta}_1, \boldsymbol{\eta}_2] = -[\boldsymbol{\eta}_2, \boldsymbol{\eta}_1]$. The space R^{2n}, together with the symplectic structure [,], is called a *symplectic vector space*. Here, a 2-form [,] on R^{2n} is called nondegenerate if, for any $\boldsymbol{\eta}_2$, $[\boldsymbol{\eta}_1, \boldsymbol{\eta}_2] = 0$ then $\boldsymbol{\eta}_1 = \boldsymbol{0}$.

In general a skew-scalar product, $[\boldsymbol{\eta}_1, \boldsymbol{\eta}_2] = \omega^2(\boldsymbol{\eta}_1, \boldsymbol{\eta}_2)$ may take a variety of forms depending on a basis in a symplectic vector space. However, if we take a suitable basis, it is represented by a very simple form

$$\omega^2 = x_1 \wedge y_1 + \cdots + x_n \wedge y_n, \tag{4.24}$$

where $(x_1, x_2, \cdots, x_n, y_1, y_2, \cdots, y_n)$ be coordinate functions on R^{2n} and this symplectic structure is called the *standard symplectic structure*. It is apparent, from the form of ω^2 in (4.24), that in this case $2n$ basis vectors $\boldsymbol{e}_{x_i}$, and $\boldsymbol{e}_{y_i}$, , $(i = 1, \cdots, n)$ satisfy the relations

$$[\boldsymbol{e}_{x_i}, \boldsymbol{e}_{x_j}] = [\boldsymbol{e}_{y_i}, \boldsymbol{e}_{y_j}] = [\boldsymbol{e}_{x_i}, \boldsymbol{e}_{y_j}] = 0, \ (i \neq j), \tag{4.25}$$

and

$$[\boldsymbol{e}_{x_i}, \boldsymbol{e}_{y_i}] = 1. \tag{4.26}$$

The basis vectors which satisfy (4.25) and (4.26) constitute a standard *symplectic basis*.

It is apparent from the definition that in the standard symplectic structure the oriented area ω^2 spanned by the two vectors $\boldsymbol{\eta}_1$ and $\boldsymbol{\eta}_2$ is projected on the n coordinate planes (x_i, y_i), $i = 1, \cdots, n$. As a result, the ω^2 is equal to the sum of the oriented areas of the parallelogram on the n coordinate planes.

It is apparent from the above discussion that CASK employs the standard symplectic basis and that matrix $\boldsymbol{K}$ in (4.19) is nothing but the expression of (4.25) and (4.26) if the order of $\boldsymbol{\Gamma}$ defined by (4.18) is even.

The geometry of a symplectic space is different from that of a Euclidean space and it is evident that the symplectic structure itself has no Euclidean metric structure. However, we can, of course, impose some Euclidean structure on the symplectic space as follows (Arnold [1]):

$$\boldsymbol{x}^t\boldsymbol{x} = \sum\nolimits_{i=1}^{n} (x_i^2 + y_i^2), \tag{4.27}$$

where

$$\boldsymbol{x} = \sum\nolimits_{i=1}^{n} \left(x_i \boldsymbol{e}_{x_i} + y\boldsymbol{e}_{y_i}\right). \tag{4.28}$$

This is equivalent to introduce a complex space C^n and define the *Hermitian scalar product*

$$\boldsymbol{\eta}_1^t \boldsymbol{\eta}_2 + i\left[\boldsymbol{\eta}_1, \boldsymbol{\eta}_2\right], \tag{4.29}$$

in the space (Arnold [1]). It is easy to see that the Hermitian scalar product defined by (4.13) is equivalent to that defined here by (4.29), if $\boldsymbol{H}$ is positive (negative) semi definite. These examinations lead to conclusion *C4*.

Conclusion *C4* suggests that DEDICOM, GIPSCAL, and HCM can take alternative forms depending upon the basis chosen. In fact, we can also choose, for example, a symplectic basis for DEDICOM and GIPSCAL. Let us first rewrite (4.4), defining an orthogonal transformation $\boldsymbol{O}$ such that

$$\boldsymbol{O}\boldsymbol{A}_s\boldsymbol{O}^t = \boldsymbol{D}_s, \tag{4.30}$$

and

$$\boldsymbol{Y} = \boldsymbol{P}\boldsymbol{O}, \tag{4.31}$$

as

$$\boldsymbol{H} = \boldsymbol{P}\boldsymbol{D}_s\boldsymbol{P}^t + i\boldsymbol{P}\boldsymbol{O}\boldsymbol{A}_{sk}\boldsymbol{O}^t\boldsymbol{P}^t. \tag{4.32}$$

If $\boldsymbol{D}_s$ is positive definite, (4.32) becomes

$$\boldsymbol{H} = \left(\boldsymbol{P}\boldsymbol{D}_s^{\frac{1}{2}}\right)\left(\boldsymbol{P}\boldsymbol{D}_s^{\frac{1}{2}}\right)^t + i\left(\boldsymbol{P}\boldsymbol{D}_s^{\frac{1}{2}}\right)\boldsymbol{D}_s^{-\frac{1}{2}}\boldsymbol{O}\boldsymbol{A}_{sk}\boldsymbol{O}^t\boldsymbol{D}_s^{-\frac{1}{2}}\left(\boldsymbol{P}\boldsymbol{D}_s^{\frac{1}{2}}\right)^t. \tag{4.33}$$

Then, as Kiers and Takane [17] have done, we can decompose the "core" matrix via a singular value decomposition into

$$\boldsymbol{D}_s^{-\frac{1}{2}}\boldsymbol{O}\boldsymbol{A}_{sk}\boldsymbol{O}^t\boldsymbol{D}_s^{-\frac{1}{2}} = \boldsymbol{T}\Sigma\boldsymbol{K}\boldsymbol{T}^t = \boldsymbol{T}\boldsymbol{J}\boldsymbol{T}^t, \tag{4.34}$$

where $\boldsymbol{T}$ is an orthogonal matrix of order $(p \times p)$. Then, (4.33) becomes

$$\boldsymbol{H} = \boldsymbol{Q}\boldsymbol{Q}^t + i\boldsymbol{Q}\boldsymbol{J}\boldsymbol{Q}^t, \tag{4.35}$$

where

$$\boldsymbol{Q} = \boldsymbol{P}\boldsymbol{D}_s^{\frac{1}{2}}\boldsymbol{T}. \tag{4.36}$$

It should be noted that if $\boldsymbol{H}$ is negative semidefinite we may merely change the sign of $\boldsymbol{H}$. To prove conclusion *C5*, we must further administer a coordinate transformation to $\boldsymbol{Q}$ in (4.35)Letting

$$\boldsymbol{R} = \boldsymbol{Q}\boldsymbol{\Sigma}^{-1/2}, \tag{4.37}$$

and noticing the relation in (4.34),

$$\boldsymbol{\Sigma K} = \boldsymbol{\Sigma}^{\frac{1}{2}}\boldsymbol{K}\boldsymbol{\Sigma}^{\frac{1}{2}} = \mathbf{J}, \tag{4.38}$$

we get

$$\boldsymbol{H} = \boldsymbol{R}\boldsymbol{\Sigma}^{-1}\boldsymbol{R}^{t} + i\boldsymbol{R}\boldsymbol{K}\boldsymbol{R}^{t}. \tag{4.39}$$

This proves conclusion *C5* for DEDICOM. The same is true for GIPSCAL.

4.3 An Interesting Property of F.D.C. Hilbert Space Models

Unlike the distance model, the Hermitian scalar product model has an interesting property in that the similarity between the pair of objects located far from the centroid of objects, say, the origin, is greater than that located near the origin, even if their distances are the same. In other words, if this model holds, then subjects are likely to underestimate the distance between the pairs of objects located far from the origin and overestimate the distance between the pairs located near the origin. To prove it, we shall first point out that for a complex pre-Hilbert space (thus, also for the f.d.c. Hilbert space), the *polar identity*

$$\begin{aligned}\varphi(\|\boldsymbol{\xi}\|, \|\boldsymbol{\tau}\|) &= \frac{1}{4}\left(\|\boldsymbol{\xi}\| + \|\boldsymbol{\tau}\|^2 - \|\boldsymbol{\xi}\| - \|\boldsymbol{\tau}\|^2\right) + \frac{1}{4}i\left(\|\boldsymbol{\xi} + i\boldsymbol{\tau}\|^2 - \|\boldsymbol{\xi} - i\boldsymbol{\tau}\|^2\right) \\ &= \frac{1}{2}\left(\|\boldsymbol{\xi}\|^2 + \|\boldsymbol{\tau}\|^2 - \|\boldsymbol{\xi} - \boldsymbol{\tau}\|^2\right) + \frac{1}{2}i\left(\|\boldsymbol{\xi}\|^2 + \|\boldsymbol{\tau}\|^2 - \|\boldsymbol{\xi} - i - \boldsymbol{\tau}\|^2\right)\end{aligned} \tag{4.40}$$

holds (Cristescu [10]), where $\varphi(\boldsymbol{\xi}, \boldsymbol{\tau})$ is a Hermitian form. Then, remembering that $h_{jk} = \varphi(\boldsymbol{v}_j, \boldsymbol{v}_k)$, which we have proved in the second section in leading to conclusion *C1*, we get

$$\begin{aligned}h_{jk} &= \frac{1}{2}\left(\left\|\boldsymbol{v}_j\right\|^2 + \|\boldsymbol{v}_k\|^2 - \left\|\boldsymbol{v}_j - \boldsymbol{v}_k\right\|^2\right) \\ &\quad + \frac{1}{2}i\left(\left\|\boldsymbol{v}_j\right\|^2 + \|\boldsymbol{v}_k\|^2 - \left\|\boldsymbol{v}_j - i\,\boldsymbol{v}_k\right\|^2\right).\end{aligned} \tag{4.41}$$

Then, in terms of (4.1), we get

$$s_{jk} = \frac{1}{2}\left(\left\|v_j\right\|^2 + \|v_k\|^2 - \left\|v_j - v_k\right\|^2\right) + \frac{1}{2}\left(\left\|v_j\right\|^2 + \|v_k\|^2 - \left\|v_j - i\,v_k\right\|^2\right). \tag{4.42}$$

Rearranging the right-hand side of (4.42), we have

$$s_{jk} = -\frac{1}{2}\left\{\left\|v_j - v_k\right\|^2 + \left\|v_j - i\,v_k\right\|^2\right\} + \left(\left\|v_j\right\|^2 + \|v_k\|^2\right). \tag{4.43}$$

Notice here that in general $\left\|v_j - iv_k\right\|^2$ is not equal to $\left\|v_k - iv_j\right\|^2$. Equation (4.43) represents the property under study, if we assume that the centroid of coordinates of objects is at or near the origin. For in this case the second term on the right-hand side of (4.43)increases as the objects j and k are away from the origin. For a real Hilbert space (i.e., a Euclidean space), it is well-known that,

$$s_{jk} = -\frac{1}{2}\left\|v_j - v_k\right\|^2 + \frac{1}{2}\left(\left\|v_j\right\|^2 + \|v_k\|^2\right). \tag{4.44}$$

It is apparent, from (4.44), that the same discussion as in the above asymmetric matrix holds for the symmetric proximity matrix.

4.4 Generalization of the Young-Householder Theorem

Conclusion *C2* in the second section motivates the following theorem. This theorem is an extension of the Young-Householder theorem into f.d.c. Hilbert space.

Theorem Let $v_j (1 \le j \le N)$ be the row vector of order n, which corresponds to the j th row of U_1, in (4.9). Put

$$d_{jk} = \left\|v_j - v_k\right\|, \quad 1 \le j \le N, \tag{4.45}$$

$$d_{jo} = \left\|v_j\right\|, \; 1 \le j \le N, \tag{4.46}$$

and

$$\bar{d}_{jk} = \left\|v_j - i\,v_k\right\|, \quad 1 \le j, \; k \le N, \tag{4.47}$$

Then the following equalities hold;

$$d_{jk} = d_{kj}, \; 1 \le j, k \le N, \; d_{jj} = 0, \; 1 \le j \le N, \tag{4.48}$$

$$d_{jo}^2 + d_{ko}^2 - \bar{d}_{jk}^2 = -\left(d_{ko}^2 + d_{jo}^2 - \bar{d}_{kj}^2\right), \; 1 \le j, k \le N, \tag{4.49}$$

$$h_{jk} = \frac{1}{2}\left(d_{jo}^2 + d_{ko}^2 - d_{jk}^2\right) + \frac{1}{2}i\left(d_{jo}^2 + d_{ko}^2 - \overline{d}_{jk}^2\right), 1 \le j, k \le N. \tag{4.50}$$

Here h_{jk} is given by (4.6), (4.7), or (4.13) and the matrix $\boldsymbol{H} = \{h_{jk}\}$, $(1 \le j, k \le N)$, is a *positive semi-definite Hermitian matrix.*

Conversely, if we are given positive real numbers $d_{jk}(1 \le j \ne k \le N)$ satisfying (4.48) and (4.49) for a suitable choice of non-negative real numbers $\overline{d}_{jk}$and$d_{jo}(1 \le j, k \le N)$, then the set $\{d_{jk}\}$ gives the mutual distances of a real (true) set of points in an f.d.c. Hilbert space if and only if $\boldsymbol{H}$ is positive semi-definite. The set of points is unique apart from an arbitrary unitary transformation.

Proof. We can prove the theorem in a similar manner to that given by Young and Householder [21]. The only difference is that in this case we must use (4.50), which is another expression of (4.41), instead of Eq. (1) of Young and Householder [21]. Equation (4.49) is merely a condition for $h_{kj} = \overline{h}_{jk}$, $\overline{h}_{jk}$nbeing the complex conjugate of h_{jk}.

The proof of the latter part of the theorem can be stated in another way. Let us first note that (4.48), (4.49), and (4.50) hold for the mutual distance of a true set of points in an f.d.c. Hilbert space. Second, if $\boldsymbol{H}$ is positive semidifinite, we can define the following distances using the row vectors $\boldsymbol{v}_j'$, $\boldsymbol{v}_k'$ corresponding to the j-th row and the k-th row of $\boldsymbol{U}_1$ in (4.9):

$$d_{jk}' = \|\boldsymbol{v}_j' - \boldsymbol{v}_k'\|, 1 \le j, k \le N, \tag{4.51}$$

$$d_{jo}' = \|\boldsymbol{v}_j'\|, 1 \le j \le N, \tag{4.52}$$

and

$$\overline{d}_{jk}' = \|\boldsymbol{v}_j' - i\boldsymbol{v}_k'\|, 1 \le j, k \le N. \tag{4.53}$$

Then, from the result of the first part of the theorem we have

$$d_{jk}' = d_{kj}', 1 \le j, k \le N, d_{jj}' = 0, 1 \le j \le N, \tag{4.54}$$

$$d_{jo}'^2 + d_{ko}'^2 - \overline{d}_{jk}'^2 = -\left(d_{ko}'^2 + d_{jo}'^2 - \overline{d}_{kj}'^2\right), \quad 1 \le j, k \le N, \tag{4.55}$$

$$h_{jk} = \frac{1}{2}\left(d_{jo}'^2 + d_{ko}'^2 - d_{jk}'^2\right) + \frac{1}{2}i\left(d_{jo}'^2 + d_{ko}'^2 - \overline{d}_{jk}'^2\right), 1 \le j, k \le N. \tag{4.56}$$

From (4.49)and (4.50), we have

$$h_{jj} = d_{jo}^2, 1 \le j \le N. \tag{4.57}$$

Moreover, from (4.55) and (4.56) we get

$$h_{jj} = d_{jo}^{\prime 2},\ 1 \leq j \leq N. \tag{4.58}$$

Then we have $d_{jo}^2 = d_{jo}^{\prime 2}$, which means that

$$d_{jo} = d_{jo}^{\prime},\ 1 \leq j \leq N. \tag{4.59}$$

Next, comparing the real parts and the imaginary parts of (4.50) and (4.56), we get

$$d_{jo}^2 + d_{ko}^2 - d_{jk}^2 = d_{jo}^{\prime 2} + d_{ko}^{\prime 2} - d_{jk}^{\prime 2},\ 1 \leq j, k \leq N, \tag{4.60}$$

$$d_{jo}^2 + d_{ko}^2 - \overline{d}_{jk}^2 = d_{jo}^{\prime 2} + d_{ko}^{\prime 2} - \overline{d}_{jk}^{\prime 2}, \quad 1 \leq j, k \leq N. \tag{4.61}$$

Equations (4.59) and (4.61) yield $d_{jk}^2 = d_{jk}^{\prime 2}$, from which we get

$$d_{jk} = d_{jk}^{\prime}. \tag{4.62}$$

Likewise, from Eqs. (4.59) and (4.61) we get

$$\overline{d}_{jk} = \overline{d}_{jk}^{\prime}. \tag{4.63}$$

The invariance of the coordinates of points over a unitary transformation is evident. ■

4.5 Discussion

Throughout this paper we have exclusively dealt with the error-free data, i. e. the error-free proximity matrices $\boldsymbol{S}$ as well as the corresponding Hermitian matrices $\boldsymbol{H}$ measured at a ratio level. However, in a practical situation, both of these matrices may be fallible and not necessarily be measured at the ratio level. Furthermore, we can neither observe nor estimate the special distances defined by (4.47), since it is not concerned with the direct diestance between two points in an f.d.c. Hilbert space in marked contrast to the distance in classical MDS. In such a case we must estimate them from the data. If the proximity judgments are measured at the ratio level, there are no missing observations, and the matrix H is positive semi-definite, we can use HCM as one such method. For in such a case HCM is also solved by a singular value decomposition and thus has a least squares (LS) property according to the Schmidt-Mirsky theorem, which is a generalization of the famous Eckart-Young theorem (Eckart & Young [11]; Stewart & Sun [20], p. 205). Otherwise, we must estimate them using some scaling procedure. GIPSCAL and its generalized version proposed recently by Kiers and Takane [17] are thought of as two such procedures, which indirectly estimate these $\boldsymbol{h}_{jk}$'s by LS criteria. Recently, Chino [7] has proposed an ML procedure for estimating and testing parameters of a general Hermitian form model, including all these discussed in this paper, given proximity or dissimilarity data measured at least at an ordinal level. The Hermitian form model discussed in this paper may also be applicable to square contingency tables, as Chino [6] has pointed out.

Acknowledgements The author wish to express their thanks to Henk A. L. Kiers, Yoshio Takane, J. O., Ramsay, and J. C. Gower for the valuable comments on earlier versions of this paper. The authors are especially indebted to Henk Kiers for substantial improvement of this paper. The authors are also grateful to anonymous reviewers for their helpful comments. Special thanks are dure to T. Fizgerald for careful proofreading. Requests for reprint should be sent to Naohito Chino, Dept. of Psychology, Aichi Gakuin University, Nissin-cho, Aichi, Japan.

Appendix

In this appendix, we shall show a MATLAB code for HFM. This code depicts the results of Example 1.24 in Chap. 1 by loading the dat (text) file named "trade2015-among-JUCR.dat" of the following form:

43480	1382	1200	55
736	189592	1161	71
1764	4832	119684	348
173	164	333	13755

The code outputs (1) the Hermitian matrix associated with the log transformed ASM (Table 1.28), (2) configurations of objects in complex dimensions, (3) eigenvalues of the Hermitian matrix, (4) eigenvectors corresponding to these eigenvalues, (4) reproduced similarity matrices for four dimensions using (4.43) which is deduced from the polar identity define by (4.41), and (5) the four digraphs associated respectively with the four configurations obtained by HFM. The log transformation of the original input data should be made only in the case when it is appropriate and necessary. One may replace the input data with any asymmetric similarity data matrix. However, the elements of this matrix must be measured at the *ratio scale* level, which is an important assumption of HFM declared in Sect. 4.1. The following is the MATLAB code:

```
%-------------------------------------------------------------
%
% Hermitian form model, HFM.
%
% file name: hfm_trade_four_nations.m
%       version 1.1 written by N. Chino, February 27, 2017.
%               revised by N. Chino, April 22, 2017.
```

```
%               revised by N. Chino, June 21, 2019.
%          version 1.1 named "hfm.m" was revised and renamed by N. Chino,
%               July 5, 2019.
%          version 1.2 named "hfm.m" was revised and renamed by N. Chino,
%                              October 5, 2019.
%
% An asymmetric multidimensional scaling model by Chino and Shiraiwa [8].
%
%-----------------------------------------------------------------
%
% Compute the configuration of objects via HFM.
%
load data/chino/trade2015-among-JUCR.dat
%
disp('0–1. Original asymmetric similarity matrix');
disp(' between Japan, America, China, and Russia in 2015');
S = trade2015_among_JUCR
N = 4;
%load data/chino/takane_voxel3.dat;
%S = takane_voxel3;
%N = 3;
%
%load data/chino/kraenke-Aberer1996.dat
% S = kraenke_Aberer1996
%
%load data/Takii14revised.mat
%U = Takii14revised;
%S = 371*U
%
% load data/Takii14origfreq.mat
% S = Takii14origfreq;
% N = 14;
%
% log transformation
for i = 1:N
    for k = 1:N
     if S(i,k)~ = 0.0 S(i,k) = log(S(i,k));
     else
          S(i,k) = –0.001;
           end;
     end;
end;
%
disp('0–2. log transformed asymmetric similarity data');
S
```

```
%
H = (S+S')/2+j*(S-S')/2;
[O,D] = eigs(H);
%
%disp('1. The Hermitian matrix computed from log asymmetric matrix, S');
disp('1. The Hermitian matrix computed from original asymmetric matrix, S');
disp(H)
%
disp('2. The configuration of objects in some complex dimensions');
disp(O)
disp('3. Eigenvalues of H via HFM');
disp(D)
%
figure(1);
O1 = O(:,1);
disp('4. The first eigenvector of H via HFM');
disp(O1)
compass(O1);
%
%Trade data (log transformed)
text(0.50, −0.03, 'J'); text(0.54, −0.06, 'A');
text(0.55, −0.01, 'C'); text(0.41,0.02, 'R');
%Takii data
%text(−0.45, −0.09, 'Ni'); text(−0.29, −0.16, 'Sn');
%text(−0.36, −0.21, 'Pd'); text(−0.31, −0.23, 'Co');
%xt(−0.14, −0.29, 'Fe'); text(−0.09, −0.29, 'Cu');
ylabel('imaginary axis');
xlabel('Fig.1 Coordinates in dimension 1');
%xlabel('Fig.1 Coordinates in dimension 1, eigenvalue = 3.53');
%print-depsc eps-files/Takii14revised.eps
pause;
%
figure(2);
O2 = O(:,2);
disp('5. The second eigenvector of H via HFM');
disp(O2)
compass(O2);
%
%Trade data (log transformed)
text(0.22,0.11, 'J'); text(0.45, −0.11, 'A');
text(−0.04, −0.07, 'C'); text(−0.91,0.00, 'R');
%Takii data
% text(0.50, −0.15, 'P'); text(0.49, −0.43, 'N'); text(0.49, −0.11, 'Co');
% text(0.23,0.0, 'Cr');
ylabel('imaginary axis');
```

```
xlabel('Fig.2 Coordinates in dimension 2');
%xlabel('Fig.2 Coordinates in dimension 2, eigenvalue = 4.44');
pause;
%
figure(3);
O3 = O(:,3);
disp('6. The third eigenvector of H via HFM');
disp(O3)
compass(O3);
%
%Trade data (log transformed)
text(−0.74,0.14, 'J'); text(0.56, −0.35, 'A');
text(−0.04,0.24, 'C'); text(0.17,0.00, 'R');
%Takii data
% text(0.50, −0.15, 'P'); text(0.49, −0.43, 'N');
% text(0.49, −0.11, 'Co'); text(0.23,0.0, 'Cr');
%Takane voxel dat
% text(0.65, 0.26, 'V1');
% text(0.54, 0.07, 'V5');
% text(0.46, 0.00, 'SPC');
ylabel('imaginary axis');
xlabel('Fig.3 Coordinates in dimension 3');
%xlabel('Fig.3 Coordinates in dimension 3, eigenvalue = 6.11');
pause;
%
% The following commands must be used for N is greater than 4.
figure(4);
O4 = O(:,4);
disp('7. The fourth eigenvector of H via HFM');
disp(O4)
compass(O4);
%
%Trade data (log transformed)
text(−0.38, −0.27, 'J'); text(−0.34,0.05, 'A');
text(0.79,0.23, 'C'); text(−0.37,0.00, 'R');
%Takii data
% text(0.50, −0.15, 'P'); text(0.49, −0.43, 'N'); text(0.49, −0.11, 'Co'); text(0.23,0.0, 'Cr');
ylabel('imaginary axis');
xlabel('Fig.4 Coordinates in dimension 4');
%xlabel('Fig.4 Coordinates in dimension 4, eigenvalue = 29.97');
pause;
%
figure(5);
%
```

```
subplot 221;
compass(O1);
text(0.50, −0.03, ‘J’); text(0.54, −0.06, ‘A’);
text(0.55, −0.01, ‘C’); text(0.41,0.02, ‘R’);
ylabel(‘imaginary axis’);
xlabel(‘a. Coordinates in dimension 1’);
%
subplot 222;
%
compass(O2);
text(0.22,0.11, ‘J’); text(0.45, −0.11, ‘A’);
text(−0.04, −0.07, ‘C’); text(−0.91,0.00, ‘R’);
ylabel(‘imaginary axis’);
xlabel(‘b. Coordinates in dimension 2’);
%
subplot 223;
compass(O3);
text(−0.74,0.14, ‘J’); text(0.56, −0.35, ‘A’);
text(−0.04,0.24, ‘C’); text(0.17,0.00, ‘R’);
ylabel(‘imaginary axis’);
xlabel(‘c. Coordinates in dimension 3’);
%
subplot 224;
compass(O4);
text(−0.38, −0.27, ‘J’); text(−0.34,0.05, ‘A’);
text(0.79,0.23, ‘C’); text(−0.37,0.00, ‘R’);
ylabel(‘imaginary axis’);
xlabel(‘d. Coordinates in dimension 4’);
print -depsc eps-files/four-configurations-of-nations.eps
pause;
%
% Compute similarity matrices in four dimensions
%
p = 1;
disp(‘8. Reproduced similarity matrices for four dimensions’);
%
[S1] = rpi(N,p,O1);
disp(‘8–1. Reproduced similarity matrix for dimension 1’);
if D(1,1)>0 [S1]
else -[S1]
end
%
[S2] = rpi(N,p,O2);
disp(‘8–2. Reproduced similarity matrix for dimension 2’);
if D(2,2)>0 [S2]
```

```
else -[S2]
end
%
[S3] = rpi(N,p,O3);
disp('8–3. Reproduced similarity matrix for dimension 3');
if D(3,3)>0 [S3]
else-[S3]
end
%
[S4] = rpi(N,p,O4);
disp('8–4. Reproduced similarity matrix for dimension 4');
if D(4,4)>0 [S4]
else -[S4]
end
%
% Depict digraphs corresponding to similarity matrices in four dimensions
%
figure(6);
% Digraphs at some points in time
% Specify the names of nodes:
names = {'J', 'A', 'C', 'R'};
%
% The special set commands in the following subplot routine enable us to
% remove coordinate axes in depicting digraphs.
subplot 221;
 G = digraph(S1,names);
 plot(G, 'k', 'Layout', 'force', 'EdgeLabel', G.Edges.Weight);
 set(gca, 'Visible', 'off', 'NextPlot', 'replaceChildren');
 text(−1.8, −2.5, 'a. digraph in dimension 1');
subplot 222;
 G = digraph(S2,names);
 plot(G, 'k', 'Layout', 'force', 'EdgeLabel', G.Edges.Weight);
 set(gca, 'Visible', 'off', 'NextPlot', 'replaceChildren');
 text(−1.8, −2.5,'b. digraph in dimension 2');
subplot 223;
 G = digraph(S3,names);
 plot(G, 'k', 'Layout', 'force', 'EdgeLabel', G.Edges.Weight);
 set(gca, 'Visible', 'off', 'NextPlot', 'replaceChildren');
 text(−1.8, −2.5,'c. digraph in dimension 3');
subplot 224;
 G = digraph(S4,names);
 plot(G, 'k', 'Layout', 'force', 'EdgeLabel', G.Edges.Weight);
 set(gca, 'Visible', 'off', 'NextPlot', 'replaceChildren');
 text(−1.8, −2.5, 'd. digraph in dimension 4');
%print -depsc eps-files/ghcde_triad_sim5_Case3-digraphs-at-four-times.eps
```

```
function [S] = rpi(N,p,Z)
% This function returns the real similarity data matrix of order N, given
% the N by p complex coordinate matrix Z, translating the these values into
% the real values via the polarized identity in the Hilbert space.
%
for a = 1:N
   for b = 1:N
     dab = abs(Z(a,:) – Z(b,:)).^2;
     daib = abs(Z(a,:) – i*Z(b,:)).^2;
     dsum = abs(Z(a,:)).^2+abs(Z(b,:)).^2;
     S([a],[b]) = dsum-(dab+daib)./2;
   end;
end;
```

Notes

This chapter is the reproduction of Chino and Shiraiwa [8]. We added a MATLAB code for analyzing empirical data by HFM in the following appendix.

As reviewed in Chap. 7, (4.1) is due to Escoufier and Grorud [12]. However, they solved the eigenvalue problem of the Hermitian matrix $\boldsymbol{H}$ in (4.1) using a well-known method which construct a double sized, real symmetric matrix shown in (7.22) of Chap. 7. Moreover, they did not refer to the space structure in which objects are embedded. By contrast, Chino and Shiraiwa have proved in this paper that objects are embedded in a Hilbert space or an indefinite metric space depending on the definiteness of $\boldsymbol{H}$.

References

1. Arnold, V. I. (1978). *Mathematical methods of classical mechanics*. Springer Verlag.
2. Chino, N. (1978). A graphical technique for representing the asymmetric relation ships between N objects. *Behaviormetrika, 5*, 23–40.
3. Chino, N. (1980). A unified geometrical interpretation of the MDS techniques for the analysis of asymmetry and related techniques. *Paper presented at the symposium on "Asymmetric multidimensional scaling" at the Spring Meeting of the Psychometric Society*, Iowa, U.S.A.
4. Chino, N. (1990). A generalized inner product model for the analysis of asymmetry. *Behaviormetrika, 27*, 25–46.
5. Chino, N. (1991a). Fundamental problems in asymmetric scaling. *Paper presented at the symposium on "Multivariate analysis of asymmetric relational data" at the 19th annual meeting of the Behaviormetric Society of Japan.* Nagoya, Japan.
6. Chino, N. (1991b). A critical review for the analysis of asymmetric relational data. *Bulletin of The Faculty of Letters of Aichi Gakuin University, 21*, 31–52.
7. Chino, N. (1992). Metric and nonmetric Hermitian canonical models for asymmetric MDS. *Paper presented at the 20th annual meeting of the Behaviormetric Society of Japan.* Tokyo, Japan.
8. Chino, N., & Shiraiwa, K. (1993). Geometrical structures of some non-distance models for asymmetric MDS. *Behaviormetrika, 20*, 35–47.

9. Constantine, A. G., & Gower, J. C. (1978). Graphical representation of asymmetric matrices. *Applied Statistics, 27*, 297–304.
10. Cristescu, R. (1977). *Topological Vector Spaces*. Noordhoff International Publishing.
11. Eckart, C., & Young, G. (1936). The approximation of one matrix by another of lower rank. *Psychometrika, 1*, 211–218.
12. Escoufier, Y., & Grorud, A., et al. (1980). Analyse factorielle des matrices carrees non symetriques [Factor analysis of square asymmetric matrices]. In E. Diday (Ed.), *Data Analysis and Informatics* (pp. 263–276). North Holland.
13. Gower, J. C. (1977). The analysis of asymmetry and orthogonality. In J. R. Barra, F. Brodeau, G. Romer, & B. van Cutsem (Eds.), *Recent developments in statistics* (pp. 109–123). North Holland.
14. Gower, J. C., & Zielman, B. (1992). *Some remarks on orthogonality in the analysis of asymmetry. Research Report RQ-92–08, Dept. of Data Theory, Univ. of Leiden, Netherlands.*
15. Harshman, R.A. (1978). Models for analysis of asymmetrical relationships among N objects or stimuli. *Paper presented at the First Joint Meeting of the Psychometric Society and the Society for Mathematical Psychology*, Hamilton, Canada.
16. Harshman, R. A., Green, P. E., Wind, Y., & Lundy, M. E. (1982). A model for the analysis of asymmetric data in marketing research. *Marketing Science, 1*, 205–242.
17. Kiers, H. A. L., & Takane, Y. (1994). A generalization of GIPSCAL for the analysis of asymmetric data. *Journal of Classification, 11*, 79–99.
18. Lancaster, P., & Tismenetsky, M. (1985). *The theory of matrices* (2nd ed.). Academic Press.
19. Schutz, B. (1980). *Geometrical methods of mathematical physics*. Cambridge University Press.
20. Stewart, G. W., & Sun, J. (1990). *Matrix perturbation theory*. Academic Press.
21. Young, G., & Householder, A. S. (1938). Discussion of a set of points in terms of their mutual distances. *Psychometrika, 3*, 19–22.

Chapter 5
How to Use HFM for Asymmetric MDS

5.1 Introduction

In this section we shall introduce HFM for the analysis of asymmetric (dis)similarity matrix, which was proposed by Chino and Shiraiwa [6]. Let us first suppose that we have an observed asymmetric similarity matrix $\boldsymbol{S} = \left[s_{jk}\right]$, whose element s_{jk} denotes the intensity of the similarity from object j to object k among N objects. Therefore, matrix $\boldsymbol{S}$ is N by N, and is in general asymmetric. We shall hereafter abbreviate the Asymmetric Similarity Matrix as ASM.

ASM can be found everywhere in our daily lives, in research laboratories, and so on. Sometimes researchers assume some ASM theoretically or hypothetically in order to explain a certain observed phenomenon. For example, one-sided love and hate among members of any informal group are typical examples which constitute such an ASM in daily lives. Trade data matrix among nations is another typical ASM. Amount of migration from one region to another is also such an example in geography.

In research laboratories, we may observe various pathways among voxels of neurons, biosynthetic pathways of nicotinamide adenine dinucleotide (NAD) (e.g., Imai & Guarente [11], Fig. 2), and so on. In field research we may observe, for example, the pecking order among a group of hens and cocks in biology (e.g., Rushen [15]). Suppose that the code is one in the case when there is a pathway or such an order, and otherwise it is zero. Then, we get an ASM whose elements are binary.

We sometimes assume some special ASM theoretically or hypothetically. For example, Sato, Akiyama, and Farmer [17] assume the payoff matrices of "rock-paper-scissors" in the context of the theory of the evolutionally stable strategy in biology. Equation (5.1) shows them.

$$\mathbf{A} = \begin{bmatrix} \varepsilon_x & -1 & 1 \\ 1 & \varepsilon_x & -1 \\ -1 & 1 & \varepsilon_x \end{bmatrix}, \mathbf{B} = \begin{bmatrix} \varepsilon_y & -1 & 1 \\ 1 & \varepsilon_y & -1 \\ -1 & 1 & \varepsilon_y \end{bmatrix}. \tag{5.1}$$

N. Chino, *Structure and Dynamics of Asymmetric Interactions*,
https://doi.org/10.1007/978-981-97-8269-7_5

The weight matrix specifically assumed in the hidden layers is also considered as a typical ASM in recurrent neural networks (e.g., Goodfellow et al. [8], Fig. 10.3). Usually, elements of the weight matrix are unknown parameters of some neural network model, and thus are estimated from data.

From the psychometrical point of view, elements of ASM are measured at various levels. For example, these elements in the affinity data discussed above are measured at an interval level. Those in trade data are measured at a ratio level. Binary data frequently observed in research laboratories, which might be considered as count data, may be viewed as measured at a ratio level. This type of data can be sometimes said to be measured at an absolute scale level. Dummy variables, 1 and − 1 in the payoff matrices might also be said to be measured at a ratio level.

For these elements of ASM, HFM assumes the ratio level of measurement. As a result, for example, affinity data matrices among members of any informal group discussed above should not be analyzed by HFM. In such a case, we may analyze such matrices using various asymmetric MDS methods which do not necessarily assume the ratio level of measurement (e.g., [13, 14]; Saburi & Chino [16]).

In this paper, we shall show how to interpret the configuration of objects obtained by applying HFM to theoretical or empirical ASMs, typical examples of which we listed above. In Sect. 5.2, we shall explain how to interpret the configuration of objects obtained by HFM. In Sect. 5.3, we show some applications of HFM to empirical ASM, and interpret the configurations of objects obtained by HFM. In Sect. 5.4, we show some applications of HFM to theoretical or hypothetical ASM, and interpret these configurations obtained by HFM. In Sect. 5.5, we shall discuss the merit of HFM, and refer to a possibility of utilizing these configurations for future research. In Sect. 5.6 we shall discuss the ASM which changes as time proceeds through the mutual interactions among objects. We introduce a certain hypothetical force acting on the state spaces of some underlying dynamical system and discuss how to interpret such a force by utilizing the theory of HFM.

5.2 Interpretation of the Configuration of Objects by HFM

In this section, we shall explain how to interpret the configuration of objects embedded in a Hilbert space, which is obtained by applying HFM to any empirical or theoretical ASM. As discussed in Chap. 4, objects are embedded in either the Hilbert space or the indefinite metric space depending on the eigenvalues of the Hermitian matrix constructed from the ASM.

However, as also discussed in that chapter, both those spaces are considered as one-dimensional Hilbert spaces if we restrict the space in one-dimension. As a result, we may interpret the n-dimensional configuration of objects in the whole space per one-dimension as if it were a one-dimensional Hilbert space, that is, a complex plane.

Remember here that in HFM the ASM, $\boldsymbol{S}$, is decomposed into the symmetric part $\boldsymbol{S}_s(= (\boldsymbol{S} + \boldsymbol{S}^t)/2)$ and the skew-symmetric part $\boldsymbol{S}_{sk}$ $(= (\boldsymbol{S} - \boldsymbol{S}^t)/2)$, first. Then, a Hermitian matrix $\boldsymbol{H} = \boldsymbol{S}_s + i\boldsymbol{S}_{sk}$ is constructed, where $i^2 = -1$. Finally, $\boldsymbol{H}$ is written as (4.11) in Chap. 4. Then, as shown in (4.33) in Chap. 4, $\boldsymbol{S}$ can be rewritten as

$$\boldsymbol{S} = \boldsymbol{X}\boldsymbol{\Omega}_s\boldsymbol{X}^t + \boldsymbol{X}\boldsymbol{\Omega}_{sk}\boldsymbol{X}^t, \tag{5.2}$$

This equation can be written, in scalar notation, as

$$s_{jk} = \sum_{l=1}^{n} \lambda_l (r_{jl}r_{kl} + c_{jl}c_{kl}) + \sum_{l=1}^{n} \lambda_l (c_{jl}r_{kl} - r_{jl}c_{kl}), \tag{5.3}$$

or

$$s_{jk} = \sum_{l=1}^{n} \lambda_l \{ (r_{jl}r_{kl} + c_{jl}c_{kl}) + (c_{jl}r_{kl} - r_{jl}c_{kl}) \}, \tag{5.4}$$

where $\boldsymbol{U}_r = [r_{jl}]$ are $\boldsymbol{U}_c = [c_{jl}]$ are defined in (4.10).

From (5.4), if we restrict the space in one-dimension we have

$$s_{jk} = \lambda_l (r_{jl}r_{kl} + c_{jl}c_{kl}) + \lambda_l (c_{jl}r_{kl} - r_{jl}c_{kl}), \tag{5.5}$$

for the lth complex plane associated with the lth eigenvalue of $\mathbf{H}$. Here, r_{jl} and c_{jl} are, respectively, the real part and the imaginary part of the complex number, v_j, corresponding to the location of object, O_j, on the lth complex plane. Since the complex plane can be identified with the two-dimensional Euclidean plane, we can also consider r_{jl} and c_{jl}, respectively, as the abscissa and the ordinate of the position of O_j, on the lth Euclidean plane.

Then, let us rewrite r_{jl} and c_{jl} as x_{j1} and x_{j2}, respectively. Moreover, let us define, respectively, the position vectors, $\boldsymbol{x}_j = [x_{j1}, x_{j2}]$ and $\boldsymbol{x}_k = [x_{k1}, x_{k2}]$ of O_j and O_k. Then, (5.5) can be rewritten as follows, if we introduce trigonometric functions:

$$s_{jk} = \lambda |\boldsymbol{x}_j| |\boldsymbol{x}_k| (\cos\theta_{jk} - \sin\theta_{jk}), \tag{5.6}$$

where θ_{jk} is the *angle* from O_j to O_k.

If we identify, respectively, *real vectors*, $\boldsymbol{x}_j$ and $\boldsymbol{x}_k$, with *complex numbers*, $v_j = x_{j1} + ix_{j2}$ and $v_k = x_{k1} + ix_{k2}$, θ_{jk} is the difference in the *arguments* θ_j and θ_k of v_j and v_k, respectively, in a complex plane. Here, the argument of the complex number z, defined for $z \neq 0$, is the *angle* which the vector originating from 0 to z makes with the positive real axis *counterclockwise*. Figure 5.1 illustrates v_j and v_k in a complex plane, assuming that v_j is on the real axis. In this case, the argument of v_j happens to be zero, and thus θ_{jk} is equal to the argument of v_k.

However, in interpreting the configuration of objects obtained by HFM, it should be noticed that the positive direction of the configuration of objects must be measured *clockwise*, considering (5.5) and (5.6). This means that the similarity from O_k associated with v_k to O_j associated.

Fig. 5.1 Configuration of two objects, O_j and O_k in the complex plane, and the argument θ_{jk} measured counterclockwise from O_j

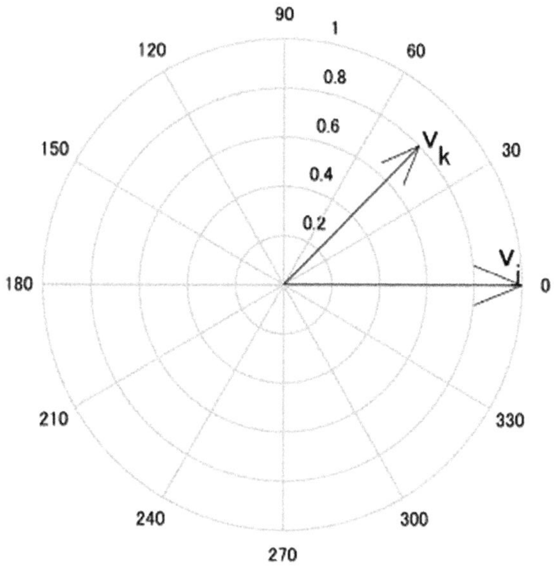

with v_j is relatively greater than that from O_j to O_k in Fig. 5.1. Remember here that the positive direction of the configuration of objects is *counterclockwise* in Chino's ASYMSCAL (Chino [1]), which is one of the asymmetric MDSs. This model is a Euclidean space model and embeds objects in the two-dimensional Euclidean space. The original model is written as

$$s_{jk} = a\left(x_{j1}x_{k1} + x_{j2}x_{k2}\right) + b\left(x_{j1}x_{k2} - x_{j2}x_{k1}\right) + c, \tag{5.7}$$

where a, b, and c are real constants.

If we identify real coordinate vector $\boldsymbol{x}_j = \left[x_{j1}, x_{j2}\right]$with the complex number $z_j = r_{j1} + ic_{j1}$ and $\boldsymbol{x}_k = [x_{k1}, x_{k2}]$ with the complex number $z_k = r_{k1} + ic_{k1}$, (5.7) can be rewritten as

$$s_{jk} = a\left(r_{j1}r_{k1} + c_{j1}c_{k1}\right) + b\left(r_{j1}c_{k1} - c_{j1}r_{k1}\right) + c. \tag{5.8}$$

Now, let us set $a = b = \lambda$. Then, (5.8) is written as

$$s_{jk} = \lambda\left(r_{j1}r_{k1} + c_{j1}c_{k1}\right) + \lambda\left(r_{j1}c_{k1} - c_{j1}r_{k1}\right) + c. \tag{5.9}$$

If we further rewrite (5.9) as

$$s_{jk} = \lambda\left(r_{j1}r_{k1} + c_{j1}c_{k1}\right) - \lambda\left(c_{j1}r_{k1} - r_{j1}c_{k1}\right) + c, \tag{5.10}$$

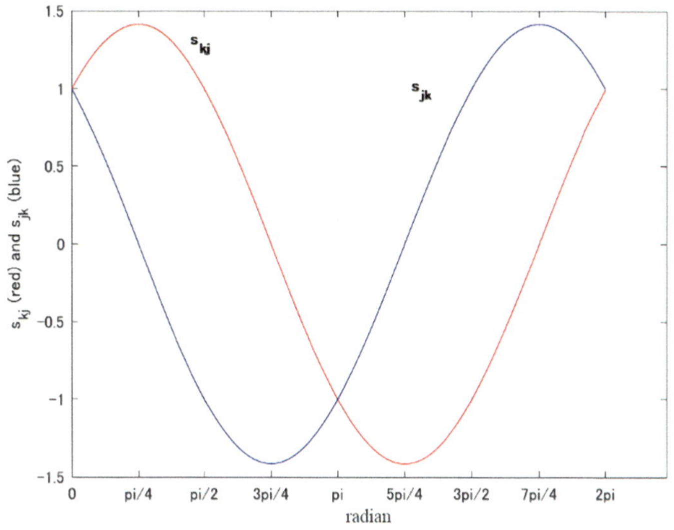

Fig. 5.2 Values of s_{jk} (blue curve) and s_{kj} (red curve) plotted against θ_{jk} when the eigenvalue of $\boldsymbol{H}$ is *positive*

and compare it with (5.5) of HFM, we see that the positive direction of the configuration5.5 of objects is opposite in the case of Chino's ASYMSCAL (Chino [1]).

In any case, s_{jk} in (5.6) of HFM depends on θ_{jk} discussed there. Figures 5.2 and 5.3 illustrate the details of s_{jk} as a function of θ_{jk}. In these figures, we set $|\boldsymbol{x}_j||\boldsymbol{x}_k|$ in (5.6) equal to 1 for simplicity. Moreover, we set λ equal to 1 and -1 for Figs. 5.2 and 5.3, respectively. These correspond to the cases when the eigenvalues of $\boldsymbol{H}$ are positive and negative, respectively.

It is apparent from Fig. 5.2 that s_{kj} is greater than s_{jk} within the range, $0 < \theta_{jk} < \pi$, while s_{jk} is greater than s_{kj} within the range, $\pi < \theta_{jk} < 2\pi$. Moreover, signs of s_{jk} and s_{kj} depend on θ_{jk}, as summarized in Table 5.1.

In contrast, it is apparent from Fig. 5.3 that s_{jk} is greater than s_{kj}.

within the range, $0 < \theta_{jk} < \pi$, while s_{kj} is greater than s_{jk} within the range, $\pi < \theta_{jk} < 2\pi$. Moreover, signs of s_{jk} and s_{kj} depend on θ_{jk}, as summarized in Table 5.2.

Finally, we shall consider a bit about the skewness between s_{jk} and s_{kj}. It is easy to show that from (5.6)

$$s_{jk} - s_{kj} = -\lambda|\boldsymbol{x}_j||\boldsymbol{x}_k| \sin\theta_{jk}. \tag{5.11}$$

Therefore, if we set $\lambda|\boldsymbol{x}_j||\boldsymbol{x}_k|$ equal to 1, we

$$s_{jk} - s_{kj} = -\sin\theta_{jk}, \tag{5.12}$$

and

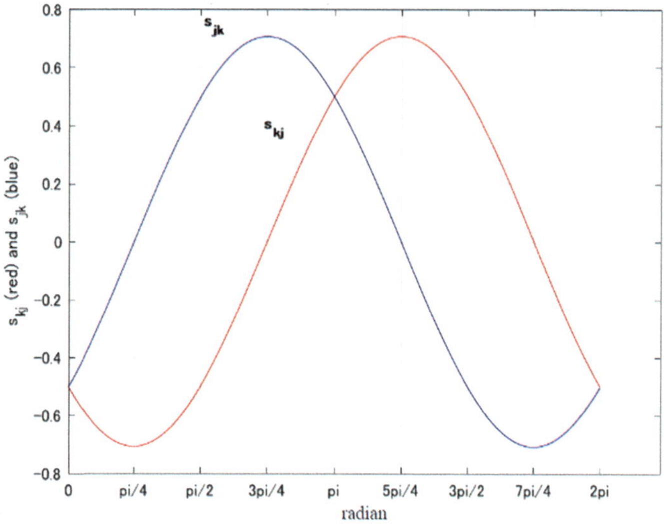

Fig. 5.3 Values of s_{jk} (blue curve) and s_{kj} (red curve) plotted against θ_{jk} when the eigenvalue of $\boldsymbol{H}$ is *negative*

Table 5.1 Summary of signs of s_{jk} and s_{kj} which depend on $\theta_{jk}\theta_{jk}$ in Fig. 5.2

θ_{jk}	$(0, \frac{\pi}{4})$	$(\frac{\pi}{4}, \frac{\pi}{2})$	$(\frac{\pi}{2}, \frac{3\pi}{4})$	$(\frac{3\pi}{4}, \pi)$	$(\pi, \frac{5\pi}{4})$	$(\frac{5\pi}{4}, \frac{3\pi}{2})$	$(\frac{3\pi}{2}, \frac{7\pi}{4})$	$(\frac{7\pi}{4}, 2\pi)$
s_{jk}	+	−	−	−	−	+	+	+
s_{kj}	+	+	+	−	−	−	−	+

Table 5.2 Summary of signs of s_{jk} and s_{kj} which depend on θ_{jk} in Fig. 5.3

θ_{jk}	$(0, \frac{\pi}{4})$	$(\frac{\pi}{4}, \frac{\pi}{2})$	$(\frac{\pi}{2}, \frac{3\pi}{4})$	$(\frac{3\pi}{4}, \pi)$	$(\pi, \frac{5\pi}{4})$	$(\frac{5\pi}{4}, \frac{3\pi}{2})$	$(\frac{3\pi}{2}, \frac{7\pi}{4})$	$(\frac{7\pi}{4}, 2\pi)$
s_{jk}	−	+	+	+	+	−	−	−
s_{kj}	−	−	−	+	+	+	+	−

$$\left|s_{jk} - s_{kj}\right| = \left|\sin\theta_{jk}\right|. \tag{5.13}$$

The green curve in Fig. 5.4 is the very amount of the skewness between the similarities, s_{jk} and s_{kj}. From this figure, it is apparent that the skewness between them takes the maximum values when θ_{jk} is $\frac{\pi}{2}$ or $3\,\frac{\pi}{2}$.

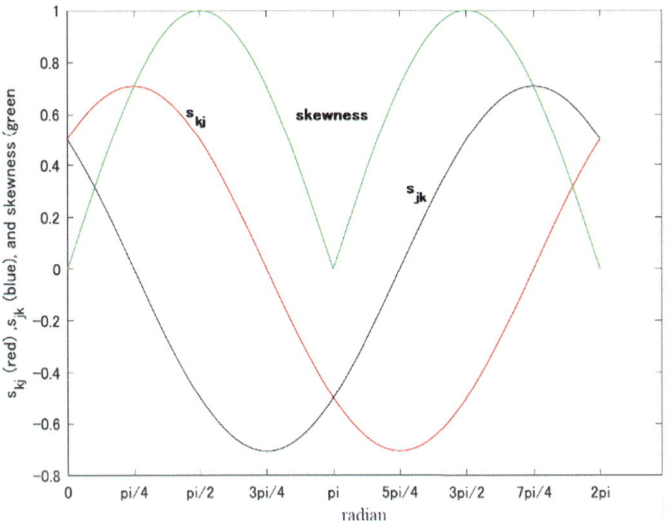

Fig. 5.4 Skewness curves (green) between s_{jk} (black curve) and s_{kj} (red curve) plotted against θ_{jk}

5.3 Applications of HFM to Empirical Asymmetric Relational Matrices

In this section we show two applications of HFM to empirical asymmetric relational data matrices. One is the international trade data among Japan, America, China, and Russia in 2015, which appeared in The Asahi News Paper in Japan. Table 5.3 shows this.

According to convention, we administered a log transformation to each element of this ASM prior to the analysis via HFM. Eigenvalues of the Hermitian matrix constructed from the transformed ASM were 29.9714, 61,145, 4.4377, and 3.5309. This means that the trade data has a *Hilbert space structure as a whole.* Of course, each of the configurations of nations on the complex planes associated with these eigenvalues is considered as embedded in a *one-dimensional Hilbert space*, as indicated in the previous section. Moreover, these configurations embedded in one-dimensional Hilbert spaces are mutually orthogonal (to be precise, *unitary*) in the whole space with the complex Hilbert space structure.

Table 5.3 The international trade data among Japan, America, China, and Russia in 2015

	1. Japan	2. USA	3. China	4. Russia
1. Japan	43,480	1382	1200	55
2. USA	736	159,592	1161	71
3. China	1764	4532	119,684	348
4. Russia	173	164	333	13,755

Figure 5.5 shows four configurations of nations corresponding to these one-dimensional Hilbert spaces. Of course, these four configurations are associated with the four eigenvalues of the Hermitian matrix discussed above.

Figure 5.5a, which is the configuration associated with the maximum eigenvalue of the Hermitian matrix, suggests that this dimension looks like a *size-factor* in PCA because four nations are compressed in a narrow region. However, if we enlarge this region, we see that major asymmetric relationships among nations observed in the ASM are reproduced in the enlarged configuration. Here, in interpreting this configuration, it should be noticed that the positive direction of this figure is *clockwise* because the eigenvalue is positive in this case. It should also be noticed that angles θ_{jk} between two nations all fall within $\pi/3$ (30 degree), as is apparent from Fig. 5.5a. Then, looking at Fig. 5.2, we see that $s_{kj} > s_{jk}$. As a result, applying this relation to nations in Fig. 5.6, the amount of export from C (China) to A (USA) is greater than that from A to C. Similar trade imbalances are indicated from C to J, and from J to A. Finally, it should be noticed that the magnitude of the skewness between s_{jk} and s_{kj}, which is defined by (5.13), increases as θ_{jk} increases within $\pi/3$, as shown in

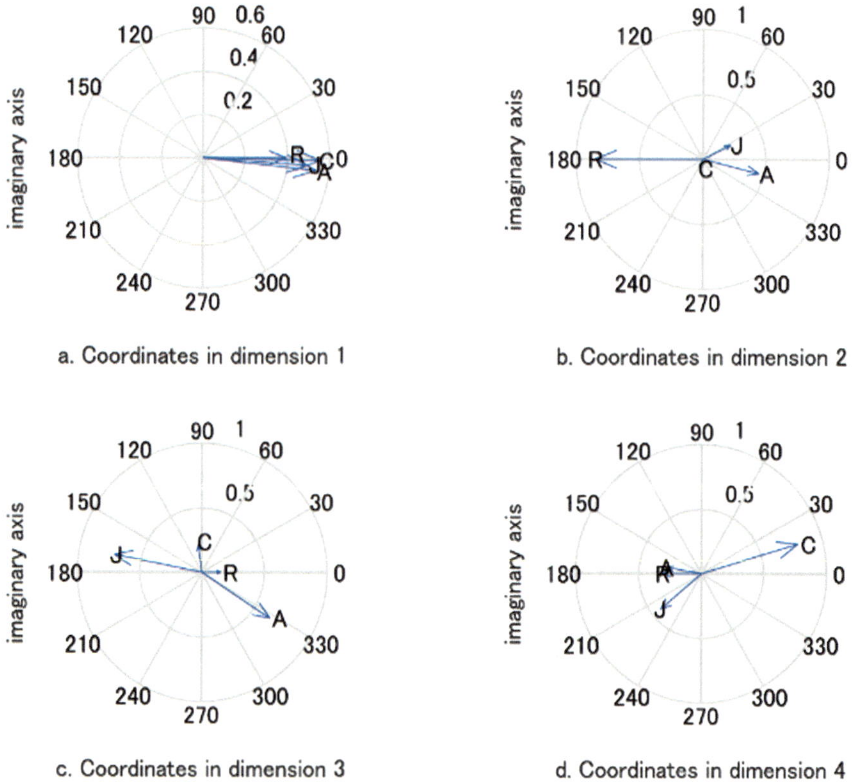

Fig. 5.5 Four configurations of nations corresponding to the one-dimensional Hilbert space. In this figure, alphabets, A, C, J, and R, indicate USA, China, Japan, and Russia

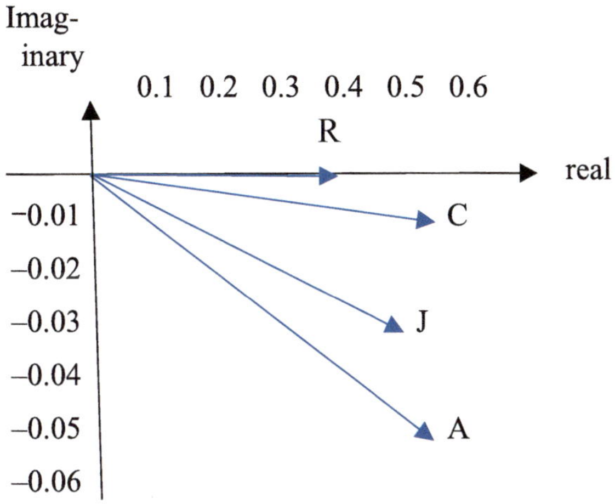

Fig. 5.6 The enlarged configuration of four nations. Here, the positive direction of this figure is clockwise because the eigenvalue is positive in this case

Fig. 5.4. Since this magnitude is also proportional to $|x_j||x_k|$ in (5.11), the magnitude of skewness between A and C might be the greatest of all the dyadic relations shown in Fig. 5.6.

The other application is concerned with the biosynthetic pathways of proteins in mammals (e.g., Imai [10]; Imai & Guarente [11]), an example of which appears in Fig. 5.2 of Imai and Guarente [11]. We reconstruct an ASM from this figure in such a way that the similarity is 1 if the directional arc exists from an element to another element of 11 substances, and otherwise it is 0. These substances are composed of tryptophan (an α-amino acid), NAD (nicotinamide adenine dinucleotide, a non-protein chemical compound), NIC (nicotinamide, an organic molecule), and so on. Table 5.4 shows this. Here, it should be noticed that in general such a pathway can be considered as a weighted digraph (directed graph) in graph theory. Since the weighted digraph accompanies a weight matrix, we say that a unique ASM is associated with any weighted digraph.

Eigenvalues of the Hermitian matrix constructed from this ASM were 1.6302, –1.6302, –1.2534, 1.2534, –0.8706, 0.8706, …. These eigenvalues mean that this data has a *holistic indefinite metric structure*. Of course, each of the configurations of the biosynthetic pathways of proteins in mammals on the complex planes associated with these eigenvalues can be embedded in a *one-dimensional Hilbert space*, as discussed previously.

Figure 5.7 shows four configurations of the biosynthetic pathways composed of these one-dimensional Hilbert spaces. These four configurations are associated with the first four eigenvalues of the Hermitian matrix discussed above.

Figure 5.7a is the configuration associated with the maximum eigenvalue of the Hermitian matrix. Although this configuration recovers the major pathways contained in the original data matrix, it does not include the important cyclic pathways, starting from O_6 (NAD) to come back to it through O_9 (NIC) and O_{10} (NMN). In contrast, Fig. 5..5.7b, which is the configuration corresponding to the maximum

Table 5.4 Biosynthetic pathways of proteins in mammals which was re-constructed from Fig. 5.2 of Imai and Guarente [11]

	1	2	3	4	5	6	7	8	9	10	11
1	0	1	0	0	0	0	0	0	0	0	0
2	0	0	0	1	0	0	0	0	0	0	0
3	0	0	0	1	0	0	0	0	0	0	0
4	0	0	0	0	1	0	0	0	0	0	0
5	0	0	0	0	0	1	0	0	0	0	0
6	0	0	0	0	0	0	1	0	1	0	0
7	0	0	0	0	0	0	0	0	0	0	0
8	0	0	0	0	0	0	0	0	0	0	0
9	0	0	0	0	0	0	0	1	0	1	0
10	0	0	0	0	0	1	0	0	0	0	1
11	0	0	0	0	0	0	0	0	0	1	0

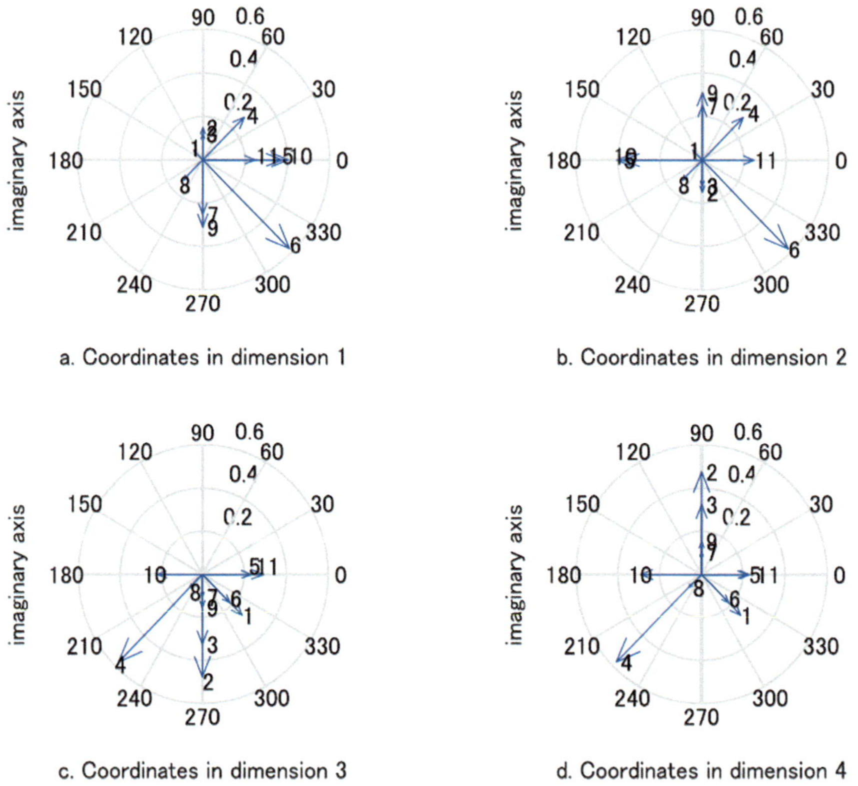

Fig. 5.7 Four configurations of the biosynthetic pathways corresponding to one-dimensional Hilbert spaces, which constitute a part of the holistic 11-dimensional Hilbert space

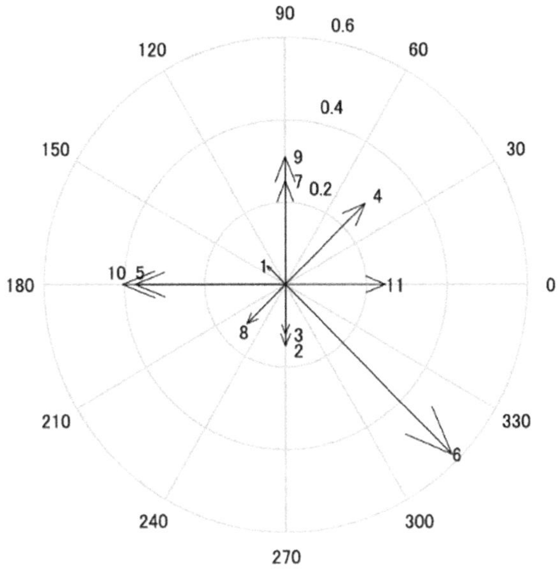

Fig. 5.8 An enlarged configuration of Fig. 5.7b

eigenvalue but with negative sign, recovers almost all the pathways contained in the original matrix.

Fig. 5.8 is an enlarged configuration of Fig. 5.7b. In interpreting this configuration, it should be noticed that its positive direction is *counterclockwise* because the eigenvalue associated with it is negative. Noticing this point, we can find the following pathways in Fig. 5.8 by following an object to another object *counterclockwise* within π radian, referring to Fig. 5.2 of Imai and Guarente [11]:

(1) O_1 (Tryptophan) $\rightarrow$ O_2 (Quinolinic acid) $\rightarrow$ O_4 (NaMN) $\rightarrow$ O_5 (deamido-NAD) $\rightarrow$ O_6 (NAD) $\rightarrow$ O_9 (NIC) $\rightarrow$ O_{10} (NMN) $\rightarrow$ O_6 (NAD)
(2) O_3 (NA) $\rightarrow$ O_4 (NaMN)
(3) O_9 (NIC) $\rightarrow$ O_8 (1-methyl-nicotinamide)
(4) O_6 (NAD) $\rightarrow$ O_7 (O-acetyl-ADP-ribose)

The first pathway includes a cycle starting from O_6 and returning to itself. The second pathway runs into the first pathway at O_4 (NaMN).

The third pathway gets away from the first pathway at O_9 (NIC). The fourth pathway also gets away from the first pathway at O_6 (NAD).

5.4 Applications of HFM to Theoretical or Hypothetical Asymmetric Relational Data Matrices

In this section we show two applications of HFM to theoretical or hypothetical asymmetric relational data matrices. Sato et al. [17] investigated the problem of learning to play the game of rock-paper-scissors, using the following set of nonlinear differential equations:

$$\begin{bmatrix} \dot{x}_i \\ \dot{y}_j \end{bmatrix} = \begin{bmatrix} x_i((\boldsymbol{Ay})_i - \boldsymbol{x}^t\boldsymbol{Ay} \\ y_j((\boldsymbol{Bx})_j - \boldsymbol{y}^t\boldsymbol{Bx} \end{bmatrix}, i = 1, \cdots, n, j = 1, \cdots, m. \tag{5.14}$$

where $\boldsymbol{x} = (x_1, \ldots, x_n)^t$ is the relative frequency vector for one population, while $\boldsymbol{y} = (y_1, \ldots, y_m)^t$ is that for the second population. This approach is based on the theory of games and the notion of *evolutionarily stable strategy* (abbreviated as ESS), of which theory was introduced by Maynard and Price [12].

In any case, in (5.14) **A** is the payoff matrix for one population, while **B** is the payoff matrix for the second population, and these two matrices are denoted as

$$\mathbf{A} = \begin{bmatrix} \varepsilon_x & -1 & 1 \\ 1 & \varepsilon_x & -1 \\ -1 & 1 & \varepsilon_x \end{bmatrix}, \mathbf{B} = \begin{bmatrix} \varepsilon_y & -1 & 1 \\ 1 & \varepsilon_y & -1 \\ -1 & 1 & \varepsilon_y \end{bmatrix} \tag{5.15}$$

where $-1 < \varepsilon_x < 1$ and $-1 < \varepsilon_y < 1$. Here, columns of these matrices are placed in the order of "rock", "paper", and "scissors". If $\varepsilon_x = -\varepsilon_y = \varepsilon$, this game is called a *zero sum game*. In matrix notation, this condition is denoted as $\boldsymbol{A} = -\boldsymbol{B}^t$. Matrices, $\boldsymbol{A}$ and $\boldsymbol{B}$, are nothing but *theoretical* examples of ASM, in that these ASMs cannot be observed and are hypothesized a priori.

Although Sato et al. [17] discuss these matrices from the viewpoint of a dynamical system, they do not discuss the *metric structure* of these matrices. Such a structure can be examined by applying HFM to them.

Let us now examine the structure of $\boldsymbol{A}$ in the case when $\varepsilon_x = 0.25$, which is one of the cases in which ε_x and ε_y are treated as a bifurcation parameter of the dynamical system described by (5.14). The eigenvalues of the Hermitian matrix constructed from this ASM were 1.9821, -1.4821, 0.2500. These eigenvalues mean that this data has a *holistic indefinite metric structure*. Figure 5.9 shows this structure. Each of the configurations shows the positions of the three elements, i.e., "1. rock", "2. paper", and "3. scissors", in a one-dimensional Hilbert space. Again, these configurations are complex orthogonal (unitary) in the holistic indefinite metric space.

Figure 5.10 shows the enlarged configuration of Fig. 5.9a. It should be noticed that the eigenvalue corresponding to it is positive. This means that the positive direction of this configuration is *clockwise*. Considering this point, we can conclude that 1 (rock) defeats 3 (scissors), 3 defeats 2 (paper), and 2 defeats 1. These relations completely reproduce the triadic relations assumed in the rock-paper-scissors game.

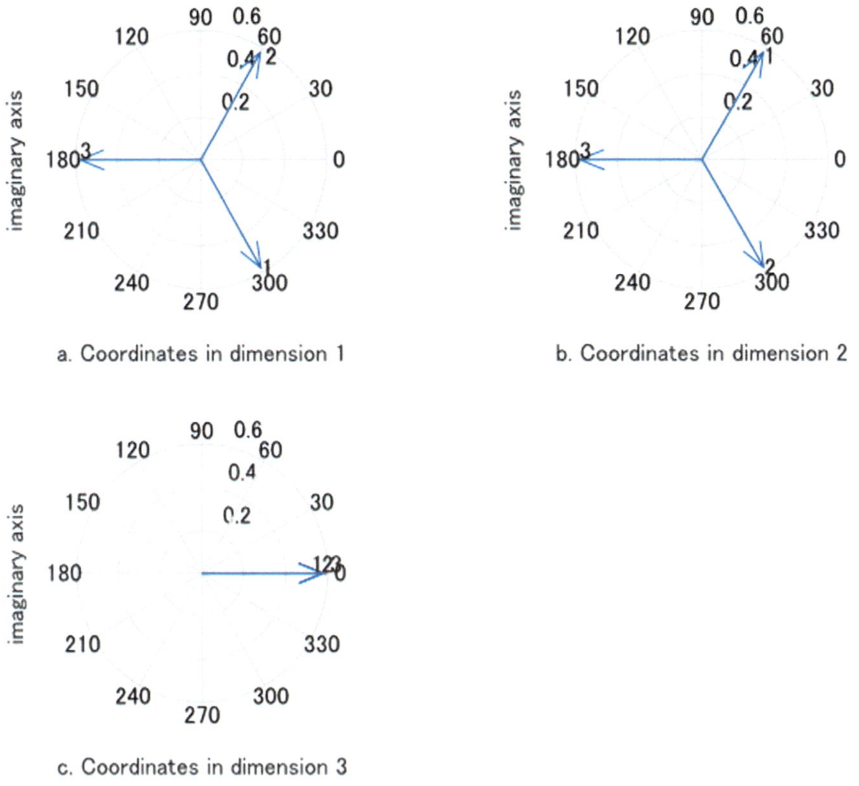

Fig. 5.9 Three configurations of the payoff matrix ***A*** corresponding to one-dimensional Hilbert spaces, which constitute the holistic 3-dimensional Hilbert space

Figure 5.11 shows the enlarged configuration of Fig. 5.9b. It should be noticed that the eigenvalue corresponding to it is *negative*. This means that the positive direction of this configuration is *counterclockwise*. Considering this point, we see that the same triadic relations as in Fig. 5.10 are reproduced.

Figure 5.9c is the configuration whose associated eigenvalue is *positive*. As a result, the positive direction of this configuration is *clockwise*. However, regardless of this information, the configuration reveals the collapse of the triadic relations.

Another application of the theoretical ASM is the following matrix,

$$\mathbf{C} = \begin{bmatrix} 1 & -1 & 1 \\ 1 & 1 & -1 \\ -1 & 1 & 1 \end{bmatrix}. \tag{5.16}$$

According to the rock-paper-scissors game, **C** is not the payoff matrix, because ε in (5.15) is out of the range defined in the payoff matrices. However, such a matrix

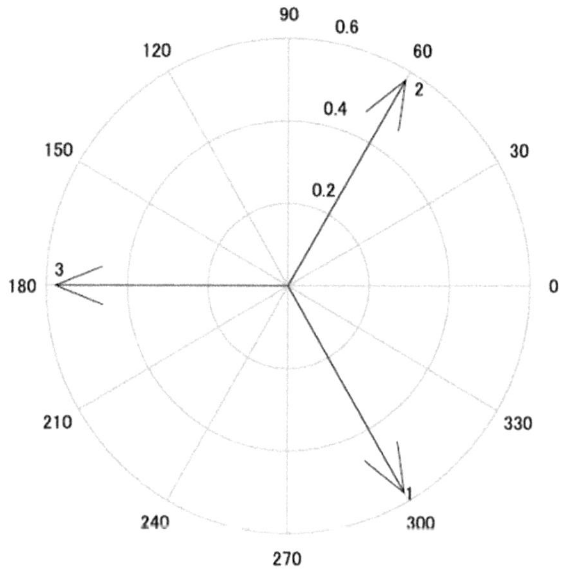

Fig. 5.10 Configuration of elements (1. Rock, 2. Paper, 3. Scissors) in the one-dimensional Hilbert space, which corresponds to the largest eigenvalue of **H** (the enlarged configuration of Fig. 5.9a)

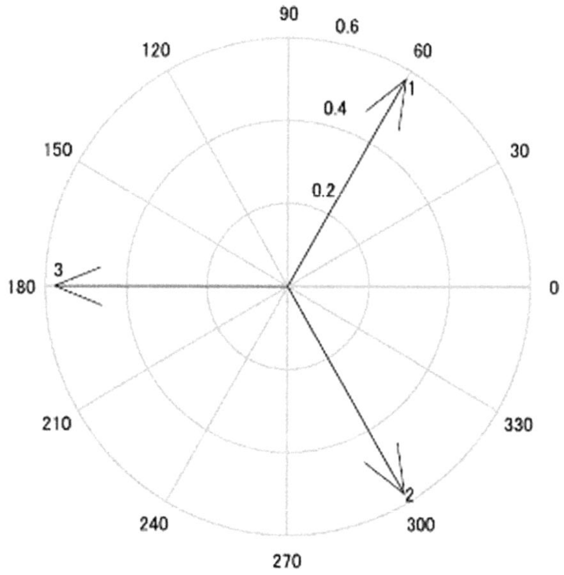

Fig. 5.11 Configuration of elements (1. Rock, 2. Paper, 3. Scissors) in the one-dimensional Hilbert space, which corresponds to the second largest eigenvalue of **H** (the enlarged configurations of Fig. 5.9b)

is plausible as a general ASM, and has already been analyzed elsewhere (e.g., Chino [2]).

In fact, eigenvalues of the Hermitian matrix constructed from this ASM were 2.7321, 1.000, and −0.7321. These eigenvalues mean that this data has a *holistic indefinite metric structure*. Figure 5.12 shows this structure. Each of the configurations shows the positions of the three elements, i.e., "1. rock", "2. paper", and "3. scissors", in a one-dimensional Hilbert space.

Figure 5.13 shows the enlarged configuration of Fig. 5.12a. At a glance, this configuration is completely different from that of Fig. 5.11. However, the sign of the eigenvalue corresponding to the configuration of.

Figure 5.11 is *negative*, while that of Fig. 5.13 is *positive*. This means that these two configurations indicate the same information about the asymmetric relations among elements.

Similarly, we see that Fig. 5.9c and Fig. 5.12b convey the same information about the asymmetric relations among elements, considering the signs of the eigenvalues corresponding to these configurations. We also find that Fig. 5.9b and Fig. 5.12c convey the same information. However, it should be noticed that the sizes of the eigenvalues for these pairs of figures are different from one another. This means that the contributions of one-dimensional Hilbert space structures to the holistic space

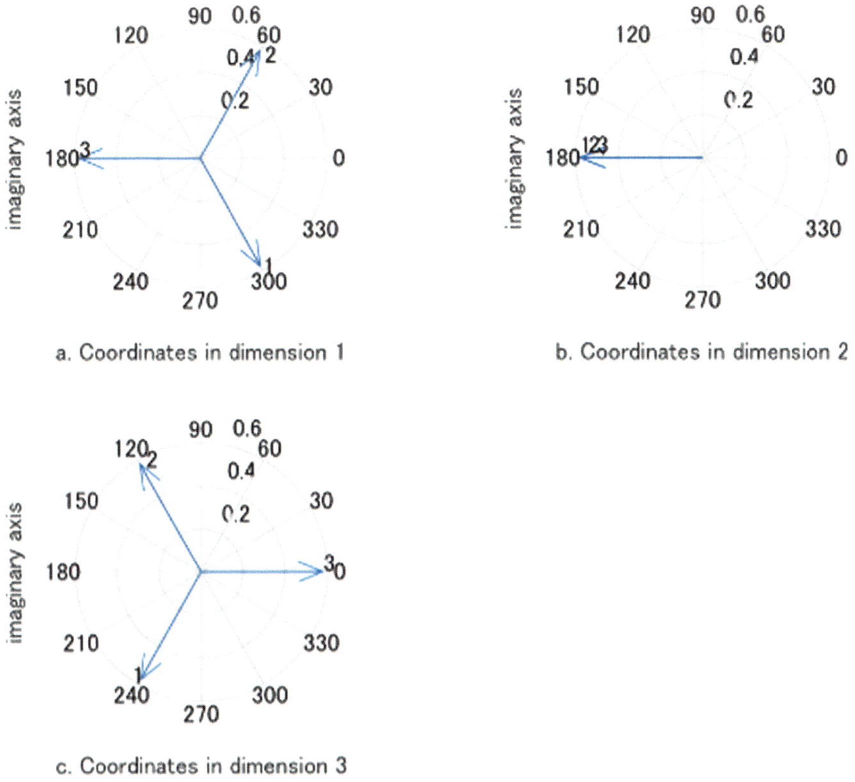

Fig. 5.12 Three configurations of the matrix **C** in (5.16) corresponding to one-dimensional Hilbert spaces, which constitute the holistic 3-dimensional Hilbert space

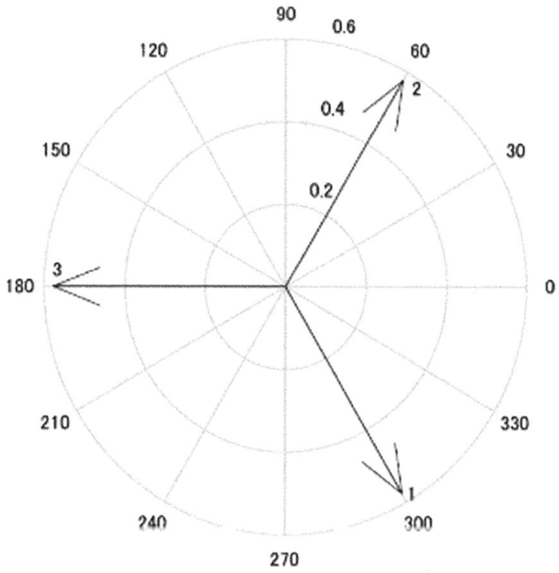

Fig. 5.13 Configuration of elements (1. Rock, 2. Paper, 3. Scissors) in the one-dimensional Hilbert space, which corresponds to the largest eigenvalue of **H** associated with **C** (the enlarged configuration of Fig. 5.12a)

structure depend on the diagonal elements, i.e., ε in (5.15), in the case of tripartite deadlock relation.

5.5 Decomposition of ASM to Elementary ASMs via HFM

In the application sections we have seen that HFM enables us to get *multiple configurations* of objects which are *mutually complex orthogonal* (i.e., unitary), given any empirical ASM or theoretical ASM. Moreover, we can recover '*real* component ASMs' from these configurations of objects embedded in a *complex* space (i.e., Hilbert space or indefinite space), using the following equation

$$s_{jk} = \|\boldsymbol{v}_j\|^2 + \|\boldsymbol{v}_k\|^2 - \frac{1}{2}\left(\|\boldsymbol{v}_j - \boldsymbol{v}_k\|^2 + \|\boldsymbol{v}_j - i\boldsymbol{v}_k\|^2\right), \tag{5.17}$$

which is a simplified version of (4.42) in Chap. 4.

For example, the theoretical ASM defined by (5.16) has three configurations of objects in the complex space, as shown in Fig. 5.12 in the previous section. If we compute *component ASMs* from these configurations using (5.17), we have

$$\boldsymbol{C}_1 = \begin{bmatrix} 0.333 & -0.455 & 0.122 \\ 0.122 & 0.333 & -0.455 \\ -0.455 & 0.122 & 0.333 \end{bmatrix}, \boldsymbol{C}_2 = \begin{bmatrix} 0.333 & -0.455 & 0.122 \\ 0.122 & 0.333 & -0.455 \\ -0.455 & 0.122 & 0.333 \end{bmatrix},$$

and

$$C_3 = \begin{bmatrix} -0.333 & -0.122 & 0.455 \\ 0.455 & -0.333 & -0.122 \\ -0.122 & 0.455 & -0.333 \end{bmatrix}. \tag{5.18}$$

Since these component ASMs are associated with the three configurations of which eigenvectors are mutually complex orthogonal, these component ASMs convey unrelated information about objects with one another. Therefore, if we apply HFM to some empirical ASM as in Table 5.4, and if we find a component configuration which is not congruent with the previous results of the experiment, it might be worthwhile to examine theoretically and/or experimentally whether such a configuration exists or not substantively.

5.6 Hypothetical Force Acting on the Hilbert Space and Its Interpretation

The ASMs we have dealt in the previous sections are matrices observed at some instant of time, or those assumed theoretically or hypothetically. However, in actual situations an ASM frequently changes as time proceeds through the *mutual interactions* among objects. In such a case, configuration spaces obtained by HFM can be thought of as *state spaces* of some underlying *dynamical system*. The state space under study is, of course, the Hilbert space or indefinite metric space.

Then, we can naturally introduce some *forces* acting on these state spaces by which objects move as time proceeds (e.g., Elaydi [7]; Hirsch & Smale [9]). As a result, the configuration of objects changes as time proceeds. If we take a snapshot of such a process at some instant in time, we can compute the *longitudinal ASMs* using (5.17) of HFM. Figure 5.14 shows an illustration of forces exerting on the configuration space in Fig. 5.6. In fact, trade imbalance changes year after year by the political and economic interactions among nations.

In Fig. 5.14, blue arrows denote the *position vectors* of the four nations, while red arrows indicate *usual vectors* denoting here the *forces* acting on the two nations, A (USA) and J (Japan). According to these forces, USA approaches Japan, while Japan does not necessarily approach America. These forces can be defined as the *mutual interaction matrix* as follows. In this case its order is four, which corresponds to the number of nations.

$$\mathbf{M} = \begin{bmatrix} m_{11} & m_{12} & m_{13} & m_{14} \\ m_{21} & m_{22} & m_{23} & m_{24} \\ m_{31} & m_{32} & m_{33} & m_{34} \\ m_{41} & m_{42} & m_{43} & m_{44} \end{bmatrix}. \tag{5.19}$$

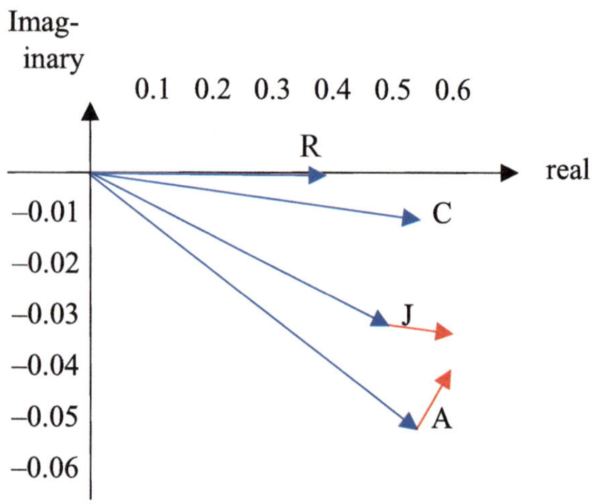

Fig. 5.14 An illustration of forces exerting on the configuration space in Fig. 5.6. Red arrows indicate the forces acting on the two nations, America and Japan. Here, the positive direction of this figure is clockwise because the eigenvalue is positive in this case

Here, each element of this matrix is a *complex number*. Moreover, subscript numbers, 1, 2, 3, and 4, denote J (Japan), A (USA), C (China), and R (Russia), respectively, as in Table 5.3. Therefore, the force emanating from USA to Japan can be expressed as, say, $m_{21} = 0.07 + 0.1i$, and that issuing from Japan to USA can be expressed as, say, $m_{12} = 0.1 - 0.01i$. We assume that the matrix **M** is defined *theoretically* or *hypothetically*, although it is also an ASM. Moreover, we assume that its elements are all *constant*.

The mutual interaction matrix is sometimes used in dynamical system models. For example, the payoff matrices hypothesized by Sato et al. [17] which was discussed in Sect. 5.4 are considered as such mutual interaction matrices. The weight matrix representing hidden-to-hidden recurrent connections assumed in *recurrent neural network models* is thought of as another example (e.g., Goodfellow et al. [8]). In this case, however, the weight matrix is *real*. Chino [3, 4] has recently been devel oping a *dynamic weighted digraph model* using a set of *nonlinear complex difference equations*, in which a mutual interaction matrix is hypothesized to explain changes in the observed ASM over time. In this case, the elements of the interaction matrix are *not real but complex*, one example of which is the matrix **M** defined in (5.19).

5.7 Conclusions

HFM is one of the MDS models which is specifically designed to analyze asymmetric (dis)similarity matrix (abbreviated as ASM), and was proposed by Chino and Shiraiwa [6]. Among various asymmetric MDS models which have been proposed up to now, HFM is a special one in that embedded spaces are *complex*, especially, a (complex) Hilbert space or an indefinite metric space, although original ASMs are always *real*. This property might sometimes be a barrier for users' understanding of HFM. Therefore, we explained in this paper how to use HFM to various ASMs which are observed in our daily lives, in research laboratories, and so on.

In the introductory section we gave some typical examples of ASMs and point out that the ratio level of measurement is required for the elements of ASM to which we apply HFM.

As is apparent from theory of HFM in Chapter 4, the keyword for introducing complex spaces is the *Hermitian form* defined in the complex space. Formulae which play important roles in relating any ASM, **S**, with this form are obtained by considering the following two elementary equations. One is the decomposition of ASM, $\boldsymbol{S}$ such that

$$\boldsymbol{S} = \boldsymbol{S}_s + \boldsymbol{S}_{sk}, \tag{5.20}$$

where $\boldsymbol{S}_s = \left(\boldsymbol{S} + \boldsymbol{S}^t\right)/2$ and $\boldsymbol{S}_{sk} = \left(\boldsymbol{S} - \boldsymbol{S}^t\right)/2$, and the other is

$$\boldsymbol{H} = \boldsymbol{S}_s + i\boldsymbol{S}_{sk}, \tag{5.21}$$

where $i^2 = -1$. Eq. (5.21) is the starting Eq. (4.1) of the theory of HFM in Chap. 4.

On the way to deducing the HFM formulas, we also get an important formula, (5.17). This equation builds a bridge between the similarity from O_j to O_k (*real*) and coordinate vectors of the two objects (*complex*).

If we apply HFM to any ASM, then we get multiple configurations of objects embedded in a (complex) Hilbert space or an indefinite metric space, depending on the signs of the eigenvalues of the Hermitian matrix $\boldsymbol{H}$ in (4.1). If the nonzero eigenvalues are all positive, we may adopt a Hilbert space as a holistic configuration space. Otherwise, we may adopt an indefinite metric space as a holistic space. Moreover, we may interpret these configurations separately per dimension, because these configurations, which are associated with the eigenvectors corresponding to these eigenvalues, are *unitary*, i.e., complex orthogonal. Furthermore, each of these configurations can be thought of as embedded in a one-dimensional Hilbert space, which can be regarded as a (real) Euclidian two-dimensional space, except the sign of the corresponding eigenvalue. If the sign is positive, we may consider that the positive direction of the configuration is *clockwise*, and if the sign is negative, we may regard it as *counterclockwise*. In Sects. 5.3 and 5.4, we pointed out two empirical ASMs and two theoretical or hypothetical ASMs, and explained how to interpret the obtained configurations of objects via HFM.

In Sect. 5.5, we focused on a theoretical or hypothetical ASM, and explained how to recover the component ASMs from the multiple configurations of objects. In Sect. 5.6, we discussed the cases in which ASM changes as time proceeds. In such cases, it will be appropriate and natural to introduce some *forces acting on the complex spaces* by which objects move as time proceeds. Then, it may be appropriate and natural to introduce a *dynamical system model* in which some *mutual interaction matrix* is assumed.

Acknowledgements The author is indebted to Gregory L. Rohe for proofreading of an earlier version of this paper.

Notes

This chapter is the reproduction of Chino [5]. We deleted Sect. 5.2 in Chino (op. cit.) which is a revisit of HFM, because HFM is fully described in Chap. 4. We corrected a few errors found in the original paper. Foremost is the (9, 10) element of the matrix shown in Table 5.4, which was set equal to 0, but must be 1. However, we did not correct the successive results associated with this error because its influences on them were negligible.

References

1. Chino, N. (1978). A graphical technique for representing the asymmetric relation ships between N objects. *Behaviormetrika, 5*, 23–40.
2. Chino, N. (2017). Dynamical scenarios of changes in asymmetric relationships on a Hilbert space. *Handout presented* at *the 45th annual meeting of the Behaviormetric Society of Japan.* 1–29.
3. Chino, N. (2018). *An elementary theory of a dynamic weighted digraph. Proceedings of the 46th annual meeting of the Behaviormetric Society of Japan.* 26–29.
4. Chino, N. (2019). An elementary theory of a dynamic weighted digraph (2). *Proceedings of the 47th annual meeting of the Behaviormetric Society of Japan.* 56–59.
5. Chino, N. (2020). How to use the Hermitian Form Model for asymmetric MDS. In T. Imaizumi, Nakayama, A., & Yokiyama, S. (Eds.), *Advanced Studies in Behaviormetrics and Data Science* (pp.19–41). Singapore: Springer Nature.
6. Chino, N., & Shiraiwa, K. (1993). A brief survey of asymmetric MDS and open problems. *Behaviormetrika, 39*, 127–165.
7. Elaydi, S. N. (1999). *An introduction to differential equations.* Springer.
8. Goodfellow, I., Bengio, Y., & Courville, A. (2016). *Deep learning*. The MIT Press.
9. Hirsch, M. W., & Smale, S. (1974). *Differential equations, dynamical systems, and linear algebra.* Academic Press.
10. Imai, S. (2011). Dissecting systematic control of metabolism and aging in the NAD World: The importance of SIRT1 and NAMPT-mediated NAD biosynthesis. *FEBS Letters, 585*, 1657–1662.
11. Imai, S., & Guarente, L. (2014). NAD^+ and sirtuins in aging and disease. *Trends in Cell Biology, 24*, 464–471.
12. Maynard Smith, J., & Price, G. R. (1973). The logic of animal conflict. *Nature, 246*, 15–18.
13. Okada, A. & Imaizumi, T. (1984). Geometric models for asymmetric similarity. *Research Reports of School of Social Relations, Rikkyo (St. Paul's) University.*
14. Okada, A., & Imaizumi, T. (1987). Nonmetric multidimensional scaling of two-mode, three-way proximities. *Journal of Classification, 14*, 195–224.
15. Rushen, J. (1982). The peck orders of chickens: How do they develop and why are they linear? *Animal Behavior, 30*, 1129–1137.
16. Saburi, S., & Chino, N. (2008). A maximum likelihood method for an asymmetric MDS model. *Computational Statistics and Data Analysis, 52*, 4673–4684.
17. Sato, Y., Akiyama, E., & Farmer, J. D. (2002). Chaos in learning a simple two-person game. *Proceedings of the National Academy of Science of the United States of America, 99*, 4748–4751.

Chapter 6
Asymmaxscal

6.1 Introduction

Asymmetric relationships are frequently observed in animal as well as human behavior. For example, pecking order is a special asymmetric relationship among a group of hens and cocks (e.g., Rushen [41]). Asymmetric relationships may become more complicated in humans than in animals. A typical example may be asymmetric sentiment relationships such as one-sided love and hate among members of informal groups in classrooms.

The asymmetric multidimensional scaling is a method which is specifically designed to analyze such asymmetric relationships among members and display them graphically by plotting each member in a certain dimensional space, given an asymmetric square data matrix $\boldsymbol{S}$ of order n whose (i, j) th element, s_{ij}, denotes the proximity of member i to member j. If we are interested in the change in these relations over time, then we have to obtain longitudinal, asymmetric square data matrices, say $\boldsymbol{S}_1, \cdots, \boldsymbol{S}_T$. In some cases s_{ij} may be dichotomous as in the pecking order data, but in other cases these may be measured at an ordinal, interval, or ratio scale level.

The asymmetric multidimensional scaling which has been developed in psychometrics since the work of Young [62] is an extension of the method called the multidimensional scaling for symmetric data, abbreviated traditionally as MDS. Hereafter we shall abbreviate asymmetric multidimensional scaling as asymmetric MDS in order to distinguish it from the traditional symmetric MDS.

The major asymmetric MDS methods include Chino [10, 11], Chino and Shiraiwa [13], Constantine and Gower [14], Escoufier and Grorud [19], Gower [22], Harshman [23], Harshman et al. [24], Kiers and Takane [27], Krumhansl [28], Okada and Imaizumi [36, 37], Rocci and Bove [40], Saito [43], Saito and Takeda [44], Sato [45], ten Berge [53], Tobler [54], Trendafilov [57], Weeks and Bentler [59], Young [62], and Zielman and Heiser [63].

N. Chino, *Structure and Dynamics of Asymmetric Interactions*,
https://doi.org/10.1007/978-981-97-8269-7_6

Although various asymmetric MDS methods have been proposed, these methods have remained to be descriptive. By contrast, we will propose a maximum likelihood method for asymmetric MDS, which extends MAXSCAL proposed by Takane [50] to asymmetric relational data. It enables us, for example, to examine not only the appropriate scaling level of the data but also the appropriate dimensionality of the configuration of members using AIC (Akaike [3]). It also enables us to perform some tests of symmetry for a given asymmetric data set. These tests enable us to verify that the data are sufficiently asymmetric.

We shall introduce two types of tests for symmetry. One type is a symmetry test to be performed prior to the application of our method to the data. It is a special test for conditional symmetry, and can be used to examine indirectly the validity of applying asymmetric MDS to the data. The other type of symmetry test pertains to a symmetry hypothesis to be tested in the scaling step. We shall introduce two kinds of tests of this type. These tests can be used to examine directly the validity of applying asymmetric MDS.

Traditionally, the symmetry hypothesis and its variants are tested using χ^2 statistics (for example, Birch [5], Bishop et al. [6], Bowker [8], Caussinus [9], Cramér [15], Goodman [21] and McNemar [34]). However, these statistics have several disadvantages. For example, their use for model comparison is limited to the case in which models to be compared constitute a hierarchical structure. Ramsay [38] shows that the asymptotic χ^2 statistic tends to be too liberal in a small sample situation. De Rooij and Heiser [18] point out that the asymptotic χ^2 statistics tend to dismiss all models except for the saturated model when the sample size is large.

Information criteria such as AIC, on the other hand, can be used to compare various models with nonhierarchical structure. Moreover, since AIC rank-orders all competing models from the best to the worst, there is no need to worry about controlling the errors in comparing these models. In the traditional goodness-of-fit tests, on the other hand, a series of tests are typically performed to identify the best model, each comparing two nested models at a time. This renders controlling the two kinds of errors (Type 1 and Type 2) extremely difficult, if not impossible. Therefore, we shall utilize information criteria rather than the asymptotic χ^2 statistic to compare the scaling levels of the data, dimensionalities of the model, and several variants of symmetry hypotheses introduced in this paper.

Basic models, the maximum likelihood criterion, and algorithms are essentially based on MAXSCAL by Takane [50], except that one of the constituent sub-models, that is, the representation model is replaced by that of Okada and Imaizumi [36]. For this reason, we shall call our method ASYMMAXSCAL-OI which means that it is a MAXSCAL applied to the Okada–Imaizumi model (abbreviated hereafter as the OI model) for asymmetric MDS.

The organization of this paper is as follows. In Sect. 6.2 we shall describe our method. In Sect. 6.3.1 we shall introduce a special test for symmetry to be performed prior to the scaling step, and in Sect. 6.3.2 two other tests to be performed in the scaling step. We shall show the asymptotic χ^2 statistics for these tests in these sections in order to clarify the distinction between the traditional symmetry hypotheses and ours. In Sect. 6.4 we shall discuss a model selection method. In Sect. 6.5 we shall show examples of application of our method to synthetic and actual data sets. Finally, in Sect. 6.6 we shall give some concluding remarks.

6.2 Method

6.2.1 *Sub-models*

In ASYMMAXSCAL as well as MAXSCAL, subjects' responses are not the usual proximities, s_{ij}'s, defined in the previous section, but rating scale categories in which subjects place certain error-perturbed proximities. For example, suppose that each subject is asked to judge the degree of friendliness of one of the governments of five nations, say, Japan to another of them, say, the USA on a 5-point rating scale with categories, "very friendly", "friendly", "neither friendly nor hostile", "hostile", "very hostile", as will be described in Sect. 6.5.2. Suppose further that a subject judges that the government of Japan is hostile to that of the USA. Then, his or her response is the fourth category, which is not the usual proximity in the context of asymmetric MDS.

Let us introduce first the population value corresponding to s_{ij} which we call the proximity model hereafter. In ASYMMAXSCAL, we assume that the proximity model is error-perturbed to give rise to a psychological value, which we call the error-perturbed proximity. We ask subjects to place it in one of the M rating scale categories, which are nothing but subjects' responses in ASYMMAXSCAL.

Subjects' responses are assumed to be determined by the following three sub-models: the representation model, the error model, and the response model, according to MAXSCAL. The representation model specifies the model of proximities (e.g., the Euclidean distance model). The error model specifies the distribution of the error-perturbed proximities (it describes how the value of the representation model is error-perturbed to generate error-perturbed proximities). In this paper we assume the normal distribution. The response model specifies the mechanism by which responses of specific form are generated. It assumes that proximities are placed in a particular category whenever error-perturbed proximities fall between the lower and the upper bounds of the category. This is similar to the law of successive categories (Torgerson [55]).

6.2.1.1 The Representation Model

The representation model specifies the representation of the proximity. Any extant model for asymmetric MDS may be chosen as the representation model in principle. In this paper, however, we will restrict our attention to the OI model (Okada and Imaizumi [36]). This model represents the symmetric part and the skew-symmetric part of the square asymmetric matrix by the Euclidean distance and the difference in the radii of circles (spheres, hyperspheres) centered at the points of the objects in that space, respectively. In the OI model, the proximity model of object i to object j, g_{ij}, is defined as

$$g_{ij} = d_{ij} - r_i + r_j, \tag{6.1}$$

where r_i is the radius of the circle attached to object i and

$$d_{ij} = \left[\sum_{a=1}^{A} \left(x_{ia} - x_{ja} \right)^2 \right]^{1/2},$$

where x_{ia} is the coordinate of object i on the ath dimension, and A is the dimensionality of the space. According to their model, an object with larger radius is more proximate to an object with smaller radius than vice versa. Okada and Imaizumi [36] introduced these parameters in order to represent the skew-symmetry contained in g_{ij}, but did not consider what kinds of psychological processes underlie these parameters. However, Okada and Imaizumi [37] interpret the term, $r_j - r_i$, as a dominance effect as discussed in Zielman and Heiser [63].

A distinguished feature of the OI model pertains to its simple form. It is a special case of the very general model by Holman [25]

$$p_{ij} = F\left[b_i^{(r)} + b_j^{(c)} + z_{ij} \right],$$

where p_{ij} is a real-valued function of several parameters. Here, F is an increasing function, while z_{ij} is a symmetric similarity function, and $b_i^{(r)}$ and $b_j^{(c)}$ are bias functions on the individual objects. In the OI model the row bias $b_i^{(r)}$ is equivalent to the negative of the column bias $b_i^{(c)}$ in the Holman model. However, one may consider such a model too simplistic because this model has no bias components for the symmetric part of g_{ij} [63].

The Holman model includes various asymmetric proximity models such as the quasi-symmetry models (Bishop et al. [6]; Caussinus [9]) and the similarity choice model (Luce [32, 33]). These models have scored a long-standing success compared with alternative models of identification confusion (e.g., Ashby and Perrin [4], Smith [48], Takane and Shibayama [52] and Townsend and Landon [56]) as Nosofsky [35] pointed out. The OI model has an advantage of visualizing the asymmetries contained in the estimated configuration of objects easily over various variants of the quasi-symmetric model as discussed in De Rooij and Heiser [17].

We also introduce a "saturated" representation model abbreviated as the SR model. This model takes $g_{ij}(i, j = 1, \cdots n)$ as parameters. It is used to test one of the symmetry hypotheses to be mentioned below. The estimates of g'_{ij} s are utilized to compute the initial values of the parameters for the representation model of interest.

6.2.1.2 The Error Model

As already discussed, the g'_{ij} s are assumed to be error-perturbed by some process to yield error-perturbed proximities. Takane [50] assumes the additive error model with the normal distribution, and the multiplicative error model with the log-normal distribution. The latter implies that g'_{ij} s are nonnegative, so that it contradicts with some representation models in which g'_{ij} s can be negative as with the OI model. Thus, we assume the following additive error model only:

$$\begin{cases} \tau_{ij}^{(k)} = g_{ij} + e_{ij}^{(k)} \\ e_{ij}^{(k)} \sim N\left(0, \sigma^2\right) \end{cases}, \tag{6.2}$$

where k indicates a subject. MAXSCAL allows for individual differences, and thus σ has subscript k. However, $\sigma'_k s$ are incidental parameters, and it is often unnatural for subjects to judge the proximities among the same set of objects many times in a short period. Therefore, we assume that subjects make one and only one proximity judgment per data set. Our additive error model corresponds to the case in which subjects are viewed as replications in MAXSCAL. The index k is suppressed hereafter.

6.2.1.3 The Response Model

Let us suppose that subjects place error-perturbed proximities in one of the M rating scale categories $C_1, \cdots, C_M$. These categories are assumed to be represented by a set of ordered intervals with upper and lower boundaries on a psychological continuum. If we denote the upper boundary of the mth category by b_m, then

$$-\infty = b_0 \le b_1 \cdots \le b_{M-1} \le b_M = \infty.$$

The probability that the error-perturbed proximity of object i to object j, τ_{ij}, falls in C_m is given by the

$$p_{ijm} = pr\left(b_{m-1} < \tau_{ij} \le b_m\right), \tag{6.3}$$

Under the additive error model of (6.2) and (6.3) can be written as

$$p_{ijm} = \int_{b_{m-1}}^{b_m} \emptyset\left(\tau_{ij}\right) d\tau_{ij}, \tag{6.4}$$

where ∅ is the density function of the normal distribution with mean g_{ij} and variance σ^2. By replacing $\left(\tau_{ij} - g_{ij}\right)/\sigma$ with z we can obtain

$$p_{ijm} = \int_{a_{ij(m-1)}}^{a_{ijm}} f(z) dz, \tag{6.5}$$

where f is the density function of the standard normal distribution, and

$$\begin{cases} a_{ij(m-1)} = \left(b_{m-1} - g_{ij}\right)/\sigma \\ a_{ijm} = \left(b_m - g_{ij}\right)/\sigma \end{cases}, \tag{6.6}$$

If we let

$$F_{ijm} = \int_{-\infty}^{a_{ijm}} f(z)dz, \tag{6.7}$$

then

$$p_{ijm} = F_{ijm} - F_{ij(m-1)}. \tag{6.8}$$

Here, b_m's ($m = 1, \cdots, M - 1$) are not necessarily equally spaced. They are unrestricted except for the implicit ordinal restriction, that is, $b_1 \le b_2 \le \cdots, b_{M-1}$. These categories constitute an ordinal scale. On the other hand, if the b_m's are assumed to be equally spaced, then we may impose the linear constraint:

$$b_m = \alpha m + \beta, (\alpha > 0; 1 \le m \le M - 1), \tag{6.9}$$

where α and β denote a scale factor and an additive constant, respectively. These categories constitute an interval scale.

6.2.2 Maximum Likelihood Criterion

Let us now introduce an indicator variable Z_{ijmk} which takes the value of one when subject k rates the proximity of object i to object j into C_m, and zero otherwise. That is

$$Z_{ijmk} = \begin{cases} 1 & when o_{ijk} \epsilon C_m \\ 0 & otherwise \end{cases}, \tag{6.10}$$

where $o_{ijk} \in C_m$ means that the proximity of object i to object j for subject k (o_{ijk}) is placed into C_m. Readers should distinguish the notation of the response o_{ijk} from the variable $\tau_{ij}^{(k)}$. The judgments for all combinations of objects are assumed to be mutually independent, so that the joint likelihood of the total observations is written as

$$L = \prod_{i}^{n} \prod_{j}^{n} \prod_{m}^{M} p_{ijm}^{Y_{ijm}}, \tag{6.11}$$

where $Y_{ijm} = \sum_{k=1} Z_{ijmk}$. We will maximize the logarithm of L in (6.11) over all the parameters in those three sub-models. It is well known that all judgments are likely to be independent under the single-judgment condition, where each subject judges only one pair of objects, with randomly sampled subjects. However, under the multiple-judgment condition, where each subject judges all pairs of objects, the judgments may or may not be independent (Bock and Jones [7], p. 16).

6.2.3 Algorithm

We use Fisher's scoring algorithm as in the original MAXSCAL, to maximize the log-likelihood. We update a vector of unknown parameters $\boldsymbol{\theta} = \left(\theta_1, \cdots, \theta_{N_p}\right)'$ by the following formula, where N_p being the total number of unknown parameters:

$$\boldsymbol{\theta}^{(q+1)} = \boldsymbol{\theta}^{(q)} + \epsilon^{(q)} \boldsymbol{I}\left(\boldsymbol{\theta}^{(q)}\right)^{-1} \boldsymbol{u}\left(\boldsymbol{\theta}^{(q)}\right), \tag{6.12}$$

where q is the index of iteration number, ε is a step-size parameter, and the vector $\boldsymbol{u}(\boldsymbol{\theta})$ and the matrix $\boldsymbol{I}(\boldsymbol{\theta})$ are defined by

$$\boldsymbol{u}(\boldsymbol{\theta}) = \left(\frac{\partial \ln L}{\partial \boldsymbol{\theta}}\right),$$

and

$$\boldsymbol{I}(\boldsymbol{\theta}) = E\left[\left(\frac{\partial \ln L}{\partial \boldsymbol{\theta}}\right)\left(\frac{\partial \ln L}{\partial \boldsymbol{\theta}}\right)^t\right] = -E\left(\frac{\partial^2 \ln L}{\partial \boldsymbol{\theta}\, \partial \boldsymbol{\theta}^t}\right).$$

The information matrix $\boldsymbol{I}(\boldsymbol{\theta})$ is singular due to the nonuniqueness of scale and location of g_{ij}'s for the SR model. Specifically, the transformation $\tilde{g}_{ij} = \psi g_{ij} + \mu$ by arbitrary constants ψ and μ in (6.6) can be compensated for by the transformations on category boundaries and σ, i.e., $\tilde{b}_m = \psi b_m + \mu$ and $\tilde{\sigma} = \psi \sigma$. The solution of the OI model is also not unique with respect to the size, the origin and the rotation of the stimulus configuration, and with respect to additive constant in the radii. Thus, we replace the regular inverse in (6.12) with the Moore–Penrose inverse.

The next task is to evaluate $\boldsymbol{u}(\boldsymbol{\theta})$ and $\boldsymbol{I}(\boldsymbol{\theta})$ in formula (6.12). It is easy to see from (6.11) that

$$\boldsymbol{u}(\boldsymbol{\theta}) = \sum_i^n \sum_j^n \sum_m^M \frac{Y_{ijm}}{p_{ijm}} \left(\frac{\partial p_{ijm}}{\partial \boldsymbol{\theta}}\right),$$

$$\boldsymbol{I}(\boldsymbol{\theta}) = \sum_i^n \sum_j^n \sum_m^M \frac{n_{ij}}{p_{ijm}} \left(\frac{\partial p_{ijm}}{\partial \boldsymbol{\theta}}\right)\left(\frac{\partial p_{ijm}}{\partial \boldsymbol{\theta}}\right)',$$

where $n_{ij} = \sum_{m=1}^{M} Y_{ijm}$. we have from (6.8)

$$\frac{\partial p_{ijm}}{\partial \boldsymbol{\theta}} = \frac{\partial F_{ijm}}{\partial \boldsymbol{\theta}} - \frac{\partial F_{ij(m-1)}}{\partial \boldsymbol{\theta}}, \tag{6.13}$$

and

$$\frac{\partial F_{ijm}}{\partial \boldsymbol{\theta}} = \frac{\partial F_{ijm}}{\partial a_{ijm}} \frac{\partial a_{ijm}}{\partial \boldsymbol{\theta}}. \tag{6.14}$$

We approximate the distribution function F_{ijm} of the standard normal distribution by that of the logistic distribution given by $F_{ijm} = 1/\left(1 + e^{-ca_{ijm}}\right)$, where c is a specific constant, since the density function of the normal distribution cannot be integrated in closed form, while that of the logistic function can, and consequently is easier to handle. We use $c = 1/0.569$ that is shown by Jeffres [26] to give a better approximation than $c = \pi/3^{1/2}$ used by Takane [50]. Then the first factor on the right-hand side of (14) is given by

$$\frac{\partial F_{ijm}}{\partial a_{ijm}} = -F_{ijm}\left(F_{ijm} - 1\right)c \tag{6.15}$$

Finally, we need to evaluate $\partial a_{ijm}/\partial \theta_l$ for $l = 1, \ldots, N_p$.

In updating $\boldsymbol{\theta}$ in (6.12), it is important to choose appropriate initial values. We may use estimates of the SR model for this purpose. We may apply an appropriate method to $G^* = \left\{g_{ij}^*\right\}$, where g_{ij}^* is the estimate of g_{ij}. The solution can be viewed as a set of reasonable initial values of these parameters. The b_m's (or α and β) and σ estimated with the SR model are also viewed as their own reasonable initial values in the OI model estimation. Therefore, it might be useful to conduct the two-step procedure, that is, to deal with the SR model estimation first, and then to use the result in the subsequent OI model estimation.

6.3 Tests for Symmetry

In this section we shall introduce a special test for conditional symmetry and two variants of symmetry tests. All of them are introduced to verify that the data are sufficiently asymmetric to apply any asymmetric MDS.

We show the asymptotic χ^2 statistics for these tests in order to clarify the distinction between the traditional symmetry hypotheses and ours. However, in the actual application we recommend to use AIC as better alternatives, for reasons which we have already pointed out in Sect. 6.1. In this case we may compare values of the criteria under these hypotheses simultaneously.

6.3.1 A Symmetry Test Prior to Fitting the Asymmetric Model

In this subsection we propose a test for conditional symmetry. It enables us to diagnose the validity of applying any asymmetric model to the set of asymmetric data under study. Since we use the method of successive categories, subjects' proximity judgments from objects to objects can be compiled into a special three-way contingency table, that is, the $n \times n \times M$ table.

Figure 6.1 shows two different arrangements of the original data, which correspond to two formally different sampling designs. Type A design assumes that each row $(Y_{ij1}, Y_{ij2}, \cdots, Y_{ijM})$ follows a multinomial distribution with parameter n_{ij} which is fixed and certain cell probabilities $p_{ij1}, p_{ij2}, \cdots, p_{ijM}$. This design is the sampling design used in our ASYMMAXSCAL as well as MAXSCAL by Takane [50].

In Type A design it is not difficult to show that under the "special" conditional symmetry hypothesis,

$$H_0^{(cs)} : p_{ijm} - p_{jim} = 0, (1 \le i < j \le n; 1 \le m \le M - 1), \tag{6.16}$$

the likelihood ratio test statistic is written as

$$G^2 = 2\sum_m^M \sum_i^n \sum_j^n Y_{ijm} \left\{ \ln Y_{ijm} - ln \left[\frac{n_{ij}(Y_{ijm} + Y_{jim})}{n_{ij} + n_{ji}} \right] \right\}, \tag{6.17}$$

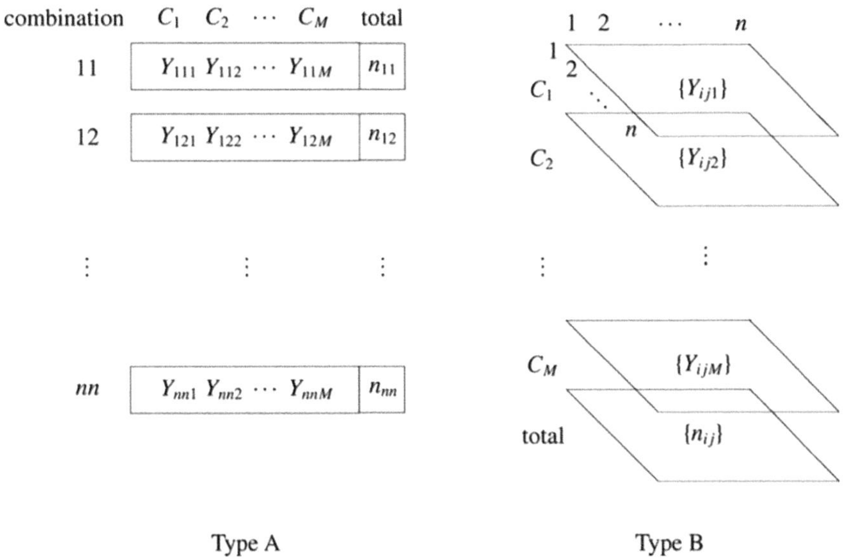

Fig. 6.1 Two types of sampling designs for the method of successive categories in ASYMMAXSCAL

which asymptotically follows the central χ^2-distribution with $\nu_{CS} = (M-1)n(n-1)/2$ degrees of freedom under $H_0^{(cs)}$.. Takane suggested this statistic (personal communication [51]).

If the conditional symmetry hypothesis stated above is accepted, we should abandon the application of any of the extant asymmetric MDS's including ASYMMAXSCAL-OI. If it is rejected, one can proceed to the application of ASYMMAXSCAL-OI to the data.

By contrast, the Type B design data which is obtained by rearranging the Type A design data looks like the data extracted from one of the traditional designs for the special three-way contingency table, that is, the $n \times n \times M$ table (Agresti [1]; Bishop et al. [6]; Read [39]), but is different from any of these traditional designs. By the traditional designs we mean (1) the multiplicative Poisson model with no fixed marginals, (2) the multinomial model with the fixed total, and (3) the product (or multiplicative) multinomial model with the layer totals fixed. We call them the Type I design, the Type II design, and the Type III design for the $n \times n \times M$ table, respectively.

In order to clarify the distinction between the Type B design and the three traditional designs, let us look at the Type I design. The Type B design follows from the Type I design by conditioning on $Y_{11\cdot} = n_{11}, \cdots, Y_{nn\cdot} = n_{nn}$, where the symbol '•' indicates the summation over all categories. It is also the case that the joint likelihood of the data, $Y_{111}, \cdots, Y_{nnM}$ of the Type B design can be converted into that of the Type A design by putting $p_{ijm} = \lambda_{ijm}/\lambda_{ij\cdot}$, where $\lambda_{ijm} = E(Y_{ijm})$ is a Poisson parameter in the Type I design and p_{ijm} is the multinomial parameter in the Type A design. Thus, the conditional symmetry hypothesis, $H_0^{CS} : p_{ijm} - p_{jim} = 0$, in the Type A design is equivalent to the hypothesis, $H_0^B : \lambda_{ijm}/\lambda_{ij\cdot} - \lambda_{jim}/\lambda_{ji\cdot}$, in the Type B design.

Furthermore, (6.17) is equivalent in form to the likelihood ratio χ^2 statistic for the traditional conditional symmetry hypothesis if and only if $n_{ij} = n_{ji}$ for all combinations of i and j. The degrees of freedom are $n(n-1)/2$ less than those for the traditional hypotheses.

The conditional symmetry test just introduced assumes none of the models for asymmetric scaling. ASYMMAXSCAL enables us to test two different symmetry hypotheses, both of which assume the error model and the response model. We shall discuss them in the next subsection.

6.3.2 Two Symmetry Tests in the Scaling Step

One symmetry test in the scaling step is the test for the symmetry hypothesis based on the SR model. This hypothesis is defined by

$$H_0^{(s/sr)} : g_{ij} - g_{ji} = 0, (1 \le i \le j \le n), \tag{6.18}$$

We test $H_0^{(s/sr)}$ using $\boldsymbol{\theta}^*$, the estimate of $\boldsymbol{\theta}$. The Wald statistic (Wald [58]; Aitchison and Silvey [2]) for $H_0^{(s/sr)}$ is written as,

$$Z_{s/sr} = \boldsymbol{h}(\boldsymbol{\theta}^*)^t \left\{ \boldsymbol{H}(\boldsymbol{\theta}^*)^t \boldsymbol{I}(\boldsymbol{\theta}^*)^{-1} \boldsymbol{H}(\boldsymbol{\theta}^*) \right\}^{-1} \boldsymbol{h}(\boldsymbol{\theta}^*), \tag{6.19}$$

which asymptotically follows the central χ^2-distribution with $\nu_{s/sr} = n(n-1)/2$ degrees of freedom under $H_0^{(s/sr)}$. Here, $\boldsymbol{h}(\boldsymbol{\theta}) = \left(h_1, \cdots, h_{\nu_{s/sr}}\right)^t$ is the vector of $g_{ij} - g_{ji}$, and $\boldsymbol{H}(\boldsymbol{\theta})$ is the $\left(N_p - 2\right) \times \nu_{s/sr}$ Jacobian matrix whose (i, j) th element is $\partial h_j / \partial \theta_i$. We remove σ and one of the b_m's or β from $\boldsymbol{\theta}^*$ in order to make $\boldsymbol{I}\left(\boldsymbol{\theta}^*\right)$ nonsingular in computing (6.19). The other symmetry test in the scaling step is the test for the symmetry hypothesis based on the OI representation model. This hypothesis is represented by

$$H_0^{(s/oi)} : r_i - r_{i+1} = 0, (1 \le i \le n-1). \tag{6.20}$$

If we use the Wald statistic as in (6.19), $Z_{s/oi}$ may be affected due to the nonuniqueness of the OI solution, and may become more complicated than that of the SR solution. An alternative method is to use the likelihood ratio statistic, $-2 \ln\left[L^{(s/oi)} / L^{(oi)}\right]$, which asymptotically follows the central χ^2 -distribution with $\nu_{s/oi} = n - 1$ degrees of freedom under $H_0^{(s/oi)}$. Here, $L^{(oi)}$ is the likelihood of the OI model, and $L^{(s/oi)}$ is that of the OI model under $H_0^{(s/oi)}$. The latter representation model is equivalent to the Euclidean distance model, $g_{ij} = d_{ij}$.

6.4 Model Selection

In this section we shall discuss the model selection method in ASYMMAXSCAL-OI which uses an information criterion. Various alternative criteria have been proposed since AIC was introduced by Akaike [3]. Which of these criteria should we choose? Winsberg and Carroll [60] chose AIC and BIC, but with symmetric MDS. In general, however, the two criteria do not necessarily lead to the same conclusion about the dimensionality. As Schwarz [47] points out, BIC leans more than AIC towards lower-dimensional models when there are 8 or more observations, and the criteria differ markedly from each other. In fact, Winsberg and Carroll confronted with the difficulty of deciding the dimensionality of their model, and examined the interpretability of their solutions to overcome it.

According to the recent literature on such a conflict between AIC and BIC, it is clear that no model selection criterion with a deterministic penalty can simultaneously enjoy the properties of AIC and BIC (Yang [61], p. 938). Thus, Yang suggests that one should keep the specific objective of inference in mind when conducting model selection. Considering that AIC is minimax-rate optimal for estimating the regression function and BIC is consistent in selecting the true model (Yang [61], p. 937), we may choose AIC if the objective is to choose the best model in the sense of prediction, and may choose BIC if it is to choose the true model. We have chosen AIC since our objective in this paper is not necessarily to choose the true model, but to choose the best model. Takane [50] uses AIC as the model selection method in MAXSCAL.

The AIC associated with a model is defined by

$$AIC = -2\ln L + 2\nu_p,$$

where L is the maximum likelihood of the model and ν_p is the effective number of parameters in the model. The model which gives the minimum AIC value is considered the best fitting model.

Table 6.1 shows the various candidate models and the corresponding hypotheses as well as ν_p's of the models. The second, fourth, and sixth models pertain to various symmetry hypotheses, while the other models do not. In ASYMMAXSCAL we shall compute AIC's of all of these models, and compare these values for choosing the most appropriate one among the models.

6.5 Example

We will show the result of a small Monte Carlo study to evaluate the performance of parameter recovery. We will also show the result of an application to the friendship data among nations in East Asia and the USA.

6.5.1 Monte Carlo Study

A small Monte Carlo study was executed in order to investigate parameter recovery with the present method for the OI model. We generated true coordinates and radii for six objects in a two-dimensional space from uniform random values, first. As in Okada and Imaizumi [36], the dispersion of constructed g_{ij}'s which are defined by (6.1) is written as

$$\sum\sum_{i\neq j}(g_{ij}-\overline{g})^2 = \sum\sum_{i\neq j}\left(d_{ij}-\overline{d}\right)^2 + 2n^2\, var(r_i), \tag{6.21}$$

where $\overline{g}$ and $\overline{d}$ are the means of g_{ij}'s and d_{ij}'s, respectively. The first and the second terms on the right-hand side of (6.21) pertain to the symmetric and skew-symmetric components of g_{ij}'s, respectively. Although the ratio of the second term to the first term might affect the parameter recovery, we adjusted the true radii so that it takes 0.5, considering Okada and Imaizumi's results.

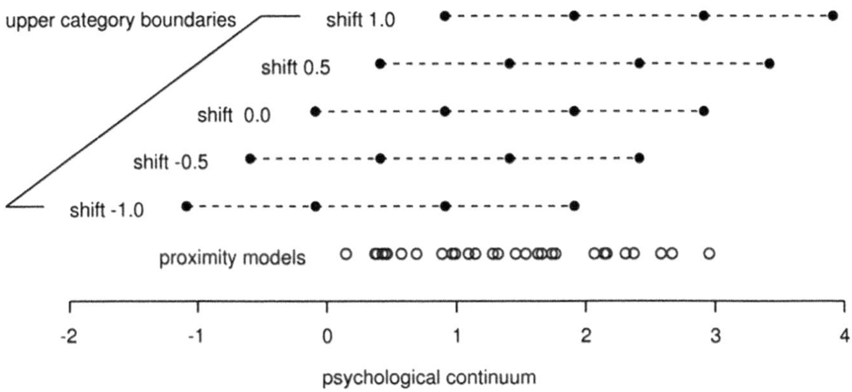

Fig. 6.2 Locations of the true category boundaries and the true proximity models on a psychological continuum used in the Monte Carlo study

Figure 6.2 illustrates the locations of the true category boundaries and the true proximity models on the psychological continuum used in the Monte Carlo study. The true category boundaries in the "shift zero" case were determined so that their mean was equivalent to $\overline{g}$ and they were equally spaced with the unit interval in both of ordinal scale trials and interval scale trials. In other cases, the true category boundaries were shifted $-$ 1.0, $-$ 0.5, 0.5, and 1.0 from the shift zero case.

We generated normal random values with mean zero and variance one (the true σ^2), and τ_{ij}'s according to (6.2). The number of samples per single proximity judgment was varied 10, 30, and 50. For each combination, thirty replications were run. We have assumed that the true dimensionality and the true scale level were known a priori.

We computed the product moment correlation coefficients between the recovered parameters and the true ones as indices of goodness of recovery. The coefficients between the g_{ij}'s constructed from the recovered parameters and the true ones were also computed. The coefficients for category boundaries were not computed in the interval scale trials, because they take the value one for any estimate. Schönemann and Carroll's [46] algorithm for a Procrustes problem was applied to the recovered coordinates in order to fit the true ones before computing the coefficients. The indeterminacies of the other parameters are concerned with their scales and/or locations, which do not affect the coefficients. The mean indices were obtained by converting these coefficients to z values, obtaining mean z, and converting back to the product moment correlation coefficient, as in Okada and Imaizumi [37].

Table 6.2 indicates these mean indices for each combination. As regard the category boundaries in the ordinal scale trials almost complete recoveries were attained. It might be due to their two-step estimation, that is, b_m's (or α and β) and σ were estimated with the SR model, first, and then their estimates were used as their initial values in the subsequent OI model estimation. The indices of the other parameters increased as the sample size increased. Assigning thirty samples per single proximity

Table 6.2 Summary of a Monte Carlo study for the OI model

	Sample size	Ordinary scale, Bounderies' shift					Interval scale, Boundaries' shift				
		− 1.0	− 0.5	0.0	0.5	1.0	− 1.0	− 0.5	0.0	0.5	1.0
Proximity models	10	#	0.956	0.960	0.956	#	#	0.964	0.955	0.965	0.958
	30	0.987	0.985	0.987	0.986	0.986	0.983	0.986	0.986	0.985	0.985
	50	0.992	0.992	0.993	0.992	0.991	0.992	0.992	0.992	0.992	0.991
Coordinates	10	#	0.970	0.976	0.976	#	#	0.978	0.986	0.980	0.972
	30	0.992	0.990	0.993	0.991	0.990	0.990	0.991	0.992	0.991	0.992
	50	0.995	0.996	0.996	0.996	0.996	0.996	0.996	0.996	0.995	0.996
Radii	10	#	0.973	0.972	0.972	#	#	0.975	0.974	0.975	0.968
	30	0.991	0.989	0.988	0.992	0.991	0.982	0.991	0.990	0.992	0.989
	50	0.994	0.994	0.994	0.995	0.993	0.993	0.995	0.993	0.993	0.993
Category boundarries	10	#	0.999	0.999	0.999	#	–	–	–	–	–
	30	0.999	1.000	1.000	1.000	1.000	–	–	–	–	–
	50	1.000	1.000	1.000	1.000	1.000	–	–	–	–	–

Note The symbol '#' indicates that inappropriate solutions were obtained

judgment seems sufficient to recover the true parameters satisfactorily. However, inappropriate solutions giving some negative p_{ijm}'s were also obtained in the special cases when the true boundaries were shifted − 1.0 and ten samples were assigned per single proximity judgment under both of the ordinal scale and the interval scale conditions, and in the case when the true boundaries were shifted 1.0 and ten samples were assigned per single proximity judgment under the interval scale condition. In such a situation, the τ_{ij}'s associated with the extreme values of g_{ij}'s on the psychological continuum might fall in C_1 or C_M in most cases, which may not provide enough information for estimation.

6.5.2 Analysis of National Friendship Data

We will illustrate the application of ASYMMAXSCAL-OI to the friendship data among nations in East Asia and the USA. Four hundred Japanese university students participated in the survey. The single-judgment sampling was chosen to make the independence assumption among all judgments more plausible. That is, each of them rated the extent to which he or she felt the government of nation A be friendly or hostile to that of nation B on a 5-point rating scale ranging from "very friendly" to "very hostile" for a randomly assigned combination of A and B among China, Japan, North Korea, South Korea, and the USA. Twenty students were assigned per single proximity judgment except for the so-called self-proximities.

Although the emphasis in this paper is not on the statistical tests for symmetry but on the diagnosis about candidate models including those with hypotheses about these tests, we first referred to the results of symmetry tests developed in this paper. Table 6.3 shows this. It is apparent that all the symmetry hypotheses are rejected at the significance level beyond the 0.01 level. These results indicate that the data are sufficiently asymmetric.

Table 6.4 shows the summary statistics for the national friendship data. According to AIC, the OI model, especially the three-dimensional OI model under the ordinal scale assumption, is the most realistic model.

Table 6.3 The results of the symmetry tests for the national friendship data

Symmetry test	Dimensionality	Ordinal scale		Interval scale	
		χ^2	d.f	χ^2	d.f
LR test for $H_0^{(s/oi)}$	1	34.99	4	33.55	4
	2	35.24	4	33.81	4
	3	37.10	4	34.87	4
	4	37.10	4	34.87	4
	–	44.46	10	43.03	10
Wald test for $H_0^{(s/sr)}$	–	χ^2		d.f	
LR test for $H_0^{(cs)}$	–	77.21		40	

Table 6.4 Summary statistics for the national friendship data

Candidate model	Dimensionality	Ordinal scale		Interval scale	
		AIC	ν_p	AIC	ν_p
The OI model	1	422	12	32.13	10
	2	6.21	15	34.33	13
	3	1.84	17	30.71	15
	4	3.84	18	32.71	16
The OI model under $H_0^{(s/oi)}$	1	31.21	8	57.68	6
	2	33.45	11	60.14	9
	3	30.95	13	57.58	11
	4	32.95	14	59.58	12
The SR model	–	3.83	23	32.13	21
The SR model under $H_0^{(s/sr)}$	–	30.96	13	57.58	11
		AIC		ν_p	
The saturated model	–	58.83		80	
The saturated model under $H_0^{(cs)}$	–	48.04		40	

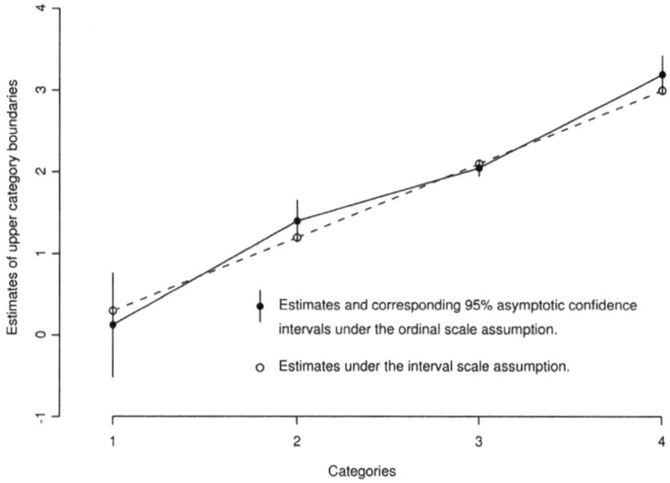

Fig. 6.3 Estimates of category boundaries with 95% asymptotic confidence intervals obtained from the three-dimensional solution for the OI model

Figure 6.3 displays the estimated category boundaries with 95% asymptotic confidence intervals in the three-dimensional solution for the OI model. The categories C2 and C4 (labeled as "friendly" and "hostile", respectively) seem to have larger intervals than that of C3 (labeled as "neither friendly nor hostile") under the ordinal scale assumption.

It should be noted that these confidence intervals are point-wise. Although none of the confidence intervals indicate significant departures, the joint departure (all of them combined) indicates a significant departure. That is, small piece-wise departures combine into a significant overall departure. This is confirmed by the fact that AIC indicates that the ordinal scale model is better than the interval scale model.

Figure 6.4 depicted the configuration and the spheres in the three-dimensional solution for the OI model under the ordinal scale assumption. The principal-axis rotation was performed on the configuration. The sphere associated with Japan is larger than those of the other nations, showing that the friendship of the Japanese government is generally unilateral. That is, Japan is perceived as more friendly to other nations than the other way round. It should be noticed that the spheres associated with all other nations but Japan are very small. These results indicate that most of the skew-symmetric part of the data seems to relate to Japan, and the relationships among the other nations are relatively symmetric.

On the first dimension, the allies of Japan and the USA are located opposite to North Korea, while South Korea and China are located about the middle but slightly closer to these allies and North Korea, respectively. This dimension seems to show the antagonism between Japan–US allies and North Korea. The second dimension seems to represent certain active opposition between the two nations, i.e., Japan and

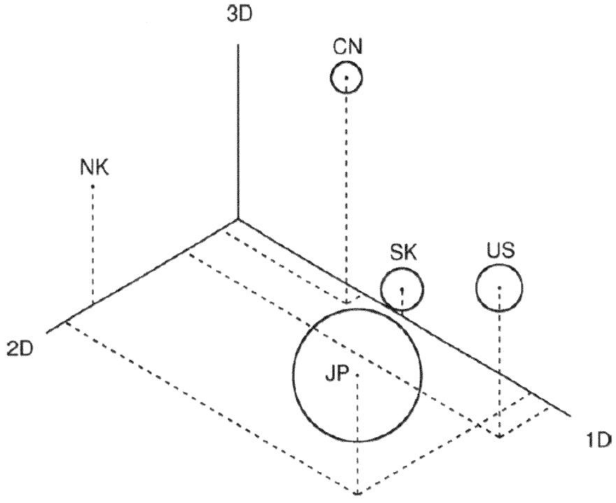

Fig. 6.4 Configuration and spheres obtained from the three-dimensional solution for the OI model under the ordinal scale assumption. Abbreviations indicate the following nations: CN = China, JP = Japan, NK = North Korea, SK = South Korea, and US = USA

North Korea, and the other nations. The third dimension seems to represent some antagonism of China toward South Korea.

Another analysis was also done using the representation model by Saito [43],

$$g_{ij} = d_{ij} + \gamma_i + \emptyset_j. \tag{6.22}$$

This model is more flexible than the OI model due to the biases which enables us to represent both of symmetric and skew symmetric components of g_{ij}, but has larger number of parameters. The minimum value of AIC among models pertaining to the Saito's model was 1005.85 with the two-dimensional case under the ordinal scale assumption. It was larger than 1001.84 with the optimal model described above, i.e., the three-dimensional OI model under the ordinal scale assumption.

6.6 Discussion

In this paper we proposed a maximum likelihood asymmetric MDS for the OI model. Our method provides an alternative to Kruskal's [29, 30] traditional nonmetric approach taken by Okada and Imaizumi [36, 37], to the asymmetric ordinal relational data. Our method can potentially be more useful than their method since it provides information other than the configuration of objects, e.g., the most appropriate dimensionality of the model based on AIC, which is more sophisticated than a descriptive statistic such as stress (Kruskal [30]; Kruskal and Carroll [31]. It also provides information as to the reliabilities of point locations in the form of confidence regions. It may be argued that such confidence regions can be derived in any descriptive method by the use of a bootstrap or some other resampling technique (e.g., De Leeuw and Meulman [16]). However, developing such a procedure for MDS is not as straightforward as in other situations. Most of the extant MDS assumes no replicated observations. Moreover, it enables us to test the validity of the assumption of asymmetry statistically using several variants of symmetry hypotheses.

The emphasis in our paper is not on hypothesis testing, but on model diagnosis using AIC. If it is of interest, however, we could also examine possible causes of asymmetry in the Type B design data, prior to fitting the asymmetric MDS model. In fact, Chino and Saburi [12] have recently utilized various tests for symmetry and related hypotheses including the quasisymmetry by Caussinus [9], the marginal homogeneity by Stuart [49], the quasi-independency by Goodman [20], and so on, given the Type B design data. Such an approach has rarely been fully discussed in the context of asymmetric MDS.

Chino and Saburi [12] have suggested that the conditional quasi-symmetry hypothesis plays a fundamental role in such an analysis. One of the major reasons is that the notion of quasi-symmetry is more general than that of symmetry. Another reason is that this notion is closely related to the basic feature of some of the extant asymmetric MDS models, as De Rooij and Heiser [17] has pointed out. Although they only deal with count data, the Type A design data are not confined to count data. Therefore, Chino and Saburi's idea can be viewed as an extension of De Rooij and Heiser [17, 18] in this respect.

We may apply the fundamental ideas and the method proposed in this paper to any of the extant asymmetric MDS models. Some unified algorithm based on these ideas and method for extant asymmetric MDS models may be fruitful and attractive as a topic for future research.

Acknowledgements The authors are indebted to Yoshio Takane, Yasunori Fujikoshi, and anonymous reviewers for their valuable suggestions and comments on earlier versions of this paper. The authors are also indebted to Gregory L. Rohe for proofreading an earlier version of this paper.

Notes

This chapter is the reproduction of Saburi and Chino [42].

References

1. Agresti, A. (2002). *Categorical Data Analysis*. Wiley.
2. Aitchison, J., & Silvey, S. D. (1960). Maximum-likelihood estimation procedures and associated tests of significance. *Journal of the Royal Statistical Society Series B, 22*, 154–171.
3. Akaike, H. (1973). Information theory and an extension of the maximum likelihood principle. In Petrov,B.N., Csaki, F. (Eds.), *The Second International Symposium on Information Theory*. Budapest: Akadémiai Kiadó, pp. 267–281.
4. Ashby, F. G., & Perrin, N. A. (1988). Toward a unified theory of similarity and recognition. *Psychological Review, 95*, 124–150.
5. Birch, M. W. (1963). Maximum likelihood in three-way contingency tables. *Journal of the Royal Statistical Society Series B, 25*, 220–233.

6. Bishop, Y. M. M., Fienberg, S. E., & Holland, P. W. (1975). *Discrete Multivariate Analysis—Theory and Practice*. The MIT press.
7. Bock, R. D., & Jones, L. V. (1968). *The Measurement and Prediction of Judgment and Choice*. Holden-Day.
8. Bowker, A. H. (1948). A test for symmetry in contingency tables. *Journal of the American Statistical Association., 43*, 572–574.
9. Caussinus, H. (1965). Contribution à l'analyse statistique des tableaux de corrélation [Contributions to the statistical analysis of correlation matrices]. *Annales de la Faculté des Sciences de l'Université de Toulouse, 29*, 77–182.
10. Chino, N. (1978). A graphical technique for representing the asymmetric relationships between N objects. *Behaviormetrika, 5*, 23–40.
11. Chino, N. (1990). A generalized inner product model for the analysis of asymmetry. *Behaviormetrika, 27*, 25–46.
12. Chino, N., Saburi, S. (2006). Tests of symmetry in asymmetric MDS. *Paper Presented at the 2nd German Japanese Symposium on Classification*. Berlin, March.
13. Chino, N., & Shiraiwa, K. (1993). Geometrical structures of some non-distance models for asymmetric MDS. *Behaviormetrika, 20*, 35–47.
14. Constantine, A. G., & Gower, J. C. (1978). Graphical representation of asymmetric matrices. *Applied Statistics, 27*, 297–304.
15. Cramér, H. (1946). *Mathematical Methods of Statistics*. Princeton University Press.
16. De Leeuw, J., & Meulman, J. (1986). A special jackknife for multidimensional scaling. *Journal of Classification, 3*, 97–112.
17. De Rooij, M., & Heiser, W. J. (2003). A distance representation of the quasi-symmetry model and related distance models. In H. Yanai, A. Okada, K. Shigemasu, Y. Kano, & J. J. Meulman (Eds.), *New Developments in Psychometrics* (pp. 487–494). Springer.
18. De Rooij, M., & Heiser, W. J. (2005). Graphical representations and odds ratios in a distance-association model for the analysis of cross-classified data. *Psychometrika, 70*, 99–122.
19. Escoufier, Y., & Grorud, A., et al. (1980). Analyse factorielle des matrices carrees non symetriques [Factor analysis of square asymmetric matrices]. In E. Diday (Ed.), *Data Analysis and Informatics* (pp. 263–276). North-Holland.
20. Goodman, L. A. (1968). The analysis of cross-classified data: Independence, quasi-independence, and interactions in contingency tables with or without missing entries. *Journal of the American Statistical Association, 63*, 1091–1131.
21. Goodman, L. A. (1985). The analysis of cross-classified data having ordered and/or unordered categories: Association models, correlation models, and asymmetry models for contingency tables with or without missing entries. *Annals of Statistics, 13*, 10–69.
22. Gower, J. C. (1977). The analysis of asymmetry and orthogonality. In J. R. Barra, F. Brodeau, G. Romer, & B. van Cutsem (Eds.), *Recent Developments in Statistics* (pp. 109–123). North-Holland.
23. Harshman, R. A. (1978). Models for analysis of asymmetrical relationships among N objects or stimuli. *Paper presented at the First Joint Meeting of the Psychometric Society and the Society for Mathematical Psychology*. Hamilton, Canada.
24. Harshman, R. A., Green, P. E., Wind, Y., & Lundy, M. E. (1982). A model for the analysis of asymmetric data in marketing research. *Marketing Science, 1*, 205–242.
25. Holman, E. W. (1979). Monotonic models for asymmetric proximities. *Journal of Mathematical Psychology, 20*, 1–15.
26. Jeffress, L. A. (1973). The logistic distribution as an approximation to the normal curve. *The Journal of the Acoustical Society of America, 53*, 1296.
27. Kiers, H. A. L., & Takane, Y. (1994). A generalization of GIPSCAL for the analysis of nonsymmetric data. *Journal of Classification, 11*, 79–99.
28. Krumhansl, C. L. (1978). Concerning the applicability of geometric models to similarity data: The interrelationship between similarity and spatial density. *Psychological Review, 85*, 445–463.

29. Kruskal, J. B. (1964). Multidimensional scaling by optimizing goodness of fit to a nonmetric hypothesis. *Psychometrika, 29*, 1–27.
30. Kruskal, J. B. (1964). Nonmetric multidimensional scaling: A numerical method. *Psychometrika, 29*, 115–129.
31. Kruskal, J. B., & Carroll, J. D. (1969). Geometrical models and badness-of-fit functions. In P. K. Krishnaiah (Ed.), *Multivariate Analysis II* (pp. 639–671). Academic Press.
32. Luce, R. D. (1959). *Individual Choice Behavior*. Wiley.
33. Luce, R. D. (1963). Detection and recognition. In R. D. Luce, R. R. Bush, & E. Galanter (Eds.), *Handbook of Mathematical Psychology* (Vol. 1, pp. 103–189). Wiley.
34. McNemar, Q. (1947). Note on the sampling error of the difference between correlated proportions or percen-tages. *Psychometrika, 12*(153–157), 35.
35. Nosofsky, R. M. (1991). Stimulus bias, asymmetric similarity, and classification. *Cognitive Psychology, 23*,
36. Okada, A., & Imaizumi, T. (1987). Nonmetric multidimensional scaling of asymmetric proximities. *Behaviormetrika, 21*, 81–96.
37. Okada, A., & Imaizumi, T. (1997). Asymmetric multidimensional scaling of two-mode three-way proximities. *Journal of Classification, 14*, 195–224.
38. Ramsay, J. O. (1980). Some small sample results for maximum likelihood estimation in multidimensional scaling. *Psychometrika, 45*, 139–144.
39. Read, C. B. (1978). Tests of symmetry in three-way contingency tables. *Psychometrika, 43*, 409–420.
40. Rocci, R., & Bove, G. (2002). Rotation techniques in asymmetric multidimensional scaling. *Journal of Compu-tational and Graphical Statistics, 11*, 405–419.
41. Rushen, J. (1982). The peck orders of chickens: How do they develop and why are they linear? *Animal Beha-vior, 30*, 1129–1137.
42. Saburi, S., & Chino, N. (2008). A maximum likelihood method for an asymmetric MDS model. *Computational Statistics and Data Analysis, 52*, 4673–4684.
43. Saito, T. (1991). Analysis of asymmetric proximity matrix by a model of distance and additive terms. *Beha-viormetrika, 29*, 45–60.
44. Saito, T., & Takeda, S. (1990). Multidimensional scaling of asymmetric proximity: Model and method. *Behaviormetrika, 28*, 49–80.
45. Sato, Y. (1988). An analysis of sociometric data by MDS in Minkowski space. In K. Matsushita (Ed.), *Statistical Theory and Data Analysis II* (pp. 385–396). North-Holland.
46. Schönemann, P. H., & Carroll, R. M. (1970). Fitting one matrix to another under choice of a central dilation and a rigid motion. *Psychometrika, 35*, 245–255.
47. Schwarz, G. (1978). Estimating the dimension of a model. *Annals of Statistics, 6*, 461–464.
48. Smith, J. E. K. (1980). Models of identification. In: Nickerson, R. (Ed.), *Attention and Performance Vol.3*. Hillsdale: Erlbaum, NJ.
49. Stuart, A. (1955). A test for homogeneity of the marginal distributions in a two-way classification. *Biometrika, 42*, 412–416.
50. Takane, Y. (1981). Multidimensional successive categories scaling: A maximum likelihood method. *Psychometrika, 46*, 9–28.
51. Takane, Y. (2005). *Personal communication*, August 26.
52. Takane, Y., & Shibayama, T. (1986). Comparison of models for stimulus recognition data. In J. de Leeuw, W. Heiser, J. Meulman, & F. Critchley (Eds.), *Multidimensional Data Analysis* (pp. 119–138). DSWO Press.
53. Ten Berge, J. M. F. (1997). Reduction of asymmetry by rank-one matrices. *Computational Statistics and Data Analysis, 24*, 357–366.
54. Tobler, W. (1976–77). Spatial interaction patterns. *Journal of Environmental Systems, 6*, 271–301.
55. Torgerson, W. S. (1958). *Theory and Methods of Scaling*. Wiley.
56. Townsend, J. T., & Landon, D. E. (1983). Mathematical models of recognition and confusion in psychology. *Mathematical Social Sciences, 4*, 25–71.

57. Trendafilov, N. T. (200). GIPSCAL revisited. A projected gradient approach. *Statistics and Computing, 12*, 135–145.
58. Wald, A. (1943). Tests of statistical hypotheses concerning several parameters when the number of observant ions is large. *Transactions of the American Mathematical Society, 54*, 426–482.
59. Weeks, D. G., & Bentler, P. M. (1982). Restricted multidimensional scaling models for asymmetric proximities. *Psychometrika, 47*, 201–208.
60. Winsberg, S., & Carroll, J. D. (1989). A quasi-nonmetric method for multidimensional scaling via an extended Euclidean model. *Psychometrika, 54*, 217–229.
61. Yang, Y. (2005). Can the strengths of AIC and BIC be shared? A conflict between model identification and regression estimation. *Biometrika, 92*, 937–950.
62. Young, F. W. (1975). An asymmetric Euclidian model for multi-process asymmetric data. Paper presented at US-Japan Seminar on MDS, San Diego, USA. pp. 79–88.
63. Zielman, B., & Heiser, W. J. (1996). Models for asymmetric proximities. *Br. J. Math. Statist. Psychol., 49*, 127–146.

Chapter 7
Brief Review of Asymmetric MDS

7.1 Introduction

Asymmetric relationships among objects are frequently observed in daily life as well as in designed experiments. They are diverse in character. One-sided love and hate among members of any informal groups are typical examples. Table 7.1 shows a sociomatrix among 10 male students in a senior-high school [25]. A group of hens and cocks shows a special asymmetric relationship called pecking order among the group (e.g., Rushen [157]). In recognition experiments the proportion of times response R_j was made when stimulus S_i was presented is not necessarily equal to the proportion of response R_i when stimulus S_j was presented (e.g., Nakatani [125]). Number of references from one journal to another also gives rise to an asymmetric data matrix (e.g., Coombs [64]). Amount of trade each nation in the world has with other nations sometimes shows huge imbalance. Table 7.2 is a set of trade data among 10 countries (including two regions) [33], which was obtained from *Statistical Yearbook 1974* [206]. Amount of migration from one region to another is a typical example in geography [197]. The number of browsing times as well as the spread of computer viruses (e.g., Balthrop et al. [10]) from one site to another on the Internet also yields asymmetric relations. The asymmetric *coupling* matrix defined in a dynamical network consisting of identical nodes is a typical example in neural networks (e.g., Amari [3]; Fukai & Shiino [67]; Kanter [96]; Kree & Zippelius [103]; Li [118]; Parisi [147]).

In general, these asymmetric relationships observed at a specific point in time or during a certain period of time are summarized in a square matrix whose elements are similarity or dissimilarity measures between objects. Such a matrix is characterized as a *one-mode, two-way square asymmetric matrix*. In contrast, if such a matrix is obtained at several points in time or during several periods of time, the data matrix is represented by a special two-mode, three-way matrix which is composed of a set of one-mode, two-way square asymmetric matrices. Sometimes such a matrix is observed per individual, and yields the same two-mode, three-way matrix.

N. Chino, *Structure and Dynamics of Asymmetric Interactions*,
https://doi.org/10.1007/978-981-97-8269-7_7

Table 7.1 Sociometric data of 10 students in a senior-high school

Rater\Ratee	1	2	3	4	5	6	7	8	9	10
1	1	3	4	3	5	5	6	4	7	7
2	2	3	4	6	7	6	5	5	4	5
3	4	4	3	5	5	4	4	3	4	5
4	4	7	6	3	7	7	4	6	4	5
5	1	7	6	7	4	7	6	6	6	5
6	4	5	4	6	5	7	4	4	4	4
7	4	5	4	4	3	3	6	6	4	6
8	2	4	4	4	5	4	5	2	4	4
9	6	r;	5	5	5	6	5	5	4	6
10	4	4	4	3	4	4	4	4	4	4

Reprinted from *Behaviormetrika*, ***39***, Table 1, Chino [33], A brief survey of asymmetric MDS and some open problems, pp. 127–165, with permission of The Behaviormetric Society

Table 7.2 Trade data between 10 countries including two regions

From\To	EC	EA	US	JN	AN	AA	ME	OA	UR	EP
EC		25.67	15.67	2.84	2.42	9.62	5.59	5.07	2.66	6.49
EA	17.99		2.43	0.70	0.35	1.33	0.76	0.82	0.78	1.51
US	16.38	2.23		8.18	1.68	1.51	2.21	6.85	1.19	0.60
JN	4.40	1.32	9.55		1.46	2.48	1.61	8.93	0.49	0.32
AN	1.70	0.12	1.14	3.08		0.08	0.16	1.01	0.24	0.13
AA	11.60	0.88	1.92	0.98	0.04		0.20	0.35	0.63	0.67
ME	10.57	1.21	1.27	5.06	0.33	0.57		2.22	0.35	0.23
OA	5.33	0.46	6.94	6.96	0.79	0.63	0.59		0.62	0.36
UR	2.88	1.04	0.19	0.84	0.01	0.66	0.67	0.45		10.03
EP	5.38	1.49	0.33	0.14	0.55	0.63	0.64	0.30	10.56	

The abbreviated country names, EC, EA, US, JN, AA, ME, OA, UR, and EP, respectively denote EEC, EFTA, USA, Japan, Australia, New Zealand, Africa, Middle East, Other Asia, USSR, and Centrally planned economics

Reprinted from *Behaviormetrika*, ***39***, Table 2, Chino [33], A brief survey of asymmetric MDS and some open problems, pp.127–165, with permission of The Behaviormetric Society

Asymmetric MDS (hereafter abbreviated as AMDS) extends symmetric MDS (hereafter abbreviated simply as MDS) to handle such asymmetric relationships among objects. As for symmetric MDS, Cox and Cox [50] give two definitions, i.e., a *narrow definition* and a *wider definition*. According to their narrow definition, MDS is a search for a low dimensional space, usually Euclidean, in which points in the space represent objects, one for each object, and such that the distances between the points in the space match as well as possible the original dissimilarities. In order to clarify the definition of MDS and also the difference between MDS and AMDS, we shall redefine their narrow definition of MDS, adding a bit stricter constraints, as follows:

The following five conditions are assumed in our narrow definition of MDS. That is,

1. Data matrix is a *one-mode, two-way square symmetric matrix* whose elements are similarity or dissimilarity measures or a special two-mode, three-way matrix which is composed of a set of one-mode, two-way square symmetric matrices.
2. Dissimilarities must be measured at the ordinal level or higher, or measured as the count data.
3. MDS searches for a low dimensional *real metric space*, *Minkowski r-metric* (a special case is Euclidean) in which points representing objects are embedded.
4. Points in the space represent the objects, each representing one object.
5. The distances between the points in the space match as well as possible the original dissimilarities.

Although Cox and Cox refer to neither the data type nor the scale level for MDS, we have added them as conditions 1 and 2. Especially, the reader should pay special attention to the term "count data" in condition 2. Although counting is sometimes viewed as an ordinal scale or higher, we shall clearly distinguish between counting and ordinal scale or higher. In this paper, following Suppes and Zinnes ([179], p.9), we shall use counting as an example of an *absolute scale*, in which there is no arbitrary choice of unit or zero available. In contrast, in the case of, say, a ratio scale, the choice of a unit is an arbitrary decision made by an individual or group of individuals.

Furthermore, if we consider an inferential procedure, the most rational analysis of count data is to treat them, not as, say, measured at an ordinal scale level, but as measured at a ratio scale level and take their natural logarithm further. The reason for this is that (1) multinomial distribution as well as Poisson distribution as standard models for count data belongs to a usual *exponential family*, and can be written in canonical form, (2) canonical parameters of these two distributions are not their population parameters themselves but the natural logarithms of their function or the natural logarithm of the population parameter, and (3) statistics corresponding to these canonical parameters have sufficient information of data (e.g., Andersen [4]).

According to their wider definition, MDS can subsume several techniques of multivariate data analysis. It covers any technique which produces a graphical representation of objects from multivariate data. It is apparent that such techniques do not necessarily satisfy all of the above conditions. Therefore, we shall adopt the narrow definition of MDS in this papar.

Two basic theorems underpinned MDS in the 1930's. These are the Eckart and Young [63] theorem and the Young and Householder [209] theorem. The former is concerned with the lower rank fit to the data matrix S, and the latter with the necessary and sufficient condition that the coordinates of objects in a multidimensional space are real points in the Euclidean space. Drawing upon these theorems, Richardson [155] proposed a method of MDS and Torgerson [199, 200] developed it further (e.g., Tucker & Messick [204]). Their method is now called *classical MDS*. There exists a large body of literature which extends classical MDS. We can divide it first into two, i.e., *descriptive MDS* and *inferential MDS*. We may further divide the latter into three. These are the so-called *special probabilistic MDS*, *maximum likelihood MDS*, and *Bayesian MDS*.

The descriptive MDS is a class of MDS which does not accompany any statistical inferences on population parameters. We shall only refer to representative ones below. The classical MDS was first extended to the case when the dissimilarities are measured at an ordinal scale level by Kruskal [107, 108], which is called nonmetric MDS. Guttman and his colleagues (e.g., Guttman [75]; Lingoes [119]) followed Kruskal, and proposed another method of nonmetric MDS called the smallest space analysis (abbreviated as SSA). Carroll and Chang [21] also extended classical MDS in such a way to handle individual differences in dissimilarity judgments, and therefore their method is called individual differences MDS. Takane et al. [189] proposed another method of individual differences MDS. In contrast with Carroll and Chang' method, their method is applicable to dissimilarity data measured on a wider class of scale levels.

As the special probabilistic MDS, we include a class of MDS methods which assumes a normal distribution on coordinates of objects to be estimated, and as a result assumes a noncentral χ^2 distributionχ^2 with some degrees of freedom and a specified noncentrality parameter on squared distances. Such an MDS technique goes back to Hefner [83], and several papers exist in the literature (e.g., Ramsay [150]; Suppes & Zinnes [179]; Zinnes & Mackay [216]).

The maximum likelihood MDS is another class of inferential MDS techniques, which assumes a normal distribution and/or log-normal distribution on observed dissimilarities (e.g., Ramsay [151–153]; Takane [180, 181, 183]; Takane & Carroll [185]). In this class of MDS methods, it is assumed that dissimilarities are obtained by the pair comparison method or by a certain rating method.

The Bayesian MDS is another inferential MDS, which is based on the Bayesian inference, and has recently been proposed by several researchers (e.g., Fong et al. [66]; Je et al. [93]; Lee [116]; Oh & Raftery [137, 138]; Okada & Mayekawa [145]; Okada & Shigemasu [146]; Park et al. [148]).

As is apparent from the brief review of MDS made above, symmetric MDS has a long history and has been almost fully developed in that both descriptive and inferential methods are now available. Furthermore, major books on MDS which thoroughly review MDS have been published (e.g., Borg & Groenen [19]; Cox & Cox [50]; Saito [159]; Takane [182]). In contrast, such books on AMDS are rare (e.g., Chino [29]), although the books on MDS mentioned above partly introduce AMDS (e.g., Borg & Groenen [19]; Cox & Cox [50]). Recently, Saito and Yadohisa [163] review AMDS fairly extensively. One possible reason for the relative paucity of literature on AMDS may be that it is still an active area of research. As a result, the notion of AMDS is still vague, and it seems that the precise definition of AMDS has not been established. Therefore, it seems appropriate and necessary to give some definition of AMDS, before we make a critical review of AMDS in this paper. In fact, depending on this definition, the history of AMDS will have to focus on different aspects. It might be possible to make a narrow definition of AMDS as well as a wider definition in a manner similar to Cox & Cox [50] for MDS. In this paper, we shall further divide the narrow definition of AMDS into three, i.e., a *narrow definition*, a *narrower definition*, and finally the *narrowest definition*.

The following six conditions are assumed for the narrowest definition of AMDS:

1. Data matrix is a one-mode, *two-way square asymmetric matrix* whose elements are similarity or dissimilarity measures or a special *two-mode, three-way matrix* which is composed of one-mode, two-way square asymmetric matrices.
2. Dissimilarities are measured at the ordinal level or higher.
3. It involves a search for a low dimensional *real metric space*, *complex metric space*, or *real asymmetric metric space*.
4. Points in the space represent the objects, each representing one object. This condition is the same as in MDS.
5. Some additional parameters other than distance are assumed in the case of the real metric space.
6. The distances between the points in the space match as well as possible the original dissimilarities. This condition is the same as condition 5 in MDS.

Condition 2 excludes a body of asymmetric methods for count data (i.e., Chino [29]). As the metric spaces in condition 3, we include the Minkowski r-metric space (the Euclidean space as a special case) for the real metric space, the Hilbert space for the complex metric space (i.e., Chino & Shiraiwa [45]; Saito & Yadohisa [163]), and the asymmetric Minkowski space for a *real asymmetric metric space* (i.e., Sato [164]). This condition excludes a familiar method for the skew-symmetric data, i.e., the *Gower diagram* (or sometimes called the canonical analysis of skew-symmetry (abbreviated as CASK by Chino [29]), because the *area property* in CASK has the symplectic but not Euclidean structure [45].

As will be discussed in detail later, CASK decomposes the skew-symmetric part of a squared asymmetric matrix into the weighted sum of a special quantity (see, (7.20) in Sect. 7.2). It is well known that this quantity is an oriented area of the parallelogram spanned by the two location vectors corresponding to two objects.

It is apparent from the above definition of AMDS that we have inherited the concept of metric space from the narrow definition of MDS discussed previously. As is well known, a metric space is a nonempty set M equipped with a positive real-valued function d: $M \times M \rightarrow R$, called the distance function, that satisfies the following axioms:

1. $\mathrm{d}(X, Y) \geq 0$, and $\mathrm{d}(X, Y) = 0 \leftrightarrow X = Y$ (positive-definite),
2. $\mathrm{d}(X, Y) = \mathrm{d}(Y, X)$ (symmetric),
3. $\mathrm{d}(X, Y) + \mathrm{d}(Y, Z) \geq \mathrm{d}(X, Z)$ (triangle inequality),

for all $X, Y, Z \in \mathrm{M}$.

Minkowski r-metric as well as Hilbert space satisfies all of these axioms, while a more general Minkowski space, which we call here the asymmetric Minkowski space, does not satisfy the second axiom. As defined elsewhere (e.g., Matsumoto [123]; Sato [164]), a Minkowski space M is a finite dimensional real vector space such that the length of a vector $\boldsymbol{x} \in M$ is given by the value $L(\boldsymbol{x})$ of a function L on M, where L is assumed to satisfy the following conditions,

1. $L(\boldsymbol{x}) \geq 0$, for any $\boldsymbol{x}$ M, $L(\boldsymbol{x}) = 0$ if and only if $\mathbf{x} = \mathbf{0}$, (nonnegativity),
2. $L(\alpha\boldsymbol{x}) = \alpha\, L(\boldsymbol{x})$, for any $\alpha > 0$, $\boldsymbol{x}$ M, (positively homogeneous of degree one),
3. $L(\boldsymbol{x} + \boldsymbol{y})$ S $L(\boldsymbol{x}) + L(\boldsymbol{y})$, for any $\boldsymbol{x}$, $\boldsymbol{y}$ M, (convexity),
4. $L(\boldsymbol{x})$ is differentiable at any non-zero $\boldsymbol{x}$.

The function L is called *Minkowski metric function*. If $L(\mathbf{x}) = L(\text{-}\mathbf{x})$ for any $\mathbf{x}$, then the metric is considered to be symmetric.

The *narrower definition* of AMDS, on the other hand, adds the count data in condition 2. This gives a criterion for checking and classifying the seemingly different approaches to an asymmetric dissimilarity data matrix using the notion of quasisymmetry in the log-linear model, as discussed in a later section.

The *narrow definition* of AMDS adds the *indefinite metric space* in condition 3. The narrow definition also adds a *one-mode, three-way square matrix* in condition 1. We shall discuss open problems on these topics in the discussion section.

According to the wider definition of AMDS, it can subsume several techniques of multivariate data analysis, and covers any techniques which produce a graphical representation of objects from multivariate data, as in MDS. It is apparent that such techniques do not necessarily satisfy all of the above conditions for AMDS. Therefore, we shall adopt the narrow definition of AMDS in this paper.

The organization of this paper is as follows. In the next section we shall briefly review a body of AMDS methods in the narrowest sense. Then, we shall briefly review a body of AMDS methods in the narrower sense, which assumes count data, and embeds objects in some metric space, and discuss the implications of checking and classifying various features of these two bodies of AMDS in the third section. In the discussion section, we shall list up the open problems on AMDS in the near future. The final section is a supplementary note on Chino [33], in which we shall list some of the recent developments in asymmetric MDS since 2012.

7.2 A Brief Review of AMDS in the Narrowest Sense

In the narrowest sense, AMDS may go back to the work of Young [208]. In his ASYMSCAL model, the squared distance from object i to object j is defined by the equation

$$d_{ij}^2 = \sum_{a=1}^{r} w_{ia}\left(x_{ia} - x_{ja}\right)^2,\ w_{ia} \geq 0, \tag{7.1}$$

where w_{ia} is the weight of object i on dimension a, x_{ia} is the coordinate of object i on dimension a, and where there are n objects and r dimensions. The distance used in this model is clearly an extension of the familiar Euclidean distance. As a result we need not only the coordinates of objects but also the object weights (which he calls the stimulus weights) in order to explain the asymmetric dissimilarity data.

A number of models have been proposed since Young proposed his ASYMSCAL. As in MDS, we can divide them into two, i.e., the *descriptive AMDS* and the *inferential AMDS*. According to the narrowest definition, almost all the extant AMDS models remain descriptive. Representative methods in this category are Borg and Groenen [19], Chino [25, 27], Chino and Shiraiwa [45], Constantine and Gower [46], Escoufier and Grorud [64], Gower [71], Harshman [77], Harshman et al. [79], Kiers and Takane [101], Krumhansl [105], Loisel and Takane [120], Okada and Imaizumi [142–144], Rocci and Bove [156], Saito [160], Saito and Takeda [162], Sato [164], ten Berge [193], Tobler [197], Trendafilov [202], Weeks and Bentler [210], Yadohisa and Niki [207], Young [208], and Zielman and Heiser [215].

Chino and Okada [41] and Chino [29] further divide the above methods (except for the models proposed after 1996) into three groups, i.e., the *augmented distance model*, the *non-distance model*, and the *extended distance model*. The augmented distance model is a family of AMDS in which some parameters are added to the metric distance between objects in order to handle asymmetry, and includes Borg and Groenen [19], some of Gower's [71] models, Krumhansl [105], Okada and Imaizumi [142–144], Saito [160], Saito and Takeda [162], Tobler [197], Weeks and Bentler [210], Yadohisa and Niki [207], Young [208], and Zielman and Heiser [214]. Young's ASYMSCAL can be classified into this family of models.

Tobler [197] proposed a unique geographical AMDS model. His *wind model* assumes a kind of wind for the observed asymmetry, and estimates it using a *vector field model*. In order to estimate the vector field from the original similarity measures, the following model is assumed

$$t_{ij} = \frac{d_{ij}}{\left(r + c_{ij}\right)}, \; c_{ij} = r\frac{\left(t_{ji} - t_{ij}\right)}{\left(t_{ij} + t_{ji}\right)}, \tag{7.2}$$

where t_{ij} is a travel effort (time, cost, etc.), and it is assumed that t_{ij} is aided by a flow c_{ij} (Tobler denotes it as $\vec{c}_{ij}$) in the direction of movement from place i to place j. Moreover, r is a rate of travel, independent of position and of direction, and is in the same units as c_{ij}. Once the vector field is estimated, it is decomposed into *divergence* and *curl-free parts*, and the *scalar* and *vector potentials* are calculated. Finally, the *gradient vectors* of the scalar potentials are drawn on the estimated configuration, which explain the travel flows. If the distances between two places are known, Tobler's method cannot be classified into AMDS in the narrowest sense, but if it is estimated from the symmetric part of t_{ij} using some appropriate MDS, his method can be viewed as an AMDS.

Yadohisa and Niki [207] proposed a vector field model similar to the Toblers' model. They assume that the locations of objects have already been determined from the symmetric part of the data via some suitable MDS method. Given the configuration of objects, they estimate vectors at those locations as well as the estimated *scalar potentials* from the skew-symmetric part of the data.

Two of the models which Gower [71] proposed, i.e., the *jet-stream model* and the *cyclone model*, are very similar to Tobler's wind model. The jet-stream model was conceived by imaging a plane flying at a constant velocity V between two towns P_i and P_j which are d_{ij} distance apart. If there is a jet-stream, velocity v, making an angle θ_{ij} with line P_i P_j then the flight times t_{ij} and t_{ji} are

$$t_{ij} = d_{ij}/\left(V + v\cos\theta_{ij}\right), t_{ji} = d_{ij}/\left(V - v\cos\theta_{ij}\right) \tag{7.3}$$

As a result, if the ratio v/V is sufficiently small to ignore v^2/V^2, the symmetric part and the skew-symmetric part of t_{ij} can be written as

$$t_{ij}(s) \doteq d_{ij}/V, t_{ji}(sk) \tag{7.4}$$

respectively. In any case, this model is formally the same as that of Tobler, if we reparameterize $c_{ij} = v\ cos\ \theta_{ij}$ and $c_{ji} = -v\ cos\ \theta_{ij}$. Although Gower was interested in analyzing symmetric parts and skew-symmetric parts separately, the jet-stream model may be thought of as an AMDS model if we analyze them simultaneously. Borg and Groenen [19] proposed a similar model to the jet-stream model in that both symmetric parts and skew-symmetric parts are analyzed simultaneously. It is called the *hill-climbing model*, and is written as

$$d^*_{ij} = d_{ij} + \left(x_i - x_j\right)^t z/d_{ij}. \tag{7.5}$$

The cyclone model also proposed by Gower [71] is similar to the jet-stream model. This model is written as

$$t_{ij} = d_{ij}/\left(V + \omega h_{ij}\right), t_{ji} = d_{ij}/\left(V - \omega h_{ij}\right) \tag{7.6}$$

As Gower points out, this model is similar to the jet-stream model except that the jet-stream is replaced by a cyclonic wind rotating about its center C at a constant angular velocity. Moreover, it is the same as Tobler's wind model, if we reparametrize $c_{ij} = \omega\ h_{ij}$ and $c_{ji} = -\omega\ h_{ij}$. As with the jet-stream model, the cyclone model may also be thought of as an AMDS model if we analyze them simultaneously. Later, we shall discuss an AMDS method proposed by Sato [164, 165] which generalizes the jet-stream model from a mathematically more sophisticated view point.

Krumhansl's *distance-density model* augments the distance d_{ij} as follows:

$$d^*_{ij} = d_{ij} + \alpha\delta(x_i) + \beta\delta\left(x_j\right), \tag{7.7}$$

where d^*_{ij} is an augmented distance, and δ (x_i) and δ (x_j) are measures of spatial density in the neighborhoods of objects i and j, respectively, while α and β are the corresponding weights applied to the densities.

By contrast, Weeks and Bentler [210] proposed an augmented distance model (the W-B model) which simplifies the distance-density model. According to the W-B model,

$$d_{ij}^{*} = bd_{ij} + c_i - c_j + a, \tag{7.8}$$

where a is an additive constant, and $b = -1$ if the data is composed of similarity mea-sures instead of dissimilarity measures. The Euclidean distance, d_{ij}, may be replaced by d_{ij}^2 in (7.8).

Okada and Imaizumi [142] proposed a more general model than the distance-density model as well as the W-B model. This model is written as

$$d_{ij}^{*} = d_{ij} + \alpha c(i, j, t) + \beta c(j, i, t), \tag{7.9}$$

where α and β are constant weight parameters, and $c(i, j, t)$ and $c(j, i, t)$ are the terms which represent the skew-symmetric component and are assumed to be positive. They consider several sub-models, one of which is the *O-I model* [143]:

$$d_{ij}^{*} = d_{ij} - r_i + r_j. \tag{7.10}$$

Okada and Imaizumi [144] extended it to the two-mode, three-way case.

Saito and Takeda [162] proposed models similar in form to the O-I model. Model 2, which is the most general one, is written as,

$$d_{ij}^{*} = d_{ij} + a\theta_i + b\theta_j + r, \tag{7.11}$$

where d_{ij} is the Minkowski r-metric, and r is an additive constant. This model is apparently a special case of (7.9) but it may be more appropriately regarded as an extension of (7.7), (7.8), and (7.10). For example, although $\delta(\mathbf{x}_i)$ and $\delta(\mathbf{x}_j)$ in (7.7) as well as r_i and r_j in (7.10) are all positive by definition, there exist no such restrictions on a, b, θ_i, and θ_j in (7.11). It is evident that the second and third right-hand terms in (7.11) are extentions of those in (7.8), although the first right-hand term of (7.11) is a special case of that in (7.8). If θ_i's are positive, they are interpreted as stimulus specific effects, analogous to spatial densities in the distance-density model.

Later, Saito [160] proposed the following model,

$$d_{ij}^{*} = d_{ij} + \theta_i + \varphi_j + \gamma, \tag{7.12}$$

which is partly a generalization of (7.8), because parameters in the second and third right-hand terms are distinct in (7.12) but they belong to the same set of parameters in (7.8). However, as for the first right-hand term, this model assumes a special case of that of the W-B model.

Most of the above models are considered as a special case of the *Holman model* [89] stated as

$$s_{ij} = F\left[m_{ij} + r_i + c_j\right], \tag{7.13}$$

where s_{ij} is a similarity between objects i and j which Holman calls the proximity data and F is some strictly increasing function, and m_{ij} is a *symmetric function*. If m_{ij} is parametrized further by the coordinates of objects in a metric space, this model can be said to be an AMDS model in the narrowest sense. In any case, the only exception for the Holman model is a generalized version of the O-I model described by (7.10). Nosofsky [136] calls the Holman model the *additive similarity and bias model*.

Zielman and Heiser [214] proposed an algorithm to fit the *slide-vector model*, which had been suggested by Kruskal in 1973. This model is written as:

$$d_{ij} = \left\{\sum_{t=1}^{r}\left(x_{it} - x_{jt} + z_t\right)^2\right\}^{1/2}, \tag{7.14}$$

where x_{it} and x_{jt}, respectively, are the coordinates of objects i and j on dimension t, and $(z_1, z_2, \cdots, z_r)$ constitutes the slide-vector $\mathbf{z}$. This model has a distinguishing feature that the diagonal elements of the model are non-zero, unlike a regular distance model. They showed how the coordinates and the slide-vector can be obtained by using an unfolding algorithm by Heiser [85]. This means that the slide-vector is a special version of the *unfolding model* originated by Coombs [47]. They also proposed a three-way generalization of the slide-vector model,

$$d_{ijk} = \left\{\sum_{t=1}^{r} u_{kt}\left(x_{it} - x_{jt} + z_t\right)^2\right\}^{1/2}, \tag{7.15}$$

However, we consider the unfolding model, especially the multidimensional unfolding models (e.g., Bennet & Hays [13]; Hays & Bennett [82]; Schönemann [166]) as they are neither a family of the augmented distance models nor a family of AMDS's in the narrowest sense, because in general the multidimensional unfolding models do not necessarily satisfy the conditions 1, 4, and 6 in our definition of AMDS in the narrowest sense. Incidentally, Zielman and Heiser [214] also proposed the *multiple slide-vector model* as well as the *row-weighted slide-vector model* as candidates for other possible generalizations.

The non-distance model is based on some quantity other than the metric distance, e.g., the inner product to the similarity measures (e.g., Chino [24, 25, 27]; Constantine & Gower [46]; DeSarbo et al. [59]; Escoufier & Grorud [64]; Gower [71]; Harshman [77]; Harshman et al. [79]; Kiers & Takane [101]; Loisel & Takane [120]; Trendafilov [202]).

Chino [24, 25] proposed an ASYMSCAL model different from Young [208]. In contrast with Young's augmented distance model, he proposed a special inner product model for AMDS, which is written as follows,

$$s_{ij} = a\left(x_{i1}x_{j1} + x_{i2}x_{j2}\right) + b\left(x_{i1}x_{j2} - x_{i2}x_{j1}\right) + c, \tag{7.16}$$

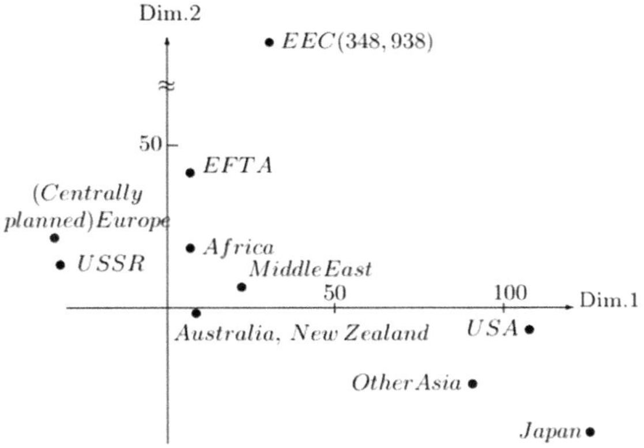

Fig. 7.1 Configuration of 10 nations including two regions obtained via Chino's ASYMSCAL (Adopted from Fig. 3.7 of Chino (1997). In this figure, the positive direction is counterclockwise

where s_{ij} is a similarity between objects i and j, and c is an additive constant, while x_{il} is the coordinate of object i on dimension l. It is apparent that the quantity in the first parentheses on the right-hand side of (7.16) and that in the second parentheses are, respectively, the inner product and the cross-product (outer product) of position vectors corresponding to two objects i and j on a two-dimensional plane. The cross-product is equivalent to the *area of the parallelogram* formed by the two position vectors.

Figure 7.1 shows the two-dimensional configuration of 10 nations including two regions for the trade data in Table 7.2. The *positive direction* in this figure is crucial in interpreting the direction of skewness in the amounts of trade between two nations. The sign of b in (7.16) determines this direction. It is apparent from the positive direction as well as the magnitudes of the parallelogram spanned by two position vectors corresponding to two nations that Japan is in a state called "unilateral love". In other words, Japan's trade surplus of exports over imports was prominent among the 10 nations.

Although Chino's ASYMSCAL was confined within the three-dimensional space, he later extended it to a generalized inner product model called GIPSCAL [27]. GIPSCAL is written in matrix form as,

$$S = aXX^t + bXL_qX^t + C1_N1_N^t, \tag{7.17}$$

where S = {s_{ij}}, a and b are constants, and c is an additive constant, while $\boldsymbol{X}$ is an $N \times q$ coordinate matrix, and $\boldsymbol{L}_q$ is a special skew-symmetric matrix (also, Chino [26]; Gower [72]).

Later, Kiers and Takane [101] pointed out that the alternating least squares (ALS) algorithm for fitting the GIPSCAL off-diagonal elements did not constitute a true ALS, and hence need not decrease the objective function value monotonically. Furthermore, they simplified the $\boldsymbol{L}_q$ in (7.17) in order to facilitate the interpretation of the skew-symmetric part of the data as follows,

$$S = a\tilde{X}\tilde{X}^t + b\tilde{X}\Delta\tilde{X}^t + C1_N 1_N^t, \tag{7.18}$$

where $\boldsymbol{\Delta}$ is a fixed matrix with singular values of the matrix $\boldsymbol{L}_{\mathrm{q}}$ in skew-symmetric 2 × 2 blocks along the diagonal. To be precise, $\boldsymbol{L}_{\mathrm{q}} = \boldsymbol{U}\boldsymbol{\Delta}\boldsymbol{U}^t$ and $XU = \tilde{X}$. Rocci and Bove [156] proposed a special case of (7.18).

Trendafilov [202] proposed another unique method for solving GIPSCAL efficiently. In his approach, the GIPSCAL problem is reformulated into an *initial value problem for matrix ordinary differential equations* on manifolds defined by the constraints of the original least-squares problems. This algorithm has been found to produce solutions which give better fits to the data than the algorithms of Chino [25] as well as Kiers and Takane [101]. Trendafilov also proposed a three-way GIPSCAL and its algorithm.

Recently, Loisel and Takane [120] proposed a fast convergent algotithm for GIPSCAL with acceleration by the minimal polynomial extrapolation. They adapted their basic algorithm to various extensions of GIPSCAL, including off-diagonal DEDICOM/GIPSCAL, and three-way GIPSCAL.

Gower [71] proposed several methods for the analysis of asymmetry, one of which includes in part the same quantity as Chino's ASYMSCAL in the two-dimensional case. The Gower diagram or CASK, referred to in the introduction section, decomposes the skew-symmetric part of the square asymmetric data matrix $\boldsymbol{S}$ via the singular value decomposition,

$$\boldsymbol{S}_{sk} = \boldsymbol{X}\boldsymbol{\Lambda}\boldsymbol{K}\boldsymbol{X}^t, \tag{7.19}$$

where $\boldsymbol{S}_{sk}$ denotes the skew-symmetric part of $\boldsymbol{S}$, while $\boldsymbol{X}$ is an $N \times N$ orthogonal matrix, $\boldsymbol{\Lambda}$ is a special diagonal matrix of singular values such that $\boldsymbol{\Lambda} = diag(\lambda_1, \lambda_1, \lambda_2, \lambda_2, \cdots, (0))$, and $\boldsymbol{K}$ is a skew identity matrix consisting of

$$\begin{pmatrix} 0 & -1 \\ 1 & 0 \end{pmatrix}$$

as diagonal blocks. In scalar form, (7.19) is written as

$$s_{ij}(sk) = \sum_t^p \lambda_t \left(x_{i,2t-1}x_{j,2t} - x_{i,2t}x_{j,2t-1} \right), \tag{7.20}$$

where p is the largest integer not exceeding $N/2$. Equation (7.20) is nothing but the second term on the right-hand side without constant b in Chino's ASYMSCAL given in (7.16). Chino [24] and Gower [71] introduced this quantity independently.

However, Gower took a philosophically different approach from Chino's. Holman points out that there exist two approaches to asymmetric proximity data. One considers the symmetric part and skew-symmetric part of the data to be inseparable parts of the same fundamental process, and the other considers the two parts to reflect different processes that can be distinguished by appropriate analysis. Chino

[24] took the former approach, while Gower [71] took the latter, although at least some other methods proposed by Gower, i.e., the jet-stream model and the cyclone model include the distance between two geographical points, as discussed earlier. In any case, we shall not regard at least CASK as an AMDS model because it possesses a *symplectic structure* which is different from that of the Euclidean metric structure (e.g., Arnold [8]; Chino & Shiraiwa [45]).

Harshman and his colleagues [77, 79] proposed a simple non-distance model called DEDICOM, which stands for the DEcomposition into DIrectional COMponents,

$$S = YAY^t, \tag{7.21}$$

where $\boldsymbol{Y}$ is the $N \times p$ loading matrix of N objects on a few basic types of objects, while $\boldsymbol{A}$ is a small asymmetric matrix of order p giving the directional relationships among the basic p types or dimensions. He called DEDICOM a non-spatial approach, and thus did not discuss its metric structure. However, Chino and Shiraiwa [45] have proven that DEDICOM has an explicit metric structure under a mild condition. Formally, Chino's ASYMSCAL is a special case of DEDICOM. Harshman et al. [79] discuss a two-mode, three-way version of DEDICOM. Takane and Kiers [186] proposed the latent class DEDICOM for square contingency tables, which is a hybrid model of a *latent class model* and a special constrained DEDICOM.

Escoufier and Grorud [64] proposed a similar model in form to Chino's ASYMSCAL from a linear algebraic point of view. Although they did not give any formal name to their model, Chino [28] called it the *Hermitian canonical model* abbreviated as HCM. HCM first decomposes the original similarity matrix $\boldsymbol{S}$ into the symmetric part $\boldsymbol{S}_s$ and the skew-symmetric part $\boldsymbol{S}_{sk}$, and compute an *Hermitian matrix* $\boldsymbol{H}$ such that $\boldsymbol{H} = \boldsymbol{S}_s + i\,\boldsymbol{S}_{sk}$, where i is a pure imaginary number. Then, they solved the eigenvalue problem of $\boldsymbol{H}$ using a well-known method which constructs a double sized, real symmetric matrix $\boldsymbol{T}$ such that

$$T = \begin{pmatrix} S_s & -S_{sk} \\ S_{sk} & S_s \end{pmatrix}, \tag{7.22}$$

It is well known that if the eigenvector corresponding to the l-th eigenvalue of $\boldsymbol{H}$ is denoted as $\boldsymbol{w}_l = \boldsymbol{u}_l + i\,\boldsymbol{v}_l$, then the eigenvalues of $\boldsymbol{T}$ come in pairs $\lambda_1, \lambda_1, \lambda_2, \lambda_2$, $\cdots$, and the eigenvectors corresponding to an identical eigenvalue λ_l are $\boldsymbol{a}_l = \left(\boldsymbol{u}_l^t, \boldsymbol{v}_l^t\right)^t$ and $\boldsymbol{b}_l = \left(-\boldsymbol{v}_l^t, \boldsymbol{u}_l^t\right)^t$ (e.g., Wilkinson [212]). Using these results, we have

$$\boldsymbol{S}_s = \sum_{l=1}^{N} \lambda_l \left(\boldsymbol{u}_l \boldsymbol{u}_l^t + \boldsymbol{v}_l \boldsymbol{v}_l^t\right), \tag{7.23}$$

and

$$\boldsymbol{S}_{sk} = \sum_{l=1}^{N} \lambda_l \left(\boldsymbol{u}_l \boldsymbol{u}_l^t - \boldsymbol{v}_l \boldsymbol{v}_l^t \right), \tag{7.24}$$

In scalar form, these are approximated as follows with only the largest eigenvalue λ_1,

$$s_{ij}(s) \fallingdotseq \lambda_1 \left(u_{i1} u_{j1} + v_{i1} v_{j1} \right),\ s_{ij}(sk) \fallingdotseq \lambda_1 \left(v_{i1} u_{j1} - u_{i1} v_{j1} \right) + c, \tag{7.25}$$

This is the result obtained by Escoufier and Grorud.

DeSarbo et al. [59] proposed a spatial MDS procedure called TSCALE based on *Tversky's contrast model*. They assume a two-mode, three-way proximity data. Letting d_{ijr}^* be the dissimilarity value on the r-th replication between the two objects i and j, for example, one of the TSCALE models is written as.

$$d_{ijr}^* = \sum_{t=1}^{T} \alpha_r \left(x_{it} - x_{jt} \right) + \sum_{t=1}^{T} \beta_r \left(x_{jt} - x_{it} \right) + \sum_{t=1}^{T} \theta_r \min \left(x_{it}, x_{jt} \right), \tag{7.26}$$

for Tversky's original linear contrast analog. Here, the function $(u - v)_+ = max(u - v, 0)$, α_r is the impact or salience that is distinctive to the first object, i, in pair ij, presented on the rth replication (e.g., subject r), β_r is the impact or salience that is distinctive to the second object, j, in pair ij, presented on the rth replication (e.g., subject r), and θ_r is the impact or salience that is common to the object pair ij presented on the rth replication. They also proposed a ratio, distinctive feature model. For details, see DeSarbo et al. [59].

The extended distance model is a family of AMDS in which some distance structure is assumed other than the traditional *real metric structure*, i.e., the Minkowski r-metric. One is the *(real) asymmetric Minkowski metric structure*, and the other the *(complex) Hilbert space structure*. Sato [164, 165] proposed to embed objects in a certain real asymmetric Minkowski space, given a set of asymmetric dissimilarity measures. Here, it should be noticed that the Minkowski space he uses is not the familiar Minkowski r-metric, but has an asymmetric metric function. The two-dimensional (asymmetric) Minkowski metric model is written as follows,

$$d_{ij}^* = d_{ij} / r_\beta(\psi | \gamma), \tag{7.27}$$

where d_{ij} is a usual Euclidean distance, and $r_\beta(\psi|\gamma)$ is a transformed *indicatrix* such that, using an original indicatrix $r(\theta|\gamma)$,

$$x_1 - \beta = r(\theta|\gamma) \cos\theta,\ x_2 - \beta = r(\theta|\gamma) \sin\theta, \tag{7.28}$$

where (x_1, x_2) is the current coordinate of point in the plane, and the original indicatrix is written as

$$r(\theta|\gamma) = \sum_{k=1}^{m} \gamma_k r_k(\theta). \tag{7.29}$$

Here, the component indicatrix in Eq. (7.29) is

$$r_k(\theta) = \mu_k + \nu_k \cos\left\{k\left(\theta - \frac{\pi}{4}\right)\right\}, \tag{7.30}$$

where $\mu_k = (k^2 + 2)/(k^2 + 3)$ and $\nu_k = 1/(k^2 + 3)$. As Sato [164] point out, if we choose the indicatrix with $k = 1$, the Minkowski metric defined by Eq. (7.27) is equivalent to the jet stream model by Gower [71]. Sato [165] extended his model to the three-dimensional case in which another asymmetric metric function, *Rander's metric function* [154], was used.

The other is to utilize the (complex) Hilbert space structure. As discussed earlier, Escoufier and Grorud [64] did not point out the metric structure in their HCM model. In contrast, Chino and Shiraiwa [45] reformulated HCM and discovered the Hilbert space structure. In contrast with Escoufier and Grorud's approach, they directly treated the eigenvalue problem of the Hermitian matrix obtained from the data matrix $\boldsymbol{S}$, and deduced the following equation:

$$H = X\Omega_s X^t + iX\Omega_{sk} X^t, \tag{7.31}$$

where

$$\Omega_s = \begin{pmatrix} \Lambda & O \\ O & \Lambda \end{pmatrix}, \Omega_{sk} = \begin{pmatrix} O & -\Lambda \\ \Lambda & O \end{pmatrix}, \tag{7.32}$$

and $\boldsymbol{\Lambda}$ is a diagonal matrix of order n which is the number of non-zero eigenvalues of $\boldsymbol{H}$, i.e., $\boldsymbol{\Lambda} = diag(\lambda_1, \lambda_2, \cdots, \lambda_n)$. The matrix $\boldsymbol{X}$ is the special real $N \times 2n$ coordinate matrix of objects, i.e., $\boldsymbol{X} = (\boldsymbol{U}_r, \boldsymbol{U}_c)$, where $\boldsymbol{U}_1 = \boldsymbol{U}_r + i\boldsymbol{U}_c$ and $\boldsymbol{U}_1$ are composed of the complex eigenvectors of $\boldsymbol{H}$ corresponding to its non-zero eigenvalues. It should be noticed that all the eigenvalues of $\boldsymbol{H}$ is real.

It is easy to show that (7.31) can be rewritten as

$$S = X\Omega_s X^t + X\Omega_{sk} X^t. \tag{7.33}$$

Chino and Shiraiwa [45] called (7.31) the *Hermitian form* model abbreviated as HFM. They showed that DEDICOM, GIPSCAL, and HCM are special cases of the Hermitian form model if their complex counterparts are considered. Furthermore, Chino and Shiraiwa [45] proved that a necessary and sufficient condition for these models to be expressible in terms of (complex) Hilbert space is the *positive semi-definiteness* of $\boldsymbol{H}$. This is an extension of the Young-Householder theorem on MDS to the case of the complex space.

Saito and Yadohisa [163] contended that there existed some unclear points of this extension, and took another look at the extension. That is, they divided the

Chino-Shiraiwa theorem into two, i.e., theorems on sufficiency and necessity, which they called theorem 7.2 and theorem 7.3, respectively. Moreover, they emphasized the role of the distance between objects j and k, d_{jk} (=$\boldsymbol{w}_j - \boldsymbol{w}_k$, where $\boldsymbol{w}_j$ and $\boldsymbol{w}_k$ are coordinate vectors corresponding to objects j and k in the Hilbert space) in discussing the extension of the Young-Householder theorem. However, in theorem 7.2 they did not discuss the other important distance $\overline{d}_{jk}$ (=$\boldsymbol{w}_j - i\boldsymbol{w}_k$, where $i^2 = -1$) in the Hilbert space, which is indispensable to the Chino-Shiraiwa theorem. For further details, see Chino and Shiraiwa [45] as well as Saito and Yadohisa [163].

All the AMDS models discussed up to now are *descriptive models* in which no statistical inference is involved. Traditionally, the AMDS in the narrowest sense might have been applied to asymmetric relational data matrices under an implicit assumption that the data are sufficiently asymmetric enough to warrant AMDS. However, the question arises: Given an $\boldsymbol{S}$, how should we ascertain whether it is sufficiently symmetric or not? Kruskal [108], for example, discusses nonsymmetry of dissimilarities in his seminal paper on nonmetric MDS. He recommends to average the (i, j) and (j, i) elements of $\boldsymbol{S}$ if they are measurements on the same underlying quantity and differ only because of random fluctuation. However, he did not provide any method for checking whether they differ only randomly. In order to answer this question empirically, it seems appropriate to develop some inferential method for it.

Recently, Saburi and Chino [158] proposed a maximum likelihood asymmetric MDS called ASYMMAXSCAL, which is an extension of MAXSCAL proposed by Takane [183]. MAXSCAL is a maximum likelihood (ML) MDS method specifically designed to analyze the symmetric similarity data measured on rating scales with a relatively small number of observation categories. ASYMMAXSCAL inherited almost all the characteristics which MAXSCAL possesses. As a result, the scale level assumed in ASYMMAXSCAL is usually higher than or equal to the ordinal level. However, as Takane [183] points out, as a special case, the same-different judgments like the confusion data can also be analyzed as if they were two-category judgments. As a result, whatever the scale level may be, the original data set for applying ASYMMAXSCAL is composed of one-dimensional frequency distributions of proximity judgments on the pairs of objects, which is called the Type A design data. By examining the estimated values of the category boundaries of the rating scale under study, one can diagnose the scale level of measurement as either the ordinal level or the interval level.

ASYMMAXSCAL provides two methods for checking whether the data are sufficiently asymmetric or not. One is a special symmetry test prior to fitting any asymmetric MDS model, and the other is some symmetry tests in a subsequent scaling step. To do the first, we rearrange the Type A data into a special $n \times n \times M$ three-way contingency table whose stratified variable is composed of the rating categories. It is called the Type B design data.

In ASYMMAXSCAL, a special conditional symmetry hypothesis, i.e.,

$$H_0^{(cs)} : p_{ijm} = p_{jim, 1 \leqq i < j \leqq N, 1 \leqq m \leqq M-1},$$

in Type A design is tested using the LR test statistic. The other method for checking the asymmetry of data in ASYMMAXSCAL is composed of two kinds of symmetry tests in the scaling step. One is the test for the symmetry hypothesis based on the saturated representation model (the SR model). Although any extant AMDS model may be chosen as the representation model in ASYMMAXSCAL, the SR model is the representation model with no structure. The other symmetry test in the scaling step is the test for the symmetry hypothesis based on a specified representation model under study. For example, Saburi and Chino [158] choose the O-I model as the specified model. As a result, the hypothesis is represented by

$$H_0^{(s/oi)} : r_i = r_{i+1}, i \leqq i \leqq N - 1.$$

In ASYMAXSCAL, several models including the above mentioned symmetry hypotheses can be compared using AIC, instead of using the statistical tests. Such a strategy has often been taken in the literature of asymmetric MDS as well as analysis of contingency tables [183, 184, 198].

Figure 7.2 shows the three-dimensional configuration and spheres for the O-I model under the ordinal scale assumption via the application of ASYMMAXSCAL to the friendship data among nations in East Asia and the USA, which was adapted from Saburi and Chino [158]. In this survey, four hundred Japanese university students participated, and the single-judgment sampling was chosen. Each subject rated the extent to which he or she felt the government of a nation as friendly or hostile to that of another nation on a 5-point rating scale. Since the sphere associated with Japan is larger than those of the other nations, it is interpreted from the O-I model that Japan is perceived as more friendly to other nations than the other way around.

7.3 A Brief Review of AMDS in the Narrowest Sense and a Criterion for Checking Features of AMDS Models

In this section, we shall first review briefly major AMDS methods for count data. As was done in the review of AMDS in the narrowest sense, we shall divide a body of AMDS in the narrower sense into the descriptive AMDS and the inferential AMDS. As was the case for MDS, the latter AMDS grew out of the descriptive methods.

In stimulus recognition experiments in psychology, typical data can be summarized in a confusion matrix $\boldsymbol{C}$, each element of which is the proportion of times response R_j was made when stimulus S_i was presented, and it is an estimate of the conditional confusion probability $P(R_j|S_i)$. According to Ashby and Perrin [9], the biased-choice model (sometimes called the similarity choice model, e.g., Nosofsky [135]) has been most successful in predicting a wide variety of confusion matrices over the last 20 years (e.g., Luce [121]; Shepard [167]). This model is written as

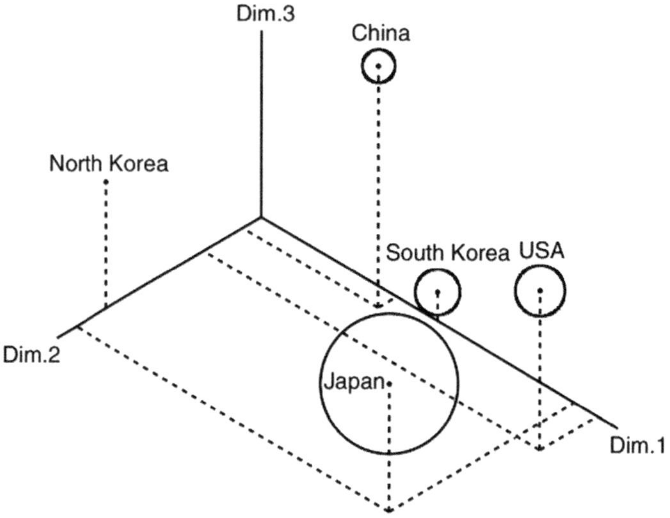

Fig. 7.2 Configuration of 5 nations and spheres surrounding each of them for the O-I model obtained via ASYMMAXSCAL (Adopted from Fig. 4 of Saburi & Chino [158])

$$P(R_j|S_i) = \frac{\beta_j \eta_{ij}}{\sum_m \beta_m \eta_{im}}, \tag{7.34}$$

where η_{ij} is the similarity of stimulus S_i to stimulus S_j, and β_j is the bias toward response R_j. Here, it is assumed that $\eta_{ij} = \eta_{ij}$ and all self-similarities are equal. Equation (7.34) is sometimes called the *unrestricted similarity-choice model* in contrast with the restricted version discussed below Takane and Shibayama [188].

The biased-choice model itself cannot be viewed as an AMDS model in the narrower sense. However, if we assume that η_{ij} is a function of the distance between the two stimuli, and the distance is parameterized by coordinates of stimuli, such a model can be considered as an AMDS model. In fact, Shepard [167] suggested replacing η_{ij} in (7.34) by the exponential decay function, $exp(-d_{ij})$, i.e.,

$$P(R_j|S_i) = \frac{\beta_j exp(-d_{ij})}{\sum_m \beta_m exp(-d_{im})}, \tag{7.35}$$

This model is called the MDS-choice model (e.g., Nosofsky [133–135]) or Shepard [168] also considered the Gaussian function, i.e., $\eta_{ij} = exp\ (-d_{ij}^2)$, and sometimes called, recpectively, the exponential MDS-choice model and the Gaussian MDS-choice model (e.g., Takane & Shibayama [188]).

Although Ashby and Perrin [9] refer to several models as versions of the MDS-choice model, some of them cannot be viewed as AMDS models in the narrower sense, while some other can be. Moreover, some of them are descriptive, but others are inferential. We shall briefly discuss the descriptive AMDS first. Shepard [167, 169] discussed not only (7.35) but also proposed a procedure for obtaining coordinates of stimuli using a version of classical MDS.

Getty et al. [68] conducted an experiment to demonstrate that the perceptual space derived from similarity judgments in a pairwise similarity-judgment task can then be used to predict behavior in an identification task. Judged (symmetric) similarity data among 8 visual stimuli were analyzed by INDSCAL [21] first, and the set of spatial coordinates for each of the stimuli were obtained. Then, the weighted Euclidean distances were computed and a set of confusion weights between stimuli were obtained using an exponential decay function. Finally, the conditional probabilities of giving the response R_j assigned to stimulus S_j when stimulus S_i was presented using a version of Luce's choice model, which assumed no differential response bias. Results indicated that this model predicts well the confusion matrices of the identification task.

Nosofsky [132] related Medin and Schaffer's [124] *context theory* to a more general theoretical framework for the modeling of choice and similarity. Their context theory assumes two rules. One is the *response-ratio rule* which states the probability of classifying any test stimuli as a member of some category, and the other the *multiplicative rule* for computing stimulus similarity. As for the response rule, Nosofsky considers it as a bias-free extension of Luce's choice theory. As regards the multiplicative rule, he points out that it arises as a special case of psychological distance between stimuli conforming to the city-block metric, and of stimulus similarity being an exponential decay function of psychological distance.

Keren and Baggen [100] proposed and tested a feature analytic model for recognition of alphanumeric characters based on Tversky's contrast model. They applied their model, to Gilmore et al.'s [69] data. However, since distances between stimuli are not resolved into coordinates of stimuli, which is consistent with the philosophy of their model, their model cannot be viewed as asymmetric MDS in the narrower sense.

Appelman and Mayzner [5] applied Krumhansl's distance-density model to the confusion matrices in three published studies which conducted a typical letter recognition experiment using a dot matrix for displaying letters. Although they examine, for example, the functional relationship between mean proportion of confusions per letter pair and the distance between letters, they do not utilize any MDS procedure to estimate coordinates of letters. For this reason, their study cannot be considered as that of asymmetric MDS in the narrower sense.

In contrast, Nakatani [125] proposed a slightly different model from the MDS-choice models discussed above. He proposed the confusion-choice model which attempts to model the recognition process with a fairly complicated perceptual process incorporating concepts from signal detection theory. His model is written as

$$P\left(R_j|S_i\right) = \sum_{k=1}^{m} e_{ik} \frac{b_j\left(r_{jk}\right)}{\sum_{l=1}^{n} b_l(r_{lk})}, \tag{7.36}$$

where b_j is the bias probability, $m = 2^n$, r_{jk} is the value of r_j for confusion state s_k, and b_j is a function of r_{jk} such that

$$b_j(r_{jk}) = \begin{cases} b_j, when r_{jk} = 1, \\ 0, when r_{jk} = 0. \end{cases}$$

Moreover, r_j and s_k are defined as follows:

First, a binary-valued mediating variable, r_j, which carries the information as to whether R_j is or is not acceptable, is defined as,

$$r_j = \begin{cases} 1, for R_j is acceptable, \\ 0, for R_j is not acceptable. \end{cases}$$

Second, a confusion state is defined as the n-tuple,

$$s_{jk} = r_{1k}, \cdots, r_{jk}, \cdots, r_{nk}.$$

Moreover, e_{ik} is called the equivocation probability

$$e_{ik} = P(s_k = r_{1k}, \cdots, r_{jk}, \cdots, r_{nk}|S_i) = \prod_{j-1}^{n} a_{ij}(r_{jk}), \tag{7.37}$$

where

$$a_{ij}(r_{jk}) = \begin{cases} a_{ij}, & when r_{jk} = 1, \\ 1 - a_{ij}, & when r_{jk} = 0. \end{cases}$$

Here, is called the *acceptance probability*,

$$a_{ij} = P(R_j acceptable|S_i) = P(q_{ij} \leq t_{j'}) = \int_{-\infty}^{t_{j'}-u_{ij}} \varphi(z)dz, \tag{7.38}$$

where $\varphi(\bullet)$ is the normal density function, $q_{ij} = u_{ij} + \varepsilon_j$, u_{ij} is the L-dimensional Euclidean distance between S_i and R_j, i.e.,

$$u_{ij} = \left\{ \sum_{l=1}^{L} (x_{il} - x_{jl})^2 \right\}^{1/2},$$

and ε_j is the channel noise, N(0, 1).

To sum up, it is evident that Nakatani's confusion-choice model is an AMDS model in the narrower sense, because the distance between each stimulus S_i and its response R_i is represented by a common point in a metric space. Shepard's MDS-choice model and Getty et al.'s weighted Euclidean distance model are also considered as the AMDS models in the narrower sense among several MDS-choice models.

In contrast with the descriptive AMDS models in the narrower sense, some researchers proposed inferential AMDS models in the 1980's, as will be discussed below: Nosofsky [135] proposed a unified quantitative approach to modeling subject's identification and categorization of multidimensional perceptual stimuli. He examined a traditional similarity choice model with the Minkowski r-metric for the former and a new categorization model which generalized the context theory of classification developed by Medin and Schaffer [124] for the latter.

It is interesting to notice here that, on the one hand, the count data obtained by an identification experiment are summarized in an $N \times N$ confusion matrix $\boldsymbol{C}$, where c_{ij} in each cell is the frequency with which stimulus i was identified as stimulus j. On the other hand, in a categorization experiment the N stimuli are classified into $m < N$ groups, each group assigned a distinct response.

Moreover, in this case the count data are summarized in an $N \times m$ confusion matrix, where c_{ij} is the frequency with which stimulus i was classified in category j. Although the new categorization model is interesting in that it takes the INDSCAL approach [21] in which selective attention is modeled by differential weighting of the component dimensions in the psychological space (Nosofsky [135], p.41), we will not consider it as an AMDS model in the narrower sense because the data matrix obtained by the categorization experiment does not satisfy the conditions, 1 and 4, of AMDS. The reason is that in Nosofsky's unified approach, he assumes the two kinds of confusion matrices.

As the function η_{ij} in (7.34) of the similarity choice model, he considers both the exponential decay function and the Gaussian function. He uses an ML criterion to estimate parameters of both models. The MDS-choice model yielded its best fits to the identification data by assuming a Gaussian function and a Euclidean metric. However, his method seems to be insufficient as an inferential method, because he neither utilize information criteria nor perform statistical tests in order to compare several models in terms of their goodness of fit (GIF).

In contrast, Takane and Shibayama [187, 188] compared various models intensively through the AIC statistic [1] by developing a standard ML estimation procedure. The likelihood of the total set of observations is stated in the same way for all the models, i.e.,

$$L = \sum_{i,j} p_{ij}^{f_{ij}},$$

where p_{ij} is the conditional probability specified by each of the models, assuming the multinomial design.

The data used in Takane and Shibayama [187] are Keren and Baggen's [100] recognition experiment data. Models compared in their analysis are the null model (the saturated model in the context of the log-linear model) (e.g., Bishop et al. [16]; Birch [15]), the unrestricted similarity-choice model (i.e., the similarity choice model), the Euclidean distance-choice model (i.e., a version of the MDS choice model), and the unique feature-choice model (a general version of Keren-Baggen's [100] model). Here, the Keren-Baggen's model is a feature analytic model for recognition of alphanumeric characters based on Tversky's [205] contrast model [100]. In any case, it is apparent that the Euclidean distance-choice model is the only inferential AMDS model in the narrower sense among the several models discussed in Takane and Shibayama [187]. Here, it should be noticed that only the Euclidean distance model among them parameterizes the distance assumed in this model by coordinates of stimuli. Takane and Shibayama [188] conducted a more extensive examination of these models and the related models using four data sets including Keren and Baggen's [100] data.

Nosofsky [136] proposed a *bias-supplemented MDS model*

$$p_{ij} = F\left[ur_i + vc_j - wd_{ij}\right] \tag{7.39}$$

where u, v, and w are positive constants. He assumes that the distance in the multidimensional space is computed using a city-block metric.

De Rooij and Herser [56, 57] proposed models called *distance-association models* for the analysis of square contingency tables, each of which might be said to be a hybrid of the log-linear model and a distance model which is closely related to MDS. In the *one-mode distance association model,*

$$\ln\left[E\left(f_{ij}\right)\right] = \lambda + \lambda_i^R + \lambda_j^C - \sum_m \left(x_{im} - x_{jm}\right)^2, \tag{7.40}$$

where f_{ij} is the observed count in the (i, j) cell of the a square contingency table, and $E\left(f_{ij}\right)$ is the expected cell value corresponding to f_{ij}. De Rooij and Heiser [56] denote $E\left(f_{ij}\right)$ by F_{ij}. They point out that it is a reduced rank version of the quasi-symmetry model. Here, at a glance, (7.40) is similar to Shepard's [168] Gaussian model (sometimes called *MDS-choice model* (e.g., Nosofsky [133], or the *Euclidean distance-choice model* [187], but the second right-hand term is slightly different in that λ_i^R in Eq. (7.40) includes no distances.

In contrast, the two-mode distance-association model [57] is no longer classified as the AMDS in the narrower sense, because it assumes two sets of coordinate matrix. As an inferential method, they apply the Pearson χ^2-test as well as the likelihood ratio (abbreviated hereafter as LR) χ^2-test (e.g., Bishop et al. [16]) in the former paper. In the latter, they utilize the *%AAF* [70] instead of the G^2, considering the fact that the traditional chi-squared distributed statistics tend to dismiss all models except the saturated model as the sample size increases indefinitely.

The AMDS models in the narrower sense also include those in the narrowest sense which were reviewed in the preceding section. Although these models are diverse in character, there seems to exist at least one criterion for checking various features of AMDS models. This is the property of *quasi-symmetry*, which was proposed by Caussinus [22] in the context of the square contingency table. As for the models for the count data, especially confusion data matrices as a typical example of such data discussed in this section, it has been pointed out (e.g., Smith [172]; Takane &

Shibayama [187]; Townsend & Landon [201]) that the similarity choice model has this property. Of course, the distance-association models proposed by De Rooij and Heiser [56, 57] also have this property, because they associate some symmetric metric function with the interaction term of the log-linear model.

As for the models assuming the ordinal level or higher in the preceding section, we can apply this criterion in checking these models if the data are obtained by a rating scale judgments as in ASYMMAXSCAL. For, in such a case, no matter what the scale level may be, we can rearrange the original Type A design data into the Type B design data, i.e., a special N × N × M contingency table, where M is the number of rating categories in the rating scale, and check whether the quasi-symmetry hypothesis holds or not. Moreover, many of the AMDS models in the narrowest sense have this property.

For example, Chino and Saburi [43] consider several extant AMDS models as the *quasi-symmetry-like AMDS models*. These are Saito's [160] model, the distance-association model, the distance-density model, the Saito and Takeda's [162] model, the slide-vector model, the W-B model, and the O-I model. The reason for labeling'quasi-symmetry-like' models is that many of these models are not necessarily designed specifically to analyze the count data. A crucial problem may be whether the quasi-symmetry property is a universal property which holds for any asymmetric phenomena. We shall discuss this problem in more detail in the discussion section.

7.4 Discussion

In this section, we shall briefly introduce problems that are presently unsolved in AMDS to facilitate its further developments of AMDS in the near future.

Problem 1 (Tests of Symmetry, Quasi-Symmetry, and Marginal Homogeneity).

As pointed out in the last paragraph of the previous section, it may be necessary and appropriate to check at least whether the data to be analyzed by AMDS has the quasi-symmetry property possibly before some AMDS model is applied to the data. On the one hand, for the count data discussed in the previous section, we can directly test the quasi-symmetry hypothesis by Caussinus [22], for example, as De Rooij and Heiser [56] did. On the other hand, we can apply the special conditional quasi-symmetry hypothesis (e.g., Bishop et al. [16]) to the Type B data discussed in the preceding section, if the data are obtained on a rating scale [158]. The reason for checking this property before any AMDS is applied is that it is logically invalid to apply any AMDS with the quasi-symmetry property to the data which have no such property.

In connection with the above method for checking the quasi-symmetry hypothesis, Saburi and Chino [158] proposed two possibilities, i.e., a symmetry test prior to fitting the AMDS model, and some symmetry tests in the scaling step. Alternatively, Chino and Saburi [42] have suggested that the conditional quasi-symmetry hypothesis plays a fundamental role in AMDS, and have proposed some sequential tests for the symmetry and related hypotheses.

Recently, Chino and Saburi have attempted to prove the following conjecture on the independence of the three hypotheses, i.e., the quasi-symmetry hypothesis for testing $H_0^{QS} : \theta \in \omega_{QS} against H_1^{QS} (Caus \sin us[22])$, a special symmetry hypothesis for testing $H_0^S : \boldsymbol{\theta} \in \omega_S against H_1^{S^*} : \theta \in \omega_{QS} - \omega_S$, and a marginal homogeneity hypothesis for testing $H_0^{MH_0} : \boldsymbol{\theta} \in \omega_{MH_0} against H_1^{MH_0^{**}} : \theta \in \omega_S - \omega_{MH_0}$. Here, the parameter spaces, Ω, ω_{QS}, ω_S, and ω_{MH_0} are defined below in Eqs. (7.41) through (7.44). It should be noticed that there exists the relation of inclusion, i.e., $\Omega \supset \omega_{QS} \supset \omega_S \supset \omega_{MH_0}$, in these spaces:

Conjecture 1. *The three LR statistics, $G^2{}_{QS}$ to test $H_0{}^{QS}$ against $H_1{}^{QS}$, $G^2{}_S$, to test $H_0{}^S$ against $H_1{}^{S^*}$, and to test against , are mutually independent stochastically, if the three hypotheses are tested seriatim and if the $H_0{}^{QS}$ is accepted first and the $H_0{}^S$ is accepted second.*

Here, parameter spaces pertaining to the above three hypotheses, using the log-linear model [15], are as follows:

First, the total parameter space for corresponding.

to an N by N cross classification table is

$$\Omega = \left\{ \left(\theta_{ij}^{(12)}, \theta_i^{(1)}, \theta_j^{(2)}, \theta^{(0)} \right), \right.$$

$$\left. -\infty < \theta_{ij}^{(12)}, \theta_i^{(1)}, \theta_j^{(2)}, \theta^{(0)} < \infty, i, j = 1, 2, \cdots, N \right\}. \tag{7.41}$$

Second, the parameter space related to the quasi-symmetry hypothesis is

$$\omega_{QS} = \left\{ \left(\theta_{ij}^{(12)}, \theta_i^{(1)}, \theta_j^{(2)}, \theta^{(0)} \right), -\infty < \theta_{ij}^{(12)} = \theta_{ji}^{(12)} < \infty, \right.$$

$$\left. -\infty < \theta_i^{(1)}, \theta_j^{(2)}, \theta^{(0)} < \infty, i, j = 1, 2, \cdots, N \right\}. \tag{7.42}$$

Third, the parameter space pertaining to the symmetry hypothesis is

$$\omega_S = \left\{ \left(\theta_{ij}^{(12)}, \theta_i^{(1)}, \theta_j^{(2)}, \theta^{(0)} \right), -\infty < \theta_{ij}^{(12)} = \theta_{ji}^{(12)} < \infty, \right.$$

$$\left. -\infty < \theta_i^{(1)} = \theta_j^{(2)} < \infty, -\infty < \theta^{(0)} < \infty\, i, j = 1, 2, \cdots, N \right\}. \tag{7.43}$$

Fourth, the parameter space pertaining to the marginal homogeneity hypothesis in terms of the log-linear model ([4], pp.208–209) is

$$\omega_{MH_0} = \left\{ \left(\theta_{ij}^{(12)}, \theta_i^{(1)}, \theta_j^{(2)}, \theta^{(0)} \right), \theta_{ij}^{(12)} = 0, \right.$$

$$-\infty < \theta_i^{(1)} = \theta_i^{(2)} < \infty, -\infty < \theta^{(0)} < \infty, i, j = 1, 2, \cdots, N \Big\}. \quad (7.44)$$

However, this conjecture has not been proven yet [43, 44]. The most difficult problem is whether the LR-statistic for testing the quasi-symmetry hypothesis, $G^2{}_{QS}$, is an *ancillary statistic* (e.g., Lehmann [117]) regarding its nuisance parameters. If $G^2{}_{QS}$ is an *ancillary statistic*, we can prove the above conjecture using the theorems by Basu [12], Lehmann [117], and Hogg and Craig [88]. Then, we can perform a sequential test for testing these three hypothesis, and can control the type 1 error completely.

Problem 2 (Multiple-Judgment Sampling).

Computation of the likelihood of the data is essential in constructing any inferential statistical method. For example, ASYMMAXSCAL assumes that each subject judges only one pair of objects, which is called the *single-judgment condition* or the *single-judgment sampling*. Computation of the likelihood as a product of likelihoods of observations is justified in such a case.

By contrast, if each subject judges all pairs of objects, it is called the *multiple-judgment condition* or the *multiple-judgment sampling* (e.g., Bock & Jones [17]). In this case, statistical dependencies among these multiple judgments are inevitable. In practice, however, it is sometimes convenient to use multiple-judgment sampling. For example, it is usual in the sociometric test to ask members of a classroom to rate all the members with a rating scale. Table 7.1 in the introductory section shows one such set of data.

Although Bock and Jones [17], for example, discuss the cases in which this requirement is relaxed somewhat in the treatment of paired comparisons, the simplest way to cope with the violation of independence of the data in the multiple-judgment condition may be to assume some appropriate statistical distribution for such data. Promising candidates for it may be the *multivariate (Bernoulli) multinomial distribution* [213] and some version of *multivariate Poisson distribution* (e.g., Johnson et al. [94]; Krummenauer [106]).

Problem 3 (Beyesian Inferences).

As mentioned in the preceding section, various Bayesian inferential MDS methods have already been developed since Oh and Raftery[138]. However, as far as we know, there had been no such method for AMDS until recently. Okada [140] has proposed such a method. Bayesian AMDS methods might be expected to overcome some shortfalls of ML methods (e.g., Oh & Raftery [138]; Price [149]). A shortfall of ML MDS methods, which was pointed out by Cox [48], is that the asymptotic theory on which ML relies may not apply in high dimensions because the number of parameters to be optimized typically grows. A well-known shortfall of ML estimation in a general setting is the so-called Neyman-Scott problem (i.e., Neyman & Scott [131]), in which the small-sample bias of some ML estimator persists as sample size increases in the case of a non-identical parent distribution.

However, the Bayesian inference cannot seem to be omnipotent. For example, as a reason for the schism within the field of statistics over the Bayesian position, Price states, "Some statisticians feel that the evidence against the frequentist approach to statistical inference and decision making is not yet sufficiently cogent to warrant discarding a large collection of procedures that by and large have worked pretty well ([149], p. 48)". Bernardo states as follows in answering a question about the axioms of probability by Irony and Singpurwalla ([92], p. 161): "These are proven existence results; they imply that the common sense 'a prior does not exist' is a mathematical fallacy: for mathematical consistency one must be a Bayesian. However, these are only existence results; they leave open the question of specifying a particular prior in each problem".

Problem 4 (Multiple Comparisons in Quasi-Symmetry).

Although the importance of the test for quasi-symmetry in the context of AMDS has already been pointed out by some researchers as discussed in the preceding section, the *multiple comparisons for the hypotheses of quasi-symmetry* have rarely been done in the context of AMDS, although several statistical theories on this problem have already been developed (e.g., Hirotsu [87]; Kastenbaum [97]). For example, the row-wise (or column-wise) multiple comparison procedure proposed by Hirotsu [87] enables us to examine in which pairs of rows (or columns) the quasi-symmetry is detected. As is well known, the quasi-symmetry hypothesis is usually defined as

$$H_0^{QS} : \theta_{ij}^{(12)} = \theta_{ji}^{(12)} \text{ for all } i \text{ and } j,$$

where $\theta_{ij}^{(12)}$ is the interaction term of the log-linear model. In contrast, Hirotsu utilizes an alternative definition of quasi-symmetry, i.e.,

$$H_0^{QS} : p_{ij}p_{jk}p_{ki} = p_{ji}p_{ik}p_{kj}$$

for the triplet *i, j, k* being all different as well as any pair of the triplet being equal, where p_{ij} is the population probability corresponding to the observed frequency f_{ij}. If H_0^{QS} is rejected, one may consider, say, the hypothesis on its row-wise partition, $H_0^{K_{ij}} : D_{ij}^t q = 0,$. where $\boldsymbol{D}_{ij}^t$ is an $n-2$ by N^2 matrix with $\boldsymbol{d}_{ijh}^t$, $h = 1, \cdots, N; h \neq i, j$ as its $N - 2$ rows, $\boldsymbol{d}_{ijh}$ being an N^2 dimensional vector defined for a triplet such that the $N(i-1)+j$th, the $N(j-1)+h$th, the $N(h-1)+i$ th elements are 1, the $N(i-1)+h$th, the $N(j-1)+i$th, and the $N(h-1)+j$th elements are $-$ 1, and all the other elements 0 if $1 \leq i < j < h \leq$ N and similarly for other permutation. Moreover, $\boldsymbol{q}$ is an N^2-dimensional vector with $ln\ p_{ij}$ as the $N(i-1)+j$th element. Such comparisons might uncover interesting features contained in the contingency table peculiar to AMDS.

Problem 5 (Positive Semidefinite Programming).

As Kumagai [110] points out, recently there has been increasing attention to the problems and applications of the *semidefinite programming* (e.g., Aakih et al. [2]; Laurent [115]; Weinberger & Saul [211]; So & Ye [173]). Kumagai [110] proposed to solve the classical MDS problem by Torgerson with asymmetric relational data in a different fashion, utilizing the semidefinite programming. His method cannot be classified into AMDS in the narrowest sense, because it does not satisfy condition 4. However, his idea might be extended further to solve various AMDS problems from the view point of semidefinite programming, because (positive) semidefiniteness plays an essential role not only in MDS but also in AMDS, as the Chino and Shiraiwa's theorem points out. Solving HFM assuming a positive semidefinite Hermitian matrix may be an easy problem to be solved in the near future.

Problem 6 (Longitudinal Asymmetric Relational Data).

In the traditional psychometric approach to a special two-mode,three-way matrix which is composed of one-mode, two-way square asymmetric matrices, the second mode usually represents individual. However, we frequently encounter a three-way matrix whose second mode denotes time. Longitudinal sociometric matrices and longitudinal trade data are typical examples. An ambitious approach to such three-way data may be to construct some vec- tor field model based on dynamical system theory in mathematics.

On the one hand, Tobler [197] and Yadohisa and Niki [207] estimate such a vector field, given a single asymmetric data matrix among objects. On the other hand, a DYNAmical System SCALing method (abbreviated as DYNASCAL) proposed by Chino and Nakagawa [40], estimates the vector field at each point in time (which is not restricted to the *gradient field of the scalar potentials*), given a set of longitudinal asymmetric relational data matrices, such as longitudinal attraction data which were gathered by Newcomb [130]. In this method, the configuration of members of a dormitory at each time is estimated from each of the longitudinal asymmetric data matrices via MULTISCALE proposed by Ramsay [153].

DYNASCAL estimates further some important features of the vector field at each time, utilizing the *qualitative theories of singularities* as well as *those of bifurcations* in interpreting the estimated vector fields. Figure 7.3 shows one of the estimated vector fields, some estimated trajectories peculiar to the vector field, and some estimated attractors (in particular, singularities and a limit cycle) of the field for the Newcomb data via DYNASCAL. These features of the vector field provide us not only with the configuration of members in a group at a specific point in time but also with changes in the local as well as the global group structure of members over time.

A limitation of these methods seems to be their assumption about the state space. For, all of these assume that it is Euclidean. By contrast, the Chino and Shiraiwa theorem suggests that the state space is a *finite-dimensional (complex) Hilbert space*. In fact, Chino [31] proposed a *complex difference equation model* of interpersonal interaction. But, it is a very restrictive model, as Chino pointed out. For example, it assumes that members will not move in his or her psychological space if there exists no skewness of sentiment between any two members. A more promising model

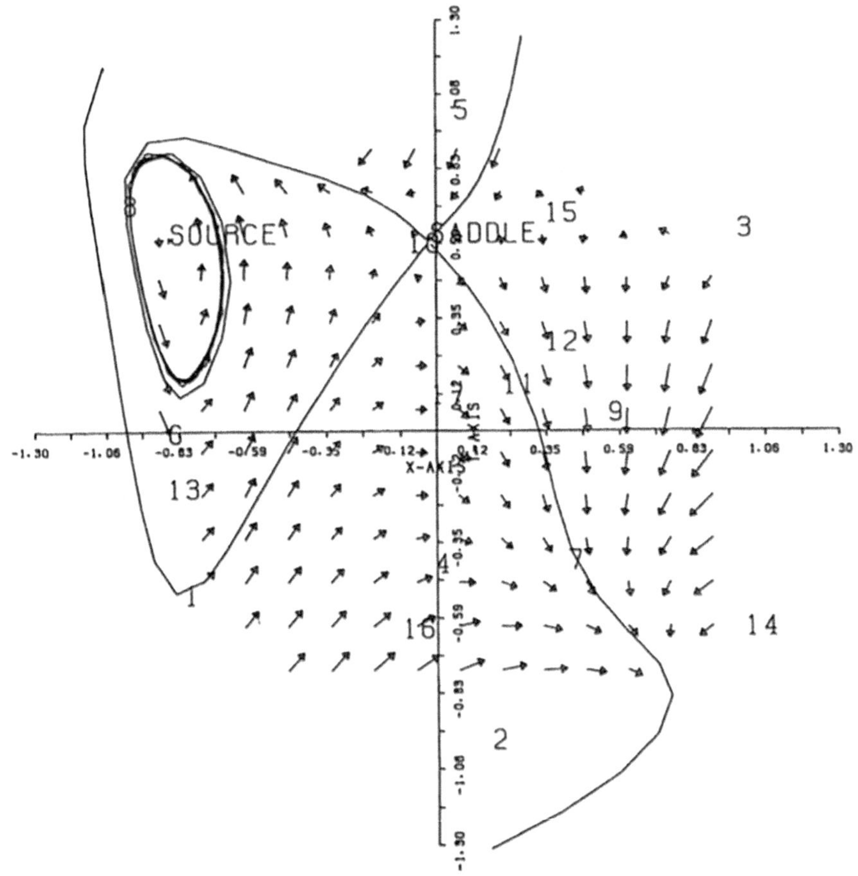

Fig. 7.3 Estimated vector field at Week 0 of Newcomb's longitudinal data. In this figure, numbers indicate the locations of members in the dormitory. (Reprinted from The Japanese Journal of Experimental Social Psychology, 2*9*, Fig. 5, Chino and Nakagawa [40], A bifurcation model of change in group structure, pp.25–38, with permission of The Japanese Group Dynamic Association)

should be constructed based on a deeper insight into the principle of interpersonal interactions and on ample empirical facts regarding interpersonal attraction.

Problem 7 (Indefinite Metric Space).

Indefinite metric space models for asymmetric MDS can generally be defined as models whose Hermitian matrix constructed directly or estimated from a one-mode, two-way square asymmetric matrix is *indefinite* (e.g., Chino & Shiraiwa [45]). Spaces with such a metric is not so familiar among social and behavioral scientists, but they are well known in physics. For example, Minkowski space in a *special relativity theory* in physics is known to possess a typical example of indefinite metric structure (e.g., Arfken & Weber [7]). In this space, the space–time interval, ds^2, is defined as

$$ds^2 = dx_0^2 - dx_1^2 - dx_2^2 - dx_3^2,$$

where $x_0 = ct$ and c is the velocity of light.

Berlin and Kac [14] proposed a mathematical model of a ferromagnet, in which it is assumed that there is a spin at each site of a regular lattice of N sites. Immediately before discussing some models for the interaction energy between neighboring spins, they examined the eigenvalue-eigenvector problem of the real *circulant matrix* (or *cyclic matrix*).

This matrix is asymmetric in general, but can be symmetric by imposing necessary conditions on the coefficients. They wrote down all the eigenvalues of $\boldsymbol{M}_c$ and the corresponding eigenvectors in the appendix.

$$M_c = \begin{pmatrix} c_1 & c_2 & c_3 & \cdots & c_{N-1} & c_N \\ c_N & c_1 & c_2 & \cdots & c_{N-2} & c_{N-1} \\ c_{N-1} & c_N & c_1 & \cdots & c_{N-3} & c_{N-2} \\ \vdots & \vdots & \vdots & \vdots & \vdots & \vdots \\ c_2 & c_3 & c_4 & \cdots & c_N & c_1 \end{pmatrix} \tag{7.45}$$

Chino [31] examined the eigenvalue-eigenvector problem of the Hermitian matrix constructed from a special 4 × 4 circulant matrix whose elements are composed of $c_1 = 1$, $c_2 = 2$, $c_3 = 3$, and $c_4 = 4$ using HFM, and showed that it has the indefinite metric structure. Kosugi (2004) considered the possible 16 patterns of the triadic asymmetric relationships among members of a group [76] from the viewpoint of the above theories by examining the eigenvalue–eigenvector structures of the Hermitian matrices constructed from the hypothetical 16 asymmetric relational matrices among three members of a group using HFM. In examining these structures, Kosugi assumes that the 3 × 3 sociomatrix has three values in which 1 denotes positive sentiment, – 1 denotes negative sentiment, and 0 indicates no such relations. Moreover, he assumes that all of the sentiments corresponding to self-similarities are positive, i.e., 1. It should be noticed that Kosugi's 16 patterns are special cases of the basic 16 patterns shown in Harary [76], because Harary's patterns assume neither the self-similarity nor the negative sentiment. In any case, Kosugi showed that 11 patterns out of 16 basic patterns have indefinite metric structures.

In contrast, Saito and Yadohisa [163] examined circle patterns of the skew-symmetric matrix constructed from a general circulant matrix $\boldsymbol{M}_c$ using SVD. It should be noticed that such circle patterns of the skew-symmetric matrix have a symplectic structure but have no Euclidean metric structures, as pointed out in the introductory section. It should also be noticed that the circulant matrix $\boldsymbol{M}_c$ does not necessarily have an indefinite metric structure. That is, if it is symmetric, then it might have a Euclidean structure or an indefinite metric structure, depending on the eigenvalue structure of the corresponding Hermitian matrix.

An interesting feature related to the asymmetric circulant matrix is the so-called *circular hierarchy*. For example, *circular triads* in dominance matrices, which is a special case of the circular hierarchy, are the cases where objects i dominates j, object *j* dominates *k*, but object *k* dominates *i*. Although there has been extensive literature on circular hierarchy in psychology, sociology, ethology, physics, and so on (e.g., Appleby [6]; Bass et al. [11]; Bodenreider [18]; Chadwick-Furman & Rinkevich [23]; De Sarbo & De Soete 58); De Vries [61]; Digby & Kempton [62]; Harshman[78]; Kendall [98]; Kendall & Smith [99]; Masure & Allee [122]; Saito [161]; Shepard [170]), it has frequently been considered as being unnatural and counterintuitive (e.g., Harshman [78]).

A more general matrix than the circulant matrix, which includes the circulant matrix as a special case is the *Toeplitz matrix* (e.g., Horn & Johnson [90]):

$$M_t = \begin{pmatrix} t_0 & t_1 & t_2 \cdots t_{N-1}\ t_N \\ t_{-1} & t_0 & t_1 \cdots t_{N-2}\ t_{N-1} \\ t_{-2} & t_{-1} & t_0 \quad t_1 \quad \cdots \quad c_{N-2} \\ \vdots & \vdots & \vdots \quad \vdots \quad \vdots \quad \vdots \\ t_{-N} & t_{-N+1} & t_{-N+2} \quad \cdots \quad t_{-1} \quad t_0 \end{pmatrix}, \tag{7.46}$$

where $t_{ij} = t_{j-i}$ for some given sequence $t_{-N}, t_{-N+1}, \cdots, t_{-1}, t_0, t_1, t_2, \cdots, t_{N-1}, t_N \in C$, and C is the field of complex number. The real version of this matrix seems to play a more important role in the analysis of circular hierarchy as well as indefinite metric space. For example, let us suppose that $N = 3$, $t_0 = t_1 = t_2 = 1$, and $t_{-1} = t_{-2} = -1$. The triadic relation with such a feature is the 9th pattern shown in Table 3. 2 of Kosugi [102]. The eigenvalues of the Hermitian matrix constructed from this matrix are $1 + \sqrt{3}$, 1 and $1 - \sqrt{3}$, which means that the above triadic relation among artificial members has an indefinite metric structure.

It will be necessary and desirable to gather empirical evidence further on circular hierarchy as well as indefinite metric structure, to examine the genetic background for such structures, and to establish substantive theories on these structures. For example, experiments conducted by Chadwick-Furman and Rinkevich [23] with a complex allorecognition system in a reef-building coral, in which circular hierarchies of overgrowth interactions were exhibited in some of the colonies of a coral reef, are interesting in that they aimed at formulating a genetic basis for their allorecognition system.

Problem 8 (One-Mode, Three-Way Asymmetric Relations).

Recently, several MDS as well as AMDS methods which assume some triadic distances have been developed, given a one-mode, three-way square data matrix. Methods for analyzing this type of triadic distance data go back to Hayashi [80]. Representative methods in this category are Cox et al. [91], Daws [52], De Rooij and Gower [54], Gower and De Rooij [73], Hayashi [80, 81], Heiser and Bennani [86], Joly and Le Calve´ [95], Nakayama [126], and Nakayama and Okada [127]. These are *descriptive MDS methods* which assume certain *symmetric, triadic distance models*. In contrast, De Rooij [53] and De Rooij and Heiser [55] pro-posed an *inferential AMDS method* which assumes some *asymmetric, triadic distance models*.

These MDS as well as AMDS methods assume some functions of *dyadic distances* except for Hayashi's (Hayashi, [80]) area model. Stated another way, neither of these models except Hayashi's deals with the triadic models which include triadic interaction (e.g., De Rooij and Gower [54]). However, there exist some triadic relational data in which it is inappropriate to reduce holistic triadic relationships into dyadic relationships. Typical examples can be found in some *balance theories* (e.g., Heider [84]; Newcomb [129]) and also in the *cognitive dissonance theory* (Festinger [65]) in social psychology.

As already discussed in Problem 7, Kosugi [102] examined the possible 16 patterns of triadic asymmetric relationships from the viewpoint of the above theories in social psychology by checking the eigenvalue-eigenvector structures of the Hermitian matrices. That is, he considers these structures from the balance theoretic viewpoint.

A possible method for handling the triadic interaction discussed above is to utilize a special skew-symmetric tensor called the *exterior form* which is usually denoted as *k-form* (or an *exterior form of degree k*). Here, a k-form is a function of k vectors which is *k-linear and antisymmetric* (e.g., Arnold [8]). Let us suppose that we embed objects in an oriented p-dimensional Euclidean space R^p. Then, the three-dimensional oriented volume of the projection of the parallelopiped with edges $\xi_1, \xi_2, \xi_3 \in R^p$ onto R^3 is a *3-form* on R^p. We may fit the square of this volume as a measure of the triadic interaction to a one-mode, three-way *square symmetric relational data*. On the other hand, for a one-mode, three-way *square asymmetric relational data*, one way to handle the skew-symmetric components of the data may be to use the *3*-form itself. In this case, it might be better to make some origin shift in defining the square of this volume to, say, the centroid of the coordinates of objects.

Of course, the idea of utilizing exterior forms is not new in the literature of MDS or, AMDS. For example, Chino's ASYMSCAL, GIPSCAL, and the Gower diagram fit area quantities to the skew-symmetric part of any one-mode, two-way square asymmetric matrix, as has already been discussed in Sect. 7.2. These area quantities are all *2*-forms. In contrast, Hayashi's area model for triadic data utilizes a special *3*-form, which is associated with the area of simplex composed of the three vertices corresponding to three objects for handling triadic data. The idea of utilizing exterior forms can naturally be extended to *MDS* as well as *AMDS for n-tuples* (e.g., Cox et al. [49]).

However, caution should be exercised when utilizing tensors in general for fitting any statistical model to data. As has already been mentioned elsewhere (e.g., De Silva & Lim [60]; Stegeman & Comon [177]), the following best rank-r approximation problem, *Approx* (A, r), has no solution in general for $r = 2, \cdots, \mathit{min}\ (d_1, \cdots, d_k)$ and $k \geq 3$, where an order-k tensor A is an element of a *tensor product* of k real vector space $V_1 \otimes \cdots \otimes V_k$, as defined in any standard algebra textbook:

Approx ($\boldsymbol{A}$, r): Given an order-k tensor determine vectors $i = 1, r$, which minimize

$$||\boldsymbol{A} - \boldsymbol{x}_1 \otimes \boldsymbol{y}_1 \otimes \cdots \otimes z_1 - \cdots - \boldsymbol{x}_r \otimes \boldsymbol{y}_r \otimes \cdots \otimes z_r||, \tag{7.47}$$

where denotes some choice of norm on

Stated another way, the best low-rank approximation problem for tensor is ill-posed for all orders (higher than 2), all norms, and many ranks, and the set of tensors that fails to have a best low-rank approximation has positive volume (De Silva & Lim, 2008). These results may be contrasted with the case when $k = 2$, i.e., the problem with the Eckart-Young theorem, which was discussed in the introductory section.

As reviewed extensively in Stegeman and Comon [177], there has already been an extensive literature on this problem and related topics, and several psychometricians have partly contributed to them (e.g., Krijnen, Dijkstra, & Stegeman [104]; Kruskal [109]; Stegeman [174–176]; Stegeman & Lathauwer [178]; Ten Berge & Kiers [194]; Ten Berge, Kiers, & De Leeuw [195]; Ten Berge, Sidiropoulos, & Rocci [196]).

It might be possible to overcome possible difficulties in introducing some appropriate tensors in the analysis of the triadic interaction problem discussed above by utilizing future methods for overcoming the ill-posedness of the lower-rank approximation problem. This is because the approach to this problem is still an active area of research, which has recently received increasing attention in statistics as well as in mathematics.

Acknowledgements The author is indebted to reviewers for their valuable suggestions and comments on earlier version of this paper. The author is also indebted to Gregory L. Rohe for proofreading of an earlier version of this paper.

Notes

This chapter is the reproduction of Chino [33]. Some of the errata in the text as well as the reference list in the original paper were corrected in this chapter.

So far, we have reviewed works on asymmetric MDS according to Chino [op. cit.]. However, asymmetric MDS is still an active area of research, and there has been a body of research since 2012 (e.g., Bove et al. [20]; Chino [34–39]; Imaizumi [91]; Kumagai [111–113]; Nakayama et al. [128]; Nakayama & Okada [127]; Okada [139]; Okada [141]; Shojima [171]; Tsuchida & Yadohisa [203]; Tanioka & Yadohisa [190–192]). A comprehensive review of these works will be done elsewhere in the near future.

Finally, a matter of crucial importance which exists in the contemporary AMDS is that the *holistic mathematical metric structure* including the symmetric part and the asymmetric part of the square asymmetric matrix is unclear except for a *Hibert space model*, HFM, by Chino and Shiraiwa [45] and a *real asymmetric Minkowski space model* by Sato [164, 165]. Such a holistic mathematical metric structure can be clarified by computing the real inner product from each of the model equations of the augmented distance models in the contemporary AMDS. The special inner product matrix thus computed are inevitably real asymmetric. If we apply HFM to this asymmetric inner product matrix, then we can examine the holistic metric structure of each of the augmented distance models. We will discuss the results of such applications elsewhere in the near future.

References

1. Akaike, H. (1974). A new look at the statistical model identification. *IEEE Transactions on Automatic Control, 19*, 716–723.
2. Alfakih, A. Y., Khandani, A., & Wolkowicz, H. (1999). Solving Euclidean distance matrix completion prob lems via semidefinite programming. *Computational Optimization and Applications, 12*, 13–30.
3. Amari, S. (1971). Characteristics of randomly connected threshold-element networks and network systems. *Proceedings of the IEEE, 59*, 35–47.
4. Andersen, E. B. (1980). *Discrete statistical models with social science applications.* North Hol land.
5. Appelman, I. B., & Mayzner, M. S. (1982). Application of geometric models to letter recognition: Distance and density. *Journal of Experimental Psychology, General, 111*, 60–100.
6. Appleby, M. C. (1983). The probability of linearity in hierarchies. *Animal Behavior, 31*, 600–6008.
7. Arfken, G. B., & Weber, H. J. J. (1995). *Mathematical methods for physicists* (4th ed.). Academic Press.
8. Arnold, V. I. (1978). *Mathematical methods of classical mechanics.* Springer Verlag.
9. Ashby, F. G., & Perrin, N. A. (1988). Toward a unified theory of similarity and recognition. *Psychological Review, 95*, 124–150.
10. Balthrop, J., Forrest, S., Newman, M. E. J., & Williamson, M. M. (2004). Technological networks and the spread of computer viruses. *Science, 304*, 527–529.
11. Bass, F. M., Pessemier, E. A., & Lehmann, D. R. (1972). An experimental study of relationships between attudes, brand preference, and choice. *Behavioral Science, 17*, 532–541.
12. Basu, D. (1955). On statistics independent of a complete sufficient statistic. *Sankhya, 15*, 377–380.
13. Bennett, J. F., & Hays, W. L. (1960). Multidimensional unfolding: Determining the the dimensionality of ranked preference data. *Psychometrika, 25*, 27–43.
14. Berlin, T. H., & Kac, C. (1952). The spherical model of ferromagnet. *The Physical Review, 86*, 821–835.
15. Birch, M. W. (1963). Maximum likelihood in three-way contingency tables. *Journal of the Royal Statistical Society, Series B, 25*, 220–233.
16. Bishop, Y. M. M., Fienberg, S. E., & Holland, P. W. (1975). *Discrete multivariate analysis-theory and practice.* The MIT press.
17. Bock, R. D., & Jones, L. V. (1968). *The measurement and prediction of judgment and choice.* Holden-Day.
18. Bodenreider, O. (2001). Circular hierarchical relationships in the UMLS: Etiology, diagnosis, treatment, com plications and prevention. In *Proceedings of the AMIA symposium*, pp. 57–61.
19. Borg, I., & Groenen, P. J. F. (2005). *Modern multidimensional scaling—Theory and applications* (2nd ed.). Springer.
20. Bove, G., Okada, A., & Vicari, D. (2021). *Methods for the analysis of asymmetric proximity data.* Springer Nature Singerpore.
21. Carroll, D. J., & Chang, J. J. (1970). Analysis of individual differences in multidimensional scaling via an N-Way generalization of "Eckart-Young" decomposition. *Psychometrika, 35*, 283–319.
22. Caussinus, H. (1965). Contribution a‘ l'analyse statistique des tableaux de corre´lation. *Annales de la Faculte´ des Sciences de l'Universite´ de Toulouse, 29*, 77–182.
23. Chadwick-Furman, N., & Rinkevich, B. (1994). A complex allorecognition system in a reef-building coral: Delayed responses, reversals and nontransitive hierarchies. *Coral Reefs, 13*, 57–63.
24. Chino, N. (1977). N-ko no taisyo kan no hitaisyo na kankei wo zusikika suru tameno ichi-giho. [A graphical technique for representing the asymmetric relationships between N objects]. In *Proceedings of the 5th annual meeting of the Behaviormetric Society of Japan*, 146–149 (in Japanese).

25. Chino, N. (1978). A graphical technique for representing the asymmetric relation ships between N objects. *Behaviormetrika, 5*, 23–40.
26. Chino, N. (1980). A unified geometrical interpretation of the MDS techniques for the analysis of asymmetry and related techniques. In: *Paper presented at the symposium on "Asymmetric multidimensional scaling" at the Spring Meeting of the Psychometric Society*, Iowa, U.S.A.
27. Chino, N. (1990). A generalized inner product model for the analysis of asymmetry. *Behaviormetrika, 27*, 25–46.
28. Chino, N. (1991). A critical review for the analysis of asymmetric relational data. *Bulletin of The Faculty of Letters of Aichi Gakuin University, 21*, 31–52.
29. Chino, N. (1997). *Hitaisho Tajigen Syakudo Koseiho [Asymmetric Multidimensional Scaling]*. Gendai Sugaku (in Japanese).
30. Chino, N. (1998). Hilbert space theory in psychology-(1) Basic concepts and possible applications. *Bulletin of the Faculty of Letters of Aichi Gakuin University, 28*, 45–66.
31. Chino, N. (2001). Complex difference system models for the analysis of asymmetry. In H. Yanai, A. Okada, K. Shigemasu, Y., Kano, and J. J. Meulman (Eds.), *New developments in psychometrics. Proceedings of the international meeting of the psychometric society IMPS2001* (pp.479–486), Osaka, Japan.
32. Chino, N. (2005). Abnormal behaviors of members predicted by a complex difference system model. *Bulletin of the Faculty of Psychological 4 Physical Science, 1*, 69–73.
33. Chino, N. (2012). A brief survey of asymmetric MDS and some open problems. *Behaviormetrika, 39*, 127–165.
34. Chino, N. (2016). A general non-Newtonian n-body problem and dynamical scenarios of solutions. In: *Paper presented at the 31th International Congress of Psychology*. Yokohama, Japan.
35. Chino, N. (2017). Dynamical scenarios of changes in asymmetric relationships over time (1). *Bulletin of The Faculty of Psychological & Physical Science, 13*, 23–31.
36. Chino, N. (2018). An elementary theory of a dynamic weighted digraph (1). *Bulletin of The Faculty of Psychological & Physical Science, 14*, 23–31.
37. Chino, N. (2019). An elementary theory of a dynamic weighed digraph (2). *Journal of the Institute for Psychological and Physical Science, 11*, 1–8.
38. Chino, N. (2020a). How to use the Hermitian form model for asymmetric MDS. In T. Imaizumi, A. Nakayama & S. Yokoyama (Eds.), *Advanced studies in behaviormetrics and data science* (pp.19–41). Springer Nature Singapore.
39. Chino, N. (2020). Hermitian symmetry on a Hilbert Space—Its applications to some asymmetric phenomena. *Natural Science, 12*, 221–236.
40. Chino, N., & Nakagawa, M. (1990). A bifurcation model of change in group structure. *The Japanese Journal of Experimental Social Psychology, 29*, 25–38.
41. Chino, N., & Okada, A. (1996). Hitaisho tajigen syakudo Koseiho to sono syuhen [Asymmetric multidimen sional scaling and related topics]. *Kodokeiryogaku, 23*, 130–152. (in Japanese).
42. Chino, N., & Saburi, S. (2006). Tests of symmetry in asymmetric MDS. In: *Paper presented at the 2nd German Japanese Symposium on Classification—Advances in data analysis and related new techniques 4 application*. Berlin, Germany, March 7–8.
43. Chino, N., & Saburi, S. (2009). Features of quasi-symmetry-like asymmetric MDS models and independence of some tests for symmetry. In: *Proceedings of the 37th annual meeting of the Behaviormetric Society of Japan* (pp. 24–25), Oita University.
44. Chino, N., & Saburi, S. (2010). Controlling the two kinds of error rate in selecting an appropriate asymmetric MDS model. *Journal of the Institute for Psychological and Physical Science, 2*, 37–42.
45. Chino, N., & Shiraiwa, K. (1993). Geometrical structures of some non-distance models for asymmetric MDS. *Behaviormetrika, 20*, 35–47.
46. Constantine, A. G., & Gower, J. C. (1978). Graphical representation of asymmetric matrices. *Applied Statistics, 27*, 297–304.
47. Coombs, C. H. (1964). *A theory of data*. Wiley.

48. Cox, D. R. (1982). Statistical significance tests. *British Journal of Clinical Pharmacology, 14*, 325–331.
49. Cox, T. F., & Cox, M. A. A. (1991). Multidimensional scaling for n-tuples. *British Journal of Mathematical and Statistical Psychology, 44*, 195–206.
50. Cox, T. F., & Cox, M. A. A. (2001). *Multidimensional scaling* (2nd ed.). Chapman & Hall/CRC.
51. Cox, T. F., Cox, M. A. A., & Branco, J. A. (1991). Multidimensional scaling for n-tuples. *British Journal of Mathematical and Statistical Psychology, 44*, 195–206.
52. Daws, J. T. (1996). The analysis of free-sorting data: Beyond pairwise co-occurrences. *Journal of Classifica tion, 13*, 57–80.
53. De Rooij, M. (2002). Studying triadic distance models under a likelihood function. In S. Nishisato, Y. Baba, H. Bozdogan, & K. Kanefuji (Eds.), *Measurement and Multivariate Analysis* (pp. 69–76). Springer.
54. De Rooij, M., & Gower, J. C. (2003). The geometry of triadic distances. *Journal of Classification, 20*, 181–220.
55. De Rooij, M., & Heiser, W. J. (2000). Triadic distance models for the analysis of asymmetric three-way prox imity data. *British Journal of Mathematical and Statistical Psychology, 53*, 99–119.
56. De Rooij, M., & Heiser, W. J. (2003). A distance representation of the quasi-symmetry model and related distance models In H. Yanai et al. (Eds.), *New Developments in psychometrics* (pp.487–494). Springer.
57. De Rooij, M., & Heiser, W. J. (2005). Graphical representations and odds ratios in a distance-association model for the analysis of cross-classified data. *Psychometrika, 70*, 99–122.
58. DeSarbo, W. S., & De Soete, G. (1984). On the use of hierarchical clustering for the analysis of nonsymmetric proximities. *Journal of Consumer Research, 11*, 601–610.
59. DeSarbo, W. S., Johnson, M., Manrai, A., & Edwards, E. (1992). TSCALE: A new multi-dimensional scaling procedure based on Tversky's contrast model. *Psychometrika, 57*, 43–70.
60. De Silva, V., & Lim, L.-H. (2008). Tensor rank and the ill-posedness of the best low-rank approximation problem. *The SIAM Journal of Matrix Analysis and Applications, 30*, 1084–1127.
61. De Vries, H. (1995). An improved test of linearity in dominance hierarchies containing unknown or tied rela tion ships. *Animal Behavior, 50*, 1375–1389.
62. Digby, R. G. N., & Kempton, R. A. (1987). *Multivariate analysis of ecological communities*. Chapman and Hall.
63. Eckart, C., & Young, G. (1936). The approximation of one matrix by another of lower rank. *Psychometrika, 1*, 211–218.
64. Escoufier, Y., & Grorud, A., et al. (1980). Analyse factorielle des matrices carrees non symetriques [Factor analysis of square asymmetric matrices]. In E. Diday (Ed.), *Data analysis and informatics* (pp. 263–276). North Holland.
65. Festinger, L. (1957). *A theory of cognitive dissonance*. Stanford University Press.
66. Fong, D. K. H., DeSarbo, W. S., Park, J., & Scott, C. J. (2010). A Bayesian vector multidimensional scaling procedure for the analysis of ordered preference data. *Journal of the American Statistical Association, 105*, 482–492.
67. Fukai, T., & Shiino, M. (1990). Asymmetric neural networks incorporating the Dale hypothesis and noise-driven chaos. *Physical Review Letters, 64*, 1465–1468.
68. Getty, D. J., Swets, J. A., Swets, J. B., & Green, D. M. (1979). On the prediction of confusion matrices from similarity judgments. *Perception & Psychophysics, 26*, 1–19.
69. Gilmore, G. C., Hersh, H., Caramazza, A., & Griffin, J. (1979). Multidimensional letter similarity derived from recognition errors. *Perception & Psychophysics, 25*, 425–431.
70. Goodman, L. A. (1971). The analysis of multidimensional contingency tables: Stepwise procedures and direct estimation methods for building models for multiple classifications. *Technometrics, 13*, 33–61.

71. Gower, J. C. (1977). The analysis of asymmetry and orthogonality. In J. R. Barra, F. Brodeau, G. Romer, & B. van Cutsem (Eds.), *Recent developments in statistics* (pp. 109–123). North Holland.
72. Gower, J. C. (1984). Multivariate analysis: Ordination, multidimensional scaling and allied topics. In E. H. Llokyd (Ed.), *Handbook of applicable mathematics* (pp.727–781). Vol. VI, Wiley.
73. Gower, J. C., & De Rooij, M. (2003). A comparison of the multidimensional scaling of triadic and dyadic distances. *Journal of Classification, 20*, 115–136.
74. Gower, J. C., & Zielman, B. (1992). Some remarks on orthogonality in the analysis of asymmetry. Research Report RQ-92–08, Dept. of Data Theory, Univ. of Leiden, Netherlands.
75. Guttman, L. (1968). A general nonmetric technique for finding the smallest coordinate space for a configura tion of points. *Psychometrika, 33*, 469–506.
76. Harary, F. J. (1968). *Graph Theory*. Massachusetts: Addison-Wesley.
77. Harshman, R. A. (1978). Models for analysis of asymmetrical relationships among N objects or stimuli. In: *Paper presented at the first joint meeting of the psychometric society and the society for mathematical psychology*, Hamilton, Canada.
78. Harshman, R. A. (1981). *DEDICOM multidimensional analysis of skew-symmetric data.* Theory. Un published manuscript.
79. Harshman, R. A., Green, P. E., Wind, Y., & Lundy, M. E. (1982). A model for the analysis of asymmetric data in marketing research. *Marketing Science, 1*, 205–242.
80. Hayashi, C. (1972). Two dimensional quantifications based on a measure of dissimilarity among three elements. *Annals of the Institute of Statistical Mathematics, 25*, 251–257.
81. Hayashi, C. (1989). Multiway data matrices and method of quantification of qualitative data as a strategy of data analysis. In R. Coppi & B. Bolasco (Eds.), *Multiway data analysis* (pp. 131–142). North-Holland.
82. Hays, W. L., & Bennett, J. F. (1961). Multidimensional unfolding: Determining configuration from complete rank order preference data. *Psychometrika, 26*, 221–238.
83. Hefner, R. A. (1958). Extensions of the law of comparative judgment to discriminable and multi-dimensional stimuli. *Doctoral dissertation*, University of Michigan.
84. Heider, F. (1946). Attitudes and cognitive organization. *The Journal of Psychology, 21*, 107–112.
85. Heiser, W. J. (1987). Joint ordination of species and sites: the unfolding technique. In P. Legendre & L. Le gendre (Eds.), *Developments in numerical ecology* (pp. 189–221). Springer Verlag.
86. Heiser, W. J., & Bennani, M. (1997). Triadic distance models: Axiomatization and least squares representation. *Journal of Mathematical Psychology, 41*, 189–206.
87. Hirotsu, C. (1983). Defining the pattern of association in two-way contingency tables. *Biometrika, 70*, 579–589.
88. Hogg, R. V., & Craig, A. T. (1956). Sufficient statistics in elementary distribution theory. *Sankhya, 17*, 209–216.
89. Holman, E. W. (1979). Monotonic models for asymmetric proximities. *Journal of Mathematical Psychology, 20*, 1–15.
90. Horn, R. A., & Johnson, C. R. (1985). *Matrix analysis*. Cambridge University Press.
91. Imaizumi, T. (2020). On detection of the unique dimensions of asymmetry in proximity data. In T. Imaizumi, A. Nakayama, & S. Yokoyama (Eds.), *Advanced studies in behaviormetrics and data science* (pp. 119–130). Singerpore: Springer Nature Singerpore.
92. Irony, T. Z., & Singpurwalla, N. D. (1997). Non-informative priors do not exist: A dialogue with Jose´ M. Bernardo. *Jouranal of Statistical Planning and Inference, 65*, 159–189.
93. Je, H., Kim, D., & Pohang, P. (2008). Bayesian multidimensional scaling for multi-robot localization. In *Proceedings of IEEE International Conference on Networking, Sensoring and Control*, pp. 926–931.
94. Johnson, N., Kotz, S., & Balakrishnan, N. (1997). *Discrete multivariate distributions*. Wiley.
95. Joly, S., & Le Calve´, G. (1995). Three-way distances. *Journal of Classification, 12*, 191–205.

96. Kanter, I. J. (1988). Inhomogeneous magnetization in dilute asymmetric and symmetric systems. *Physical Review Letters, 60*, 1891–1894.
97. Kastenbaum, M. A. (1960). A note on the additive partitioning of chi-square in contingency tables. *Biometrics, 16*, 416–422.
98. Kendall, M. G. (1962). *Rank correlation methods* (3rd ed.). Charles Griffin.
99. Kendall, M. G., & Smith, B. B. (1940). On the method of paired comparisons. *Biometrika, 31*, 324–345.
100. Keren, G., & Baggen, S. (1981). Recognition models of alphanumeric characters. *Perception & Psychophysics, 29*, 234–246.
101. Kiers, H. A. L., & Takane, Y. (1994). A generalization of GIPSCAL for the analysis of asymmetric data. *Journal of Classification, 11*, 79–99.
102. Kosugi, K. (2004). Shinri-Ronri to Taido-riron he no Suri-apurouchi [Mathematical Approaches to Interpersonal Psycho-Logic and Attitude Theory]. *Unpublished doctoral dissertation*, Kwansei-Gakuin University, Nishinomiya.
103. Kree, R., & Zippelius, A. (1995). Asymmetrically diluted neural networks. In E. Domany, J.L. van Hemmen & K. Schulten (Eds.), *Models of Neural Networks I* (2nd ed.) (pp.201–220).
104. Krijnen, W. P., Dijkstra, T. K., & Stegeman, A. (2008). On the non-existence of optimal solutions and the occurrence of "degeneracy" in the CANDECOMP/PARA- FAC model. *Psychometrika, 73*, 431–439.
105. Krumhansl, C. L. (1978). Concerning the applicability of geometric models to similarity data: The interrelationship between similarity and spatial density. *Psychological Review, 85*, 445–463.
106. Krummenauer, F. (1998). Limit theorems for multivariate discrete distributions. *Metrika, 47*, 47–69.
107. Kruskal, J. B. (1964). Multidimensional scaling by optimizing goodness of fit to a nonmetric hypothesis. *Psychometrika, 29*, 1–27.
108. Kruskal, J. B. (1964). Multidimensional scaling: A numerical method. *Psychometrika, 29*, 115–129.
109. Kruskal, J. B. (1989). Rank, decomposition, and uniqueness for 3-way and N-way arrays. In R. Coppi & S. Bolasco (Eds.), *Multiway Data Analysis* (pp. 7–18). North Hol-land.
110. Kumagai, A. (2010). Kotenteki Tajigen Schakudo Koseiho ni motozuku Hitaisyo Kanrensei deta no Bunseki [Analysis of asymmetric relational data based on classical multidimensional scaling]. *Nihon Oyo Sugakukai Ronbunshi, 20*, 57–66. (in Japanese).
111. Kumagai, A. (2014). Extension of classical MDS to treat dissimilarities not satisfying axioms of distance. *Japan Journal of Industrial and Applied Mathematics, 31*, 111–124.
112. Kumagai, A. (2015). Multidimensional scaling in dually flat spaces. *Japan Journal of Industrial and Applied Mathematics, 32*, 51–63.
113. Kumagai, A. (2016). Semilocal properties of canonical divergences in dually flat spaces. *Japan Journal of Industrial and Applied Mathematics, 33*, 417–426.
114. Kuriki, S. (1991). A construction of simultaneous confidence regions and multiple comparison procedures concerning the paired comparison model. *Ouyo Tokeigaku[Applied Statistics], 20*, 127–137 (in Japanese).
115. Laurent, M. (2001). Matrix completion problems. Encyclopedia of. *Optimization, 3*, 221–229.
116. Lee, M. D. (2008).Three case studies in the Bayesian analysis of cognitive models. *Psychonomic Bulletin 4 Review, 15*, 1–15.
117. Lehmann, E. L. (1983). *Theory of Point Estimation*. Wiley.
118. Li, Z. (2008). Exponential stability of synchronization in asymmetrically coupled dynamical net-works. *Chaos, 18*, 1–11.
119. Lingoes, J. C. (1973). *The Guttman-Lingoes nonmetric program series*. Mathesis Press.
120. Loisel, S., & Takane, Y. (2011). Generalized GIPSCAL re-revisited: A fast convergent algorithm with acceleration by the minimal polynomial extrapolation. *Advances in Data Analysis and Classification, 5*, 57–75.
121. Luce, R. D. (1963). Detection and recognition. In R. D. Luce, R. R. Bush, & E. Galanter (Eds.), *Handbook of mathematical psychology* (pp. 103–189). Wiley.

122. Masure, R. H., & Allee, W. C. (1934). The social order in flocks of the common chicken and the pigeon. *The Auk, 51*, 306–327.
123. Matsumoto, M. (1986). *Foundation of finsler geometry and special finsler space*. Kai-seisha.
124. Medin, D. L., & Schaffer, M. M. (1978). Context theory of classification learning. *Psychological Review, 85*, 207–238.
125. Nakatani, L. H. (1972). Confusion-choice model for multidimensional psychophysics. *Journal of Mathematical Psychology, 9*, 104–127.
126. Nakayama, A. (2005). A multidimensional scaling model for three-way data analysis. *Behaviormetrika, 32*, 95–110.
127. Nakayama, A., & Okada, A. (2012). Reconstructing one-mode three-way asymmetric data for multidimensional scaling. In A. WolfgangGaul, A. Andreas Geyer-Schulz & J. K. Lars Schmidt-Thieme (Eds.), *Challenges at the interface of data analysis, computer science, and optimization* (pp.133–141). Springer
128. Nakayama, A., Tsurumi, H., & Okada, A. (2014). A symmetry test for one-mode three-way proximity data. In W. Gaul, A. Geyer-Schulz, Y. Baba, & A. Okada (Eds.), *German-Japanese interchange of data analysis results—Studites in classification, data analysis, and knowledge organization* (pp. 125–134). Springer International Publishing.
129. Newcomb, T. M. J. (1953). An approach to the study of communicative acts. *Psychological Review, 60*, 393–404.
130. Newcomb, T. M. (1961). *The acquaintance process*. Holt. Rinehart and Winston.
131. Neyman, J., & Scott, E. L. (1948). Consistent estimates based on partially consistent observations. *Econometrika, 16*, 1–32.
132. Nosofsky, R. M. (1984). Choice, similarity, and the context theory of classification. *Journal of Experimental Psychology, 10*, 104–114.
133. Nosofsky, R. M. (1985a). Luce's choice model and Thurstone's categorical judgment model com-pared: Kornbrot's data revisited. Perception 4 *Psychophysics, 37*, 89–91.
134. Nosofsky, R. M. (1985b). Overall similarity and the identification of separable-dimension stimuli: A choice model analysis. Perception 4 *Psychophysics, 38*, 415–432.
135. Nosofsky, R. M. (1986). Attention, similarity, and the identification-categorization of integral stimuli. *Journal of Experimental Psychology: Learning, Memory, and Cognition, 13*, 87–108.
136. Nosofsky, R. M. (1991). Stimulus bias, asymmetric similarity, and classification. *Cognitive Psychology, 23*, 94–140.
137. Oh, M.-S., & Raftery, A. E. (2001). Bayesian multidimensional scaling and choice of dimension. *Journal of the American Statistical Association, 96*, 1031–1044.
138. Oh, M.-S., & Raftery, A. E. (2007). Model-based clustering with dissimilarities: A Bayesian approach. *Jouranal of Computational and Graphical Statistics, 16*, 559–585.
139. Okada, A. (2012). Analysis of car switching data by asymmetric multidimensional scaling based on singular value decomposition. In W. Gaul, A. Geyer-Schulz, L. Schmidt-Thieme & J. Kunze (Eds.), *Challenges at the interface of data analysis, computerscience, and optimization* (pp.143–150). Springer-Verlag.
140. Okada, K. (2011). Beizu suitei ni yoru hitaisyo MDS. [Asymmetric MDS by Bayesian estimation]. In *Proceedings of the 39th annual meeting of the Behaviormetric Society of Japan*. Okayama Science University (in print).
141. Okada, K. (2012). A Baysian approach to asymmetric multidimensional scaling. *Behaviormetrika, 39*, 49–62.
142. Okada, A., & Imaizumi, T. (1984). Geometric models for asymmetric similarity. *Research Reports of School of Social Relations*, Rikkyo (St. Paul's) University.
143. Okada, A., & Imaizumi, T. (1987). Nonmetric multidimensional scaling of asymmetric proximities. *Behaviormetrika, 21*, 81–96.
144. Okada, A., & Imaizumi, T. (1997). Asymmetric multidimensional scaling of two-mode, three-way proximities. *Journal of Classification, 14*, 195–224.
145. Okada, K., & Mayekawa, S. (2011). Bayesian nonmetric successive categories multidimensional scaling. *Behaviormetrika, 38*, 17–32.

146. Okada, K., & Shigemasu, K. (2010). Bayesian multidimensional scaling for the estimation of a Minkowski exponent. *Behavior Research Methods, 42*, 899–905.
147. Parisi, G. (1986). Asymmetric neural networks and the process of learning. *Journal of Physics A: Mathematical and General, 19*, L675–L680.
148. Park, J., DeSarbo, W. S., & Liechty, J. (2008). A hierarchical Bayesian multidimensional scaling methodology for accommodating both structural and preference heterogeneity. *Psychometrika, 73*, 451–472.
149. Price, S. J. (1989). Bayesian statistics—Principles, models, and applications. Wiley
150. Ramsey, J. O. (1969). Some statistical considerations in multidimensional scaling. *Psychometrika, 34*, 167–182.
151. Ramsay, J. O. (1977). Maximum likelihood estimation in multidimensional scaling. *Psychometrika, 42*, 241–266.
152. Ramsay, J. O. (1978). Confidence regions for multidimensional scaling analysis. *Psychometrika, 43*, 145–160.
153. Ramsay, J. O. (1982). Some statistical approaches to multidimensional scaling data. *The Journal of the Royal Statistical Society,Series A (General), 145*, 285–312.
154. Randers, G. (1941). On an asymmetrical metric in the four-space of general relativity. *Physical Review, 59*, 195–199.
155. Richardson, M. W. (1938). Multidimensional psychophysics. *Psychological Bulletin, 35*, 659.
156. Rocci, R., & Bove, G. (2002). Rotation techniques in asymmetric multidimensional scaling. *Journal of Computational Statistics and Graphical Statistics, 11*, 405–419.
157. Rushen, J. (1982). The peck orders of chickens: How do they develop and why are they linear? *Animal Behavior, 30*, 1129–1137.
158. Saburi, S., & Chino, N. (2008). A maximum likelihood method for an asymmetric MDS model. *Computational Statistics and Data Analysis, 52*, 4673–4684.
159. Saito, T. (1980). *Tajigen Shakudo Kouseiho [Multidimensional Scaling]*. Tokyo: Asakura Book Company (in Japanese).
160. Saito, T. (1991). Analysis of asymmetric proximity matrix by a model of distance and additive terms. *Behaviormetrika, 29*, 45–60.
161. Saito, T. (2002). Circle structure derived from decomposition of asymmetric data matrix. *Journal of the Japanese Society of Computational Statistics, 15*, 1–18.
162. Saito, T., & Takeda, S. (1990). Multidimensional scaling of asymmetric proximity: Model and method. *Behaviormetrika, 28*, 49–80.
163. Saito, T., & Yadohisa, H. (2005). *Data analysis of asymmetric structures—Advanced approaches in computational statistics*. Marcel Dekker.
164. Sato, Y. (1988). An analysis of sociometric data by MDS in Minkowski space. In K. Matsusita (Ed.), *Statistical Theory and Data Analysis II* (pp. 385–396). North Holland.
165. Sato, Y. (1989). Minkowski keiryo o mochiita hitaisyo ruijisei no kyori hyogen ni tsuite. [On the distance representation of asymmetric dissimilarities by Minkowskimetric function]. *Computational Statistics, 2*, 35–45. (in Japanese).
166. Schönemann, P. H. (1970). On metric multidimensional unfolding. *Psychometrika, 35*, 349–366.
167. Shepard, R. N. (1957). Stimulus and response generalization: A stochastic model relating generalization to distance in psychological space. *Psychometrika, 22*, 325–345.
168. Shepard, R. N. (1958). Stimulus and response generalization: Deduction of the generalization gradient from a trace model. *Psychological Review, 65*, 242–256.
169. Shepard, R. N. (1958). Stimulus and response generalization: Tests of a model relating generalization to distance in psychological space. *Journal of Experimental Psychology, 55*, 509–523.
170. Shepard, R. N. (1964). Circularity in judgments of relative pitch. *Journal of the Acoustical Society of America, 36*, 2346–2353.
171. Shojima, K. (2012). Asymmetric triangulation scaling. Asymmetric multidimensional scaling for visualizing interr-item dependency structure. *Behaviormetrika, 39*, 27–48.

172. Smith, J. E. K. (1982). Recognition models evaluated: A commentary on Keren and Baggen. *Perception & Psychophysics, 31*, 183–189.
173. So, A.M.-C., & Ye, Y. (2007). Theory of semidefinite programming for sensor network localization. *Mathematical Programming, 109*, 367–384.
174. Stegeman, A. (2006). Degeneracy in CANDECOMP/PARAFAC explained for p x p x 2 arrays of rank p + 1 or higher. *Psychometrika, 71*, 483–501.
175. Stegeman, A. (2007). Degeneracy in CANDECOMP/PARAFAC explained for several three-sliced arrays with a two-valued typical rank. *Psychometrika, 72*, 601–619.
176. Stegeman, A. (2008). Low-rank approximation of generic p x q x 2 arrays and diverging com-ponents in the CANDECOMP/PARAFAC model. *SIAM Journal on Matrix Analysis and Applications, 30*, 988–1007.
177. Stegeman, A., & Comon, P. (2010). Subtracting a best rank-1 approximation may increase tensor rank. *Linear Algebra and its Applications, 433*, 1276–1300.
178. Stegeman, A., & De Lathauwer, L. (2009). A method to avoid diverging components in the CAN-DECOMP/PARAFAC model for generic Ix J x 2 arrays. *SIAM Journal on Matrix Analysis and Applications, 30*, 1614–1638.
179. Suppes, P., & Zinnes, J. L. (1963). Basic measurement theory. In R. D. Luce et al. (Eds.), *Handbook of mathematical psychology*, Vol.1, Wiley.
180. Takane, Y. (1978). A maximum likelihood method for nonmetric multidimensional scaling: 1. The case in which all empirical pairwise orderings are independent-theory. *Japanese Psychological Research, 20*, 7–17.
181. Takane, Y. (1978). A maximum likelihood method for nonmetric multidimensional scaling: 2. The case in which all empirical pairwise orderings are independent- Evaluation. *Japanese Psychological Research, 20*, 105–114.
182. Takane, Y. (1980). *Tajigen Shakudoho [Multidimensional Scaling]*. Tokyo University Press. (in Japanese).
183. Takane, Y. (1981). Multidimensional successive categories scaling: A maximum likelihood method. *Psychometrika, 46*, 9–28.
184. Takane, Y. (1987). Analysis of contingency tables by ideal point discriminant analysis. *Psychometrika, 52*, 493–513.
185. Takane, Y., & Carroll, J. D. (1981). Nonmetric maximum likelihood multidimensional scaling from directional rankings of similarities. *Psychometrika, 46*, 389–405.
186. Takane, Y., & Kiers, H. A. L. (1997). Latent class DEDICOM. *Journal of Classification, 14*, 225–247.
187. Takane, Y., & Shibayama, T., et al. (1986). Comparison of models for stimulus recognition data. In J. de Leeuw (Ed.), *Multidimensional Data Analysis* (pp. 119–148). DSWO Press.
188. Takane, Y., & Shibayama, T. (1992). Structures in stimulus identification data. In F. G. Ashby (Ed.), *Multidimensional models of perception and cognition* (pp. 335–362). Lawrence Erlbaum Associates.
189. Takane, Y., Young, F. W., & de Leeuw, J. (1977). Nonmetric individual differences multi-dimensional scaling: An alternative least squares method with optimal scaling features. *Psychometrika, 42*, 1–68.
190. Tanioka, K., & Yadohisa, H. (2016). Discriminant coordinates for asymmetric dissimilarity data based on radius model. *Behaviormetrika, 43*, 1–17.
191. Tanioka, K., & Yadohisa, H. (2018). Unfolding models for asymmetric dissimilarity data with external information based on path structures. *International Journal of Software Innovation, 6*(3), 53–66.
192. Tanioka, K., & Yadohisa, H. (2018). Asymmetric MDS with categorical external information based on radius model. *Procedia Computer Science, 140*, 284–291.
193. Ten Berge, J. M. F. (1997). Reduction of asymmetry by rank-one matrices. *Computational Statistics & Data Analysis, 24*, 357–366.
194. Ten Berge, J. M. F., & Kiers, H. A. L. (1999). Simplicity of core arrays in three-way principal com-ponent analysis and the typical rank of p x q x 2 arrays. *Linear Algebra and its Applications, 294*, 169–179.

195. Ten Berge, J. M. F., Kiers, H. A. L., & De Leeuw, L. (1988). Explicit CA-NDECOMP/PARAFAC solutions for a contrived 2 x 2 x 2 array of rank three. *Psychometrika, 53*, 579–584.
196. Ten Berge, J. M. F., Sidiropoulos, N. D., & Rocci, R. (2004). Typical rank and INDSCAL dimensionality for symmetric three-way arrays of order I x 2 or I x 3 x 3. *Linear Algebra and its Applications, 388*, 363–377.
197. Tobler, W. (1976–77). Spatial interaction patterns. *Journal of Environmental Systems, 6*, 271–301.
198. Tomizawa, S. (1992). Multiplicative models with further restrictions on the usual symmetry model. *Communications in Statistics—Theory and Methods, 21*, 693–710.
199. Torgerson, W. S. (1954). A law of categorical judgment. *The American Psychologist, 9*, 483.
200. Torgerson, W. S. (1958). *Theory and Methods of Scaling*. Wiley.
201. Townsend, J., & Landon, D. E. (1982). An experimental and theoretical investigation o the constant-ratio rule and other models of visual letter confusion. *Journal of Mathematical Psychology, 25*, 119–162.
202. Trendafilov, N. T. (2002). GIPSCAL revisited. A projected gradient approach. *Statistics and Computing, 12*, 135–145.
203. Tsuchida, J., & Yadohisa, H. (2016). Asymmetric multidimensional scaling of N-mode M-way categorical data using a log-linear model. *Behaviormetrika, 43*, 103–138.
204. Tucker, L. R., & Messick, S. (1963). An individual differences model for multidimensional scaling. *Psychometrika, 28*, 333–367.
205. Tversky, A. (1977). Features of similarity. *Psychological Review, 84*, 327–352.
206. United Nations. (1975). *Statistical Year Book 1974*. United Nations Publication.
207. Yadohisa, H., & Niki, N. (1999). Vector field representation of asymmetric proximity data. *Comunications in Statistics, Theory and method, 28*, 35–48.
208. Young, F.W. (1975). An asymmetric Euclidean model for multi-process asymmetric data. In: *Paper presented at U.S.-Japan Seminar on MDS*, San Diego, U.S.A.
209. Young, G., & Householder, A. S. (1938). Discussion of a set of points in terms of their mutual distances. *Psychometrika, 3*, 19–22.
210. Weeks, D. G., & Bentler, P. M. (1982). Restricted multidimensional scaling models for asymmetric proximities. *Psychometrika, 47*, 201–208.
211. Weinberger, K. Q., & Saul, L.K. (2004). Unsupervised learning of image manifolds by semidefinite programming. *Proceedings of the 2004 IEEE Computer Society Conference on Computer Vision and Pattern Recognition. 70*, 77–90.
212. Wilkinson, J. H. (1965). *The algebraic eigenvalue problem*. Clarendon Press.
213. Wishart, J. (1949). Cumulants of multivariate multinomial distributions. *Biometrika, 36*, 47–58.
214. Zielman, B., & Heiser, W. J. (1993). Analysis of asymmetry by a slide vector. *Psychometrika, 58*, 101–114.
215. Zielman, B., & Heiser, W. J. (1996). Models for asymmetric proximities. *British Journal of Mathematical and Statistivbal Psychology, 49*, 127–146.
216. Zinnes, J. L., & MacKay, D. B. (1983). Probabilistic multidimensional scaling: Complete and incomplete data. *Psychometrika, 48*, 27–48.

Part II
The Dynamics of Asymmetric Interactions

In Part II of this book, which includes Chaps. 8 and 9, the dynamics of asymmetric interactions is discussed from the viewpoint of *dynamical systems* in mathematics, especially *qualitative theoriesof dynamical systems*. Chapter 8 is a reproduction of a paper by the author and one of his colleagues from 1990. We added notes and an appendix to this paper. In this chapter, we introduce a *bifurcation model* of changes in group structure over time, which is called the *dynamical* system *scaling* (abbreviated as DYNACSCAL). This model is a two-dimensional *nonlinear differential equation* model and estimates the *vector field* as well as the basic *trajectories* of the field at each point in time, given a set of observed longitudinal sociomatrices between members of a group. Neither *chaos* nor *fractal* appears in the estimated vector field at each time, since this model is a set of nonlinear differential equations in R^2. However, various *singularities* and *limit cycles* appear theoretically, and we show the estimated orbits of these special points and cycles by applying DYNASCAL to an empirical data. As for basic concepts, stability notions, and bifurcations of singular points of the vector field described by differential equations, we shall address those in Chap. 12 of Part III.

Chapter 9 introduces an elementary theory of the *dynamical* (or dynamic) *weighted digraph* (abbreviated as DWD), which the author has been developing since 2000. In the earlier versions, we called it a complex difference equation model. Since there is a correspondence between a *weighted digraph* and an asymmetric similarity matrix (abbreviated as *ASM*), we can embed vertices of the observed weighted digraph in a *Hilbert space* or in an indefinite metric space depending on the nature of ASM, which we have already discussed in Chap. 4 of Part I. In this theory, we assume the configuration of vertices of a digraph embedded in a *Hilbert space* at an instant of time to be an initial state space of a set of *complex polynomial difference equations*. Depending on the parameters of this set of equations, the qualitative patterns of the solution curves of this system involve fixed points, fractals, and so on, theoretically. We further assume an *asymmetric interaction matrix* among vertices, which is composed of a part of these parameters. Qualitative behaviors of the solution curves are determined by the patterns of the eigenvalues of this matrix in the case when the system is linear. In the case when the system is nonlinear with parameters including those of the asymmetric interaction matrix, various curious qualitative behaviors such as fractals are seen. We shall recommend readers who are not familiar with differential and difference equations to look through Chaps. 10 through 12 of Part III of this book before they move on to Chaps. 8 and 9.

Chapter 8
A Bifurcation Model of Changes in Group Structure

8.1 Introduction

About ninety years ago, Lewin [20, 21] introduced various dynamic concepts such as *social force field*, *equilibrium*, and *barrier*, to explain the psychological influence of environment on the behavior and development of children, tried to develop a new geometry which he termed *hodological space*, and proposed a *topological psychology* (Lewin [22, 23]).

As Deutsch [15] pointed out, these concepts were indeed helpful in giving insight into behavior in conflict situations and in understanding the difference between escape and goal-seeking behavior. However, according to Rosenberg [32], "these topological concepts were, in the final analysis, primarily a way of enriching English vocabulary and syntax; … Specific predictions about behavior were rarely possible from Lewin's mathematical expressions or from his topological representations of psychological situations". In brief, there were few operational definitions of these concepts and thus his field theory permitted no verification based on data.

Lewin was motivated to develop the new geometry, i.e., the hodological space, because he felt that the topology, which was *qualitative* in nature, did not permit determination of direction and distance which are *quantitative* in nature (Lewin [23]). In other words, he aimed at developing a new geometry which is both qualitative and quantitative in nature.

In this paper, we introduce a method which enables us to deal with both aspects without formulating a new geometry and show an example of application, technical difficulties in applications of the method and way of overcoming these difficulties. The method is called DYNASCAL (DYNamical system SCALing) and has been developed by Chino and Nakagawa (Chino [6–9]; Chino & Nakagawa [10–13]).

DYNASCAL deals with those processes in which individuals accumulate or dissolve into subgroups via complicated interactions among them. For example, freshmen in a class become acquainted with each other gradually and informal groups are formed in due course. A big political group once formed sometimes dissolves into a few formal subgroups by oppositions of political beliefs and strategies.

N. Chino, *Structure and Dynamics of Asymmetric Interactions*,
https://doi.org/10.1007/978-981-97-8269-7_8

To model such a process in mathematical terms, DYNASCAL presupposes a set of longitudinal configurations of individuals. By the longitudinal configurations we mean the coordinates of individuals at each time in a certain space. This space may be considered as a *life space* in Lewin's terminology. We define the longitudinal configurations operationally as the configurations obtained by the repeated applications of some appropriate multidimensional scaling (MDS) to each of longitudinal dissimilarity judgements among individuals such as longitudinal sociomatrices. DYNASCAL estimates the vector field at each time, given the set of longitudinal configurations.

In estimating the vector field at each time, DYNASCAL utilizes qualitative theories of *dynamical system* especially those of *singularities*, *structural stability*, and *bifurcations of vector field*. These theories teach us a body of knowledge about qualitative features of the estimated vector fields. For example, theories of singularities of vector field tell us what kinds of local features of a vector fields will be possible. Theories of structural stability teach us what kinds of features of a vector field can be observed and cannot be observed *in principle*. Theories of bifurcations of vector field tell us what sort of change in the qualitative features of a vector field will happen. We shall discuss these theories in more detail in later sections.

Our idea of vector field is essentially the same as that proposed by Lewin [20]. The major distinction between his method and ours is that the former does not utilize the qualitative theories of dynamical system, some of which had never been developed at his age. These are the notion of structural stability (Andronov & Pontrjagin [3]), theory of global structural stability (Peixoto & Peixoto [29]), and recent developments in bifurcation theories.

Utilization of the qualitative theories of singularities is not new in constructing mathematical models of social behavior but hardly any literature can be found which utilizes the theories of structural stability or those of bifurcation (Chino [8, 9]; Nakagawa & Chino [25]). For example, Simon and Guetzkow [33] formalize some of Festinger's propositions (Festinger [16]) about communication processes in small groups and organize them into a differential equation model. Variables specifying the state of the system are the perceived discrepancy of opinion on an issue among members of a group and the cohesiveness of the group. They examine the stability of the system at hypothesized equilibriums, but do not refer to the structural stability of the system.

Cervin and Henderson [5] propose differential equation models of persuasive interaction between two individuals. In this case variables specifying the state of the system are momentary probabilities of the two individuals holding a positive opinion on a topic. The fit of Model 2, which is a special case of Model 1, was fitted to experimental data. Abelson [2] examines qualitative behaviors of the systems in some special cases of Model 1 including Model 2. However, none of them realizes the important point that the system described by Model 2 is structurally unstable and cannot be observed in principle. Abelson [1] extends the models into multi-person cases and refers to the possibility of obtaining a stable equilibrium which means an outcome situation with zero opinion discrepancy.

In the next three sections we shall briefly introduce fundamental concepts about differential equation and qualitative theories of dynamical system to help readers to interpret the results obtained by DYNASCAL. In the second section, we shall make mathematical definitions of the notions which are fundamental in fitting a differential equation to any real system. Then we shall discuss the theories of singularities and structural stability in the third section. In the fourth section we shall introduce the theories of bifurcations. In the fifth section we shall show an example of application of DYNASCAL to Newcomb's data. We shall also refer to technical difficulties in applying the method to empirical data and show ways of overcoming them. In the final section we shall discuss how our model can be potentially applicable to various areas of research, and how it can be developed further from the point of view of self-organization to which there has been an increasing attention in various branches of science.

8.2 Specification of the Differential Equations

In fitting a differential equation to any real system that changes in time, one must specify two things: quantity which specifies the instantaneous state of a system and form of the differential equation which specifies the manner in which the system can change in time. The former is called the *state variables* of the system, which constitute a *state space*. In our case, the state variables are coordinates of points in a certain dimensional state space, which may correspond to a *life space* in Lewin's terminology.

Before we specify our differential equation, a few general remarks are in order. Suppose in general that the number of dimensions is p. Then, the instantaneous state of our system is denoted by the state vector, $\boldsymbol{x} = \left(x_1, x_2, \cdots, x_p\right)^t$, whose components are coordinates of a point in the space.

Since velocity is a rate at which the state variable x_i (in our case, a coordinate of a point) changes in time, any velocity vector is defined in terms of x as

$$\dot{\boldsymbol{x}} = \frac{d\boldsymbol{x}}{dt} = \left(\frac{dx_1}{dt}, \frac{dx_2}{dt}, \cdots, \frac{dx_p}{dt}\right)^t.$$

Qualitative behaviors of the system whose dimension in higher than two are very complex in general. On the contrary, the qualitative behaviors of the two-dimensional system are relatively well known. For this reason, we choose the following equation:

$$\begin{pmatrix} dx_1/dt \\ dx_2/dt \end{pmatrix} = \begin{pmatrix} f_1(x_1, x_2, b) \\ f_2(x_1, x_2, b) \end{pmatrix} \tag{8.1}$$

Equation (8.1) is two-dimensional, nonlinear, and is called *autonomous* since the right-hand side does not include time t. The reason for including a parameter b in the right-hand side of the equation is based on the assumption that the qualitative features of the vector field will change as a certain parameter changes continuously. In bifurcation theory, this parameter is called a bifurcation parameter.

Since the qualitative behaviors of the system described by (8.1) depend upon both the state variables (x_1, x_2) and a parameter b, we first examine the behaviors of the following nonlinear autonomous system in the next section:

$$\begin{pmatrix} dx_1/dt \\ dx_2/dt \end{pmatrix} = \begin{pmatrix} f_1(x_1, x_2) \\ f_2(x_1, x_2) \end{pmatrix} \tag{8.2}$$

The qualitative behaviors of the system described by (8.1) can then be stated as changes in the behaviors of (8.2) over the parameter value b.

In the next section we shall examine what kind of qualitative behaviors will appear locally as well as globally for the vector field described by (8.2).

8.3 Behaviors of the Two-Dimensional Nonlinear Autonomous System

In examining the qualitative behaviors of any system, we must consider two aspects: one is the *local behavior* of the system and the other the *global behavior*. By the former we mean the qualitative feature of the family of solution curves of the system within some restricted region near a point in the state space. In contrast we mean by the latter the qualitative feature of the family of the solution curves in the whole state space. We shall first discuss the local behavior of the nonlinear autonomous system described by (8.2), and then discuss the global one.

It is well known that most of the qualitative behaviors of the nonlinear system described by (8.2) can be examined locally by *linearizing the system* at a special point $\boldsymbol{x}^*$ called *singularity* or the *singular point* defined as follows (Hirsch & Smale [19]):

$$\begin{pmatrix} dx_1/dt \\ dx_2/dt \end{pmatrix} = J\left(x^*\right) \begin{pmatrix} x_1 \\ x_2 \end{pmatrix} \tag{8.3}$$

where

$$J\left(x^*\right) = \begin{pmatrix} \partial f_1/\partial x_1 & \partial f_1/\partial x_2 \\ \partial f_2/\partial x_1 & \partial f_2/\partial x_2 \end{pmatrix}_{x=x^*} \tag{8.4}$$

Singularity is sometimes called equilibrium, which is well-known to social scientists. In general, a singularity is the point at which first-derivatives all vanish. In the case of (8.2), the $\boldsymbol{x}^*$ is a solution of

$$\begin{pmatrix} dx_1/dt \\ dx_2/dt \end{pmatrix} = \begin{pmatrix} f_1(x_1, x_2) \\ f_2(x_1, x_2) \end{pmatrix} = \begin{pmatrix} 0 \\ 0 \end{pmatrix} \tag{8.5}$$

Equation (8.5) has generally more than one solution. The matrix $\boldsymbol{J}(\boldsymbol{x}^*)$ in (8.4) is called the *Jacobian* at singularity $\boldsymbol{x}^*$.

Since the Jacobian at point $\boldsymbol{x}^*$ is the matrix whose components are all constants, (8.3) is considered as a *general linear autonomous system*. Qualitative behaviors of the system of this kind are classified in terms of singularities of the vector field. Precisely, if the determinant of the Jacobian is not zero, qualitative behaviors of the system can be classified by the eigenvalues of the Jacobian, which determine the types of singularities.

In the next section we shall discuss the case in which the determinant is zero at a singularity. Such a singularity is called the *degenerate singularity*. As will be explained there, the degenerate singularity cannot be observed in principle. On the other hand, the singularity cannot be observed in principle. On the other hand, the singularity whose Jacobian has nonzero determinant is called *nondegenerate singularity*.

Figure 8.1 shows fundamental nondegenerate singularities and qualitative behaviors of the system near these singularities. As already noted, by the qualitative behaviors we mean qualitative features of the family of solution curves for the system. From now on, we call the solution curves *orbits*. Singularities except *saddle* and *center* in Fig. 8.1 are sometimes referred to simply as *nodes*. For the details of the classification, we shall refer to Hirsch & Smale [19].

We shall now show that all the qualitative behaviors of the system near these singularities are successfully interpreted psychologically as well as sociologically. Let us supposed that the individuals are located in a plane and let us imagine that the individuals are moving around a singularity on the plane under the influence of the force field described by Eq. (8.3). Consider first the cased of *sink* and *inward node*. As shown in Fig. 8.1b and c, respectively, the force acting on every point in these fields attracts individuals to each singularity. Therefore, we can regard the qualitative behavior of the system near the singularity as that peculiar to a *group formation process*. On the other hand, the force acting repels individuals from singularities in case of *source* and *outward node* shown in Fig. 8.1d and e. Therefore, we can regard the behavior as that peculiar to a *group dissolution process*.

The qualitative behavior near saddle is completely different from those stated above. The line formed by the two orbits which go into saddle is considered as a kind of boundary which divides a psychological or sociological region into two, since, as shown in Fig. 8.1a, forces acting on both sides of this boundary are opposite. In other words, the boundary acts as a *threshold* or *barrier*.

The qualitative behavior near the singularity called center is interesting. Suppose, for example, that for an individual the point is attractive in a sense but repulsive at the same time in another sense. In such an ambivalent situation, it may be that he or she goes around it in a psychological or sociological space. But, as will be discussed later, few can observe no center for a theoretical reason.

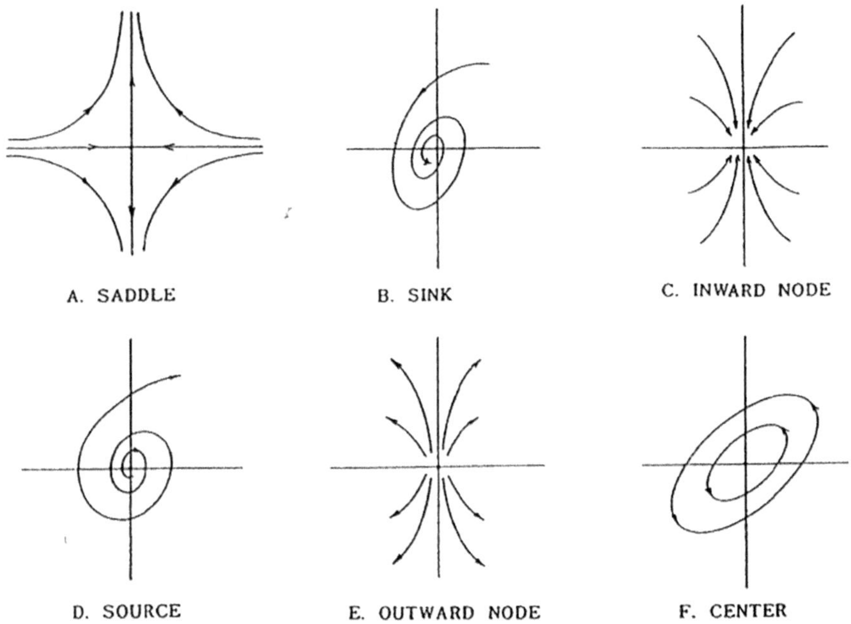

Fig. 8.1 Fundamental kinds of nondegenerate singularities

We can thus examine the local qualitative behavior of the system near the singularities of the vector field described by Eq. (8.2) and interpret them. The nonlinear differential equation, however, exhibits another sort of behavior frequently. It is exhibited by a kind of closed orbit called the *limit cycle*. Figure 8.2 shows two fundamental limit cycles.

As shown in Fig. 8.2a, *α-limit cycle* includes a sink inside the closed orbit and the behavior outside the closed orbit is that of a source. On theω-limit cycle contrary, *ω-limit cycle* in Fig. 8.2b has a source inside the closed orbit and the behavior outside

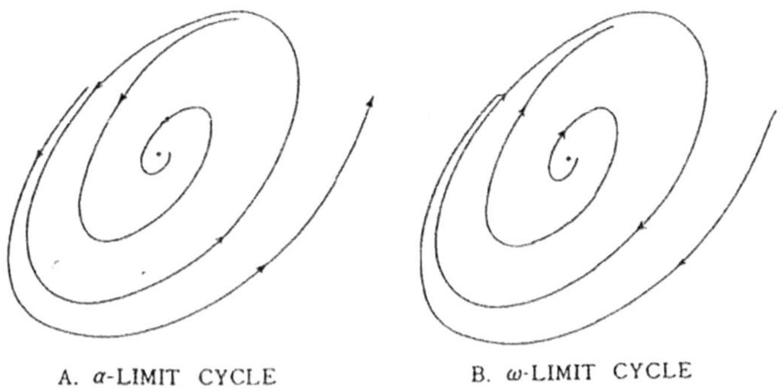

Fig. 8.2 Fundamental kinds of limit cycles

the closed orbit is that of a sink. We can interpret these behaviors of limit cycles as those peculiar to a *closed group*. This is because the closed orbit is considered as a psychological or social *barrier* which separates members located inside the orbit from those located outside.

Singularities and limit cycles discussed so far exhibit typical local qualitative behaviors of the nonlinear autonomous system described by Eq. (8.2). Of course, there exist, at least theoretically, other kinds of local qualitative behaviors of this system. However, they cannot be observed in principle (Guckenheimer & Holmes [17]). This is concerned with the notion of *structural stability*. Here, a vector field described by an autonomous differential equation is said to be structurally stable when orbital properties do not change by a small perturbation of the field.

According to the Hartman & Grobman theorem, the field is *locally structurally stable* if the Jacobian of the system at a singularity has non-zero real part, that is, no zero or purely imaginary eigenvalues. Here, the singularity which satisfies the condition is called *hyperbolic*. Only the center shown in Fig. 8.1f has zero real parts. It follows that center is structurally unstable. Orbital property of a structurally unstable system changes by a small perturbation and therefore can not be observed by any method which accompanies measurement errors or computation errors. This is the reason why center is unobservable in principle. On the other hand, the other hand, the other singularities shown in Fig. 8.1 as well as the limit cycles shown in Fig. 8.2 are both observable since they all are hyperbolic.

In order to discuss the global behaviors of the system described by (8.2) we must specify the state space of the system further. Let the space be a closed region whose boundary is a simple smooth curve. Then, according to Peixoto's theorem (Peixoto & Peixoto [29]), under some general conditions, a vector field defined in the closed region is *globally structurally stable* if and only if:

(1) the number of singularities and closed orbits is finite and each is hyperbolic,
(2) there are no orbits connecting saddle points,
(3) the nonwandering set consists of singularities of singularities and closed orbits alone.

Here, the third condition means that, in a structurally stable two-dimensional system, things that characterize orbital properties are singularities and limit cycles only.

We are now in a position to examine the qualitative behaviors of the system described by (8.1), which can be described by (8.2) over the parameter value b.

8.4 Change in Behaviors of the System Over a Parameter Value

Let us suppose that there exists a certain factor which influences the force field under study and the value of the factor changes continuously by the complicated interactions among individuals or some influence from outside the system. Mathematically, such a factor is represented by a bifurcation parameter. Equation (8.1) is an autonomous system with the simplest bifurcation parameter. Qualitative behavior of the system changes as the parameter value varies.

As we did for the system described by (8.2) in the previous section, we must examine changes in the local behaviors as well as the global behaviors of the system described by (8.1). The former is concerned with the *local bifurcation* and the other the *global bifurcation*. Four kinds of bifurcations are the best known bifurcations which depend on a single parameter in a two-dimensional system (Guckenheimer & Holmes [17]). These are *saddle-node*, *pitchfork*, *transcritical*, and the *Hopf bifurcations*. For space limitation we shall discuss the saddle-node and the Hopf bifurcations only. For further discussions, see Chino [9].

The saddle-node bifurcation is the process in which, for instance, there exists no singularity at the outset and then a *degenerate singularity* is born to bifurcate into a *saddle* and a *sink*. The reverse process is also a saddle-node bifurcation. In this case, a saddle and a sink approach gradually and are suddenly unified into a degenerate singularity. Figure 8.3a shows the former case.

As indicated earlier, degenerate singularities cannot be observed in principle since they are structurally unstable (Guckenheimer & Holmes [17]). Instead we might observe a saddle and a node as if they suddenly appeared or disappeared.

It is well known that all bifurcations of one-parameter families at a singularity with a zero eigenvalue can be perturbed to saddle-node bifurcations. In this sense, the saddle-node bifurcation is a typical bifurcation. Thus, we frequently encounter a saddle-node bifurcation in the case of a singularity with a zero eigenvalue. Otherwise, there may be something special which prevents that bifurcation from occurring.

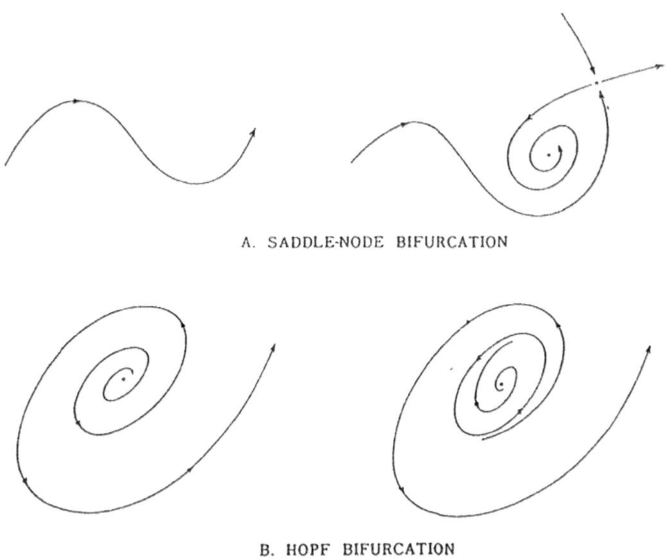

Fig. 8.3 Two examples of local bifurcations

The Hopf bifurcation is the case where eigenvalues of the Jacobian at a singularity are pure imaginary. In this case, either an α-limit cycle or an ω-limit cycleω appears before or after the bifurcation. Figure 8.3b shows the case in which the α-limit cycle appears.

We shall now show that these local bifurcations can be interpreted psychologically or sociologically. First, the saddle-node bifurcation is interpreted as a process of *order formation* or *dissolution*. For example, Fig. 8.3a is considered as an order formation process in which some local group formation process by a sink begins with a barrier by a saddle, which roughly divides the space into two. The reverse process is regarded as a dissolution process. If the node of the bifurcation is a source, it is also interpreted as a process of order formation or dissolution. In this case, by the order formation process we mean, for example, the process in which a barrier formed by a saddle appears suddenly with a point which repels every object, that is, a source.

The interpretation of the Hopf bifurcation is somewhat complicated. We can show four patterns of this bifurcation (Chino [9]). Each of them is considered as either a group formation process or a group dissolution process. For space limitation, we will discuss only one of them, as is explained below.

In the case of Fig. 8.3b, a small α-limit cycle with a sink appears in the center of a source and develops. Since the sink is considered as a group formation process, and since the original global qualitative features of the source is destroyed gradually from the center by the sink, this kind of Hopf bifurcation is considered as a group formation process.

We shall now discuss the global bifurcations. They are relatively poorly understood in mathematics compared with the local ones and this global bifurcation theory is still an active area of research. *Saddle loop* and *saddle connection* are a few well known examples. Figure 8.4 illustrates them. In the case of a saddle loop, one of the two orbits which go out of a saddle enters the saddle. On the other hand, in the case of a saddle connection, two orbits, one of which goes out of a saddle and the other enters the other saddle, connect to form an orbit. Psychological or sociological interpretation of these global bifurcations is also possible (see Chino [9]).

We have so far examined the qualitative behaviors of the system described by (8.1). Examination has shown that these behaviors can be classified compactly using some bifurcation theories of the vector field. Moreover, we have found that many of those behaviors can be interpreted psychologically as well as sociologically.

Before we interpret the results obtained by DYNASCAL according to the theories discussed above, a remark is in order. The point is that we will not use (8.1) directly as the model equation of DYNASCAL. The reason is that in general we find it difficult to identify the single bifurcation parameter b in (8.1). To overcome this difficulty we replace the parameter with time t. This makes the system *nonaoutonomous*. However, one can think of successive vector fields described by the nonautonomous system as changes in the vector field described by an autonomous system over time. Then, no problem arises in applying the theories of dynamical system to the estimated system by DYNASCAL.

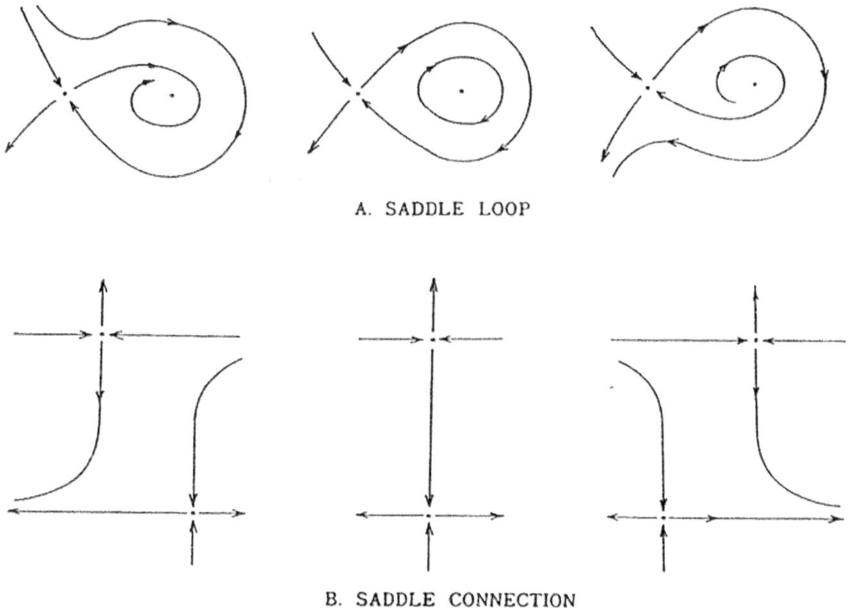

Fig. 8.4 Examples of global bifurcations

8.5 Application

In this section we shall show results of applying DYNASCAL to a set of longitudinal dissimilarity data. First, we shall refer to a set of longitudinal sociomatrices (see, Nordlie [28]), which was gathered by Newcomb [26], and illustrate how group formation process is analyzed by our bifurcation model. Second, we shall show other empirical data to discuss technical difficulties of DYNASCAL.

Newcomb observed an acquaintance process of 17 previously unacquainted male students, who lived together in a fraternity-style house, expenses paid. The obtained self-reports of complete rank orderings of the other 16 by attraction during each of the 16 weeks of experiment are longitudinal dissimilarity matrices. He conducted this type of experiment twice. For convenience, we shall choose the second experiment in Year 2, which is a replication of the first experiment in Year 1.

As a preprocessing of original ranking data, SSA2 (Guttman [18]) is applied to each of the 15 sociomatrices (Week 9 was missing in Newcomb's data) in this case and longitudinal configurations are obtained. Application of DYNASCAL yields 15 estimated vector fields corresponding to 15 weeks (Chino [8]). However, we will show only three of them. Figures 8.5 through 8.7 show the vector fields at Week 0, 1, and 15, respectively. We select them because the earliest and latest periods are most crucial for noting changes in vector field.

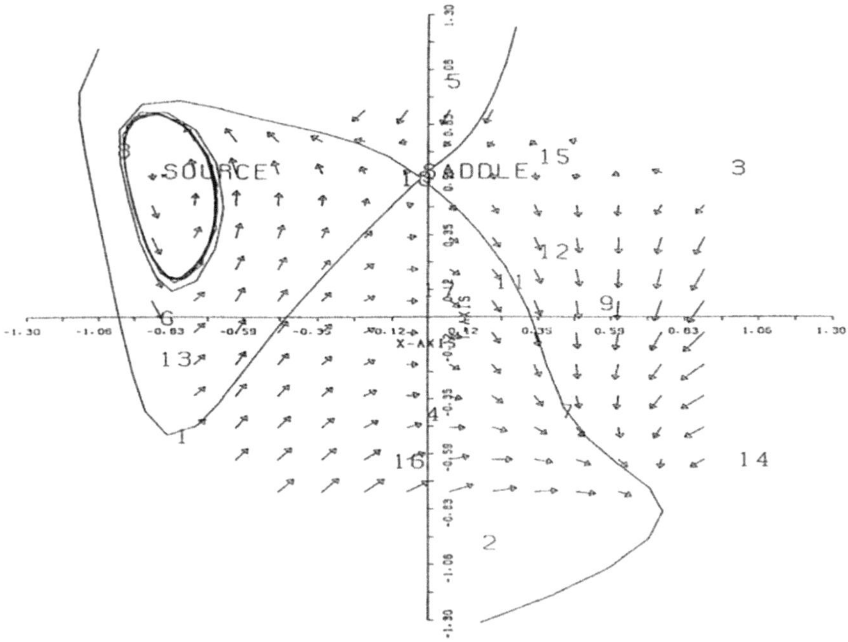

Fig. 8.5 Estimated vector field at week 0 for Newcomb's data

In these figures numbers indicate the locations of members, and vector show the estimated velocity vectors at some grid points. Some orbits of the estimated system are shown to facilitate understanding of the global qualitative features of each of the vector fields.

In Fig. 8.5, we have two singularities, a saddle and a source. The saddle clearly divides the initial psychological space into two, that is the left upper part and the right lower part of the space. The former includes members 1, 5, 6, 8, 13, although member 5 is located far from members 1,6, 8, 13. The latter includes almost all remaining members except member 10. These results almost coincide with those obtained by depicting traditional sociograms for this set of data (Chino [7]; Newcomb [26]).

An important point is that estimated vector fields in general convey much more dynamic information about each individual as well as group structure than traditional sociograms. For example, the vector field in Fig. 8.5 tells us that member 10 is under an unstable circumstance in the sense that only a small change in location leads him either to the left upper part or to the right lower part of the psychological space. This reflects the instability of the singularity, saddle.

Another dynamic information is obtained by noting the source in the left upper region of the field. As was remarked in the third section, a source indicates a group dissolution process. Readers may question whether such a dissolution process should take place so early at a period of acquaintance process. However, if we note the existence of an ω-limit cycle which surrounds the source and the fact that this limit cycle disappears from the vector fields at Week 1 in Fig. 8.6, we can conclude that a group formation process due to a once born sink has been degenerating by the appearance of the limit cycle. Such an inference is possible if we remember processes of the Hopf bifurcation discussed in the fourth section. We can also make an inference, from locations of members 1, 6, 8, 13, in the left upper region, that these members were involved in the group formation process. This inference is verified by examining the sociograms at Weeks 0 and 1.

In the vector field at Week 1, we find five singularities including three saddles, an outward node, and a source. However, we shall neglect singularities located near the boundary of the vector field because they merely play a peripheral role in interpreting properties of the vector field. This space is roughly divided into two by the saddle located near the origin. The source located in the fourth quadrant indicates a dissolution process of the members placed around it.

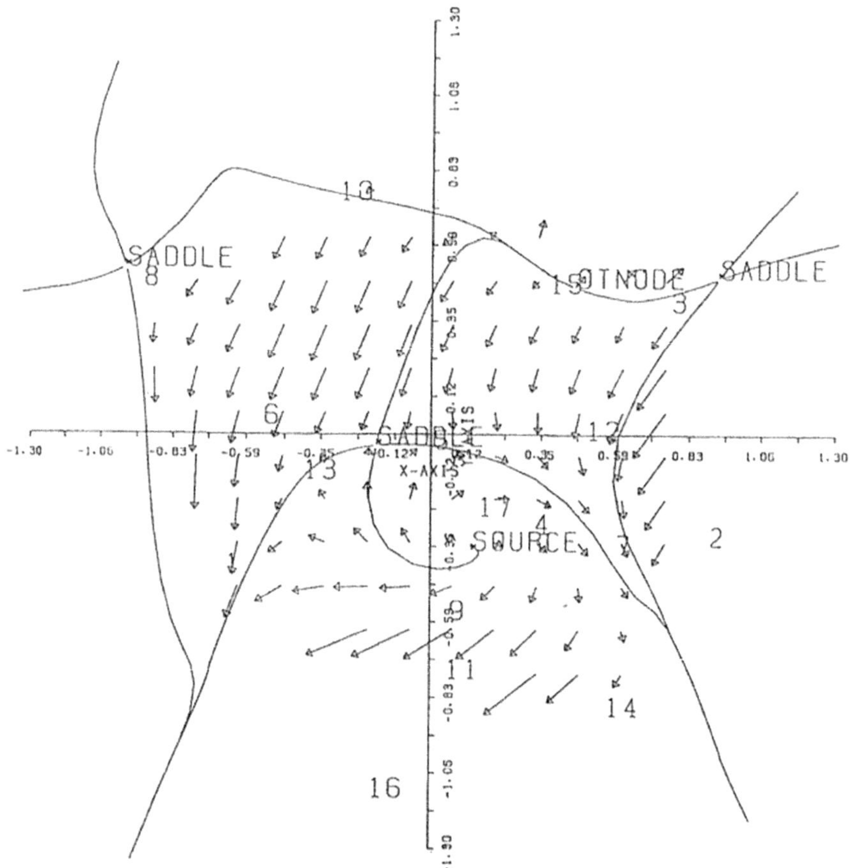

Fig. 8.6 Estimated vector field at week 1 for Newcomb's data

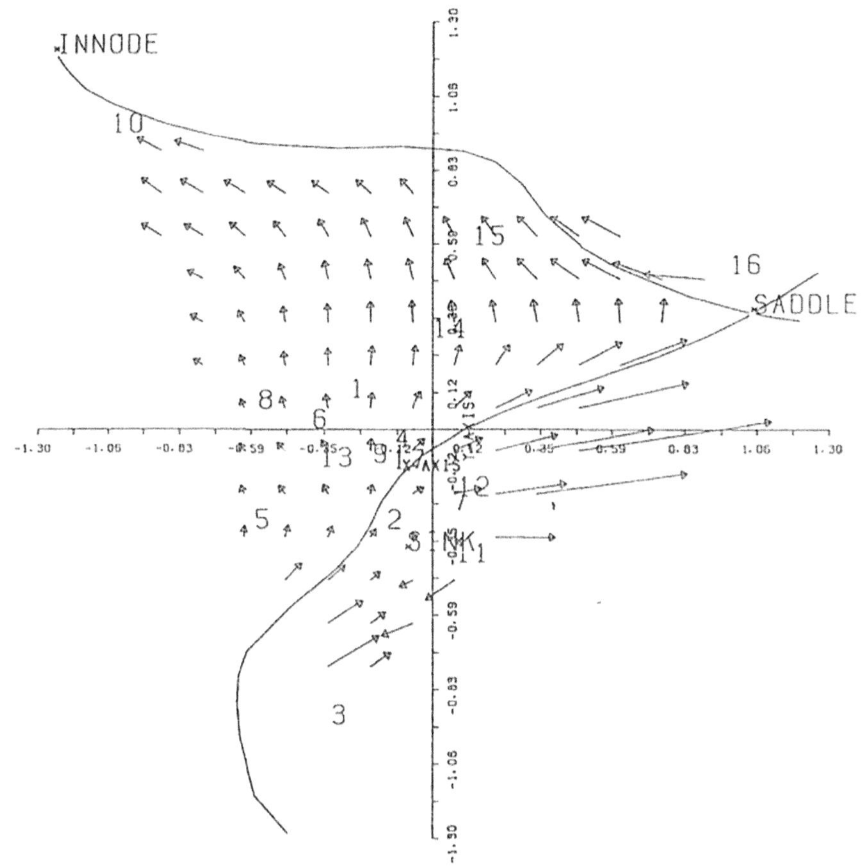

Fig. 8.7 Estimated vector field at week 15 for Newcomb's data

There is no obvious singularity except the peripheral part of the vector field shown in Fig. 8.7. However, it is evident, from the global feature of velocity vectors, that this field is dissipating all the members of this experimental group. It is probable, at this final week of the experiment, that some global group dissolution process is going on.

As we illustrated above, DYNASCAL vividly reveals underlying dynamical processes of social interactions among individuals, given a set of longitudinal dissimilarity judgements among individuals. However, there exist some technical difficulties in applying our model to empirical data.

Most crucial one is the influence of judgmental error on the recovery of the system. For, as reported elsewhere (Chino & Nakagawa [12, 13]), simulation studies indicate that DYNASCAL cannot recover assumed qualitative orbital properties of true vector fields if the standard error of dissimilarity judgment exceeds a small constant.

Fortunately, however, there is a way to overcome this difficulty. To do this, we can utilize a theorem on *maximum likelihood estimators*. This theorem states that under fairly general regularity conditions, the asymptotic distribution of the maximum likelihood estimator is multivariate normal with covariance matrix which equals the inverse of *information matrix*. As the information matrix is proportional to number of samples or replications, we can increase the accuracy of estimation of parameters by increasing the number of samples drawn from an identical parent population, even if the standard error of dissimilarity judgment exceeds a critical value.

In our case, these parameters are coordinates of individuals at each time, which can be estimated by some appropriate maximum likelihood MDS. MULTISCALE, which is one of such MDS's, depicts *confidence regions* of coordinates of individuals, assuming that the dissimilarity judgments obey a lognormal distribution (Ramsay [30, 31]).

Figure 8.8 shows 95% confidence ellipses of coordinates of 8 individuals (characters) at the earliest period of time in a video movie (Chino & Nakagawa [12]). In this experiment, 62 university students judged the intensities of the relations among 8 characters appeared in the movie on a seven-point rating scale. The content of the movie was divided into six parts and each of them was shown once a week. This figure indicates that a large sample size is necessary to attain a high accuracy of estimation of coordinates. Unfortunately, Newcomb's data discussed above do not meet the requirement of high accuracy, because in this case the sample size of each dissimilarity judgment is one. Nevertheless, we cite them because orbital properties of the estimated vector fields coincide very well with those inferred from sociograms as well as common sense.

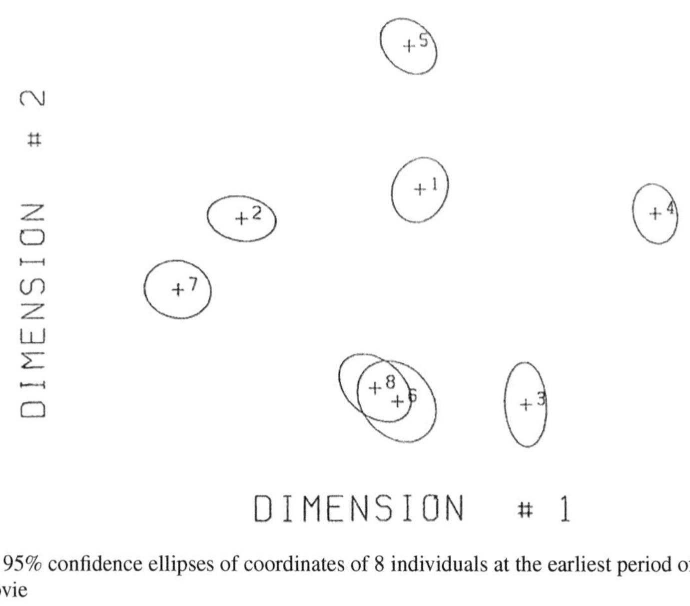

Fig. 8.8 95% confidence ellipses of coordinates of 8 individuals at the earliest period of time in a video movie

Simulation studies indicate that number of individuals as well as number of points in time are also factors which influence the recovery of the systems by DYNASCAL. For example, the number of individuals in the video experiment seems to be not enough to recover true fields, although accuracy of estimation of the coordinates almost reached a required value by the sample size of observers. In future simulation studies, we have to determine the critical numbers of them enough to guarantee the recovery of the system by DYNASCAL.

8.6 Discussion

In this paper, we proposed a bifurcation model of change in group structure in order to proceed further the idea of social force field initiated by K. Lewin some fifty years ago. The social force field under study is specified as that generated by complex interactions among individuals in any formal or informal group. Moreover, it is assumed, for theoretical simplicity, that the fields is described by a system of two-dimensional nonlinear differential equations, which is estimated based on empirical data.

We introduced a method called DYNASCAL for defining operationally the social force field based on the data. Our method is an amalgamation of knowledge about multidimensional scaling developed in psychometrics and various qualitative theories of dynamical system in mathematics, especially, theories of singularities, structural stabilities, and bifurcations of the vector field. DYNASCAL can provide us with estimated vector fields and their qualitative properties at each time, given a set of longitudinal dissimilarity judgments such as longitudinal sociomatrices.

It was shown that.

(1) qualitative properties of any vector fields described ty the system of differential equations can be classified into a strikingly few categories and
(2) almost all the qualitative properties which are observable in principle can be interpreted psychologically as well as sociologically.

Furthermore, it was demonstrated by an illustrative example of application, that our method reveals vividly the underlying group dynamics under study, although there exist some difficulties in applying the method to empirical data.

Bifurcation theory in general provides a natural approach to the problem of the emergence of new patterns and structure although we only use it to model change in group structure in time. In fact, the theory has recently been playing a fundamental role in understanding the problem of *self-organization* in physical and biological system Nicolis & Prigogin [27]; Thompson & Stewart [35]. We believe that this theory as well as the paradigm of self-organization will get a citizenship in psychology in the near future.

Appendix

In this appendix we shall discuss how to estimate the vector field at each week, given a set of longitudinal sociomatirces, which we did not discuss in this paper.

The following is the steps of estimating these vector fields:

Let us assume first that N is the number of members, which is also the order of each of these sociomatrices, while T is the number of times, at which these matrices are observed.

Step 1.

Apply SSA2 (Guttman [18] and Lingoes [24] to each of the longitudinal sociomatrices, as a preprocessing of original ranking data. SSA2, particularly, the *row solutions* give us the longitudinal configurations of members, in a Euclidean space, as pointed out in the text. As is well known, SSA2 has two types of solutions, that is, *row solution* and *column solution*. We have chosen the row solutions, because the rows indicate the raters and the columns indicate the rateees. We have specified two-dimensional configurations in the application since the model Eq. (8.3) is two-dimensional. Let these configurations be, $\boldsymbol{X}_1, \boldsymbol{X}_2, ..,$

$\boldsymbol{X}_T$, each of whose order is $N \times T$.

Step 2.

Perform the *orthogonal Procrustes rotation* (e.g., Ten Berge [34]) to each pair of configurations obtained in Step 1, seriatim.

To be precise, find the orthogonal transformation matrix $\boldsymbol{O}_1$ of order 2 which minimize $f(O_1) = tr(X_2O_1 - X_1)^t(X_2O_1 - X_1)$, and compute $X_2^{(r)} = X_2O_1$, first. Then, find the orthogonal transformation matrix $\boldsymbol{O}_2$ of order 2 which minimize $f(O_2) = tr\left(X_3O_2 - X_2^{(r)}\right)^t\left(X_3O_2 - X_2^{(r)}\right)$, and compute $X_3^{(r)} = X_3O_2$, second. Then, find the transformation matrix $\boldsymbol{O}_3$ of order 2 which minimize $f(O_3) = tr\left(X_4O_3 - X_3^{(r)}\right)^t\left(X_4O_3 - X_3^{(r)}\right)$, and compute $X_4^{(r)} = X_4O_3$, third. By continuing these steps seriatim, find $X_1, X_2^{(r)}, X_3^{(r)}, \cdots, X_T^{(r)}$, finally.

Step 3.

Let $\boldsymbol{Y}_1, \boldsymbol{Y}_2, \boldsymbol{Y}_3, \cdots, \boldsymbol{Y}_T$ be respectively, $X_1, X_2^{(r)}, X_3^{(r)}, \cdots, X_T^{(r)}$, and let $\boldsymbol{y}_{ik}$ be the ith row vector of $Y_k, t = 1, 2, \cdots, T$. Then, the sequence $\{\boldsymbol{y}_{i1}, \boldsymbol{y}_{i2,\cdots}, \boldsymbol{y}_{iT}\}$ is considered as the *estimated solution curve* of the set of nonlinear differential equations in DYNASCAL associated with the ith member of a group, where i ranges from 1 to N. In Step 3, fit a one-dimensional *spline function* (of order 3) (e.g., Atkinson [4]) to each of the N estimated solution curves, and compute the *first derivative* of this function at each point of members in the Euclidean plane.

Step 4.

Let the sequence $\{\mathbf{v}_{i1}, \mathbf{v}_{i2}, \cdots, \mathbf{v}_{iT}\}$ be composed of the first derivative vectors associated with the ith member of a group computed in Step 3, where $v_{it} = (v_{ixt}, v_{iyt})$, and v_{ixt} and v_{iyt} are respectively the derivative components in the horizontal direction and vertical direction of the plane. In Step 4, smooth the N sequences $\{\mathbf{v}_{i1}, \mathbf{v}_{i2}, \cdots, \mathbf{v}_{iT}\}, i = 1, 2, \cdots, N$, associated with the sequences $\{\mathbf{y}_{i1}, \mathbf{y}_{i2,\cdots,}\mathbf{y}_{iT}\}, i = 1, 2, \cdots, N$, using a two-dimensional *smoothing spline function*.

Let the sequences $\{\hat{\mathbf{v}}_{i1}, \hat{\mathbf{v}}_{i2}, \cdots, \hat{\mathbf{v}}_{iT}\}, i = 1, 2, \cdots, N$ be the smoothed sequences of $\{\mathbf{v}_{i1}, \mathbf{v}_{i2}, \cdots, \mathbf{v}_{iT}\}, i = 1, 2, \cdots, N$, associated with the coordinate sequences $\{\mathbf{y}_{i1}, \mathbf{y}_{i2,\cdots,}\mathbf{y}_{iT}\}, i = 1, 2, \cdots, N$. We use this set of the smoothed sequences associated with the coordinate sequences per time instant t, as an input data, and estimate the singular points of the estimated vector field at each point in time, utilizing the qualitative theory of the vector field discussed in Sects. 8.3 and 8.4.

Notes

This chapter is the reproduction of Chino and Nakagawa [14]. Although DYNASCAL can reveals the underlying group dynamics under study vividly as discussed above, there is a limitation of DYNASCAL in that it cannot describe chaotic behaviors because it is restricted to the *two-dimensional real nonlinear system* described by (8.1). It is expected that DYNASCAL can be extended to a higher dimensional system. There is another limitation of DYNASCAL in that it utilizes a Procrustes rotation to measure the nearness of the neighboring configurations of members since the configurations obtained by applying SSA2 to the longitudinal asymmetric ranking matrices are *free up to rotations*. Such rotations of configurations are, at a glance, rational. However, we might not be able to guarantee mathematically that any of the neighboring points of the solutions to a differential equation have the property of nearness in the sense of the Procrustes rotations. A third limitation of DYNASCAL is that each of the configurations of members is *origin free* since SSA2 is a *distance model*. To overcome this difficulty, we may use HFM described in Chap. 4 since it is an *inner product model* (to be precise, *Hermitian inner produce model*). However, at present, the ordinal version of HFM has not been developed. But, even if we use HFM we cannot avoid the rotation free problem of configurations.

References

1. Abelson, R. P. (1964). Mathematical models of the distribution of attitudes under controversy. In N. Frederikson and H. Gulliksen (Eds.), *Contributions to mathematical psychology*. Holt, Rinehart & Winston.
2. Abelson, R. P. (1967). Mathematical models in social psychology In G. Lindzey and E. Aronson (Eds.) *Handbook of social psychology*. Addison-Wesley.
3. Andronov, A. A., & Pontrjagin, L. S. (1937). Systems grossiers. *Doklady Akademii Nauk SSSR, 14*, 247–250.
4. Atkinson, K. E. (1978). *An introduction to numerical analysis*. John Wiley & Sons.

5. Cervin, V. B., & Henderson, G. P. (1961). Statistical theory of persuasion. *Psychological Review, 68*, 157–166.
6. Chino, N. (1981). Some trials for a unified geometrical interpretation of several methods for the analysis of asymmetry and related techniques—*A letter from N. Chino to Dr. T. M. Newcomb.*
7. Chino, N. (1984a). Spatio-temporal structure of Newcomb's data. *Bulletin of The Faculty of Humanities of Aichigakuin University, 14*, 84–100 (in Japanese).
8. Chino, N. (1984b). Toward a theory of dynamical system in group dynamics. *Working paper of the specific researches in education and psychology by the aid of The Ministry of Education in Japan*, Nagoya University.
9. Chino, N. (1988). A bifurcation model of changes in interdependence structure among objects. *Bulletin of The Faculty of Humanities of Aichigakuin University, 17*, 85–109.
10. Chino, N., & Nakagawa, M. (1983). A vector field model for sociometric data. *Paper presented at the 11th annual meeting of the Behaviormetric Society of Japan.*
11. Chino, N., & Nakagawa, M. (1986). ML-Dynascal (theory). *Paper presented at the 14th annual meeting of the Behaviormetric Society of Japan*, Tokyo, August.
12. Chino, N., & Nakagawa, M. (1987). Measurement error in the DYNASCAL. *Paper presented at the 15th annual meeting of the Behaviormetric Society of Japan*, Fukuoka, August.
13. Chino, N., & Nakagawa, M. (1988). An application of the DYNASCAL to empirical data. *Paper presented at the 16th annual meeting of the Behaviormetric Society of Japan*, Chiba, August.
14. Chino, N., & Nakagawa, M. (1990). A bifurcation model of change in group structure. *The Japanese Journal of Experimental Social Psychology, 29*, 25–38 (an English issue).
15. Deutsch, M. (1968). Field theory in social psychology. In G. Lindzey & E. Aronson (Eds.), *The handbook of social psychology*, 2nd ed., 1, 412–487.
16. Festinger, L. (1950). Informal social communication. *Psychological Review, 57*, 271–282.
17. Guckenheimer, J., & Holmes, P. (1983). *Nonlinear oscillations, dynamical systems, and bifurcations of vector fields*. Springer-Verlag.
18. Guttman, L. (1968). A general nonmetric technique for finding the smallest coordinate space for a configuration of points. *Psychometrika, 33*, 469–506.
19. Hirsch, M. W., & Smale, S. (1974). *Differential equations, dynamical systems, and linear algebra.* Academic Press.
20. Lewin, K. (1933). Environmental forces in child behavior and development. In Murchison, C. (Ed.), *Handbook of child psychology*. Clark University Press.
21. Lewin, K. (1935). *A dynamic theory of personality*. McGraw-Hill.
22. Lewin, K. (1936). *Principles of topological psychology*. McGraw-Hill.
23. Lewin, K. (1938). The conceptual representation and measurement of psychological forces. *Contribution to Psychological Theory, 1*, 4.
24. Lingoes, J. C. (1973). *The Guttman-Lingoes Nonmetric Program Series*. Mathesis Press.
25. Nakagawa, M., & Chino, N. (1987). A dynamical model of social interaction. *The Japanese Journal of Psychonomic Science, 6*, 11–20.
26. Newcomb, T. M. (1961). *The acquaintance process*. Holt. Rinehart and Winston.
27. Nicolis, G., & Prigogin, I. (1977). *Self-organization in nonequilibrium systems—From dissipative structures to order through fluctuations*. Wiley-Interscience.
28. Nordlie, P. G. (1958). A longitudinal study of interpersonal attraction in a natural group setting. *A doctoral dissertation submitted to University of Michigan.*
29. Peixoto, M. C., & Peixoto, M. M. (1959). Structural stability in the place with enlarged boundary condition. *Anais da Academia Brasileia de Ciencias, 31*, 135–160.
30. Ramsay, J. O. (1977). Maximum likelihood estimation in multidimensional scaling. *Psychometrika, 42*, 241–266.
31. Ramsey, J. O. (1982). Some statistical approaches to multidimensional scaling data. *The Journal of the Royal Statistical Society Series A (General), 145*, 285–312.
32. Rosenberg, S. (1968). Mathematical models of social behavior. In G. Lindzey & E. Aronson (Eds.), *The Handbook of Social Psychology*, 2nd ed. 1, 179–244.

33. Simon, H. A., & Guetzkow, H. (1955). A model of shrt- and long-run mechanisms involved in pressures toward uniformity in groups. *Psychological Review, 62*, 56–68.
34. Ten Berge, J. M. F. (1977). Orthogonal Procrustes rotation for two or more matrices. *Psychometrika, 42*, 267–276.
35. Thompson, J. M. T., & Stewart, H. B. (1986). *Nonlinear dynamics and chaos—Geometrical methods for engineers and scientists*. Wiley.

Chapter 9
Elementary Theory of the Dynamical Weighted Digraph

9.1 Introduction

In Chap. 8, we introduced a differential equation model called DYNASCAL, which was proposed with one of my colleagues in 1990. In that model we utilized *a set of nonlinear differential equations* to model the dynamics of asymmetric interactions among objects. However, DYNASCAL has several shortfalls, which were pointed out in Chap. 8. Moreover, there exist other serious problems in general in modelling a set of longitudinal asymmetric similarity matrices. One is concerned with the number of instants of time at which such matrices can be observed. For example, we cannot gather sociomatrices so frequently in a short period of time because in such a situation subjects might not rate other members independently. Another is concerned with the nature of the phenomenon for which we want to observe. Most of the dynamic changes in the phenomena observed in the behavioral and life sciences seem to occur in a discrete time interval. In such a phenomenon it seems to be natural and appropriate to utilize a *difference equation* model. Particularly, if we deal with longitudinal asymmetric similarity matrices among objects, it seems to be appropriate to utilize a difference equation model whose space domain is *comple*x, which is a rational consequence of HFM discussed in Chap. 4, which was proposed by Chino and Shiraiwa [19] in 1993. We shall recommend readers who are not familiar with difference equations to look through Chap. 12 in Part III of this book before they look at the subsequent sections in this chapter.

In this chapter we shall introduce an elementary theory of the *dynamical (or dynamic) weighted digraph* (abbreviated as *DWD*) for the analysis of asymmetric interactions between objects or nodes. A precursor of DWD had been called *complex difference system model* [5–11, 18] and did not include the notion of *weighted digraph*. The state space of this model is assumed to be a finite-dimensional (complex) *Hilbert space* or an *indefinite metric space* depending on the definiteness of the *hypothetical Hermitian matrix* which has a one-to-one correspondence to an *observed asymmetric similarity matrix* between objects. This assumption is based on the

N. Chino, *Structure and Dynamics of Asymmetric Interactions*,
https://doi.org/10.1007/978-981-97-8269-7_9

Chino-Shiraiwa theorem [19], which has been discussed in Chaps. 4 and 7. The reason why we choose a Hilbert space (or an indefinite metric space) is that it is one of the most appropriate spaces to grasp the holistic structure of asymmetric similarity matrix between objects among various extant space models mathematically, as discussed in Chap. 4. We shall introduce this model in Sect. 9.2 as an earlier version of the complex difference system model.

In 2014, Shiraiwa (personal communication, February) pointed out a shortfall of the above complex difference system model in that it is not *holomorphic*, the notion of which we shall introduce in Chap. 10. As a result, we cannot utilize the theory of complex dynamical systems in mathematics [15], although it bears numerical iterations and simulations of the model. Since then, we have discarded one of the equations in the above model [12–18]. We shall introduce this simplified version in Sect. 9.3.

In 2017, the author has realized that an asymmetric similarity matrix is considered as a *weight matrix of digraph*, and has added the notion of weighted digraph in the simplified version discussed above since then [12–18]. We shall introduce this new version of the complex difference system model in Sect. 9.3.2. This version may be called *DWD* as another name.

In Sect. 9.4, we show applications of several DWD models to an empirical data which has already been shown in Table 5.3 of Chap. 5. In Sect. 9.5, we shall list some possible future works on DWD. These include pure mathematical works and applications of DWD to various empirical data.

Before proceeding, it will be necessary to distinguish between the notion of *similarity* in psychometrics and that of similarity in fractal geometry. In psychometrics, an element of a similarity matrix has been called similarity, proximity, affinity, and so on, as discussed, for example, in Chaps. 4, 7, and 8. By contrast, in fractal geometry, the word, "similarity", is concerned with the similarity dimension as discussed in Chap. 10, and *self-similarity* is a key concept in fractal geometry. Therefore, hereafter in this chapter we shall assign the word, *proximity*, to an element of an asymmetric similarity matrix $\boldsymbol{S} = \{s_{jk}\}$ between objects (or members). As a result, we shall call a nondiagonal element s_{jk} *proximity*, while we shall call a diagonal element s_{jj} *self-proximity*.

9.2 Earlier Version of the Complex Difference System Model

In an earlier version of the complex difference system model, we proposed a class of complex difference system models [5, 6]. Let us assume that in pairwise relations the following two axioms hold:

1. If one has a positive sentiment to the other, then one will move toward the other in a sociopsychological space,

2. If one has a negative sentiment to the other, then one will move away from the other in the space.

Second, let us assume that the state space of a small group of size N is a finite-dimensional (complex) *Hilbert space* or an *indefinite metric space*. This assumption is based on the Chino-Shiraiwa theorem [19], which has already been discussed in Chaps. 3, 4, 7 and 8. Let us denote the coordinates of member j at discrete point in time n in this space by

$$z_{j,n} = \left(z_{j,n}^{(1)}, z_{j,n}^{(2)}, \cdots, z_{j,n}^{(p)}\right)^t. \tag{9.1}$$

Third, let us assume that the interpersonal sentiment structure can be observed only at discrete time intervals, and that the strength of the interpersonal sentiment of a member toward others at discrete time n + 1 can be predicted by that of the sentiment at discrete time n. Then, a simple and plausible class of models [5, 6] may be a complex difference system model, which is called a simple linear model composed of the following three equations,

$$z_{j,n+1} = z_{j,n} + \sum_{k \neq j}^{N} \boldsymbol{D}_{jk,n}\left(z_{k,n} - z_{j,n}\right),\ j = 1, 2, \cdots, N, \tag{9.2}$$

where

$$\boldsymbol{D}_{jk,n} = diag\left(w_{jk,n}^{(1)}, w_{jk,n}^{(2)}, \ \ldots, w_{jk,n}^{(p)}\right), \tag{9.3}$$

and

$$w_{jk,n}^{(l)} = a_n^{(l)} r_{j,n}^{(l)} r_{k,n}^{(l)} sin\left(\theta_{k,n}^{(l)} - \theta_{j,n}^{(l)}\right),\ l = 1,\ 2, \cdots, p. \tag{9.4}$$

Here, the first right-hand side factor of (9.4), $a_n^{(l)}$, is a known constant which characterizes the speed of change in the configuration of objects on the lth complex plane in the state space at time n. The second right-hand side factor, $r_{j,n}^{(l)}$, is the modulus of object j on dimension l at time n. Angles $\theta_{j,n}^{(l)}$ and $\theta_{k,n}^{(l)}$, respectively, denote usual counterclockwise oriented angles of object j and object k from the real axis in the lth complex plane at time n. It is not difficult to see that each of the N equations constituting this model described by (9.2) is a nonhomogeneous linear difference system that is nonautonomous, or time-variant (e.g., Elaydi [20]). From (9.4), we have $w_{kj,n}^{(l)} = -w_{jk,n}^{(l)}$, since the sinusoidal function is odd.

It is not difficult to see from (9.4) that all the members cease moving when all the configurations of members on the complex planes become any of the following three states:

(1) All the members concentrate on a point.
(2) All the members concentrate on a line which passes through the origin,

(3) Members are located on a regular polygon whose center is the origin.

For further details, see Chino [5, 6].

A more complicated model composed of the following four equations may be constructed by converting the linear term of (9.2) into nonlinear one:

$$z_{j,\,n+1} = z_{j,n} + \sum_{m=1}^{q}\sum_{k\neq j}^{N} \boldsymbol{D}_{jk,n}^{(m)} \boldsymbol{f}^{(m)}\left(z_{j,n} - z_{k,n}\right) + z_0,\; j = 1,\,2,\cdots,N, \tag{9.5}$$

$$\boldsymbol{f}^{(m)}\left(z_{j,n} - z_{k,n}\right) = \left(\left(z_{j,n}^{(1)} - z_{k,n}^{(1)}\right)^m, \left(z_{j,n}^{(2)} - z_{k,n}^{(2)}\right)^m, \ldots, \left(z_{j,n}^{(p)} - z_{k,n}^{(p)}\right)^m\right)^t, \tag{9.6}$$

$$\boldsymbol{D}_{jk,n}^{(m)} = diag\left(w_{jk,n}^{(1,m)}, w_{jk,n}^{(2,m)}, \ldots, w_{jk,n}^{(p,m)}\right), \tag{9.7}$$

and

$$w_{jk,n}^{(l,m)} = a_n^{(l,m)} r_{j,n}^{(l,m)} r_{k,n}^{(l,m)} sin\left(\theta_{k,n}^{(l,m)} - \theta_{j,n}^{(l,m)}\right),\; l = 1, 2,\cdots,p,\; m = 1,\,2,\cdots,r \tag{9.8}$$

Here, m denotes the degree of the vector function $\boldsymbol{f}^{(m)}\left(z_{j,n} - z_{k,n}\right)$, which is assumed to have the maximum value q. This model is very general and might enable us to describe various possible changes in asymmetric relationships among members over time. The term z_0 is a complex constant vector. The Eqs. (9.7) and (9.8) are the same as those in the simple linear model except that the superscript parameter m is added to it.

9.2.1 Linear Model

In this section we shall examine properties of the solution curves of some linear models which are the special cases of a class of complex difference system models discussed in Sect. 9.2.

Example 9.1 The following is the simplest model of the complicated linear model discussed above, which describes the change in dyadic relationships between two members over time. In this model, we omit the superscripts corresponding to m and p in this model since both m and p equal 1.

$$\begin{cases} z_{j,n+1} = z_{j,n} + w_{jk,n}\left(z_{j,n} - z_{k,n}\right) \\ z_{k,n+1} = z_{k,n} + w_{jk,n}\left(z_{k,n} - z_{j,n}\right) \\ w_{jk,n} = c r_{j,n} r_{k,n} sin\left(\theta_{k,n} - \theta_{j,n}\right) \end{cases}, \tag{9.9}$$

where c is a special case of $a_n^{(l,m)}$ in (9.8). Figure 9.1 shows the four configurations

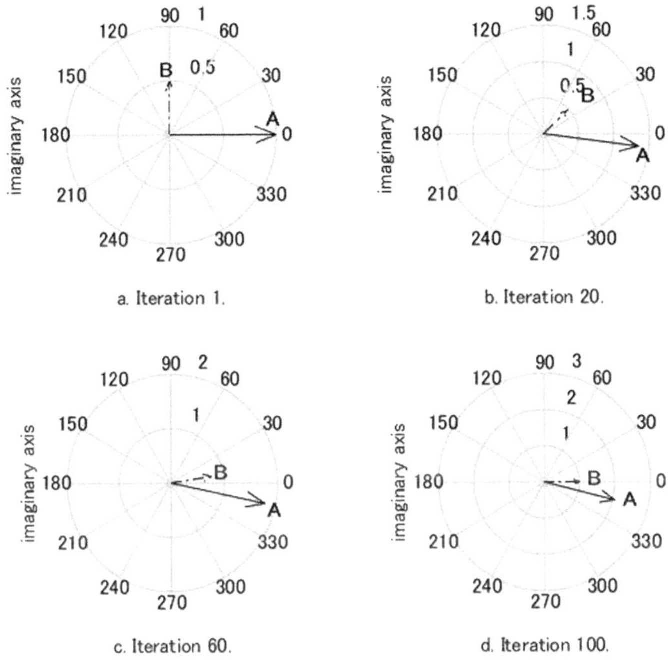

Fig. 9.1 Four configurations of members between iteration 1 and iteration 100, where the horizontal axis and the vertical axis denote the real axis and the imaginary axis, respectively

of members between iteration 1 (discrete point in time $n = 1$) and iteration 100 (discrete point in time $n = 100$). In this figure, we denote the initial points $z_{j,1}$.

and $z_{k,1}$ by A and B, respectively. These points A and B were set equal to 1 and $i/2$, where i^2 is -1 in the *one-dimensional Hilbert space*, which can be identified as a *complex plane C*.

Figure 9.2 shows the orbits of the two members j and k on the same plane, each of which depicts the orbit from iteration 1 to iteration 200. The orbit of member j denoted by A starting from A_1 is depicted by a thick red line, while that of member k denoted by B starting from B_1 by a thin black line. The relatively large dot indicates the origin.

At a glance the two orbits are on the same line from the initial point where A starts. On the other hand, one can see from Fig. 9.3c that the angle from A to B decreases gradually as time proceeds.

Example 9.2 The following is another linear model, which describes the change in triadic relationships between three members over time:

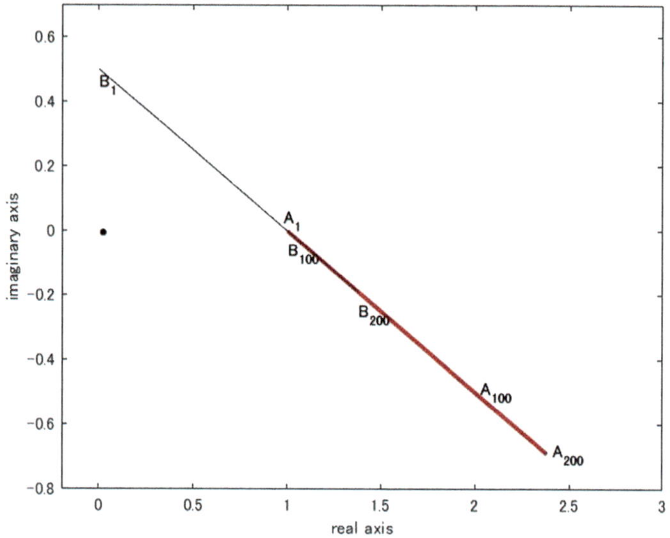

Fig. 9.2 Orbits (solution curves) of the two members in the complex plane

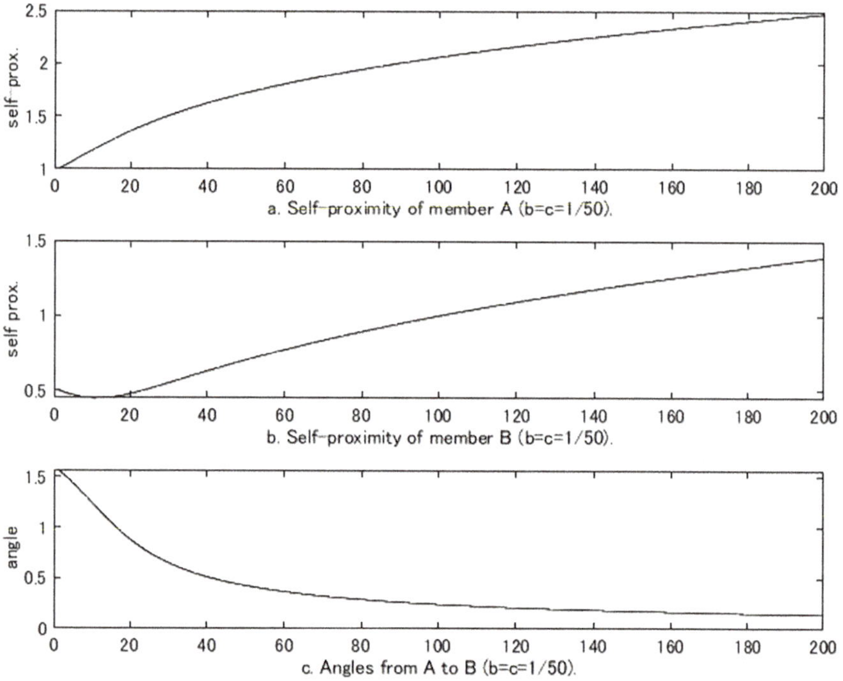

Fig. 9.3 Change in the self-proximities as well as the angle between the two members over time

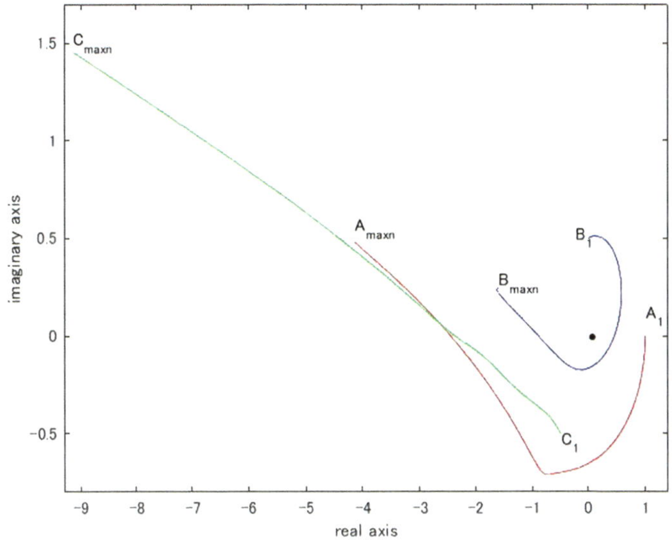

Fig. 9.4 The orbits of the three members in a complex plane

$$\begin{cases} z_{i,n+1} = z_{in} + w_{ij,n}\left(z_{i,n} - z_{j,n}\right) + w_{ik,n}\left(z_{i,n} - z_{k,n}\right) \\ z_{j,n+1} = z_{jn} + w_{jk,n}\left(z_j, n - z_{k,n}\right) + w_{ji,n}\left(z_{j,n} - z_{i,n}\right) \\ z_{k,n+1} = z_{kn} + w_{ki,n}\left(z_{k,n} - z_{i,n}\right) + w_{kj,n}\left(z_{k,n} - z_{j,n}\right) \\ w_{ij,n} = cr^d_{i,n} r^d_{j,n} \sin\left(\theta_{j,n} - \theta_{i,n}\right) \\ w_{ik,n} = cr^d_{i,n} r^d_{k,n} \sin\left(\theta_{k,n} - \theta_{i,n}\right) \\ w_{jk,n} = cr^d_{j,n} r^d_{k,n} \sin\left(\theta_{k,n} - \theta_{j,n}\right) \end{cases}, \tag{9.10}$$

where c is a special case of $a_n^{(l,m)}$ in (9.8), $r^d_{j,n}$ and $r^d_{k,n}$ are special cases of $r_{j,n}^{(l,m)}$ and $r_{k,n}^{(l,m)}$ in (9.8), respectively. Moreover, $w_{ji,n} = -w_{ij,n}$, $w_{ki,n} = -w_{ik,n}$, and $w_{kj,n} = -w_{jk,n}$, since the sinusoidal function is odd. Figure 9.4 shows the orbits of the three members in a complex plane. The initial points A_1, B_1, andC_1 of three members were set equal to 1, $i/2$, and $-0.5 - 0.5i$, respectively, while both c and d were set equal to 1/50. The final points $A_{\max n}$, $B_{\max n}$, and$C_{\max n}$ are those points at iteration 600.

9.2.2 Nonlinear Model

In this section we shall examine the property of a simple nonlinear model, which is a special case of the nonlinear model described by (9.5) through (9.8).

Example 9.3 The following is a special case of the complicated difference system model described by (9.5) through (9.8), in which the order of the polynomial is 2 and no linear term is included,

$$\begin{cases} z_{j,n+1} = z_{jn} + w_{jk,,n}\left(z_{j,n} - z_{k,n}\right)^2 + z_0 \\ z_{k,n+1} = z_{kn} + w_{kj,,n}\left(z_{k,n} - z_{j,n}\right)^2 + z_0 \\ w_{jk,n} = cr_{j,n}^d r_{k,n}^d sin\left(\theta_{k,n} - \theta_{j,n}\right) \end{cases}, \tag{9.11}$$

where c is a special case of $a_n^{(l,m)}$ in (9.8), $r_{j,n}^d$ and $r_{k,n}^d$ are special cases of $r_{j,n}^{(l,m)}$ and $r_{k,n}^{(l,m)}$ in (9.8), respectively. Moreover, $w_{kj,n} = -w_{jk,n}$,, from the characteristics of the third equation in (9.11). The constant term z_0 was set equal to $0.01i$.

Figure 9.5 shows the changes in self-proximities of two members as well as the change in the angle between the two members over time, while Fig. 9.6 depicts orbits of the two members in the same complex plane. The initial points A_1 and B_1 of these members were set equal to 1 and $i/2$, respectively. The maximum number of iterations was set equal to 1000.

Figures 9.5a and b show that the self-proximities of the two members diverge as time proceeds, while Fig. 9.5c indicates that the angle between the two members converges to π as time proceeds. Divergence of the self-proximity is not a realistic phenomenon in daily life.

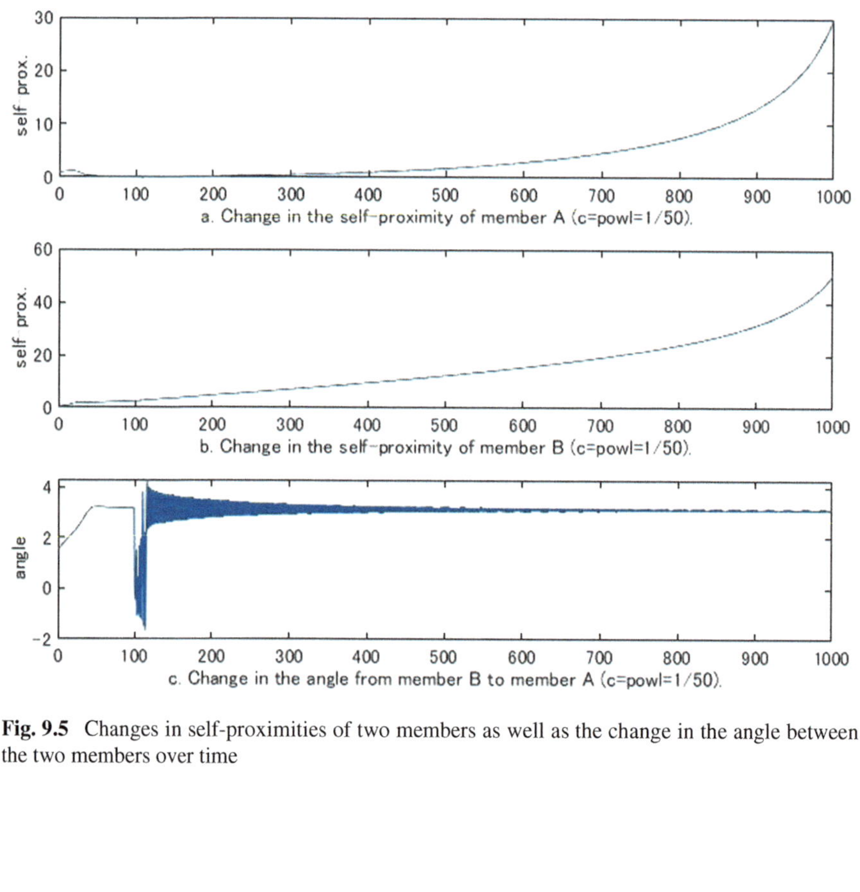

Fig. 9.5 Changes in self-proximities of two members as well as the change in the angle between the two members over time

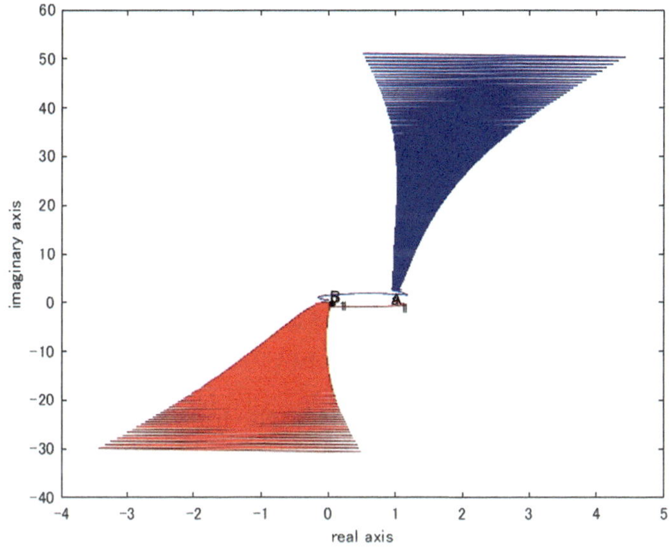

Fig. 9.6 Orbits of the two members in the same complex plane

9.3 Recent Versions of the Complex Difference System Model

As introduced in Sect. 9.1, the earlier version of our complex difference system models has a shortfall in that it is not holomorphic, and therefore we cannot utilize the theory of complex dynamical systems in mathematics in the earlier version. Recent version described by the following equations is a revised version of the complex difference system models discussed in Sect. 9.2.

$$z_{j,n+1} = z_{j,n} + \sum_{m=1}^{q} \sum_{k \neq j}^{N} \boldsymbol{D}_{jk,n}^{(m)} \boldsymbol{f}^{(m)}\left(z_{j,n} - z_{k,n}\right) + \boldsymbol{g}\left(\boldsymbol{u}_{j,n}\right) + z_0,\ j = 1, 2, \cdots, N, \tag{9.12}$$

$$\boldsymbol{f}^{(m)}\left(z_{j,n} - z_{k,n}\right) = \left(\left(z_{j,n}^{(1)} - z_{k,n}^{(1)}\right)^m, \left(z_{j,n}^{(2)} - z_{k,n}^{(2)}\right)^m, \ldots, \left(z_{j,n}^{(p)} - z_{k,n}^{(p)}\right)^m\right)^t, \tag{9.13}$$

and

$$\boldsymbol{D}_{jk,n}^{(m)} = diag\left(w_{jk,n}^{(1,m)}, w_{jk,n}^{(2,m)}, \ldots, w_{jk,n}^{(p,m)}\right). \tag{9.14}$$

Here, $z_{j,n}$ denotes the coordinate vector of member j at time n in a p-dimensional Hilbert space or a p-dimensional indefinite metric space. Moreover, m denotes the degree of the vector function $\boldsymbol{f}^{(m)}\left(z_{j,n} - z_{k,n}\right)$, which is assumed to have the maximum value q. Moreover, $w_{jk,n}^{(1,m)}, w_{jk,n}^{(2,m)}, \ldots, w_{jk,n}^{(p,m)}$ are *complex constants* in this model. Therefore, we shall hereafter omit the subscript n to express clearly that these parameters are constants, unless otherwise stated. This model is very general and might enable us to describe various possible changes in asymmetric relationships among members over time. $\boldsymbol{g}(\boldsymbol{u}_{j,n})$ is a *control* (e.g., Elaydi [21]; Ott et al. [30]) and z_0 is a complex constant vector.

The recent version includes two kinds of complex difference system models. One is the difference system model which imposes certain constraints on the complex constants, $w_{jk}^{(1,m)}, w_{jk}^{(2,m)}, \ldots, w_{jk}^{(p,m)}$. To be precise, these constraints are $w_{kj}^{(1,m)} = -w_{kj}^{(1,m)}$, $m = 1, 2, \cdots, q$, each of which is a skew-symmetric constraint. The other is the difference system model which imposes no constraints on these complex constants.

As indicated in Sect. 9.1, the recent version also includes the notion of *weighted digraph* associated with the complex difference system. Since the state space in which objects are embedded is assumed to be a (complex) Hilbert space or an indefinite metric space depending on the definiteness of the hypothetical Hermitian matrix, *polarization identity* (or polar identity) holds for every pair of the coordinate vectors of objects in the state space (see, Theorems 2.8a and b in Chap. 2; (7.40) in Chap. 7):

In fact, from (7.41) in Chap. 7, we have

$$h_{jk} = \frac{1}{2}\left(\|z_j^2\| + \|z_k^2\| - \|z_j\| - \|z_k^2\right)\| + \|\frac{1}{2}i\left(z_j^2\| + \|z_k^2\| - \|z_j\| - \|i\,z_k^2\|\right), \tag{9.15}$$

where h_{jk} is the $(j,\ k)$ element of the Hermitian matrix constructed from an asymmetric similarity matrix between objects. Moreover, we have

$$s_{jk} = \frac{1}{2}\left(\|z_j^2\| + \|z_k^2\| - \|z_j\| - \|z_k^2\right)\| + \|\frac{1}{2}\left(z_j^2\| + \|z_k^2\| - \|z_j\| - \|iz_k^2\|\right), \tag{9.16}$$

where s_{jk} is the (j, k) element of the (real) asymmetric similarity matrix, $\boldsymbol{S}$.

Therefore, given a configuration of objects $z_{j,n}$ $(\equiv z_j)$, $j = 1, \cdots, N$, at iteration n in a Hilbert space, we can compute $\boldsymbol{S}$ at any point of iteration. As a result, we can construct the digraph associated with $\boldsymbol{S}$ at any point of iteration. This is the reason why we call our complex difference system model for the analysis of asymmetric interactions between objects or nodes *dynamical weighted digraph* (DWD), since the weighted digraph is associated with $\boldsymbol{S}$.

In Sect. 9.3.1, we shall examine properties of the solution curves of the former difference system model. In Sect. 9.3.2, we shall examine those of the latter difference system model.

Before proceeding, we shall consider a similar model equation to (9.12), which cannot be included in (9.12) when the power m in (9.13) is even. This model can be written as follows:

$$z_{j,\,n+1} = z_{j,n} + \sum_{m=1}^{q} \sum_{k \neq j}^{N} \boldsymbol{D}_{jk,n}^{(m)} \boldsymbol{f}\left(\left(z_{j,n}\right)^{(m)} - \left(z_{k,n}\right)^{(m)} \right) + \boldsymbol{g}\left(\boldsymbol{u}_{j,n}\right) + z_0,\ j = 1, 2, \cdots, \mathrm{N}, \tag{9.17}$$

$$\boldsymbol{f}\left(\left(z_{j,n}\right)^{(m)} - \left(z_{k,n}\right)^{(m)} \right) = \left(\left(z_{j,n}^{(1)}\right)^{m} - \left(z_{k,n}^{(1)}\right)^{m}, \ldots, \left(z_{j,n}^{(\mathrm{p})}\right)^{m} - \left(z_{k,n}^{(\mathrm{p})}\right)^{m} \right)^{t}, \tag{9.18}$$

and $\boldsymbol{D}_{jk,n}^{(m)}$ is the same as (9.14). In this model, m is assumed to be even.

We shall point out two examples of this model in Examples 9.6 and 9.7.

Remark Having not studied this model for several years, we had not noticed the difference between (9.12) and (9.17) until recently. M. Nakagawa (personal communication, February 29, 2024) pointed out this difference.

9.3.1 *Complex Difference System Models with Skew-Symmetric Constraints on* $\boldsymbol{w_{jk}^{(1,m)}, w_{jk}^{(2,m)}, \ldots, w_{jk}^{(p,m)}}$

In this section we shall examine properties of the solution curves of the complex difference system models which imposes skew-symmetric constraints on $w_{jk}^{(1,m)}, w_{jk}^{(2,m)}, \ldots, w_{jk}^{(p,m)}$.

(a) Linear Model

Example 9.4 The following is a dyadic linear model with skew-symmetric constraints on w_{jk}, which is a special case of $w_{jk,n}^{(1,m)}$ in (9.14).

$$\begin{cases} z_{j,n+1} = z_{j,n} + w_{jk}\left(z_{j,n} - z_{k,n}\right) \\ z_{k,n+1} = z_{k,n} + w_{kj}\left(z_{k,n} - z_{j,n}\right) \end{cases}, \tag{9.19}$$

where $w_{kj} = -w_{jk}$ and $w_{jk} = 0.01(1 + i)$.

Figure 9.7 shows the changes in self-proximities of two members as well as the change in the angle between the two members over time, while Fig. 9.8 depicts orbits of the two members in the same complex plane. The initial points A_1 and B_1 of these members were set equal to 1 and $i/2$, respectively. The maximum number of iterations was 200. It should be noticed from these figures that the angle from members A to B approaches zero as time proceeds although the orbits of the two members continue to be parallel as time proceeds.

Example 9.5 The following is another linear model, which describes the change in *tripartite deadlock* relationships between three members over time.

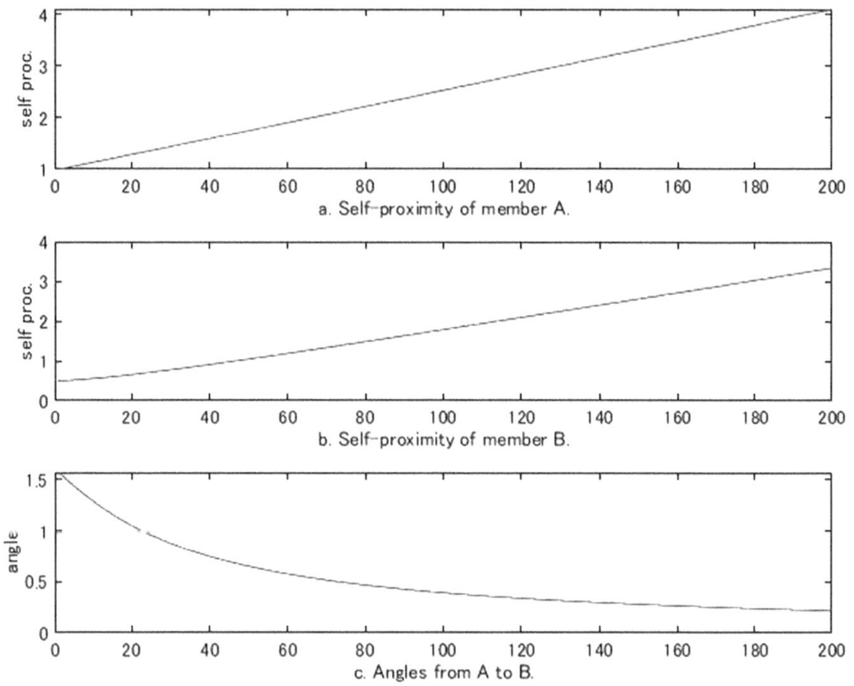

Fig. 9.7 Changes in self-proximities of two members as well as the change in the angle between the two members over time in (9.19)

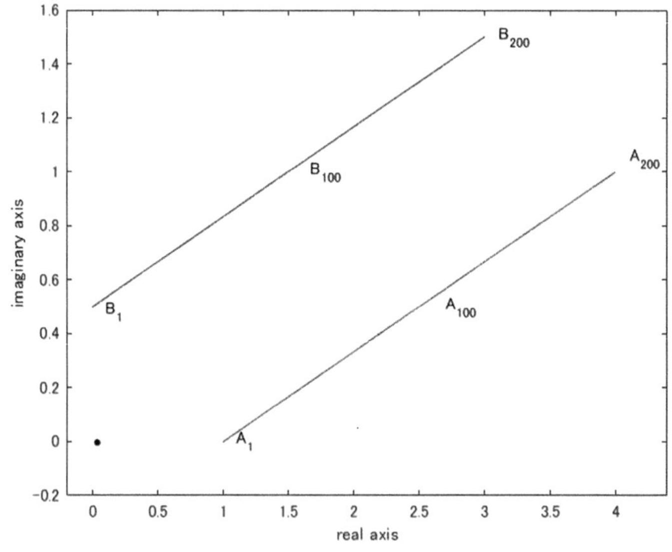

Fig. 9.8 Orbits of the two members in (9.19) in the same complex plane

$$\begin{cases} z_{i,n+1} = z_{i,n} + w_{ij}\left(z_{i,n} - z_{j,n}\right) + w_{ik}\left(z_{i,n} - z_{k,n}\right) \\ z_{j,n+1} = z_{j,n} + w_{jk}\left(z_{j,n} - z_{k,n}\right) + w_{ji}\left(z_{j,n} - z_{i,n}\right) \\ z_{k,n+1} = z_{k,i} + w_{ki}\left(z_{k,n} - z_{i,n}\right) + w_{kj}\left(z_{k,n} - z_{j,n}\right) \end{cases}, \quad (9.20)$$

where $w_{ji} = -w_{ij}$, $w_{ki} = -w_{ik}$, and $w_{kj} = -w_{jk}$, while $w_{ij} = w_{jk} = w_{ki} = 0.01$.

Figure 9.9 shows the four configurations of members at iterations, 1, 20, 60, and 100. Figure 9.10 shows the changes in self-proximities of three members over time, while Fig. 9.11 shows the orbits of the three members in a complex plane. The initial points A_1, $B_{1,}$andC_1 of three members were set equal to 1, $\cos\frac{2\pi}{3} + i\sin\frac{2\pi}{3}$, and, $\cos\frac{2\pi}{3} - i\sin\frac{2\pi}{3}$ respectively, which was not shown in Fig. 9.11. The final points $A_{\max n}$, $B_{\max n,}$and$C_{\max n}$ are those points at iteration 600.

It is apparent from Fig. 9.9 that the tripartite deadlock relationship is invariant up to the rotation of the configuration. It is also apparent from Fig. 9.10 that the self-proximities of the three members increase gradually as time proceeds.

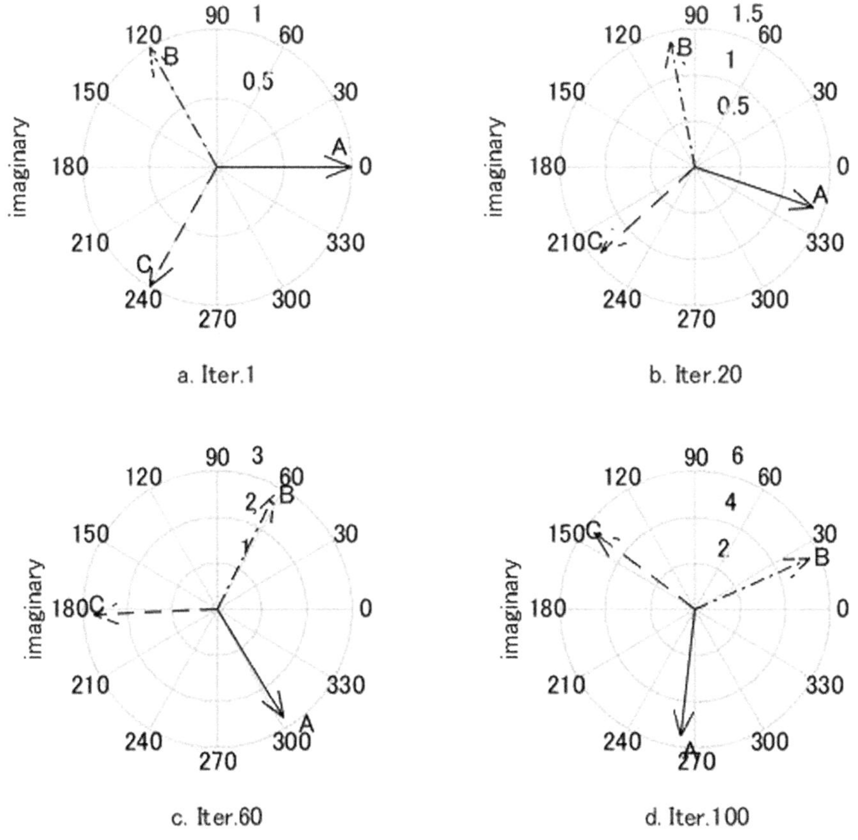

Fig. 9.9 Four configurations of members at iterations, 1, 20, 60, and 100

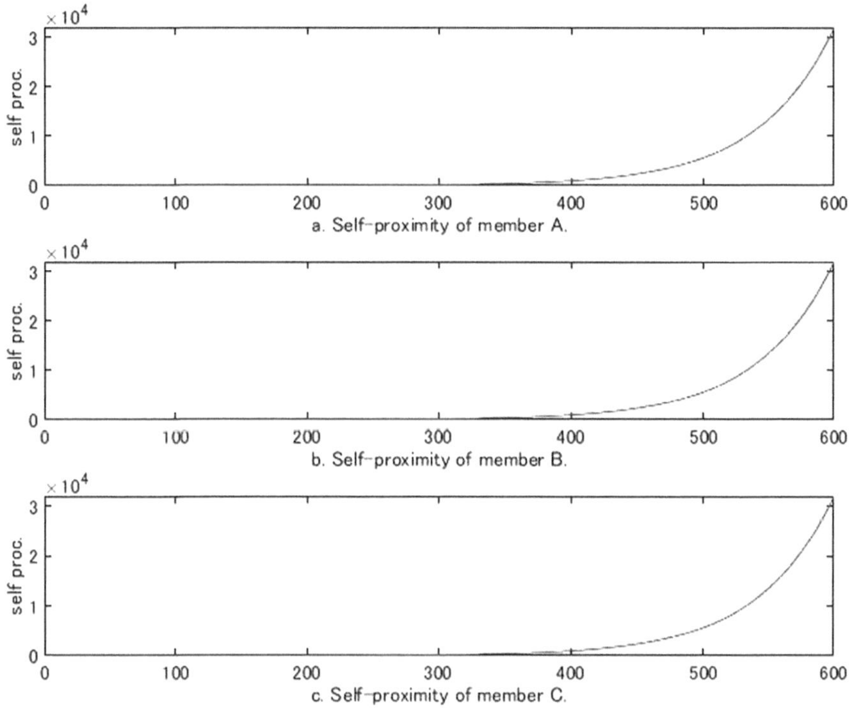

Fig. 9.10 Changes in self-proximities of the three members over time

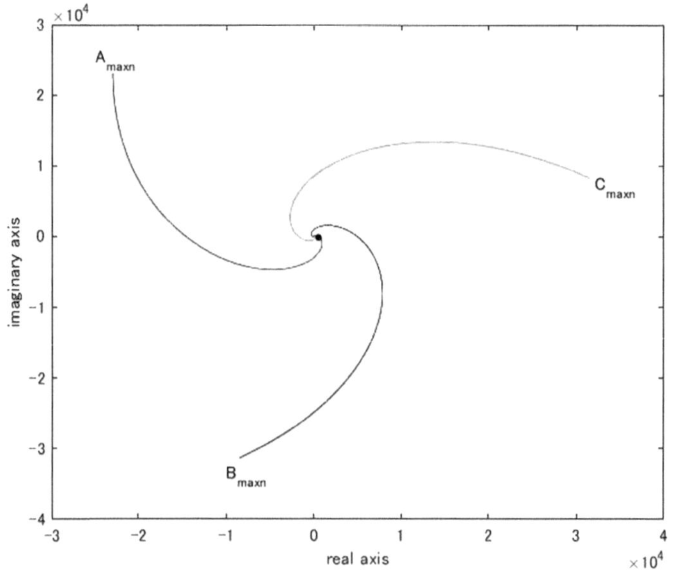

Fig. 9.11 Orbits of the three members in a complex plane in (9.17)

From Fig. 9.11 one can see that the orbits of the three members of this system depicted on the same plane form a spiral curve and diverge, as time proceeds.

(b) Nonlinear Model

Example 9.6 The following is a special case of the complex difference system models described by (9.17), in which the order of the polynomial is 2 and no linear term is included,

$$\begin{cases} z_{j,n+1} = z_{j,n} + w_{jk}\left(z_{j,n}^2 - z_{k,n}^2\right) + z_0 \\ z_{k,n+1} = z_{k,n} + w_{kj}\left(z_{k,n}^2 - z_{j,n}^2\right) + z_0 \end{cases}, \tag{9.21}$$

where $w_{kj} = -w_{jk}$, and $w_{jk} = 0.01$. The constant term z_0 in (9.12) was set equal to $0.01 + 0.01i$.

Figure 9.12 shows the changes in self-proximities of the two members and the angle between the two members, while Fig. 9.13 depicts orbits of the two members on the same complex plane. The initial points A_1andB_1 of these members were set equal to 1 and $i/2$, respectively. The constant term z_0 was set equal to $0.01i$. The maximum number of iterations was set equal to 240.

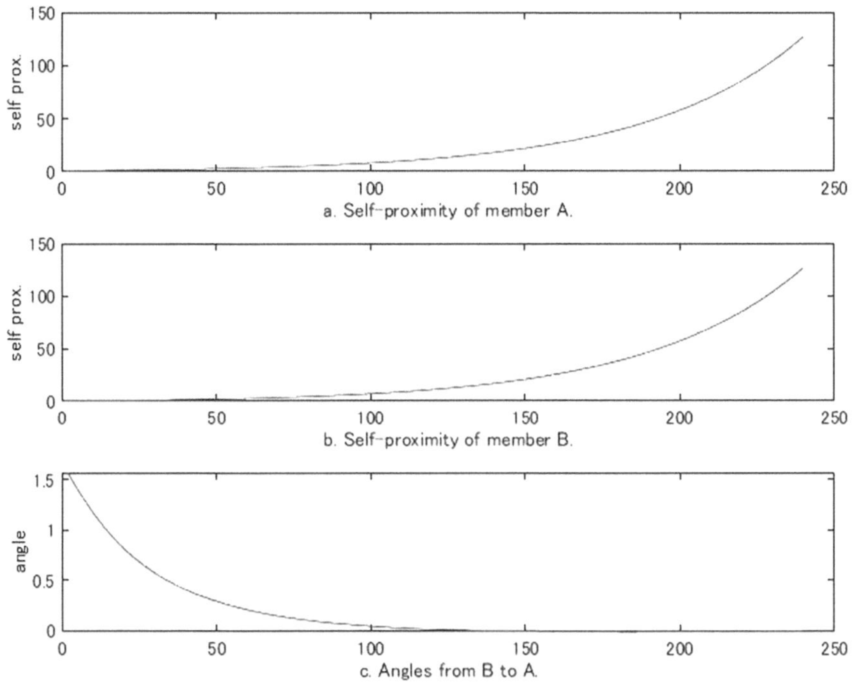

Fig. 9.12 Changes in self-proximities of the two members and the angle between the two members

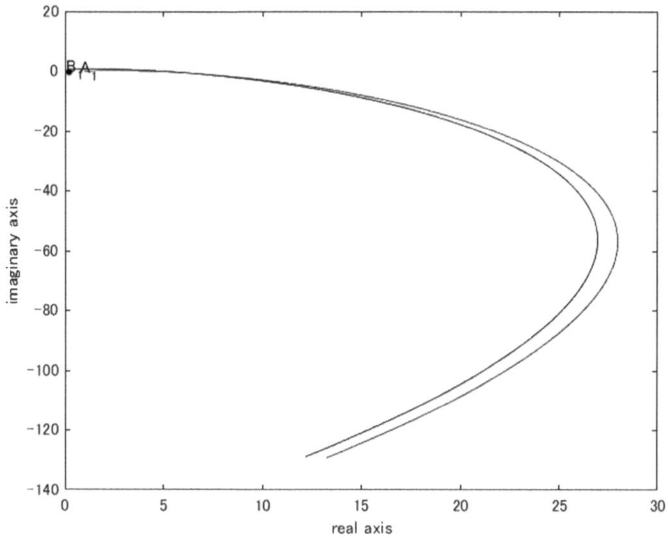

Fig. 9.13 Orbits of the two members on the same complex plane

Example 9.7 The following is another nonlinear model described by (9.17), which describes triadic relationships among members over time. In this case, the order of the polynomial is 2 and no linear term is included,

$$\begin{cases} z_{i,n+1} = z_{i,n} + w_{ij}\left(z_{i,n}^2 - z_{j,n}^2\right) + w_{ik}\left(z_{i,n}^2 - z_{k,n}^2\right) \\ z_{j,n+1} = z_{j,n} + w_{jk}\left(z_{j,n}^2 - z_{k,n}^2\right) + w_{ji}\left(z_{j,n}^2 - z_{i,n}^2\right) \\ z_{k,n+1} = z_{k,n} + w_{ki}\left(z_{k,n}^2 - z_{i,n}^2\right) + w_{kj}\left(z_{k,n}^2 - z_{j,n}^2\right) \end{cases}, \tag{9.22}$$

where $w_{ji} = -w_{ij}$, $w_{kj} = -w_{jk}$, $w_{ik} = -w_{ki}$, and $w_{ij} = w_{jk} = w_{ki} = 0.01$. The constant term z_0 in (9.12) was set equal to $0.01i$.

Figure 9.14 shows the changes in self-proximities of the three members, *A*, *B*, and *C*. Figure 9.15 shows changes in angles from A to B, B to C, and C to A. Figure 9.16 depicts orbits of the three members on the same complex plane. The initial points A_1, B_1*and* C_1 of these members were set equal to $1, i/2$, and $-0.5-0.5i$, respectively. The maximum number of iterations was set equal to 8000.

At a glance, each of the changes in the self-proximities of the three members give rise to irregular transcient bursts as time proceeds. As will be examined soon, the time series in Fig. 9.14a, for example, indicates a *chaos*. On the other hand, changes in angles from *A* to *B*, *B* to *C*, and *C* to *A* in Fig. 9.15 are reminiscent of a *typical chaos*. In fact, for example, the largest Lyapunov exponent of the time series in Fig. 9.15a is positive, which means that this time series is characterized by a typical chaos, as will be examined soon. Since asymmetric proximities change as the angles change according to HFM (see Chaps. 7 and 8), this means that the asymmetric proximities in

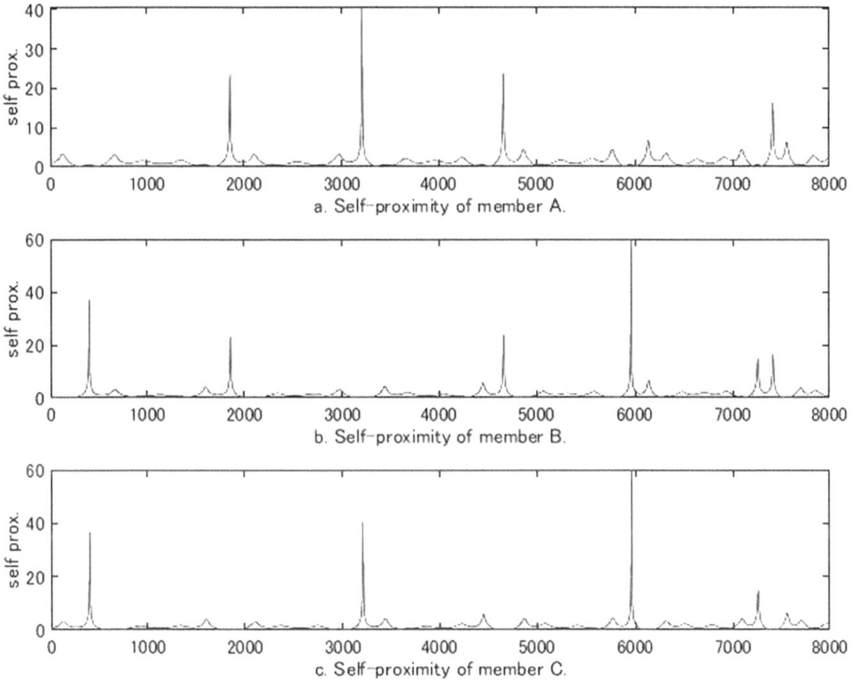

Fig. 9.14 Changes in self-proximities of the three members, *A*, *B*, and *C*

this triadic nonlinear model change chaotically through the complicated asymmetric interactions between three members.

In Fig. 9.16, the red curve, the blue curve, and green curve, respectively, show the orbits of members *A*, *B*, and *C*.

Figure 9.17 shows the estimated Lyapunov exponents for the time series of changes in self-proximities of member *A* depicted in 9.14a. They were 0.0211, −0.0014, and − 0.0258. This result indicates that the changes in self-proximities of member *A* exhibit a chaos.

Figure 9.18 shows the estimated Lyapunov exponents for the angles between members *A* and *B* over time shown in 9.15a. They were 0.1598, −0.0054, and − 0.0620. The largest Lyapunov exponent also indicates that the changes in the angle from members *A* to *B* exhibit a chaos.

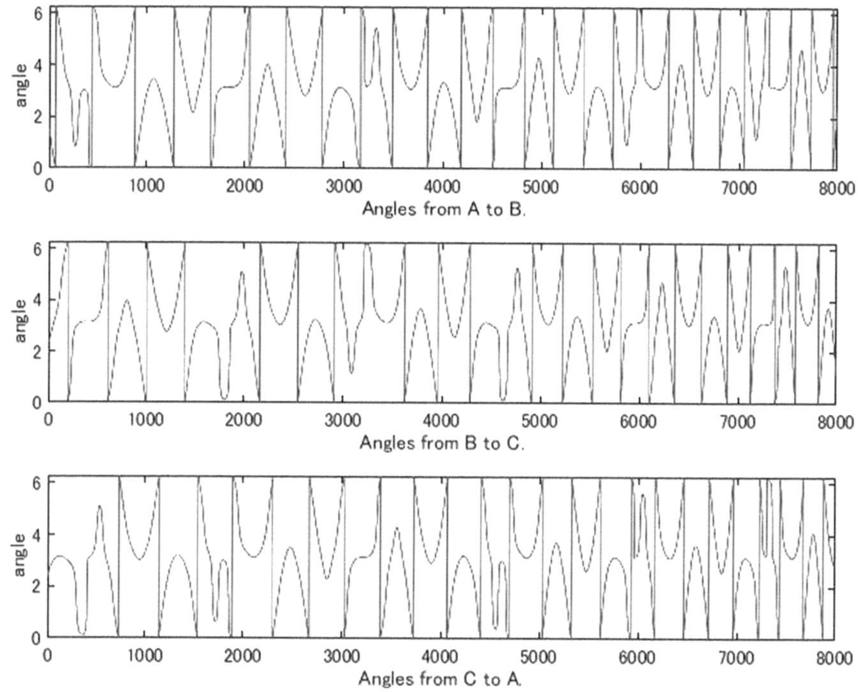

Fig. 9.15 Changes in angles between A and B, B and C, and C and A

9.3.2 *Complex Difference System Models with no Constraints on* $w_{jk}^{(1,m)}, w_{jk}^{(2,m)}, \ldots, w_{jk}^{(p,m)}$

In this section we shall examine properties of the solution curves of the complex difference system models which imposes no constraints on $w_{jk}^{(1,m)}, w_{jk}^{(2,m)}, \ldots, w_{jk}^{(p,m)}$.

(a) Linear Model

Let us consider a *dyadic* linear model as the simplest model of the complex difference system models which impose no constraints on $w_{jk}^{(1,m)}, w_{jk}^{(2,m)}, \ldots, w_{jk}^{(p,m)}$, first. In such a model, one of the superscript parameter m of w_{jk} equals 1. Moreover, let us assume that the state space of this system is one-dimensional Hilbert space. Then, the other superscript parameter p of w_{jk} is also 1. In such a model, we can omit all the superscript parameters of w_{jk}, and we may write $w_{jk}^{(1,m)}$ simply as w_{jk}. In any case, this model is the same as that described by (9.19). However, in this case, we impose no constraints on w_{jk}. This enables us to discuss the linear model from a general framework concerning the w_{jk}.

In fact, we can classify the solution curves of the linear models of this type according to the *eigenvalues* of the following special matrix composed of the w_{jk}'s.

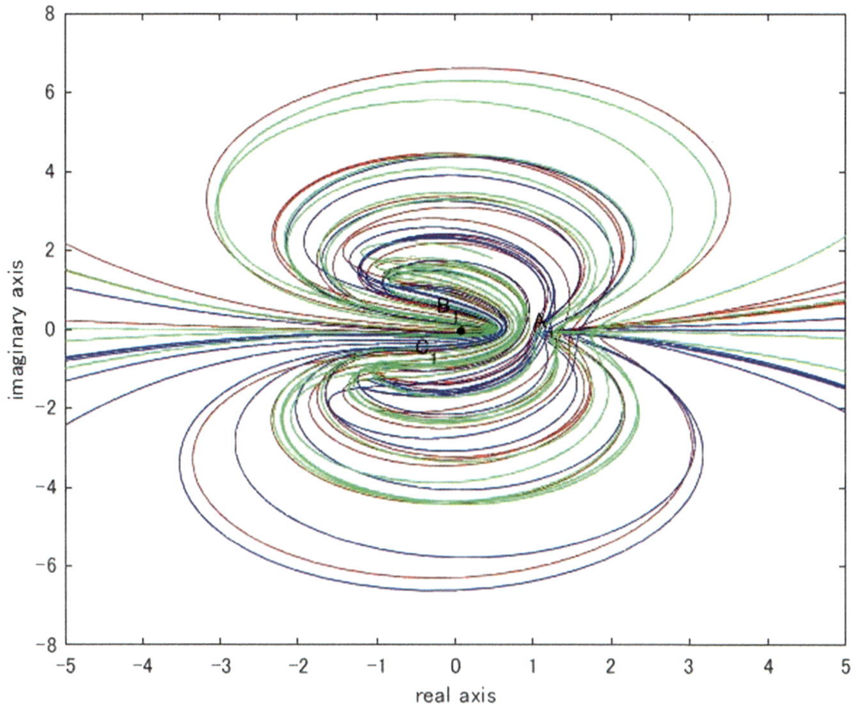

Fig. 9.16 Orbits of the three members on the same complex plane

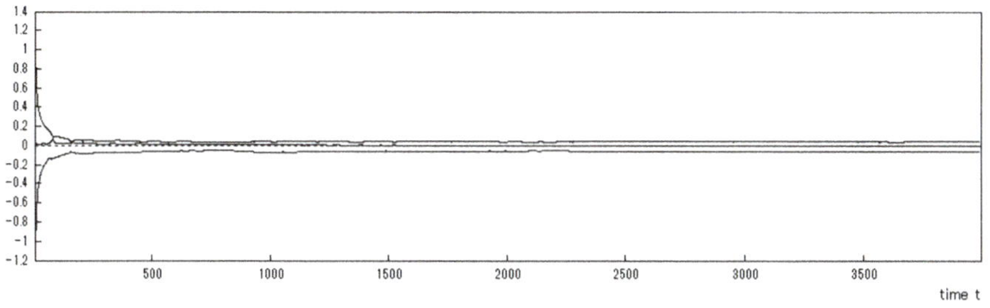

Fig. 9.17 Lyapunov exponents for the self-proximity of member A computed by SundayChaos Times

Before proceeding, let us denote $z_n = \left(z_{j,n}, z_{k,n}\right)^t$, and let us rewrite (9.19) using vector notation as

$$z_{n+1} = A_2 z_n, \tag{9.23}$$

where

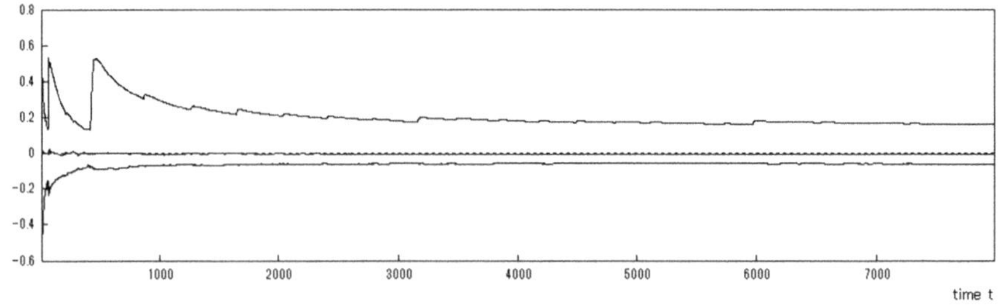

Fig. 9.18 Lyapunov exponent for the angle between members A and B computed by SundayChaos Times

$$\boldsymbol{A}_2 = \begin{pmatrix} 1+w_{jk} & -w_{jk} \\ -w_{kj} & 1+w_{kj} \end{pmatrix}. \tag{9.24}$$

Here, z_n should not be confused with $z_{j,n}$ defined by (9.2) earlier. The eigenvalues of $\boldsymbol{A}_2$ are 1 and $1 + w_{jk} + w_{kj}$.

It is not difficult to prove that in the dyadic case, dynamical scenarios of the solution curve have the three patterns depending on the absolute value of the eigenvalue, $1 + w_{jk} + w_{kj}$, which is shown in the next theorem.

Theorem 9.1 Let a dyadic linear system be described by (9.23) with a constant matrix $\boldsymbol{A}_2$ defined by (9.24). Then, the solution curve of (9.23)

$$\begin{cases} \textit{diverges}, & \textit{if } \left|1+w_{jk}+w_{kj}\right| > 1 \\ \textit{diverges on two lines}, & \textit{if } \left|1+w_{jk}+w_{kj}\right| = 1 \\ \textit{converges}, & \textit{if } \left|1+w_{jk}+w_{kj}\right| < 1 \end{cases}. \tag{9.25}$$

This theorem can be proved if we solve the solution curves of this system by applying the *Putzer algorithm* (e.g., Elaydi [21], p. 106–109) to the linear system described by (9.23). In this case we may identify (9.23) with the linear two-dimensional difference equation with real coefficients, because the real two-dimensional space can be identified with a complex plane.

From this theorem, we can say that the dyadic linear model shown in Example 9.4 is the special case in which $\left|1+w_{jk}+w_{kj}\right|$ equals 1. In fact, the orbit, i.e., the solution curve of the system described by (9.19) corresponds to the second case of Theorem 9.1.

Example 9.8 Let us consider another dyadic linear model described by (9.23) in which $w_{jk} = 0.02(1+i)$ and $w_{kj} = -0.01(1+i)$. The initial values $z_{j,1}$and$z_{k,1}$ were set equal to 1 and $i/2$, respectively. Then, the eigenvalue $1+w_{jk}+w_{kj} = 1.01+0.01i$. Therefore, $\left|1+w_{jk}+w_{kj}\right| \doteq 1.01$, and thus from Theorem 9.1 we can expect that the solution curve diverges as time proceeds. Figure 9.19 shows the orbits of two members in the same plane. We denote $z_{j,1}$and$z_{k,1}$ by $A_1 and B_1$, respectively, in this figure.

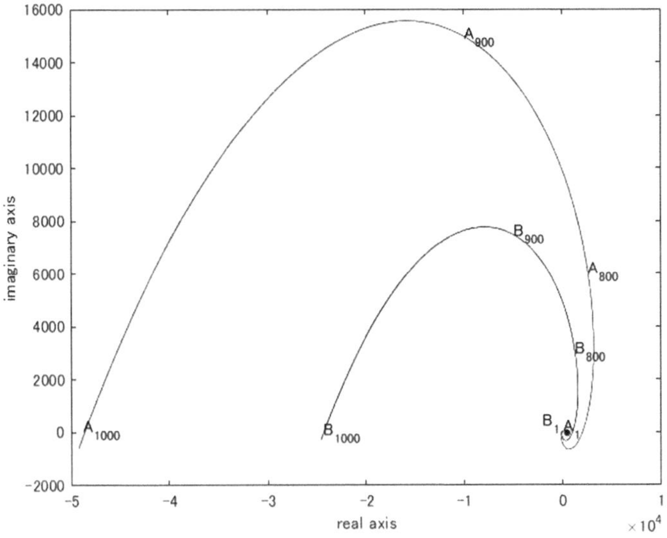

Fig. 9.19 Orbits of the two members A and B in the same complex plane. Initial points are denoted by A_1 and B_1

Example 9.9 Let us consider a third dyadic linear model described by (9.23), in which in which $w_{jk} = 0.01(1+i)$ and $w_{kj} = -0.02(1+i)$. The initial values $z_{j,1}$and$z_{k,1}$ were set equal to 1 and $i/2$, respectively, as in Example 9.8. Then, the eigenvalue $1 + w_{jk} + w_{kj} = 0.99 - 0.01i$. Therefore, $\left|1 + w_{jk} + w_{kj}\right| \fallingdotseq 0.9901$, and thus from Theorem 9.1 we can expect that the solution curve converges as time proceeds. Figure 9.20 shows the orbits of two members in the same plane. We denote $z_{j,1}$and$z_{k,1}$ by $A_1 and B_1$, respectively, in this figure.

Example 9.10 Let us now consider the *triadic* linear model. This model is the same as that defined in (9.20), but has no constraints on w_{ij}'s. Let us now rewrite (9.20) as

$$z_{n+1} = \boldsymbol{A}_3 z_n, \tag{9.26}$$

where

$$\boldsymbol{A}_3 = \begin{pmatrix} 1 + w_{jk} + w_{jl} & -w_{jk} & -w_{jl} \\ -w_{kj} & 1 + w_{kl} + w_{kj} & -w_{kl} \\ -w_{lj} & -w_{lk} & 1 + w_{lj} + w_{lk} \end{pmatrix}. \tag{9.27}$$

Here, $z_n = \left(z_{j,n}, z_{k,n}, z_{l,n}\right)^t$. The eigenvalues of $\boldsymbol{A}_3$ are 1 and

$$1 + \frac{1}{2}\sum_{g \neq h}^{3}\sum_{h}^{3} w_{gh} \pm \sqrt{D}, \tag{9.28}$$

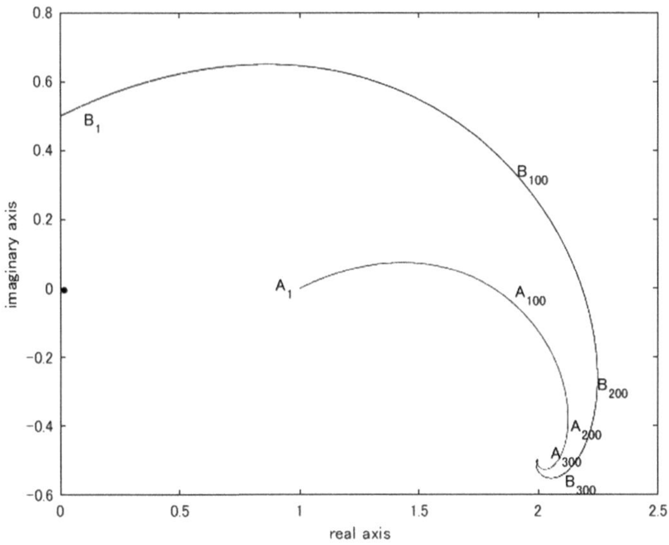

Fig. 9.20 Orbits of the two members A and B in the same complex plane. Initial points are denoted by A_1 and B_1

which determine the patterns of solution curves, where D is a special quadratic function of w_{gh}'s. We can solve the solution curves of this system by applying the Putzer algorithm to the linear system described by (9.26) in the same manner as in the dyadic linear system. However, we will not describe them here because they are very lengthy and complicated.

Example 9.5 discussed earlier is considered as a special case of (9.26). Here, in Example 9.10 we show a triadic linear model described by (9.26), in which $w_{jk} = -0.01(1+i)$, $w_{kj} = -0.02(1+i)$, $w_{jl} = 0.01(1+i)$, $w_{lj} = -0.02(1+i)$, $w_{kl} = 0.01(1+i)$, and $w_{lk} = -0.02(1+i)$. The initial values were the same as those of Example 9.5, which represent a *tripartite deadlock*. Figure 9.21 shows the orbits of three members in the same plane. After 1000 iterations, three orbits almost converged to a point which is denoted as A, B, C_{1000} in the figure. In this example, the eigenvalues of $\boldsymbol{A}_3$ were 1, $0.98+0.02i$, and$0.97+0.03i$, and their absolute values were 1.00, 0.9901, and 0.9851.

(b) Nonlinear Model

Let us consider now nonlinear complex difference system models which impose no constraint on $w_{jk}^{(1,m)}, w_{jk}^{(2,m)}, \ldots, w_{jk}^{(p,m)}$. We shall first consider such models with quadratic terms,

$$\begin{cases} z_{j,n+1} = z_{j,n} + w_{jk}^{(1)}\left(z_{j,n} - z_{kn}\right) + w_{jk}^{(2)}\left(z_{j,n} - z_{k,n}\right)^2 \\ z_{k,n+1} = z_{k,n} + w_{kj}^{(1)}\left(z_{k,n} - z_{jn}\right) + w_{kj}^{(2)}\left(z_{k,n} - z_{j,n}\right)^2 \end{cases}. \tag{9.29}$$

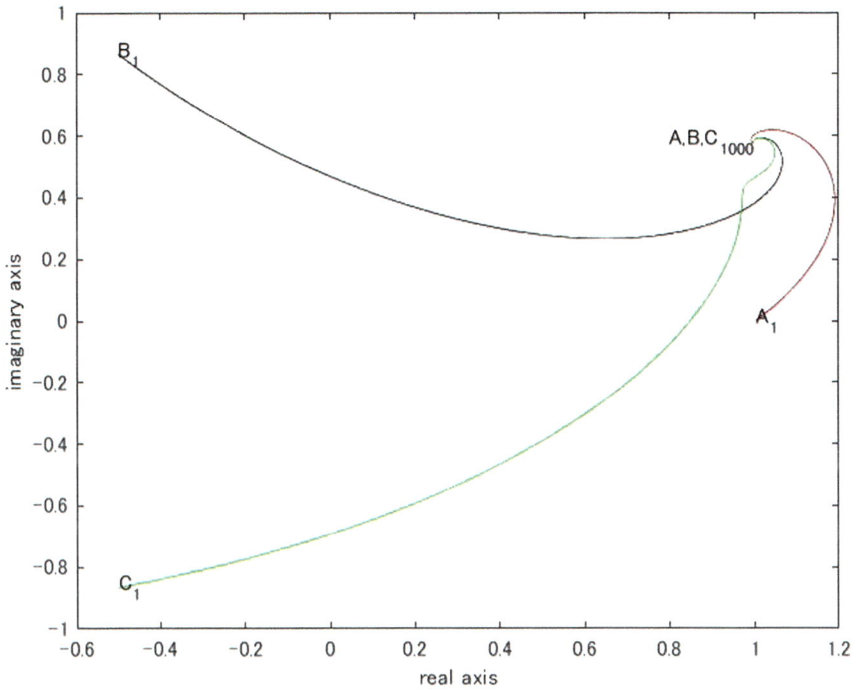

Fig. 9.21 Orbits of three members in the same plane

This type of system has a very desirable property in that we can utilize the heritage of the theory of the *complex dynamical system* developed in mathematics directly in classifying its solution curves. In fact, we can rewrite (9.29) in the form of *Mandelbrot set* as follows (Shiraiwa, K., personal communication, [32]):

First, let us compute the difference between $z_{j,n+1}$ and $z_{k,n+1}$ in (9.29) as

$$z_{j,n+1} - z_{k,\,n+1} = \left(1 + w_{jk}^{(1)} + w_{kj}^{(1)}\right)\left(z_{j,n} - z_{kn}\right) + \left(w_{jk}^{(2)} - w_{kj}^{(2)}\right)\left(z_{j,n} - z_{k,n}\right)^2. \tag{9.30}$$

Second, let us define a new variable,

$$u_{jk,n} = z_{j,n} - z_{k,n}. \tag{9.31}$$

Then, (9.28) can be written as

$$u_{jk,n+1} = \beta_{jk}^{(1)} u_{jk,n} + \beta_{jk}^{(2)} u_{jk,n}^2, \tag{9.32}$$

where

$$\beta_{jk}^{(1)} = 1 + w_{jk}^{(1)} + w_{kj}^{(1)},\ \beta_{jk}^{(2)} = w_{jk}^{(2)} - w_{kj}^{(2)} \tag{9.33}$$

Now, let us define another new variable,

$$Z_{jk,n} = \beta_{jk}^{(2)} u_{jk,n} + \frac{1}{2}\beta_{jk}^{(1)}. \tag{9.34}$$

Then, $Z_{jk,n+1}$ can be written as

$$Z_{jk,n+1} = Z_{jk,n}^{2} + \gamma_{jk}^{(1)}, \tag{9.35}$$

where

$$\gamma_{jk}^{(1)} = \frac{1}{2}\beta_{jk}^{(1)} - \frac{1}{4}\left[\beta_{jk}^{(1)}\right]^{2}, \tag{9.36}$$

If $c = \gamma_{jk}^{(1)}$ satisfies the subset defined later in Theorem 12.24 in Chap. 12, the complex system defined by (9.35) describes the Mandelbrot set.

It is interesting to note that $\gamma_{jk}^{(1)}$ in (9.36) is the function which includes only $\beta_{jk}^{(1)}$. Stated another way, $\gamma_{jk}^{(1)}$ is composed of coefficients of the linear term of (9.29).

Example 9.11 In this example, we shall examine the behaviors of the dyadic nonlinear system described by (9.29), in which $w_{jk}^{(1)} = w_{jk}^{(2)} = 1$, and $w_{kj}^{(1)} = w_{kj}^{(2)} = 1.5375 - 0.9564i$. Initial values of $z_{j,1}$and$z_{k,1}$ were set equal to $0.50094 + 0.0059i$ and -0.5, respectively.

Figure 9.22 shows the configurations of the two members at iterations, 1, 20, and 1000. In the figure, $A_{\max n}$, for example, indicate A_{1000}.

Figure 9.23 shows the orbit of member B after iteration 1000. Since this orbit is packed too close to distinguish between neighboring loci, we shall zoom up the orbit up to the first 100 iterations in Fig. 9.25. In Fig. 9.24, we shall cover this orbit with yellow, and depict the orbit of member A in the same plane. It is apparent from Fig. 9.25 that the orbit of member B is composed of dynamic cycles in marked contrast to that composed of tiny cycles of member A.

Figure 9.24 shows the orbits of two members in the same plane after iteration 1000.

To check whether the two kinds of cycles shown in Fig. 9.25 are chaotic or not, we computed the Lyapunov exponents of these cyclic orbits. Figures 9.26 and 9.27, respectively shows Lyapunov exponent of the orbits of the self-proximities of member A and member B, which were embedded in a three-dimensional space. The largest Lyapunov exponents of these orbits indicate that they have a chaotic feature, although the patterns of Lyapunov exponents of these orbits are different.

However, the Lyapunov exponent of the orbit of the angle from member B to member A, which was embedded in a three-dimensional space, tells us that it does not have a chaotic feature. Figure 9.28 shows this. That is, all the Lyapunov exponent were negative.

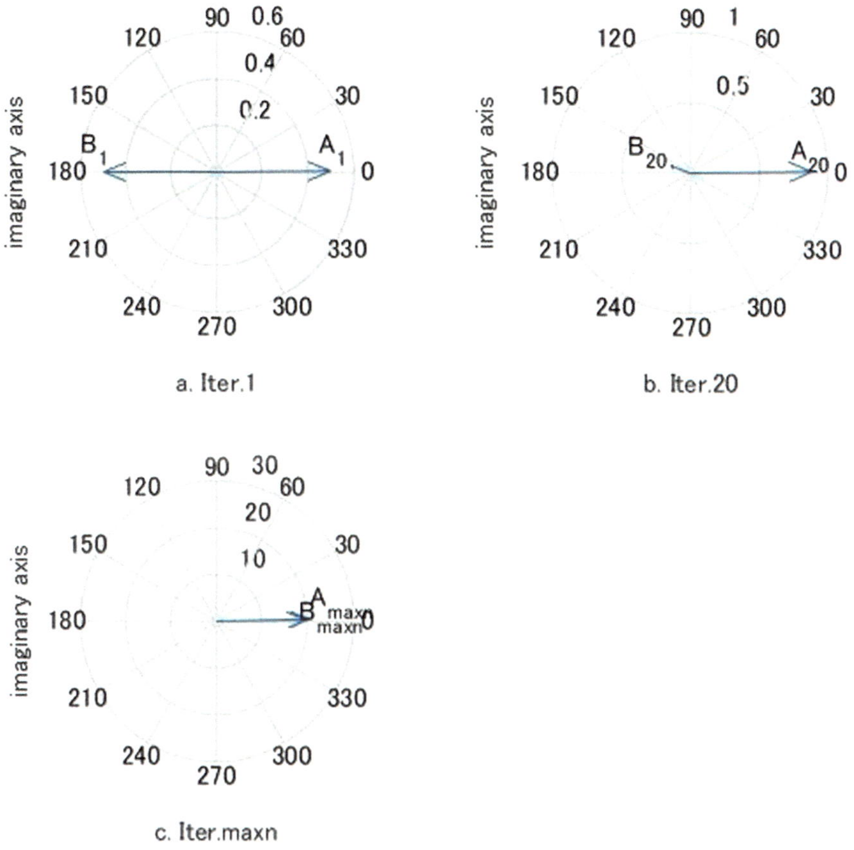

Fig. 9.22 Changes in configurations of the two members at iterations, 90, and manx = 1000

Then, it would be natural to ask whether the holistic orbital property of the self-proximities of two members is chaotic or not. To answer this question, we computed the Lyapunov exponents of these proximities. In this case, we embedded these time series in a two-dimensional space. As shown in Fig. 9.29, the largest Lyapunov exponent was not positive. This means that the holistic orbital property of the self-proximities of two members has not a chaotic feature.

So far, we have examined the orbital properties of (9.29) in the case of Example 9.11. Now we shall examine the orbital properties of (9.35) associated with (9.29) in the case of this example. In this case, $\beta_{jk}^{(1)}$ in (9.33) is 2.5475–0.9564i, and thus $\gamma_{jk}^{(1)}$ in (9.36) is about $-0.900138 + 0.7400145i$. Therefore, (9.35) is expressed as

$$Z_{jk,n+1} = Z_{jk,n}^2 - 0.900138 + 0.7400145i. \tag{9.37}$$

This map belongs to the *Mandelbrot set* and is almost the same as the famous *Julia set* in Fig. 12.25, which has innumerous *buds*, in Chap. 12. In other words, the

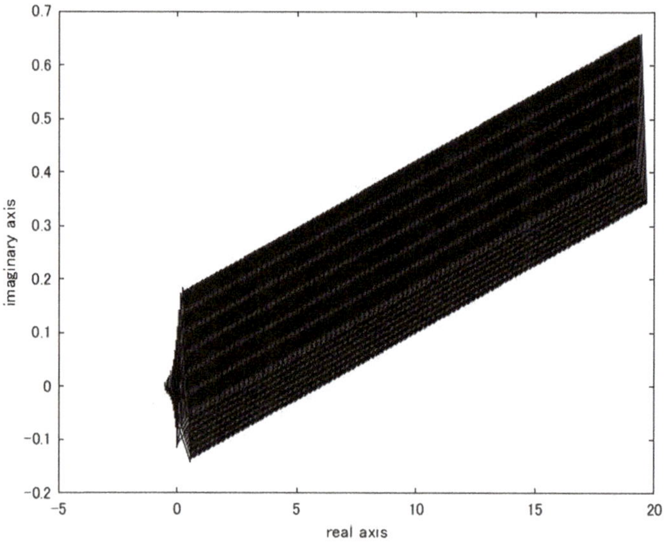

Fig. 9.23 Orbit of member *B* after iteration 1000

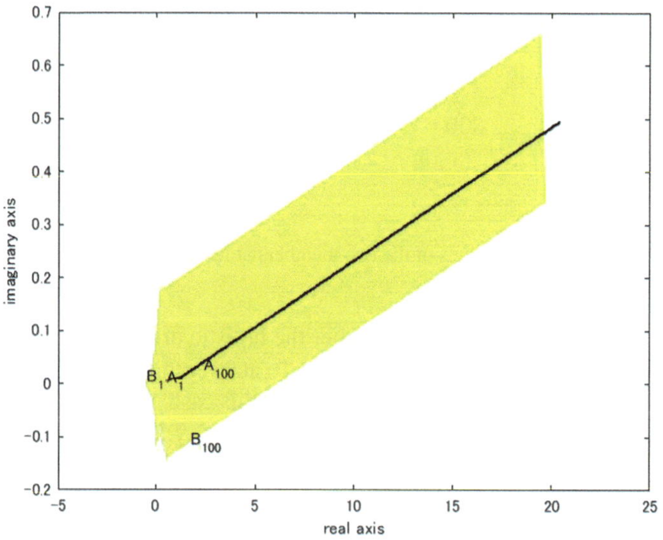

Fig. 9.24 Orbits of two members in the same plane after 1000 iterations

two-dimensional nonlinear difference system described by (9.29) can be regarded as a one-dimensional nonlinear difference system which belongs to a Mandelbrot set in the case of Example 9.11, and has a *fractal structure*.

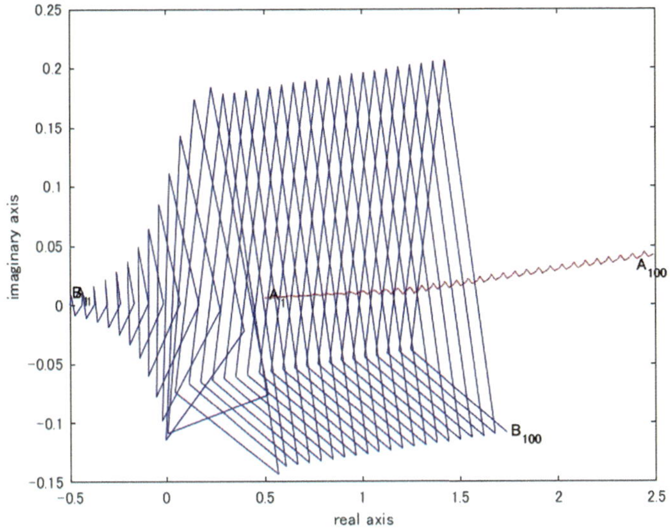

Fig. 9.25 Orbits of two members in the same plane after the first 100 iterations

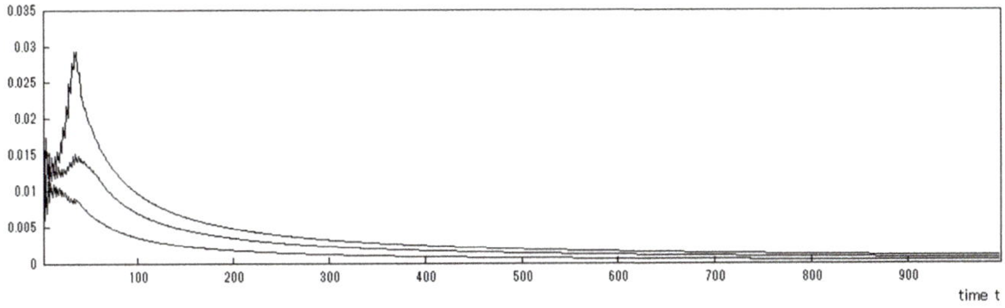

Fig. 9.26 Lyapunov exponent of the orbit of member A which was embedded in a three-dimensional space

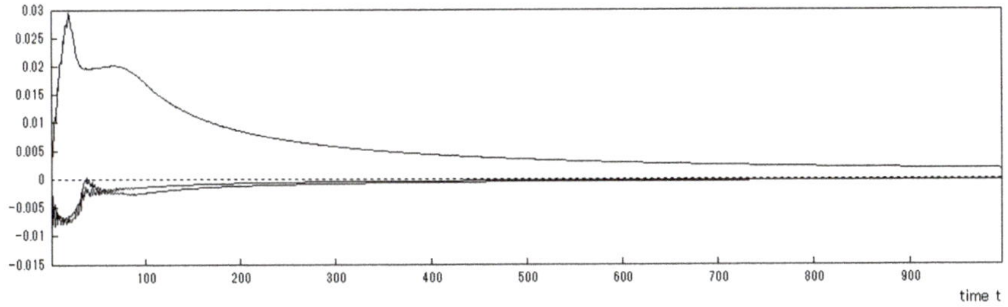

Fig. 9.27 Lyapunov exponent of the orbit of member B which was embedded in a three-dimensional space

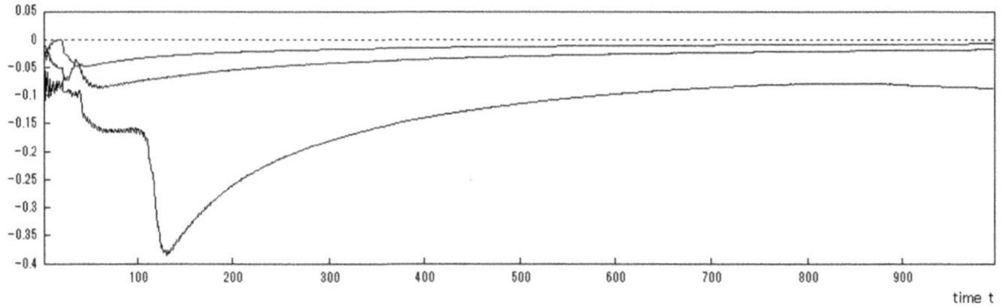

Fig. 9.28 Lyapunov exponent of the orbit of the angle from member B to member A which was embedded in a three-dimensional space

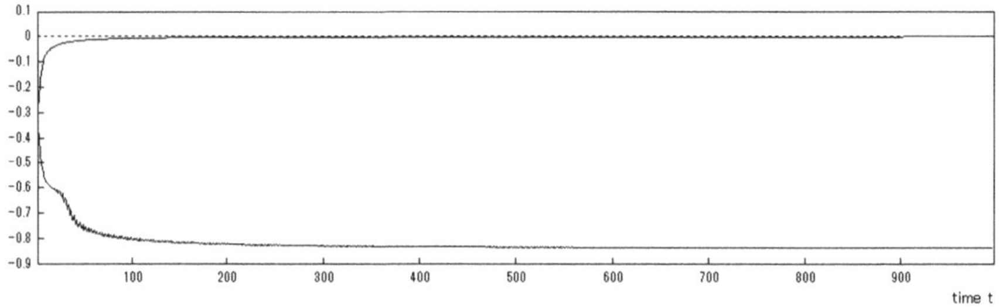

Fig. 9.29 Lyapunov exponents of the orbits of the self-proximities of two members which were embedded in a two-dimensional space

Example 9.12 In this example, we shall examine the behaviors of the dyadic nonlinear system described by (9.29), in which $w_{jk}^{(1)} = 0.9564i$, and $w_{kj}^{(1)} = -1 - 0.9564i$. and $w_{jk}^{(2)} = 0.01$, and $w_{kj}^{(2)} = 1.5375 - 0.9564i$ Initial values of $z_{j,1}$and$z_{k,1}$ were set equal to $-0.205537992601818 - 0.470295604664077i$ and -0.5, respectively. Computation was performed using the high precision format called ‘format long’ in MATLAB. The maximum number of iterations were set equal to 50,000.

Figure 9.30 shows changes in the self-proximities of members A and B, the proximity from members A to B, and the angle from B to A. Figure 9.31 shows changes in configurations of two members at iterations, 9, and 50,000, where A_m and B_m indicate the points of members A and B at iteration m ($=$ 50,000), while *maxn* indicates 50,000.

Figures 9.32 , 9.33 and 9.34 show the three *weighted digraphs* corresponding to the three configurations of members depicted in Fig. 9.31, which were obtained by applying (9.16) to these configurations. Dynamical changes in these weighted digraphs over time are the results of complex asymmetric interactions between members of some group described by an underlying complex difference system model. This is the very reason why we call our model discussed in this chapter DWD.

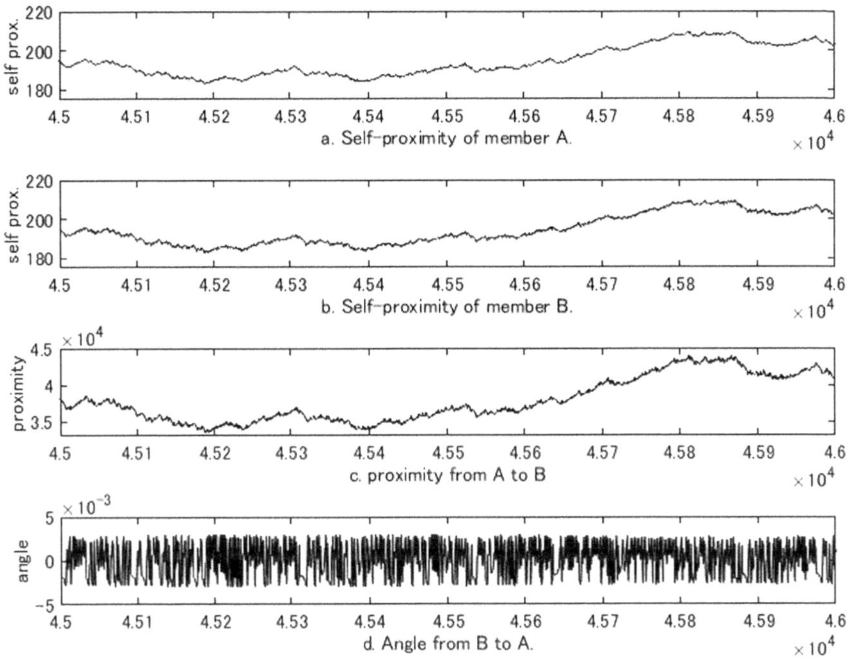

Fig. 9.30 Changes in the self-proximities from members A and B, the proximity from member A to member B, and the angle from member B to member A

Figures 9.35 and 9.36, respectively, show the changes in coordinates of members A and B after 50,000 iterations. In Fig. 9.35, A_1 indicates the point at iteration 1 and $A_{50,000}$ the point at iteration 50,000. Similarly, in Fig. 9.36, B_1 indicates the point at iteration 1 and $B_{50,000}$ the point at iteration 50,000. In contrast, Fig. 9.37 shows the simultaneous plots of the orbits depicted in Figs. 9.35 and 9.36 using different colors, yellow and black. The orbit of member A is drawn by yellow, while that of member B is drawn by black.

Then, it would be natural to ask whether the holistic orbital property of the self-proximities of two members shown in Fig. 9.30a, b is chaotic or not, as in Fig. 9.23. To answer this question, we computed the Lyapunov exponents of these proximities. Here, in computing them, we confined these orbits within iterations between 42,000 and 50,000, since the size of the original timeseries is too large to compute them by SundayChaos Times. As in the case of Fig. 9.29, we embedded these time series in a two-dimensional space. As shown in Fig. 9.38, the largest Lyapunov exponent was positive (0.5654). This means that the holistic orbital property of the self-proximities of two members seems to have a chaotic feature. However, it will be better not to conclude for the time being that this set of time series has a chaotic structure, since the largest Lyapunov exponent is small and it depends on embedding parameters.

So far, we have examined the orbital properties of (9.29) in the case of Example 9.11. Now we shall examine the orbital properties of (9.35) associated with (9.29)

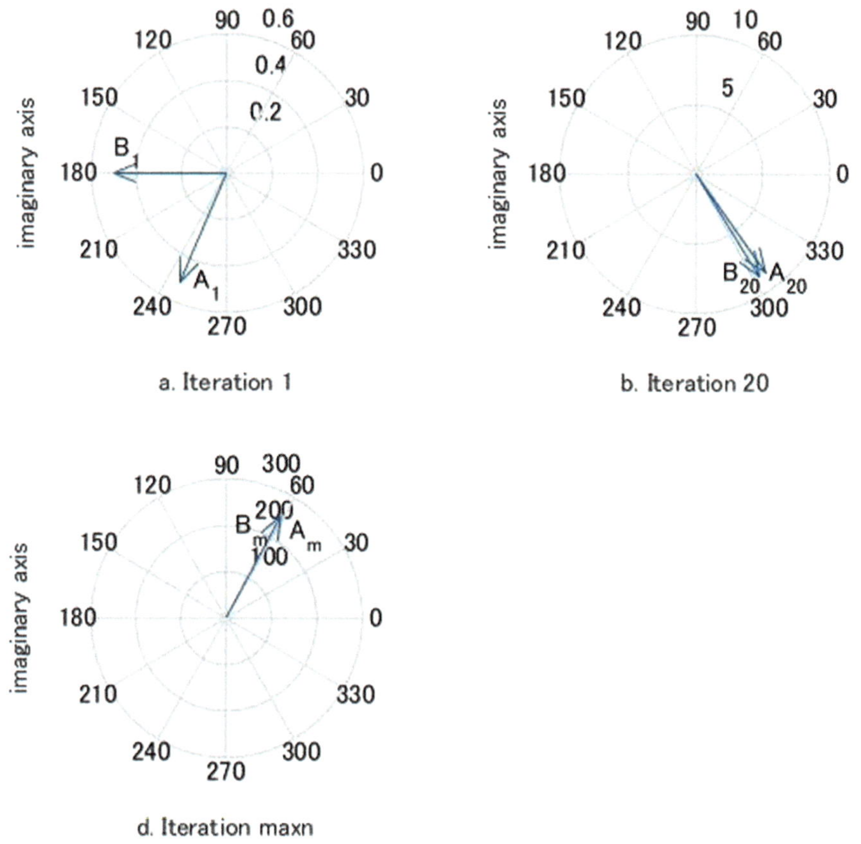

Fig. 9.31 Changes in configurations of two members at iterations, 9, and 50,000, where A_m and B_m indicate the points of members A and B at iteration m(=50,000) which are almost the same, while *maxn* indicates 50,000

in the case of this example. In this case, β^1_{jk} in (9.33) is 0, and thus $\gamma^{(1)}_{jk}$ in (9.36) is also 0. Therefore, (9.35) is expressed as

$$Z_{jk,n+1} = Z^2_{jk,n}. \tag{9.38}$$

This map belongs to the *Mandelbrot set* and has a *Julia set*. In other words, the one-dimensional nonlinear difference system in Example 9.12, which is associated with the two-dimensional nonlinear difference system described by (9.29), a has a Julia set. It is well known that the Julia set of this system is the unit circle with origin 0 in C, where C is the complex plane. Moreover, it is easy to show that this system has two fixed points, 1 and 0 (the origin), and these points are repelling and superattracting, respectively. As a result, when we start from an arbitrary point on the unit circle, the trajectory moves chaotically on the circle, at least theoretically.

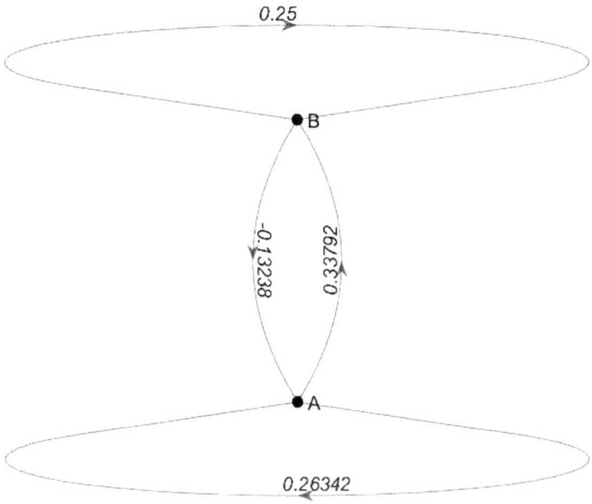

Fig. 9.32 Digraph of two members at iteration 1

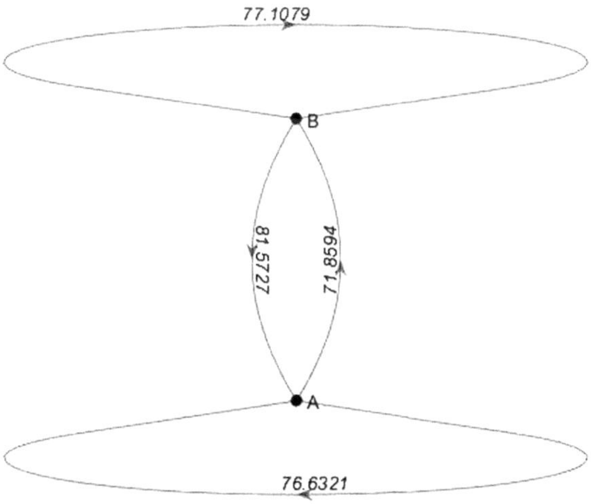

Fig. 9.33 Digraph of two members at iteration 20

Example 9.13 In this example, we shall examine the behaviors of the dyadic nonlinear system described by (9.29), in which $w_{jk}^{(1)} = 0.9564i$, and $w_{kj}^{(1)} = -0.9564i$ and $w_{jk}^{(2)} = 0.01$, and $w_{kj}^{(2)} = 1.5375 - 0.9564i$. Initial values of $z_{j,1}$and$z_{k,1}$ were set equal to $-0.205537992601818 - 0.470295604664077i$ and -0.5, respectively, as in Example 9.12. Computation was performed using the high precision format called 'format long' in MATLAB. The maximum number of iterations were set equal to 50,000.

To check whether the holistic orbital property of the self-proximities of two members shown in Fig. 9.39a, b is chaotic or not, as in Fig. 9.37, we computed the Lyapunov exponents of these proximities. Here, in computing them, we confined

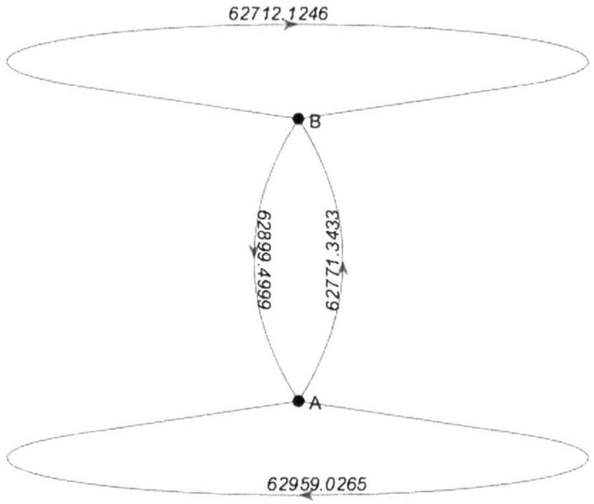

Fig. 9.34 Digraph of two members at iteration 50,000

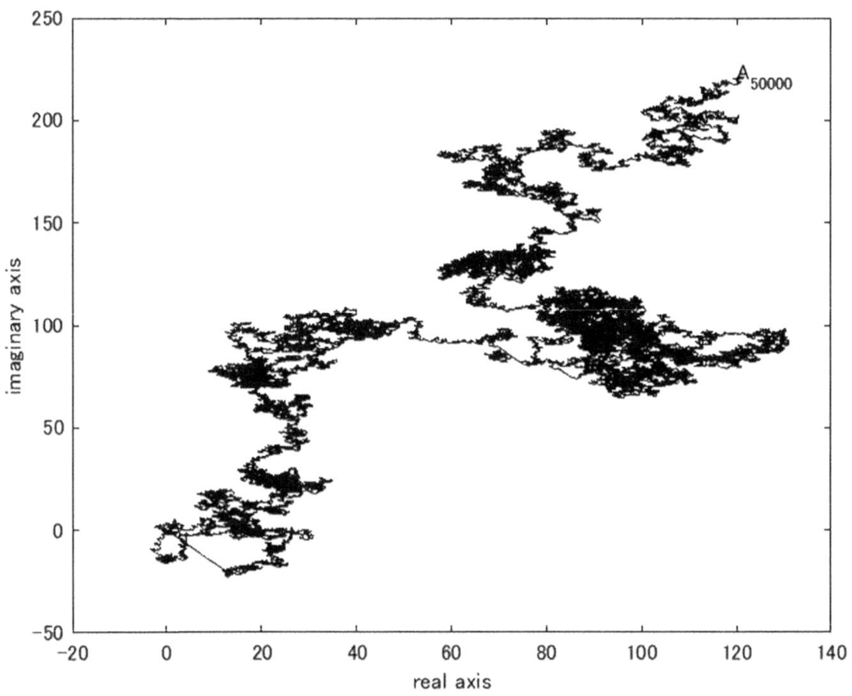

Fig. 9.35 Orbit of member *A* after 50,000 iterations

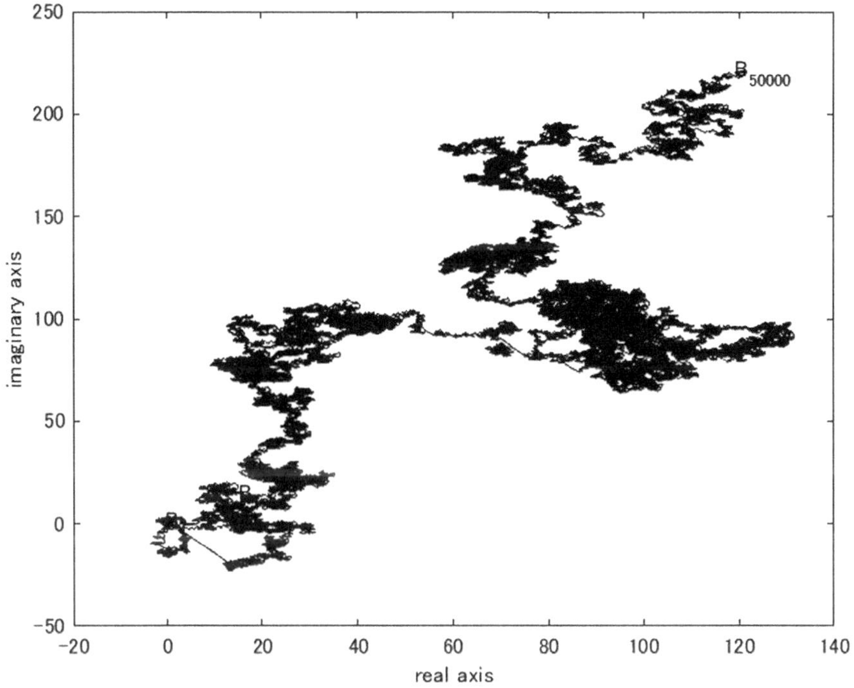

Fig. 9.36 Orbit of member *B* after 50,000 iterations

these orbits within iterations between 42,000 and 50,000, since the size of the original time series is too large to compute them by SundayChaos Times. As in the case of Fig. 9.38, we embedded these time series in a two-dimensional space. As a result, the Lyapunov exponents were 0.5643 and − 0.3564. Since the largest Lyapunov exponent was positive, the holistic orbital property of the self-proximities of two members may be said to have a chaotic feature (Figs. 9.40, 9.41 and 9.42).

Finally, to check whether the orbits of the self-proximities in Figs. 9.39a and b as well as the orbit of the proximity in Fig. 9.39c have fractal features, we computed the power spectra corresponding to these orbits. Figures 9.43, 9.44 and 9.45 show these spectra. From Fig. 9.43, the power α in power spectrum $p(f) \propto f^{-\alpha}$, which will be introduced in the note of Definition 12.22 in Chap. 12, is roughly inspected as 1.0 in high frequencies. Therefore, the Hausdorff dimension (or fractal dimension) D_H, which is computed by Berry's equality $D_H = (5 - \alpha)/2$, is about 2. Berry [4] calls it *extreme fractal*. Usually, this case is called $1/f$ noise.

From Fig. 9.44, the power α of the orbit of the self-similarities of member B is roughly inspected as 0.75 in high frequencies. Therefore, D_H is about 2.13. As a result, this orbit can be said to have a fractal dimension proposed by Mandelbrot [27].

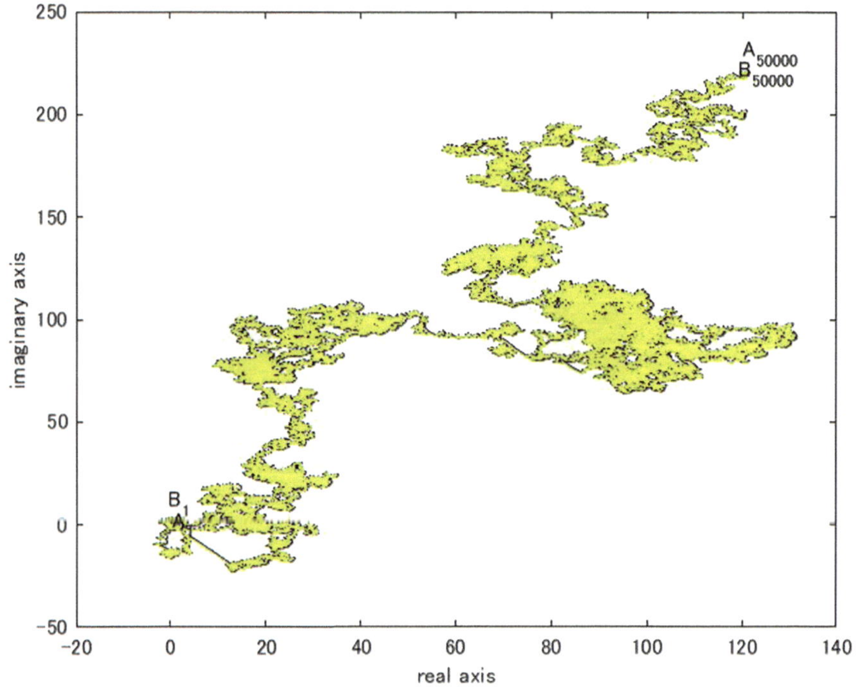

Fig. 9.37 Orbits of the two members in the same plane

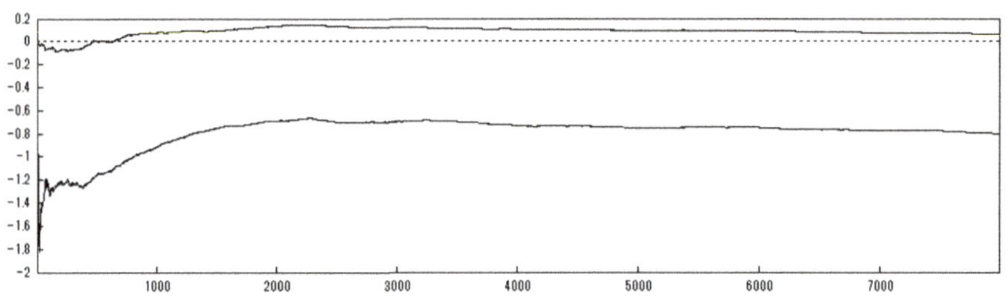

Fig. 9.38 Lyapunov exponents of the orbits of the self-proximities of two members which were embedded in a two-dimensional space

From Fig. 9.45, the power α of the orbit of the proximity from member A to 55frequency domain is temporally discontinuous. Therefore, this case may be roughly considered as an extreme fractal, as in Fig. 9.43.

So far, we have examined the orbital properties of (9.29) in the case of Example 9.13. Now we shall examine the orbital properties of (9.35) associated with (9.29) in the case of this example. In this case, β_{jk}^{1} in (9.33) is 1, and thus $\gamma_{jk}^{(1)}$ in (9.36) is 0.25. Therefore, (9.35) is expressed as

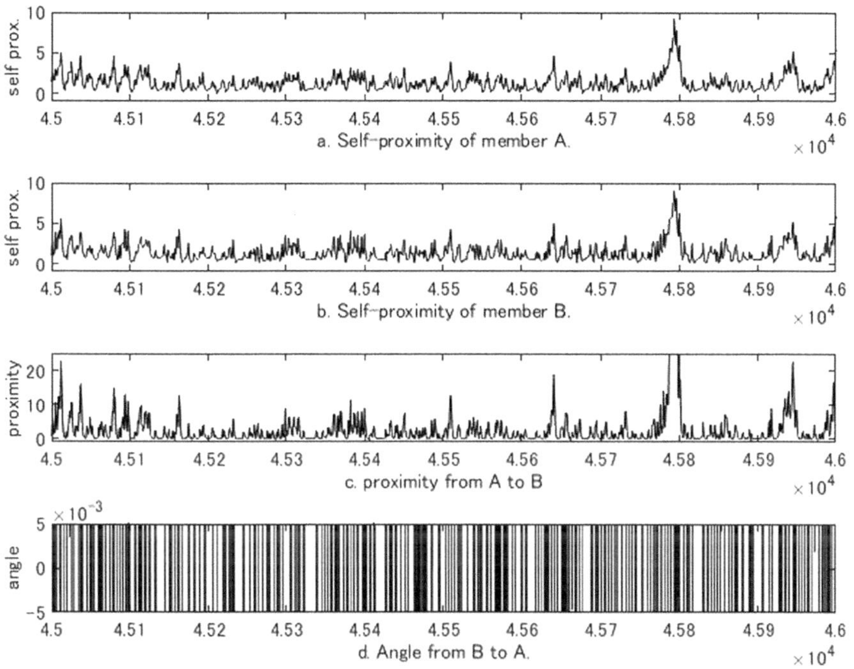

Fig. 9.39 Changes in the self-proximities from members A and B, the proximity from member A to member B, and the angle from member B to member A, in Example 9.13

$$Z_{jk,n+1} = Z_{jk,n}^2 + 0.25. \tag{9.39}$$

This map belongs to the *Mandelbrot set* and is the same as the famous *Julia set* in Fig. 12.29, which is called *cauliflower*, in Chap. 12. In other words, in this case the two-dimensional nonlinear difference system described by (9.29) can be regarded as a one-dimensional nonlinear difference system which belongs to a Mandelbrot set in Fig. 12.28 in Chap. 12, and has a *fractal structure*. For further details, see also Examples 12.12 through 12.14 in that chapter.

9.4 Applications of DWD to Empirical Data

We shall now show some applications of DWD to an empirical data which has already been shown in Table 5.3 in Sect. 5.3 of Chap. 5, which we shall reproduce here as Table 9.1.

One way of applying DWD to empirical data is to use the *configuration of objects* obtained by applying HFM to empirical *asymmetric similarity matrix* (abbreviated as *ASM*) which is obtained at an instant of time, as *initial values* of a set of complex difference equations specified by DWD.

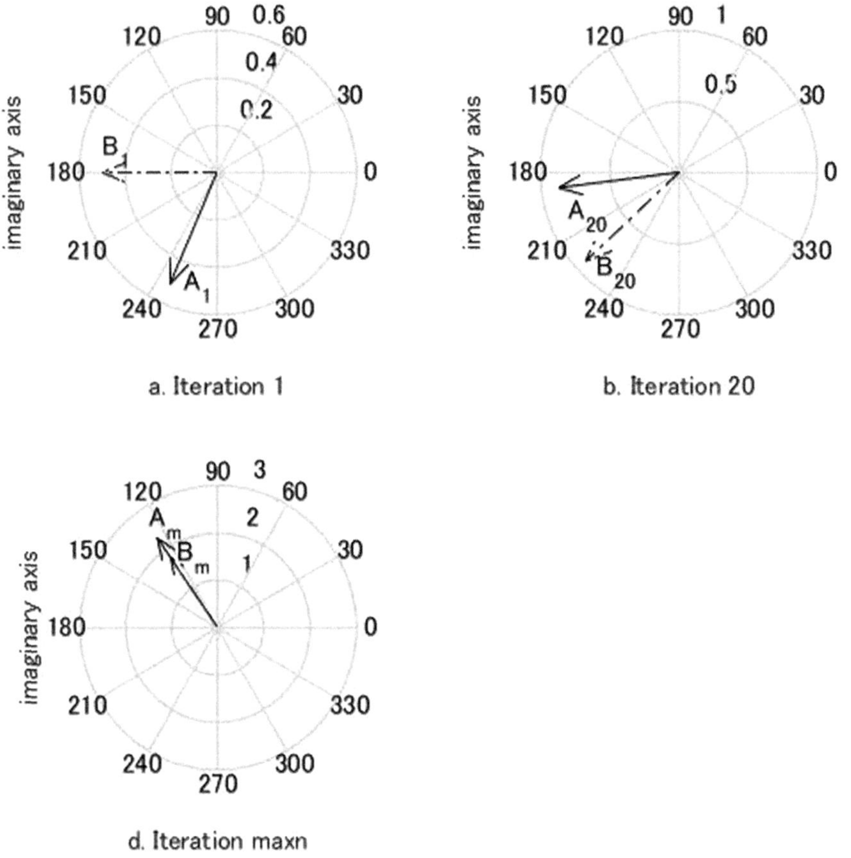

Fig. 9.40 Changes in configurations of two members at iterations, 9, and 50,000, where A_m and B_m indicate the points of members *A* and *B* at iteration m(= 50,000), while *maxn* indicates 50,000, in Example 9.13

As discussed in that section, we get the four configurations of nations which are associated with the four dimensions in a four-dimensional Hilbert space. For simplicity let us choose the configuration of nations corresponding to the largest eigenvalues, which is depicted in Fig. 5.6. We shall reproduce it as Fig. 9.46 in the following. In this figure, *J*, *A*, *C*, and *R*, are the abbreviations of Japan, USA, China, and Russia, respectively. We shall sometimes use these full names of nations instead of the abbreviated forms in the figures to be shown later.

Now, let us suppose generally that we are interested in the following problem:

Problem. How the asymmetric relations among objects will change through the *mutual asymmetric interactions* among them? Here, we mean by the change in the asymmetric relations

(1) Changes in the configuration of objects over time.

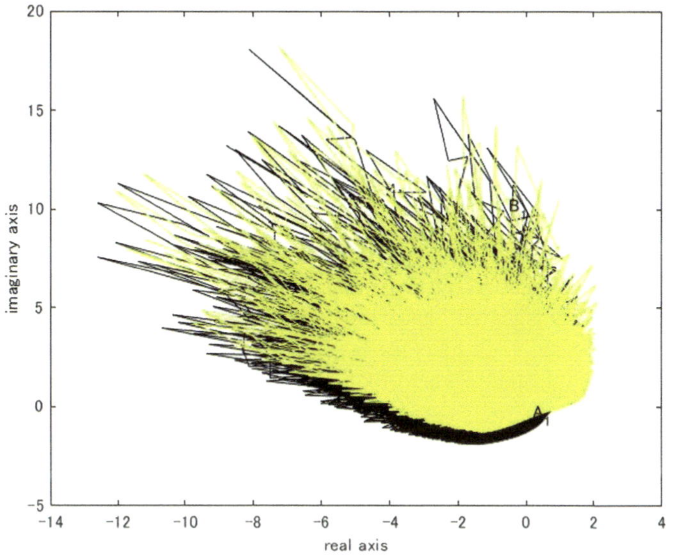

Fig. 9.41 Orbits of the two members in the same planes in Example 9.13

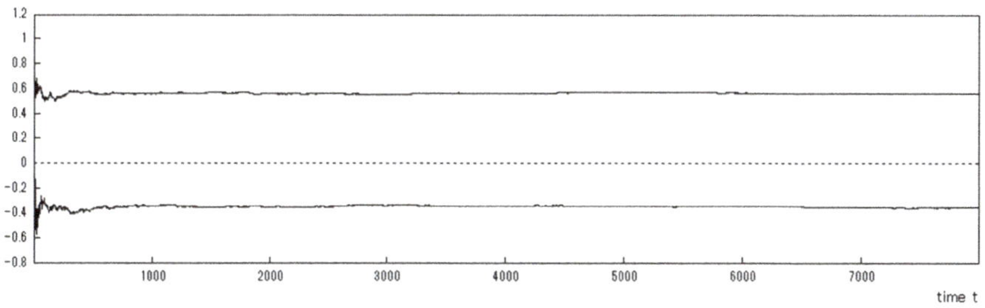

Fig. 9.42 Lyapunov exponents of the orbits of the self-proximities of two members which were embedded in a two-dimensional space in Example 9.13

(2) Changes in ASM over time.
(3) Changes in the digraph corresponding to the ASM over time.

Changes in the configuration of objects over time are predicted by a specified complex difference equation of DWD which is generally defined by (9.12) introduced in Sect. 9.3. We have already shown various examples of these changes using *hypothetical data* in Sects. 9.2 and 9.3. It should be noted that major interests in these changes have not been *quantitative* but *qualitative*.

Changes in ASM over time are computed by (9.16) which is deducted from a *polarization identity* defined, for example, by (9.15) in Sect. 9.3 in this chapter and by (4.42) in Sect. 4.3 of Chap. 4. The polarization identity is a key equation in HFM [19].

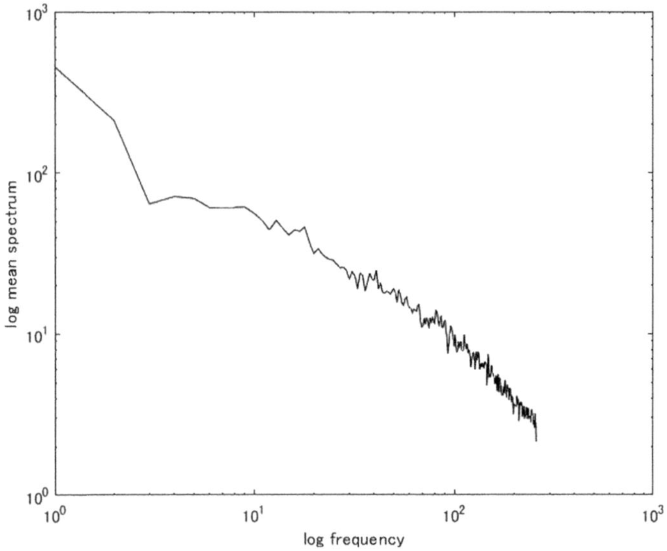

Fig. 9.43 Mean power spectrum corresponding to the orbit of the self-proximities of member *A* shown in Fig. 9.39a

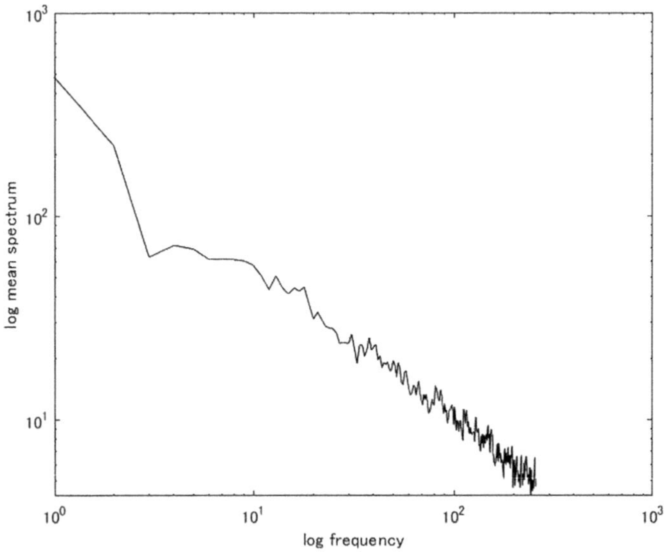

Fig. 9.44 Mean power spectrum corresponding to the orbit of the self-proximities of member *B* shown in Fig. 9.39b

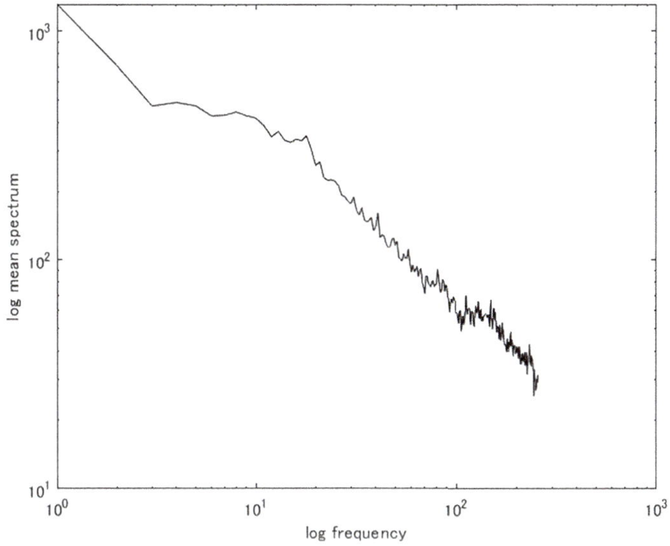

Fig. 9.45 Mean power spectrum corresponding to the orbit of the proximity from member *A* to member *B* shown in Fig. 9.39c

Table 9.1 The international trade data among Japan, America, China, and Russia in 2015

	1. Japan	2. USA	3. China	4. Russia
1. Japan	43,480	1382	1200	55
2. USA	736	159,592	1161	71
3. China	1764	4532	119,684	348
4. Russia	173	164	333	13,755

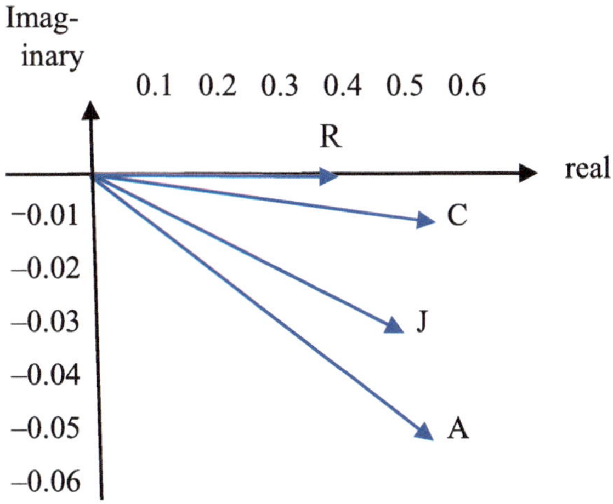

Fig. 9.46 The enlarged configuration of four nations. Here, the positive direction of this figure is clockwise because the eigenvalue is positive in this case

Changes in the digraph corresponding to the ASM over time can be depicted by using the estimated ASM, as shown in Sect. 9.3 in the case of hypothetical data. Figures 9.32, 9.33 and 9.34 in Sect. 9.3.2 are examples of such digraphs in the case of a hypothetical data.

We can solve this problem according to the following four steps:

Step 1. Apply HFM to the *empirical ASM* of order N and check whether the objects can be embedded in a *Hilbert space* or an *indefinite metric space* by computing the *eigenvalues* of the *Hermitian matrix*, $\boldsymbol{H}$, defined by (4.1) of Chap. 4 corresponding to the empirical ASM, which must be *measured at a ratio level of measurement*. For convenience we shall reproduce this equation in the following as (9.40).

$$\boldsymbol{H} = \boldsymbol{S}_s + i\boldsymbol{S}_{sk}, \tag{9.40}$$

where $\boldsymbol{S}_s$ and $\boldsymbol{S}_{sk}$ are the symmetric part and the skew-symmetric part of the ASM, respectively.

Step 2. Obtain the configuration of objects in a one-dimensional Hilbert space corresponding to the first eigenvalue or the second eigenvalue if all the eigenvalues are positive, as in the result of the trade data in Sect. 5.3. By contrast, if the eigenvalues include positive and negative ones, then we may choose the dimension with the largest absolute value of these eigenvalues. Such a dimension may be regarded as a one-dimensional Hilbert space except for the sign of the eigenvalue. The configuration obtained here is used as the *initial values* in solving the complex differential equation model of DWD under study.

Step 3. Specify a *hypothetical mutual interaction matrix* of the complex set of difference equations of DWD and compute the matrix $\boldsymbol{A}_N$ associated with this interaction matrix in the case of a *linear model.* Specify a *Jacobian matrix* $\boldsymbol{J}_N$ in the case of a *nonlinear model,* where N is the number of objects in the empirical ASM. As regards the Jacobian matrix of a set of nonlinear difference equations, see Sect. 12.4 in Chap. 12.

Then, in the former case, compute the eigenvalues of $\boldsymbol{A}_N$ and examine the qualitative features of the solution curves of the linear model by applying some traditional time series analysis such as *power spectra* as well as applying some nonlinear analyses such as *Lyapunov exponents* to each of or all of these curves.

In the latter case, compute the eigenvalues of $\boldsymbol{J}_N$ at a *fixed point* (or each of the *fixed points* if any) and examine the local qualitative features of the nonlinear model.

Step 4. Solve the linear or nonlinear model by specifying the *initial values of the configuration of objects* obtained in Step 2 as well as *weight parameters* of the model which is under study. Generally, the weight parameters of (9.12) can be specified freely. Then, examine the three features of the model in the above problem.

In the following sections we shall show several examples of applications using the empirical data in Table 9.1 which represents a *tetradic relationship* among four nations (i.e., $N = 4$).

9.4.1 Linear Model

The following is a simple linear model of tetradic relationship:

Example 9.14

$$\begin{cases} z_{i,n+1} = z_{i,n} + w_{ij}\left(z_{i,n} - z_{j,n}\right) + w_{ik}\left(z_{i,n} - z_{k,n}\right) + w_{il}\left(z_{i,n} - z_{l,n}\right) \\ z_{j,n+1} = z_{j,n} + w_{jk}\left(z_{j,n} - z_{k,n}\right) + w_{jl}\left(z_{j,n} - z_{l,n}\right) + w_{ji}\left(z_{j,n} - z_{i,n}\right) \\ z_{k,n+1} = z_{k,n} + w_{kl}\left(z_{k,n} - z_{l,n}\right) + w_{ki}\left(z_{k,n} - z_{i,n}\right) + w_{kj}\left(z_{k,n} - z_{j,n}\right) \\ z_{l,n+1} = z_{l,n} + w_{li}\left(z_{l,n} - z_{i,n}\right) + w_{lj}\left(z_{l,n} - z_{j,n}\right) + w_{lk}\left(z_{l,n} - z_{k,n}\right) \end{cases}, \tag{9.41}$$

where $w_{ji} = -w_{ij}$, $w_{ki} = -w_{ik}$, $w_{li} = -w_{il}$, $w_{kj} = -w_{jk}$, $w_{lj} = -w_{jl}$, and $w_{lk} = -w_{kl}$, while $w_{ij} = w_{ik} = w_{il} = w_{jk} = w_{jl} = w_{kl} = 0.01$.

The mutual asymmetric interactions will be specified generally by the *mutual interaction matrix* [14] introduced in (5.19) in Sect. 5.6 of Chap. 5 in the case of a *linear model* of DWD. We shall reproduce it as (9.42) in the following.

$$\mathbf{M} = \begin{bmatrix} m_{11} & m_{12} & m_{13} & m_{14} \\ m_{21} & m_{22} & m_{23} & m_{24} \\ m_{31} & m_{32} & m_{33} & m_{34} \\ m_{41} & m_{42} & m_{43} & m_{44} \end{bmatrix}. \tag{9.42}$$

This matrix is obtained by rearranging the coefficients of the right-hand side of the linear model of DWD in (9.41) row by row by a rule along the corresponding row of the matrix $\boldsymbol{M}$. This rule is such that we set the element m_{ab} in (9.42) equal to the coefficient w_{ab} in (9.41) except that we set all the diagonal elements $m_{aa} = 1$. Then, $\boldsymbol{M}$ can be written as,

$$\boldsymbol{M} = \begin{bmatrix} 1 & w_{ij} & w_{ik} & w_{il} \\ w_{ji} & 1 & w_{jk} & w_{jl} \\ w_{ki} & w_{kj} & 1 & w_{kl} \\ w_{li} & w_{lj} & w_{lk} & 1 \end{bmatrix}. \tag{9.43}$$

A more important matrix than $\boldsymbol{M}$ which plays a crucial role in examining the properties of the *qualitative features* of the linear model of DWD is the matrix $\boldsymbol{A}_4$ defined by (9.44).

$$\boldsymbol{A}_4 = \begin{bmatrix} 1 + w_{i,sum} & -w_{ij} & -w_{ik} & -w_{il} \\ -w_{ji} & 1 + w_{j,sum} & -w_{jk} & -w_{jl} \\ -w_{ki} & -w_{kj} & 1 + w_{k,sum} & -w_{kl} \\ -w_{li} & -w_{lj} & -w_{lk} & 1 + w_{l,sum} \end{bmatrix}, \tag{9.44}$$

where $w_{i,sum} = \sum_{b \neq i} w_{ib}$, $w_{j,sum} = \sum_{b \neq j} w_{jb}$, $w_{k,sum} = \sum_{b \neq k} w_{kb}$, and $w_{l,sum} = \sum_{b \neq l} w_{lb}$, in the one-dimensional Hilbert space. If we use this matrix, then we can obtain

$$z_{n+1} = A_4 z_n, \tag{9.45}$$

where $z_n = [z_{in}, z_{jn}, z_{kn}, z_{ln}]^t$ and $z_{in}, z_{jn}, z_{kn}, z_{ln}$ are the coordinates of the four objects in a one-dimensional Hilbert space at an instant in time n. Equation (9.45) is an alternative expression of (9.41). The reader should notice that there are several differences in the arrangements of the elements of $\boldsymbol{M}$ and $\boldsymbol{A}_4$. The reason for this is that $\boldsymbol{A}_4$ must satisfy (9.45), while $\boldsymbol{M}$ need not.

As is the case with $\boldsymbol{A}_2$ and $\boldsymbol{A}_3$ which have already been introduced in Sect. 9.3.2, the qualitative features of the solution curve of (9.41) can be determined by the features of the eigenvalues of $\boldsymbol{A}_4$, if we identify it as a real matrix with real elements in a real two-dimensional Cartesian space and apply the *Putzer algorithm* (e.g., Elaydi [21] to (9.41) as in Theorem 9.1.

Returning to the matrix $\boldsymbol{M}$ for a while, we shall consider the meaning of the elements of this matrix. For example, let us pick up m_{21} and m_{12} of this matrix. These elements are considered as two *free vectors* with red initiated from the countries J (Japan) and A (USA) in the Hilbert space, as shown in Fig. 5.14, which we shall reproduce in Fig. 9.47. Here, the red arrow emanated from J (Japan, whose number in Table 9.1 is 1) is associated with, m_{12}, where subscript 12 indicates the $(1, 2)$ element of $\boldsymbol{M}$. Therefore, m_{12} denotes the direction of the mutual association from Japan to USA in the one-dimensional Hilbert space since the second subscript 2 represents USA in Table 9.1.

Remark Figure 9.47 clearly implies the difference between the *two complex asymmetric matrices*, one being the *empirical Hermitian matrix* computed by (9.40) and the other being the *mutual asymmetric interaction matrix* defined by (9.43). Although the four *blue vectors* which constitute the configuration of nations are computed as the four eigenvectors of the *empirical* Hermitian matrix defined by (9.40), the two

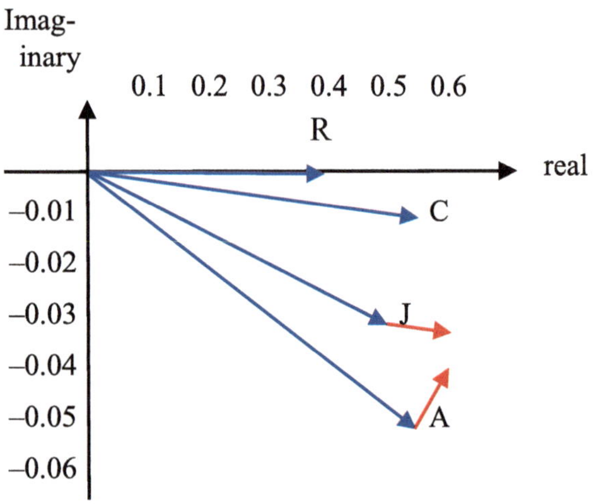

Fig. 9.47 An illustration of forces exerting on the configuration space in Fig. 5. 6. Red arrows indicate the forces acting on the two nations, America and Japan. Here, the positive direction of this figure is clockwise because the eigenvalue is positive in this case

elements of the mutual asymmetric interaction matrix which are denoted by *red vectors* are specified as *hypothetical* quantities.

Now, we shall apply the simple linear model for the tetradic relationship defined by (9.41) to the ASM in Table 9.1. The matrix $\boldsymbol{A}_4$ for this table can be specified, for example, as follows, if we set

$$\begin{aligned}
&w_{ij} = 0.1 - 0.01i,\ w_{ik} = 0.07 + 0.1i,\ w_{il} = -0.1i,\\
&w_{ji} = 0.07 + 0.1i,\ w_{jk} = -0.2i,\ w_{jl} = 0.4i,\\
&w_{ki} = 0.1 - 0.1i,\ w_{kj} = 0.07 - 0.4i,\ w_{kl} = 0.2 + 0.2i,\\
&w_{li} = 0.1,\ w_{lj} = 0.4i,\ w_{lk} = 0.2 - 0.1i :
\end{aligned}$$

$$\boldsymbol{A}_4 = \begin{bmatrix} 1.17 - 0.01i & -0.1 + 0.01i & -0.07 - 0.1i & 0.1i \\ -0.07 - 0.1i & 1.07 + 0.3i & 0.2i & -0.4i \\ -0.1 + 0.1i & -0.07 + 0.4i & 1.37 - 0.3i & -0.2 - 0.2i \\ -0.1 & -0.4i & -0.2 + 0.1i & 1.30 + 0.3i \end{bmatrix}. \tag{9.46}$$

The eigenvalues of this matrix are 1.4310–0.3329i, 1.2406+0.7084i, 1.2383–0.0855i, and 1.0000. It is expected from these values that the solution curves *diverge* in this case from the analogy of the discussions on $\boldsymbol{A}_2$ and $\boldsymbol{A}_3$ in Sect. 9.3.2 as well as from the theories which we shall introduce in Sects. 12.4.2, 12.4.3, and 12.5.1.

If we compute the solution curves *algebraically* via the *Putzer algorithm* using these eigenvalues and the corresponding eigenvectors, we can examine the qualitative features of these curves theoretically. Generally, this job is very laborious if the order p of the matrix $\boldsymbol{A}_p$ is higher than 3. However, if we may proceed to the next step without solving this problem algebraically, then we can obtain the solution curves *computationally*.

Figure 9.48 depicts the changes in self-proximities of the four nations over 2000 iterations. It is apparent that these changes are very slow until about 1800 iterations, and after that self-proximities of all nations increase rapidly.

Figure 9.49 draws changes in angles between all pairs of the four nations.

These time series seem to change almost regularly, and thus we may conclude that they do not exhibit any chaotic behavior. In fact, the system under consideration is a linear model, and we cannot expect such a chaotic behavior at least theoretically.

Figure 9.50 depicts the changes in configurations of the four nations, j, k, l, and m at iterations 1, 20, 60, and 100.

Figure 9.51 shows the orbits of four nations in the same complex plane after the 2000 iterations. It is evident that the qualitative feature of this system is considered as a *source*.

In this way, we can choose the elements of the hypothetical mutual interaction matrix freely and examine what kind of the scenario on the change in asymmetric relations among nations will appear through the specified mutual interaction.

As another possible application, we shall consider a problem of how to starve a target nation into surrender efficiently by inhibiting the trades of the nation under

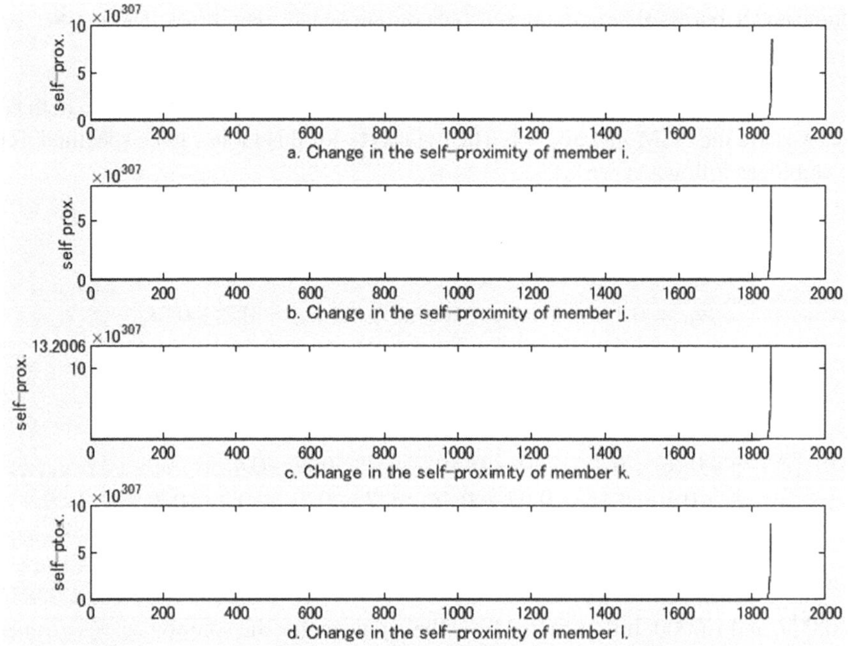

Fig. 9.48 Changes in the self-proximities of four tetrad nations over time

a circumstance that there exist one group of friendly nations and the other group of unfriendly nations to the target nation. Moreover, suppose that both groups have trade relationships with the target nation.

If we have a trade data like Table 9.1 among all of these nations at an instant of time, then we may proceed to *Step 1* and *Step 2* of the problem discussed in this section and obtain one-dimensional configuration of nations in a Hilbert space. Then, specify a hypothetical mutual interaction matrix and proceed to *Step 3* and *Step 4.* Finally, check the obtained scenario of the solution curves. If that scenario is not efficient, then consider another hypothetical mutual interaction matrix and return to *Step 3.*

Example 9.15 In this example, we shall show another example A_4 for Table 9.1 and the results of the linear model of DWD, which is specified by the *linear part* of the *nonlinear model* of DWD to be shown in the following section, Sect. 9.4.2. In contrast to the results of Example 9.14, orbits of the solution curves exhibit a fixed point which *converge* to a point.

The following is a set of the weights in such a linear part of the nonlinear model:

$$
\begin{aligned}
&w_{ij} = -0.3 - 0.01i,\ w_{ik} = -0.3 + 0.1i,\ w_{il} = -0.1i,\\
&w_{ji} = 0.07 + 0.1i,\ w_{jk} = -0.02i,\ w_{jl} = 0.01i,\\
&w_{ki} = -0.1 - 0.1i,\ w_{kj} = -0.02 - 0.01i,\ w_{kl} = -0.02 + 0.02i,
\end{aligned}
$$

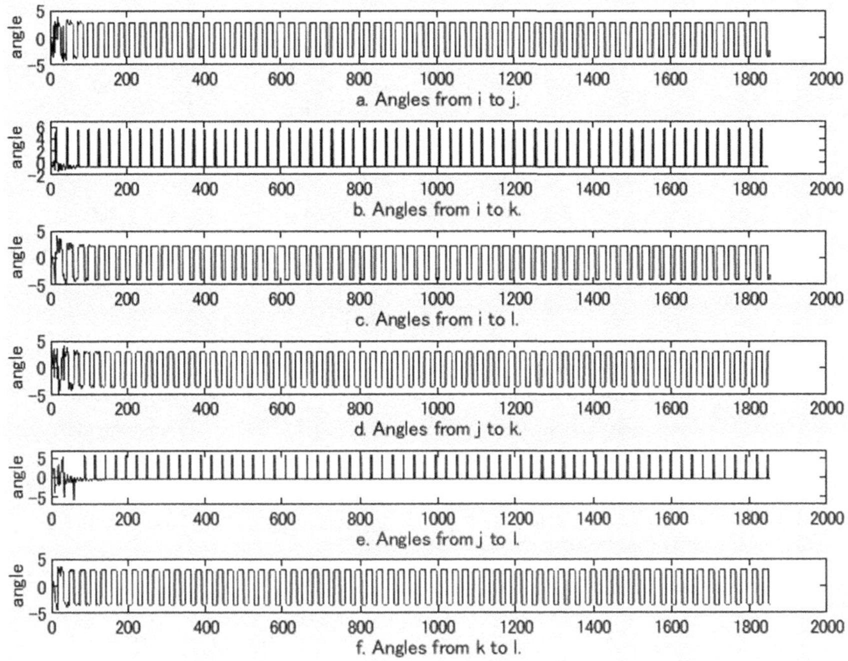

Fig. 9.49 Changes in the angles between four tetrad nations over time

$$w_{li} = -0.1,\ w_{lj} = 0.01i,\ w_{lk} = -0.02 - 0.1i.$$

Initial values of the configuration of the four nations of this model are the same as those of the linear model whose $\boldsymbol{A}_4$ is defined by (9.46). We chose the above weight parameters so that the absolute values of the diagonal elements of the resultant matrix $\boldsymbol{A}_4$ become relatively smaller than those in the previous example of the linear model. As a result, we have

$$\boldsymbol{A}_4 = \begin{bmatrix} 0.4 - 0.01i & 0.3 + 0.01i & 0.3 - 0.1i\ 0.1i \\ -0.07 - 0.1i & 1.07 + 0.09i & 0.02i\ \ -0.01i \\ 0.1 + 0.1i & 0.02 + 0.01i & 0.86 - 0.09i\ \ 0.02 - 0.02i \\ 0.1 & -0.01i & 0.02 + 0.1i\ \ 0.88 - 0.09i \end{bmatrix}. \tag{9.47}$$

The eigenvalues of this matrix are 1.0, 0.9515+0.0220i, 0.9029 – 0.0741i, and 0.3557 – 0.0479.

The following Fig. 9.52 shows changes in the self-proximities of the four nations over time. In contrast to the changes in those self-proximities in the previous example, each of the self-proximity of four nations converges to the same value after a short time interval.

The following figure depicts changes in angles between four nations over time.

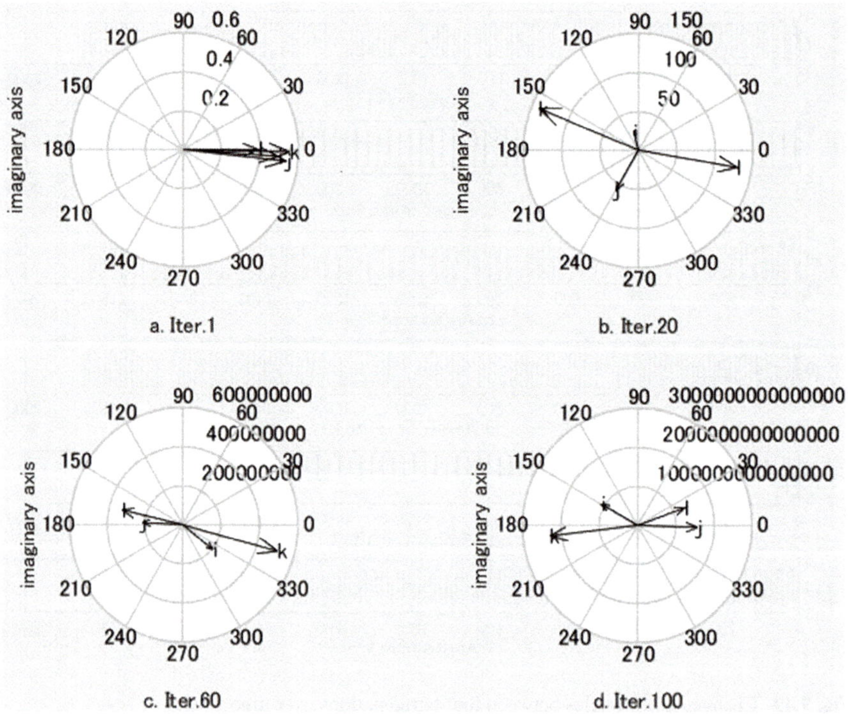

Fig. 9.50 Changes in the configurations of four tetrad nations over time

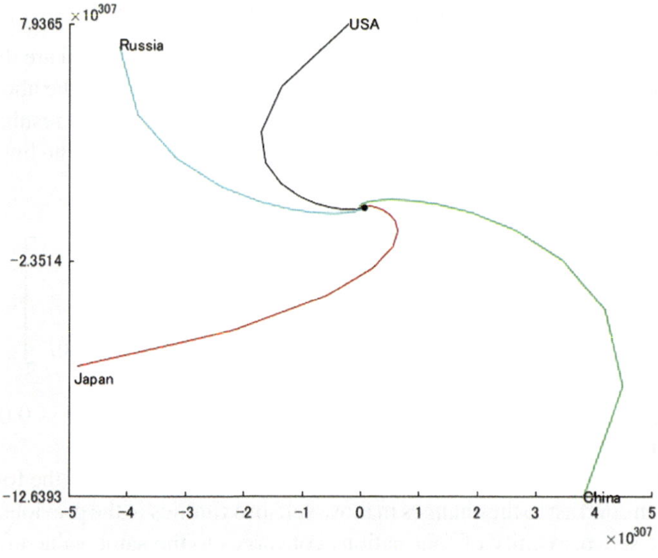

Fig. 9.51 Orbits of four tetrad nations in the same complex plane. The origin of this figure is depicted by a large dot with color, black

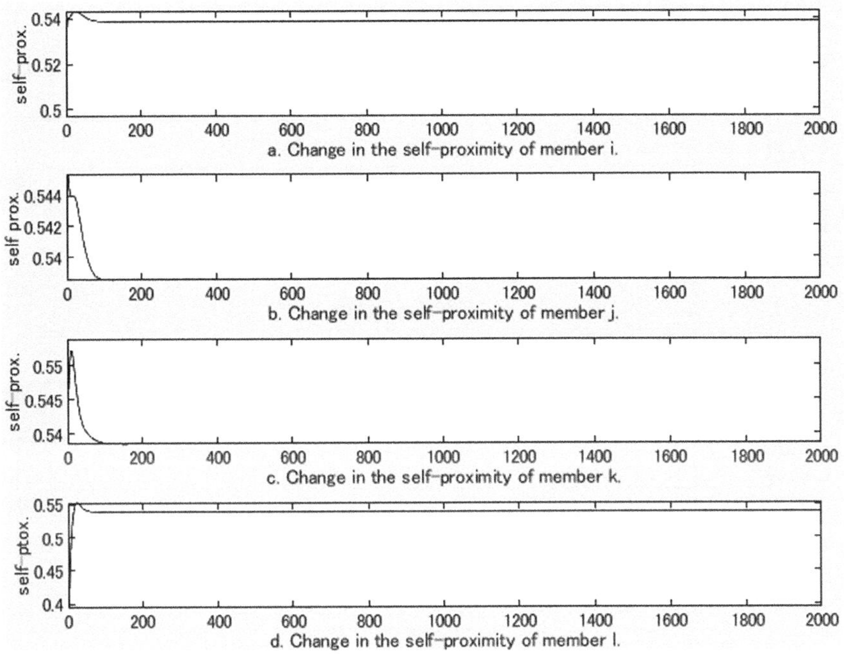

Fig. 9.52 Change in the self-proximities of four tetrad nations over time in Example 9.15

As in the case of self-proximities, each of the angles between nations converges to a constant in a short time interval, in contrast to those in the previous linear model (Fig. 9.53).

The following figure draws the orbits of the four nations in the same complex plane after 2000 iterations. The reader may realize that *small changes* in the weight parameters of DWD produce a *radical change* in the *qualitative feature* of the orbital properties of the DWD model.

9.4.2 Nonlinear Model

In this section, we shall apply a nonlinear model of DWD, which is similar to (9.29), to an empirical data shown in Table 9.1. Since this data is an ASM of order 4, the nonlinear model is defined as follows. This model is a special case of (9.12) in Sect. 9.3 in which p = 1, m = 2, $\boldsymbol{g}(\boldsymbol{u}_{j,n}) = 0$, and $z_0 = 0$.

Example 9.16

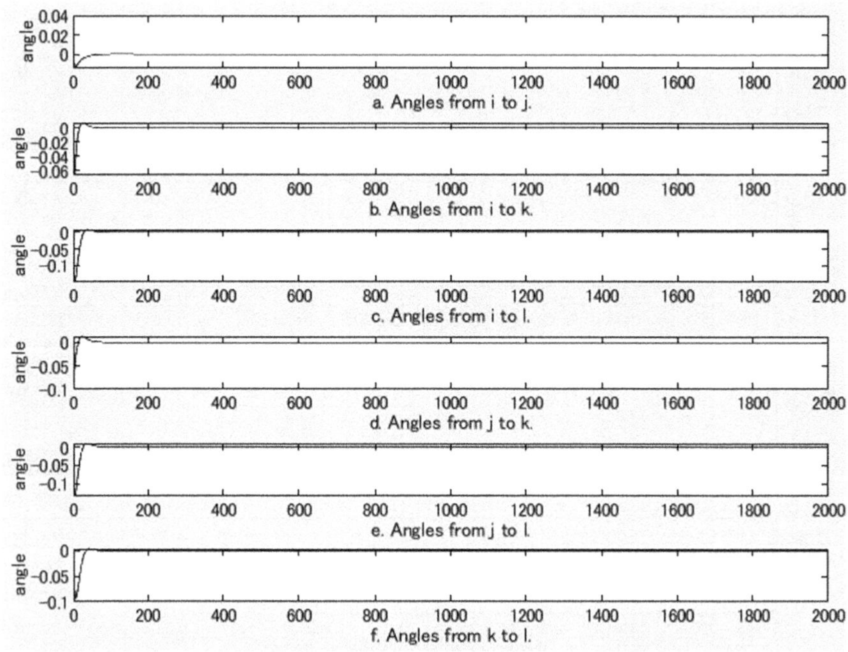

Fig. 9.53 Changes in the angles between four tetrad nations over time in Example 9.15

$$\begin{cases} z_{i,n+1} = z_{i,n} + w_{ij1}\left(z_{i,n} - z_{j,n}\right) + w_{ik1}\left(z_{i,n} - z_{k,n}\right) + w_{il1}\left(z_{i,n} - z_{l,n}\right) \\ \quad +w_{ij2}\left(z_{i,n} - z_{j,n}\right)^2 + w_{ik2}\left(z_{i,n} - z_{k,n}\right)^2 + w_{il2}\left(z_{i,n} - z_{l,n}\right)^2, \\ z_{j,n+1} = z_{j,n} + w_{jk1}\left(z_{j,n} - z_{k,n}\right) + w_{jl1}\left(z_{j,n} - z_{l,n}\right) + w_{ji1}\left(z_{j,n} - z_{i,n}\right) \\ \quad w_{jk2}\left(z_{j,n} - z_{k,n}\right)^2 + w_{jl2}\left(z_{j,n} - z_{l,n}\right)^2 + w_{ji2}\left(z_{j,n} - z_{i,n}\right)^2, \\ z_{k,n+1} = z_{k,i} + w_{kl1}\left(z_{k,n} - z_{l,n}\right) + w_{ki1}\left(z_{k,n} - z_{i,n}\right) + w_{kj1}\left(z_{k,n} - z_{j,n}\right)), \\ \quad w_{kl2}\left(z_{k,n} - z_{l,n}\right)^2 + w_{ki2}\left(z_{k,n} - z_{i,n}\right)^2 + w_{kj2}\left(z_{k,n} - z_{j,n}\right)^2, \\ z_{l,n+1} = z_{l,n} + w_{li1}\left(z_{l,n} - z_{i,n}\right) + w_{lj1}\left(z_{l,n} - z_{j,n}\right) + w_{lk1}\left(z_{l,n} - z_{k,n}\right) \\ \quad w_{li2}\left(z_{l,n} - z_{i,n}\right)^2 + w_{lj2}\left(z_{l,n} - z_{j,n}\right)^2 + w_{lk2}\left(z_{l,n} - z_{k,n}\right)^2 \end{cases} \tag{9.48}$$

The following is a set of the weights in this nonlinear model in this example:

$$\begin{aligned} & w_{ij1} = -0.3 - 0.01i,\ w_{ik1} = -0.3 + 0.1i,\ w_{il1} = -0.1i, \\ & w_{ij2} = -0.02 - 0.01i,\ w_{ik2} = -0.02 + 0.1i,\ w_{il2} = 0.02 - 0.1i, \\ & w_{ji1} = 0.07 + 0.1i,\ w_{jk1} = -0.02i,\ w_{jl1} = 0.01i, \\ & w_{ji2} = -0.01i,\ w_{jk2} = 0.1 - 0.02i,\ w_{jl2} = -0.03i, \\ & w_{ki1} = -0.1 - 0.1i, w_{kj1} = -0.02 - 0.01i,\ w_{kl1} = -0.02 + 0.02i, \\ & w_{ki2} = -0.1 + 0.01i, w_{kj2} = -0.01i,\ w_{kl2} = 0.1 + 0.02i, \\ & w_{li1} = -0.1,\ w_{lj1} = 0.01i, w_{lk1} = -0.02 - 0.1i. \end{aligned}$$

$$w_{li2} = -0.03,\ w_{lj2} = -0.03 + 0.1i,\ w_{lk2} = -0.01 - 0.02i.$$

The Jacobian matrix $\boldsymbol{J}_4$ is equal to the $\boldsymbol{A}_4$ defined by (9.47) in this model. Thus, we have

$$\boldsymbol{J}_4 = \begin{bmatrix} 0.4 - 0.01i & 0.3 + 0.01i & 0.3 - 0.1i\ 0.1i \\ -0.07 - 0.1i & 1.07 + 0.09i & 0.02i\ -0.01i \\ 0.1 + 0.1i & 0.02 + 0.01i & 0.86 - 0.09i\ 0.02 - 0.02i \\ 0.1 & -0.01i & 0.02 + 0.1i\ \ 0.88 - 0.09i \end{bmatrix}. \tag{9.49}$$

These eigenvalues are of course the same as those of $\boldsymbol{A}_4$ in (9.47). As a result, the qualitative feature of the solution curves of this model is expected to be similar to that of the model shown in Example 9.15, in which the linear part of this model is specified.

The Fig. 9.55 draws the changes in the self-proximities of four tetrad nations over time in Table 9.1. In this example, all the self-proximities of these nations converge to a constant in merely 30 iterations.

Figure 9.56 depicts changes in the angles between four tetradic nations over time. All of these angles also converge to a fixed point, that is, zero, in short iterations.

Figure 9.57 shows the orbits of four tetrad nations over time in the same complex plane. As expected, the qualitative feature of these orbits is basically equivalent to that in Fig. 9.54.

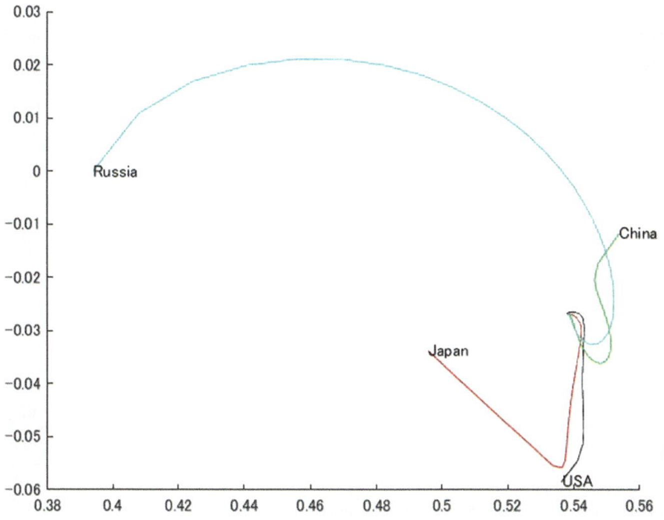

Fig. 9.54 Orbits of four tetrad nations in the same complex plane. A fixed point may be located near the point to which all the orbits of these nations converge in this figure

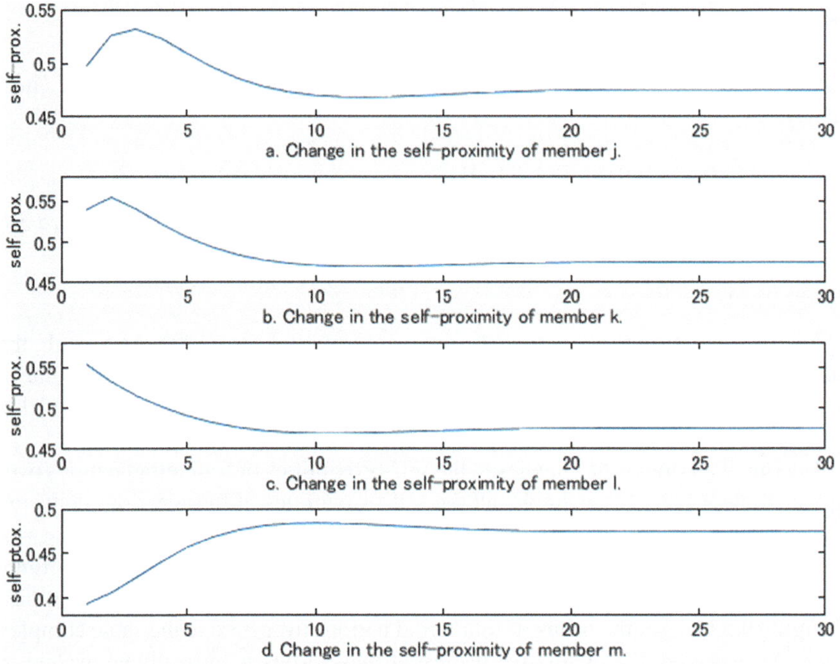

Fig. 9.55 Change in the self-proximities of four tetrad nations over time in Example 9.16

Example 9.17 In this example, we consider another nonlinear model in the following which is a special case of the model defined in (9.17).

$$\begin{cases} z_{i,n+1} = z_{i,n} + w_{ij}\left(z_{i,n}^2 - z_{j,n}^2\right) + w_{ik}\left(z_{i,n}^2 - z_{k,n}^2\right) + w_{il}\left(z_{i,n}^2 - z_{l,n}^2\right) \\ z_{j,n+1} = z_{j,n} + w_{jk}\left(z_{j,n}^2 - z_{k,n}^2\right) + w_{jl}\left(z_{j,n}^2 - z_{l,n}^2\right) + w_{ji}\left(z_{j,n}^2 - z_{i,n}^2\right) \\ z_{k,n+1} = z_{k,n} + w_{kl}\left(z_{k,n}^2 - z_{l,n}^2\right) + w_{ki}\left(z_{k,n}^2 - z_{i,n}^2\right) + w_{kj}\left(z_{k,n}^2 - z_{j,n}^2\right) \\ z_{l,n+1} = z_{l,n} + w_{li}\left(z_{l,n}^2 - z_{i,n}^2\right) + w_{lj}\left(z_{l,n}^2 - z_{j,n}^2\right) + w_{lk}\left(z_{l,n}^2 - z_{k,n}^2\right) \end{cases}, \tag{9.50}$$

where

$$w_{ji} = -w_{ij},\ w_{ki} = -w_{ik},\ w_{li} = -w_{il},\ w_{kj} = -w_{jk},\ w_{lj} = -w_{jl},$$
$$w_{lk} = -w_{kl}. \tag{9.51}$$

In this example, we specified the weights such that

$$w_{ij} = w_{ik} = w_{il} = w_{jk} = w_{jl} = w_{kl} = 0.01.$$

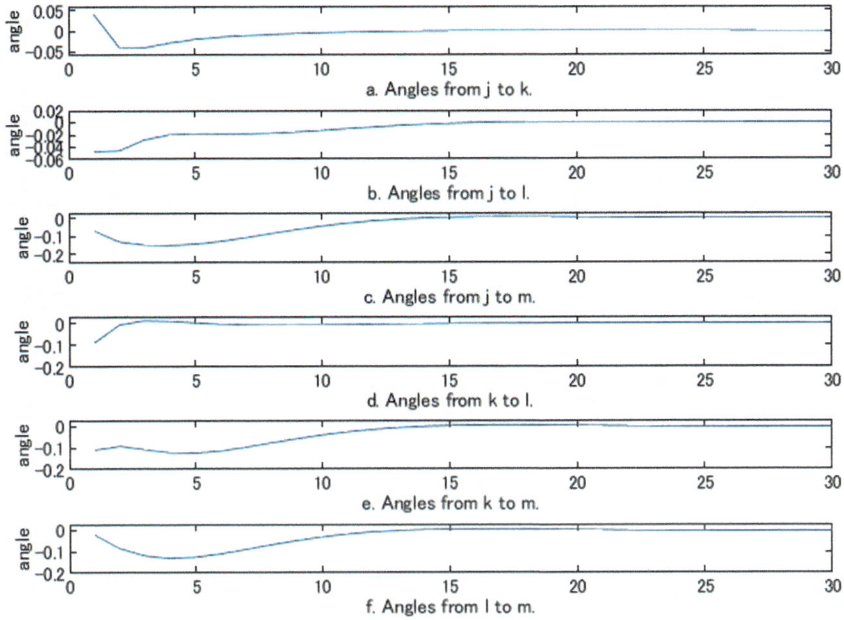

Fig. 9.56 Changes in the angles between four tetrad nations over time in Table 9.1

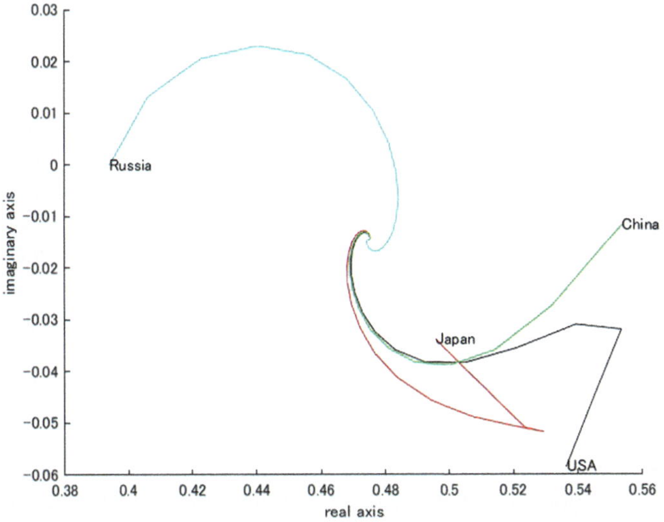

Fig. 9.57 Orbits of four tetrad nations in the same complex plane. A fixed point may be located near the point to which all the orbits of these nations converge in this figure

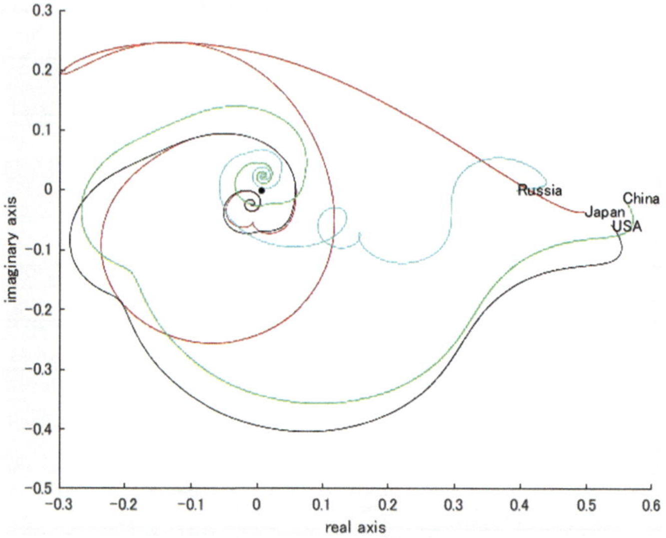

Fig. 9.58 Orbits of four tetrad nations in the same complex plane, where the DWD model is composed of the nonlinear part. Two fixed points may be located near the origin to which two of the orbits of Japan and USA and the other two of the orbits of China and Russia converge to different fixed points

Initial values of the configuration of the four nations of this model are the same as that shown in Fig. 9.46.

Figure 9.58 shows the orbits of the four nations in the same complex plane. As is apparent from this figure that there appear *two fixed points* near the origin. Moreover, it is evident that China and Russia converge to a fixed point located above the origin, while Japan and USA converge to the other fixed point located below the origin.

Generally, even in dynamics in one complex variable such as Mandelbrot set (see, for example, Sect. 12.5.2 in Chap. 12), we frequently encounter multiple fixed points. Therefore, it seems to be natural to observe such two fixed points in our model introduced above. It seems to be important to notice here that such a qualitative feature of the orbits of four nations is not determined only by a specific *mutual interaction between the two nations* but is determined by the *mutual interactions among the four nations*.

Example 9.18 The final example is the nonlinear model of (9.50) with no skew-symmetric restrictions described by (9.51). In this case, weight parameters of this example are as follows:

$$w_{ij} = 0.01 + 0.2i,\ w_{ik} = -0.2i,\ w_{il} = 0.3 + 0.2i,$$
$$w_{jk} = 0.02i,\ w_{jl} = 0.01 + 0.2i,\ w_{kl} = 0.01.$$

The initial configuration of nations is the same as in the previous examples.

Figure 9.59 depicts changes in the self-proximities of four nations over 8,000 iterations. In the case of this model, self-proximities converge to some constant values. By contrast, changes in the angle between four nations sometimes oscillate and exhibit *chaotic behaviors*, as will be checked below. Figure 9.60 draws changes in the angles between four nations over time. It is apparent that the second and the third time series in Fig. 9.60 oscillate until about 6000 iterations and suddenly stop oscillating and converge to a constant. Such a phenomenon is reminiscent of the *intermittency*, which will be discussed in Sect. 12.3 in Chap. 12. In general, *intermittent chaos* goes with at least one *bursting phase* which is characterized by an intense oscillation and one *laminar phase which is characterized by a quiet sequence.* For example, the time series in the first 6,000 iterations in the change in angles of the second time series of Fig. 9.60 is possibly considered as a bursting phase and the remaining series after 6,000 iterations may be considered as a laminar phase. Such an intermittent chaos in difference equations has been found, for example, in the *modified Bernoulli map* (Aizawa & Kohyama [1]) and in a neural network model by Rulkov [31].

Figure 9.61 is the orbits of four tetrad nations in the same complex plane in Example 9.18. It is evident that the four nations converge to two fixed points near the origin. Japan (A) approaches to a fixed point at the left-side of the origin, while USA (B), China (C) and Russia (D) approach to the other fixed point at the right-side of the origin.

Figure 9.62 shows *Lyapunov exponent* for the angle between B (USA) to C (China) in the first 2,000 iterations. Since the largest Lyapunov exponent of this time series is positive, the second time series of Fig. 9.60 may be said to include a chaos in the first 2,000 iterations.

9.5 Possible Future Works on DWD

In this chapter we have introduced an elementary theory of the dynamical weighted digraph which has been developed by the author since 2000. We have examined the qualitative behaviors of the solution curves of some of the special cases of the complex difference system models described by (9.5) through (9.11) in Sect. 9.2. We have also examined those described by (9.12) through (9.18) in Sect. 9.3. These examinations have been restricted to *dyadic* and *triadic* asymmetric interaction models. In Sect. 9.4, we have applied DWD to an empirical data which is shown in Table 9.1. In that section, we have shown several applications of a *tetradic* interaction model to an empirical data.

However, there is plenty of room for further development of DWD which has been introduced in this chapter. For example, it will be interesting to examine the pentadic asymmetric interaction models which can be assumed in our models, especially those described by (9.12) through (9.14).

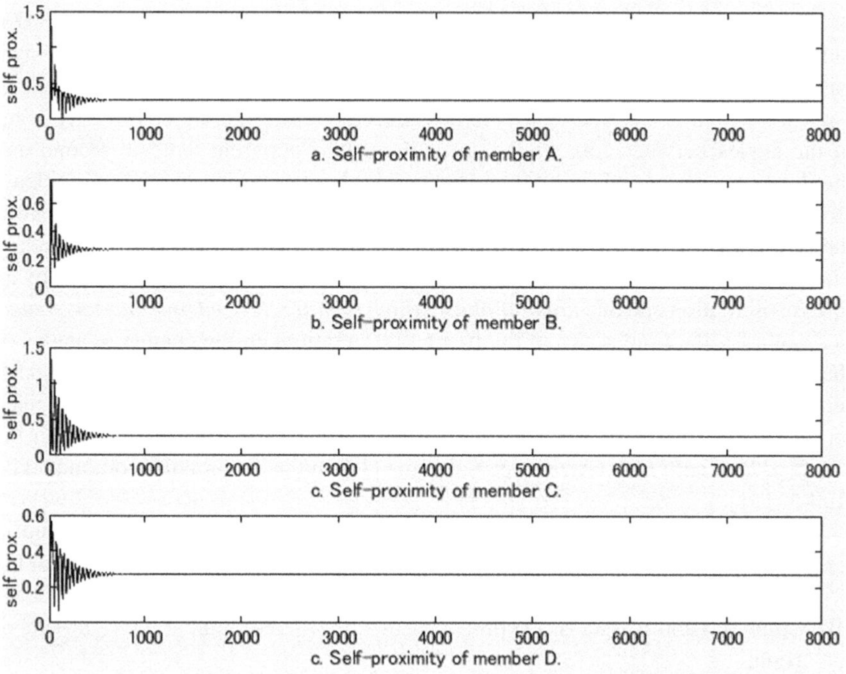

Fig. 9.59 Change in the self-proximities of four nations over time in Example 9.18

As for the degree of the polynomials in our model, that is, *m*, in (9.12), our examination has been restricted within 2. It will be interesting to examine our model in the case when polynomial degree *m* is larger than 2.

As regards the dimensions of the Hilbert space, that is, *p*, in (9.14), our theoretical and numerical considerations have been restricted to 1. In mathematics, a famous textbook on complex dynamical system written by Milnor [28] is restricted to one complex variable, which may be roughly equivalent to the complex plane in one-dimensional Hilbert space as a special case of our *p*-dimensional Hilbert space. Recently, Ebeling [20] has extended the complex dynamics to functions of several complex variables. However, the assumption of the several variables seems to be slightly different from that of our *p*-dimensional model. For, our model is based on the *congruent transformation* (e.g., Lancaster & Tismenetsky [26]) of a Hermitian inner product which is the one-to-one correspondence to an observed real ASM, and the dimensions of the configuration of members to be embedded in a complex space are *unitary*, that is, *complex orthogonal* to each other, while the complex variables assumed in Ebeling do not seem to fulfill this requirement. As for the *unitary matrix* we have already discussed in Chap. 4 and we shall introduce it in Chap. 11 in more detail.

As for the applications of DWD to an observed digraph, we can examine changes in digraph over time as shown in Figs. 9.32, 9.33 and 9.34 by applying (9.16) to the

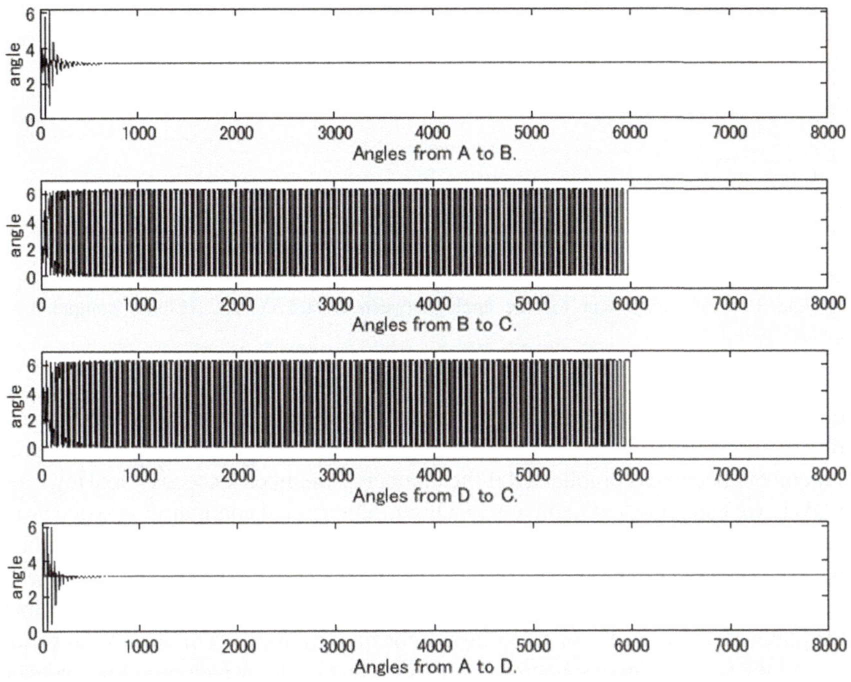

Fig. 9.60 Changes in the angles between four nations over time in Example 9.18

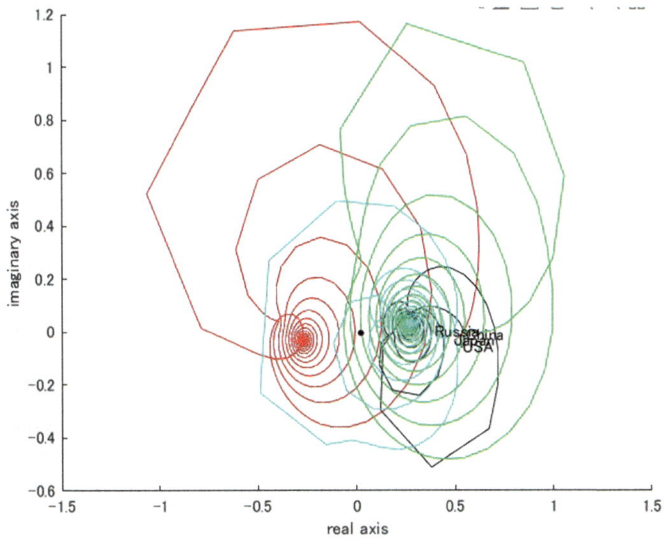

Fig. 9.61 Orbits of four tetrad nations in the same complex plane in Example 9.18

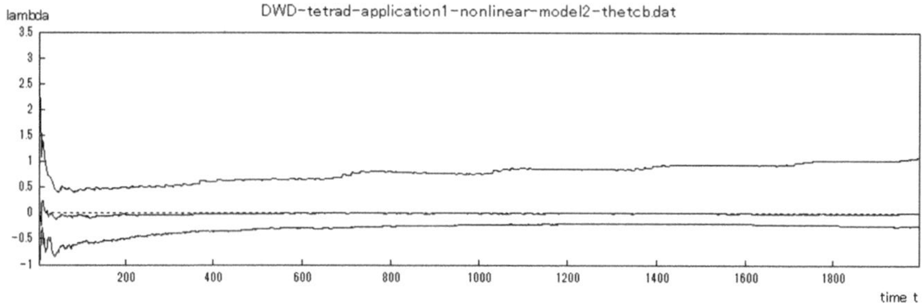

Fig. 9.62 Lyapunov exponent for the angle between B (USA) to C (China) computed by SundayChaos Times

configurations of members in a Hilbert space, which can be obtained by DWD at any point of iterations. However, there exists a shortfall in digraph in that a representation of members becomes complicated if the number of members exceeds five. However, in DWD we can represent members in a high-dimensional configuration space as an alternative way of representing members, which enables us to represent members visually far simpler.

DWD can be applied to empirical data as well as hypothetical data in the various disciplines of science. As for the complex constant z_0 in (9.12) of our DWD, which is a *control term*, we may set some arbitrary (vector) value which might be deduced theoretically or hypothetically in the corresponding discipline of science.

As another example, DWD can possibly be applied to neural network data, because we think that neural network has in principle asymmetric connections between neurons. As will be referred to in Chapter 11, a *phasor neural network* model is known in which Hebbian postulate is extended to complex case (e.g., Aizenberg et al. [2]; Aizenberg & Aizenberg [3]; Jankowski et al. [25]; Hirose [24]; Noest [29]). In this model, the weight matrix composed of synaptic strengths defined by (11.128) is assumed to be complex. By contrast, in DWD, we assume that the state space in which members of a group is embedded is a p-dimensional Hilbert space (or an indefinite metric space) and that the motions of members through asymmetric interactions in the Hilbert space are described by a set of complex polynomial difference equations.

Moreover, in DWD, the initial configuration of members in a Hilbert space can be estimated from the observed real asymmetric similarity matrix between members at an instant in time. Therefore, if we can observe, for example, the mutual strengths of connections between boxels in the brain at an instant in time, we can embed boxels in a Hilbert space (or indefinite metric space) by applying HFM in Chap. 4. Then, we can apply DWD to this data, and we can simulate the qualitative features of the motions of boxels by setting an arbitrary set of parameters $w_{jk}^{(1,m)} \cdots , w_{jk}^{(p,m)}$ which is a special set of parameters defined in (9.14). Here, it should be noted that, in *deep learning*, it is usual that we use *recurrent neural network model* which is described by a set of *real difference equations*, an example of which is defined as

$$\begin{cases} \boldsymbol{h}^{(n)} = tanh\left(\boldsymbol{a}^{(n)}\right) \\ \boldsymbol{a}^{(n)} = \boldsymbol{b} + \boldsymbol{W}\boldsymbol{h}^{(n-1)} + \boldsymbol{U}\boldsymbol{x}^{(n)} \\ \boldsymbol{o}^{(n)} = \boldsymbol{c} + \boldsymbol{V}\boldsymbol{h}^{(n)} \end{cases}, \tag{9.52}$$

where $\boldsymbol{h}^{(n)}$ is a vector composed of the hidden layer activations, $\boldsymbol{W}$ is a weight matrix which represents the hidden-to-hidden connections, $\boldsymbol{U}$ is another weight matrix which represents the input-to-hidden connections, $\boldsymbol{x}^{(n)}$ is a vector input sequence, $\boldsymbol{o}^{(n)}$ is composed of specific output values, and $\boldsymbol{c}$ is a constant vector (Goodfellow et al. [22]).

In applying DWD to empirical data, emphasis is placed on the classification of the *qualitative features* of the solution curves (or orbits) instead of their *quantitative features*. To do so, we utilize *qualitative theories of dynamical systems* such as the theory of the singular points of the vector fields of differential equations as well as that of the fixed points of the difference equations (e.g., Elaydi [21]; Guckenheimer and Holms [23]; Milnor [28]). However, we have not completed the classifications of the fixed points in tetradic models and higher order linear models of DWD as well as the *local* and *global qualitative features* of the nonlinear models of DWD yet. In any case, it should be noticed here that, by focusing on the qualitative features of the solution curves, DWD is released from the problem of *statistical errors* of each of the coordinate points of members estimated via HFM since the estimation errors of initial estimates of the configuration of members are not essential for the scenarios of the solution curves of DWD. In this sense, we hope that DWD will play a crucial role in opening up a new horizon of analysis in applied sciences in marked contrast to most of the traditional statistical analyses in the social, behavioral, and life sciences.

Finally, we have not developed any systematic method for obtaining an efficient set of weight parameters of our complex difference equation models in DWD so that we can obtain a desirable scenario of the solution curves (i.e., a desirable classification of the solution curves). At present, we specify these parameters arbitrarily or by intuition. However, it is likely that there exist multiple solutions to this problem because the desirable scenario of such solution curves is merely *qualitative*.

Appendix

In this appendix, we shall show a MATLAB code which depicts the results of Example 9.13.

```
%-------------------------------------------------------------------
%
% file name: ghcde_diad_sim2_Case4_reducted.m
% version 1.1 written by N. Chino, February 28, 2023.
% revised by N. Chino, March 2023.
%
%-------------------------------------------------------------------
```

```
%
format long;
%
% Parameter values of DWD
%
P = 1, m = 2, maxn = 50,000, oN = 2,
%where p is the number of dimensions of the Hilbert space,
%     m is the degree of polynomial equation,
%     maxn is the number of iterations,
%     oN is the number of dimensions of the difference system, which
%     is equal to the number of members or nodes which interact with
%     each other.
%
% Initial values of the orbit
za(1) = -0.205537992601818–0.470295604664077j,
zb(1) = -0.5
% thinning parameter values
lowit = 42,000, intvl = 1, upit = 49,999
% initial values of the weights in this model
alp1jk = 0.9564j, alp1kj = -0.9564j
alp2jk = 0.01, alp2kj = 1.5375–0.9564j
%
Beta1jk = 1 + alp1jk + alp1kj
beta2jk = alp2jk-alp2kj
gam1jk = beta1jk/2-beta1jk*beta1jk/4
%
angba(1) = angle(zb(1))-angle(za(1))
for n = 1:maxn
% It should be noticed here that positive direction is clockwise in HFM.
difzab(n) = za(n)-zb(n);
difzba(n) = -difzab(n);
difzs(n) = difzab(n)*difzab(n);
zabplus(n) = alp1jk*difzab(n) + alp2jk*difzs(n);
zbaplus(n) = alp1kj*difzba(n) + alp2kj*difzs(n);
zjkdef(n) = beta2jk*difzab(n) + beta1jk/2;
if n = = 1
    zjk(1) = zjkdef(1)
    abszjk(1) = abs(zjk(1));
    angba(1) = angle(zb(1))-angle(za(1));
end
za(n + 1) = za(n) + zabplus(n);
zb(n + 1) = zb(n) + zbaplus(n);
zjk(n + 1) = zjkdef(n)*zjkdef(n) + gam1jk;
abszjk(n + 1) = abs(zjk(n + 1));
if abszjk(n + 1) ~ = 1
```

```
      za(n + 1) = za(n + 1)/abszjk(n + 1);
      zb(n + 1) = zb(n + 1)/abszjk(n + 1);
  end
  angba(n + 1) = angle(zb(n + 1))-angle(za(n + 1));
  end;
  %
  disp('1. Iteration history of coordinates, self-similarities, and angles');
  t = [1:maxn];
  %
  disp(' 1–0. Coordinates et al. of object a at some points in time')
  dx = [' n      zaplus(n)','            za(n)'];
  disp(dx)
  %
    for n = lowit:intvl:upit
    x = [zabplus(n),za(n)];
  n,disp(x)
  end;
  %
  for n = lowit:intvl:upit
    x = [zbaplus(n),zb(n)];
    n,disp(x)
  end;
  %
  disp(' 1–1. Coordinates of points at some points in time')
  dx = ['       n','       za(n)','       zb(n)'];
  disp(dx)
  %
  for n = lowit:intvl:upit
    x = [n,za(n),zb(n)]; disp(x)
  end;
  %
  disp(' 1–2. self-similarities of points at some points in time')
  dx = [' n',' abs(za(n))',' abs(zb(n))'];
  disp(dx)
  for n = lowit:intvl:upit
    x = [n,abs(za(n)),abs(zb(n))]; disp(x)
  end;
  %
  disp('1-3. Angle between objects a and b at some points in time')
  dy = ['       n','   angba(n)'];
  disp(dy)
  for n = lowit:intvl:upit
    y = [n,angba(n)]; disp(y)
  end;
  %
```

```
disp('2. Plots of the Iteration history of self-similarities');
ssa = abs(za(t)); ssb = abs(zb(t));
for n = lowit:intvl:upit
   nn = n-lowit + 1;
   zap(nn) = za(n);
   zbp(nn) = zb(n);
   ssap(nn) = abs(zap(nn));
   ssbp(nn) = abs(zbp(nn));
end;
%
Z = [za(1);zb(1)];
[Sf] = rpi(oN,p,z);
[S1] = [Sf];
Sall = Sf;
% Compute the next similarity matrix and stack it rowwise.
for n = 2:maxn
   z = [za(n);zb(n)];
    [Sn] = rpi(oN,p,z);
   if n = = 20.
    [S20] = [Sn];
   end
   %
   if n = = 21
      [S21] = [Sn];
   end
   %
   if n = = 22
      [S22] = [Sn];
   end
   %
   %
   if n = = 23
    [S23] = [Sn];
   end
   %
   if n = = maxn
    [Smaxn] = [Sn];
   end
   Sall = [Sf;Sn];
   Sf = Sall;
end;
%
for n = 1:maxn
   nij = 1 + 2*(n-1);
   sij(n) = Sall(nij,2);
```

```
end;
%
for n = lowit:intvl:upit
   nn = n-lowit + 1;
   sijp(nn) = sij(n);
   simijp(nn) = sijp(nn);
end;
%
Cp = [ssap;ssbp];
ssapt = ssap'; ssbpt = ssbp'; simijpt = simijp'; Cpt = Cp';
save('data/chino/ghcde-dyad-sim2-case4-ssap.dat', 'ssapt', '-ASCII');
save('data/chino/ghcde-dyad-sim2-case4-ssbp.dat', 'ssbpt', '-ASCII');
save('data/chino/ghcde-dyad-sim2-case4-simijp.dat', 'simijpt', '-ASCII');
save('data/chino/ghcde-dyad-sim2-case4-Cp.dat', 'Cpt', '-ASCII');
%
figure (1);
subplot 411;
plot(ssa,'k');
axis([45000 46,000 -1 10]);
ylabel('self prox.');
xlabel('a. Self-proximity of member A.');
%
subplot 412;
plot(ssb,'k');
axis([45000 46,000 -1 10]);
ylabel('self prox.');
xlabel('b. Self-proximity of member B.');
%
subplot 413;
disp('2. Plots of the iteration history of the proximity between 2 objects');
simij = sij(t);
plot(simij,'k');
axis([45000 46,000 -1 25]);
ylabel('proximity');
xlabel('c. proximity from A to B');
%
disp('3. Plots of the iteration history of the angle from a to b');
subplot 414;
thetba = angba(t);
plot(thetba,'k');
axis([45000 46,000 -0.005 0.005]);
ylabel('angle');
xlabel('d. Angle from B to A.');
pause;
%
```

```
disp('4. Plots of the configurations of objects at each time');
disp('In these figures, positive direction is clockwise.');
%
figure (2);
subplot 221;
z = [za(1);zb(1)];
compass(z);
text(-0.15,-0.48,'A_1'); text(-0.50,0.09,'B_1');
ylabel('imaginary axis');
xlabel('a. Iteration 1');
disp(' (1) Asymmetric similarity matrix at t(1)');
[S] = rpi(oN,p,z).
%
subplot 222;
z = [za(20);zb(20)];
compass(z);
text(-0.8,-0.3,'A_{20}'); text(-0.6,-0.7,'B_{20}');
ylabel('imaginary axis');
xlabel('b. Iteration 20');
disp(' (3) Asymmetric similarity matrix at t(20)');
[S] = rpi(oN,p,z).
%
subplot 223;
z = [za(maxn);zb(maxn)];
compass(z);
text(-1.5,2.2,'A_{m}'); text(-0.9,1.5,'B_{m}');
ylabel('imaginary axis');
xlabel('d. Iteration maxn');
disp(' (6) Asymmetric similarity matrix at t(maxn)');
[S] = rpi(oN,p,z)
disp(' (7) Angle between objecs A and B');
disp(angba(maxn))
%
% Reproduce the configuration of objects at time t(maxn) via HFM.
%
If S(1,2) < 1000000
   H = (S + S')/2 + j*(S–S')/2;
    [O,D] = eigs(H);
   disp('5. The Hermitian matrix reproduced from t_maxn configuration');
   disp(H)
end
%
figure (3);
names = {'A','B'};
%
```

```
  G = digraph(S1,names);
  plot(G,'k','Layout','force','EdgeLabel',G.Edges.Weight);
  set(gca,'Visible','off','NextPlot','replaceChildren');
  text(-0.04,-0.7,'a. digraph at Iter.1');
pause;
%
figure (4);
  G = digraph(S20,names);
  plot(G,'k','Layout','force','EdgeLabel',G.Edges.Weight);
  set(gca,'Visible','off','NextPlot','replaceChildren');
  text(-0.04,-0.7,'b. digraph at Iter.20');
  pause;
%
figure (5);
  G = digraph(Smaxn,names);
  plot(G,'k','Layout','force','EdgeLabel',G.Edges.Weight);
  set(gca,'Visible','off','NextPlot','replaceChildren');
  text(-0.04,-0.7,'c. digraph at Iter.50,000');
pause;
%
disp('Coordinates of points at some points in time')
dx = ['      n','      za(n)','      zb(n)'];
disp(dx)
zapart = za(1:maxn);
zareal = real(zapart);
zaimag = imag(zapart);
%
figure (6);
set(0,'DefaultAxesColorOrder',[0,0,0],...
    'DefaultAxesLineStyleOrder','-|-.|--|:').
%plot(za,'-k');
plot(zapart,'-k');
text(0.2,-0.47,'A_1');
%text(16, 12,'A_{1000}');
text(120, 221,'A_{50,000}');
xlabel('real axis');
ylabel('imaginary axis');
pause;
zbpart = zb(1:maxn);
zbreal = real(zbpart);
zbimag = imag(zbpart);
%
figure (7);
set(0,'DefaultAxesColorOrder',[0,0,0],...
'DefaultAxesLineStyleOrder','-|-.|–|:').
```

```
%plot(zb,'-k');
plot(zbpart,'-k');
text(-0.5,0,'B_1');
text(15, 13,'B_{1000}');
text(119, 220,'B_{50000}');
ylabel('imaginary axis');
xlabel('real axis');
pause;
%
figure (8);
set(0,'DefaultAxesColorOrder',[0,0,0],...
'DefaultAxesLineStyleOrder','-|-.|--|:').
plot(zbpart,'-k');
hold all
plot(zapart,'-y');
text(0.2,-0.47,'A_1');
%text(-0.5, 0.1,'B_1');
text(-0.5,10,'B_1');
text(119, 220,'B_{50000}');
text(120, 230,'A_{50000}');
ylabel('imaginary axis');
xlabel('real axis');
print -depsc eps-files/takens/ghcde/ghcde-sim2-case4-1-chaotic-trajectories.eps
pause;
%
% Spectral analysis of ssap, ssbp, and simijp.
%
Speca = abs(spectrogram(ssap,500));
Specb = abs(spectrogram(ssbp,500));
Specs = abs(spectrogram(simijp,500));
slast = size(Speca);
sss = [1:slast];
%
figure (9);
loglog(sss,Speca);
ylabel('log spectrum');
xlabel('log frequency');
%
figure (10);
yma = mean(Speca,2);
loglog(yma);
ylabel('log mean spectrum');
xlabel('log frequency');
%
figure (11);
```

```
loglog(sss,Specb);
ylabel('log spectrum');
xlabel('fog frequency');
%
figure (12);
ymb = mean(Specb,2);
loglog(ymb);
ylabel('log mean spectrum');
xlabel('log frequency');
%
figure (13);
loglog(sss,Specs);
ylabel('log spectrum');
xlabel('log frequency');
%
figure (14);
yms = mean(Specs,2);
loglog(yms);
ylabel('log mean spectrum');
xlabel('log frequency');
function [S] = rpi(N,p,Z)
% This function returns the real similarity data matrix of order N, given
% the N by p complex coordinate matrix Z, translating the these values into
% the real values via the polarized identity in the Hilbert space.
%
for a = 1:N
  for b = 1:N
     dab = abs(Z(a,:)-Z(b,:)).^2;
     daib = abs(Z(a,:)-i*Z(b,:)).^2;
     dsum = abs(Z(a,:)).^2 + abs(Z(b,:)).^2;
     S([a],[b]) = dsum-(dab + daib)./2;
  end;
end;
```

References

1. Aizawa, Y., & Kohyama, T. (1984). Asymptotically non-stationary chaos. *Progress of Theoretical Physics, 71*, 847–850.
2. Aizenberg, N. N., Ivaskiv, Y. L., & Pospelov, D. A. (1971). A certain generalization of threshold functions. *Dokrady Akademii Nauk SSSR, 196*, 1287–1290.
3. Aizenberg, N. N. & Aizenberg, I. N. (1992). CNN based on multi-valued neuron as a model of associative memory for grey scale images. In *CNNA '92 Proceedings of 2nd International Workshop on Cellular Neural Networks Applications*, Munich, Germany, pp. 36–41.
4. Berry, M. V. (1979). Diffractals. *Journal of Physics A: Mathematical and General, 12*, 781–797.

5. Chino, N. (2000). Complex space models for the analysis of asymmetry. *Paper presented at The International Conference on Measurement and Multivariate Analysis*. Banff, Canada.
6. Chino, N. (2001). Complex difference system models for the analysis of asymmetry. In H. Yanai, . Okada, K. Shigemasu, Y. Kano, & J. J. Meulman (Eds.), *New Developments in Psychometrics* (pp. 479–486). Tokyo: Springer-Verlag.
7. Chino, N. (2002). Complex space models for the analysis of asymmetry. In S. Nishisato, Y. Baba, H. Bozdogan, & K. Kanefuji (Eds.), *Measurement and Multivariate Analysis* (pp. 107–114). Springer-Verlag.
8. Chino, N. (2003). Fitting complex difference system models to longitudinal asymmetric proximity matrices. *Paper presented at the 13th International Meeting and the 68th Annual Meeting of the Psychometric Society*. Cagliari, Sardinia, Italy.
9. Chino, N. (2005). Abnormal behaviors of members predicted by a complex difference system model. *Bulletin of the Faculty of Psychological & Physical Science, 1*, 69–73.
10. Chino, N. (2006). Asymmetric multidimensional scaling and related topics. *Manuscript for the invited talk at the Weierstrass Institute for Applied Analysis and Stochastics*, Berlin, Germany.
11. Chino, N. (2014). A general non-Newtonian n-body problem and dynamical scenarios of solutions. *Proceedings of the 42th Annual Meeting of the Behaviormetric Society of Japan* (pp. 48–51). Sendai, Japan.
12. Chino, N. (2016). A general non-Newtonian n-body problem and dynamical scenarios of solutions. In *Paper presented at the 31th International Congress of Psychology*. Yokohama, Japan.
13. Chino, N. (2017). Dynamical scenarios of changes in asymmetric relationships over time (1). *Bulletin of The Faculty of Psychological & Physical Science, 13*, 23–31.
14. Chino, N. (2018). An elementary theory of a dynamic weighted digraph (1). *Bulletin of The Faculty of Psychological & Physical Science, 14*, 23–31.
15. Chino, N. (2018b). Dynamical Scenarios of changes in asymmetric relationships over time (2). *Journal of the Institute for Psychological and Physical Science, 10*, 7–14
16. Chino, N. (2019). An elementary theory of a dynamic weighed digraph (2). *Journal of the Institute for Psychological and Physical Science, 11*, 1–8.
17. Chino, N. (2020). How to use the Hermitian form model for asymmetric MDS. In T. Imaizumi, A. Nakayama, and S. Yokoyama (Eds.), *Advanced Studies in Behaviormetrics and Data Science* (pp. 19–41). Springer Nature Singapore.
18. Chino, N. (2024). Complex difference system models for asymmetric interaction. In E. Beh, , R. Lombardo, and J. Clavel (Eds.) *Advanced Studies in Behaviormetrics and Data Science—Essays in Honor of Sizuhiko Nishisato*. Springer Nature Singerpore.
19. Chino, N., & Shiraiwa, K. (1993). Geometric structures of some non-distance models for asymmetric MDS. *Behaviormetrika, 20*, 35–47.
20. Ebeling, W. (2007). *Functions of Several Complex Variables and Their Singularities*. Friedr. Vieweg & Sohn.
21. Elaydi, S. N. (1999). *An introduction to difference equations* (2nd ed.). Springer-Verlag.
22. Goodfellow, I., Bengio, Y., & Courville, A. (2016). *Deep learning*. The MIT Press.
23. Guckenheimer, J., & Holmes, P. (1983). *Nonlinear oscillations, dynamical systems, and bifurcations of vector fields*. Springer-Verlag.
24. Hirose, A. (2003). *Complex-valued neural networks—Theories and applications*. World Scientific.
25. Jankowski, S., Lozowski, A., & Zurada, J. M. (1996). Complex-valued multistate neural associative memory. *IEEE Transactions on Neural Networks, 7*, 1491–1496.
26. Lancaster, P., & Tismenetsky, M. (1985). *The theory of matrices*. Academic Press.
27. Mandelbrat, B. B. (1977). *The fractal geometry of nature*. W. H. Freeman and Company.
28. Milnor, J. (2000). *Dynamics in one complex variable—Introductory lectures* (2nd ed.). Friedr. Vieweg & Sohn.
29. Noest, A. J. (1988). Discrete-state phasor neural networks. *Physical Review A, 38*, 2196–2199.
30. Ott, E., Grebogi, C., & Yorke, J. A. (1990). Controlling Chaos. *Physical Review Letters, 64*, 1196–1199.

31. Rulkov, N. F. (2002). Modelling of spiking-bursting neural behavior using two-dimensional map. *Physical Review E, 65*, 041922.
32. Shiraiwa, K. (2014). Personal communication. April 25.

Part III
Supplementary Chapters

Part III of this book is composed of supplementary chapters, which include Chaps. 10 through 12. In these chapters, definitions, theorems, and theories are introduced which are related to the notions, theorems, and theories that appeared in Chaps. 1 through 9. Chapter 10 introduces the reader to the definitions and theorems useful for the understanding of the structure of asymmetric interactions.

The set theory and notions of group in group theory in mathematics play crucial roles in the fundamental scales in psychological measurement. These are discussed in Sects. 10.2 and 10.3 seriatim. All the models discussed in this book assume a certain scale level of measurement. Notions of complex number and the notion of Hilbert space and related theorems that are essential in HFM are introduced in Chap. 4 of Part I. These notions are also indispensable in DWD, introduced in Chap. 9 of Part II. These are discussed in Sections 10.3 and 10.4 seriatim.

Chapter 11 introduces the definitions and theories useful for the understanding of the structure of asymmetric interaction. These include matrix theory, asymmetric MDS, graph and spectral graph theories, graph embedding, and neural networks. Chapter 12 introduces definitions and theorems useful for the dynamics of asymmetric interactions. These include those related to differential equation as well as difference equation, which are discussed in Sects. 12.2, 12.4, and 12.5. The notions of chaos and related topics are common to differential and difference equations and are discussed in Sect. 12.3.

The purpose of providing these supplementary chapters is to make this book self-contained as far as possible for the readers whose fields belong to various disciplines of science.

Chapter 10
Definitions and Theorems Useful for the Understanding of the Structures of Asymmetric Interactions

10.1 Introduction

In this chapter, we shall present various definitions and theorems useful for the understanding of the contents discussed in the previous chapters mainly in Part I of this book. This is in order to make this book as self-contained as possible. We shall omit all the proofs of the theorems listed in this chapter but give references to proofs at the end of each theorem. One reason for this is that in this book we assume the majority of readers to be applied scientists and not mathematicians. Another reason is to avoid making this book too big.

One method for analyzing the structure of asymmetric interactions among elements is to represent them as points in a certain dimensional *vector space*. For example, in *symmetric MDS* which was developed in psychometrics, we frequently embed elements in an r-dimensional *Euclidean space* or more general *Minkowski spaces* which include the Euclidean space as a special case, given a *symmetric* similarity or dissimilarity matrix among elements (e.g., Torgerson [54]; Kruskal [32, 33]). In such cases, it is usual to define various *metric functions* on an r-dimensional vector space. We sometimes define an inner product in that space. In this case, a certain norm is induced by the inner product, which has a metric property.

In *asymmetric MDS* which has also been developed mainly in psychometrics, we frequently embed elements in a certain r-dimensional *augmented* Euclidean space, given an *asymmetric* similarity or dissimilarity matrix among elements, as reviewed in Chap. 7. We sometimes embed elements in a *Hilbert space* which includes the Euclidean space as a special case.

In *spectral graph theory*, nodes (or vertices) of a graph (usually an *undirected graph*) are embedded in an r-dimensional Euclidean space, by solving the eigenvalue problem of a *Laplace matrix* (or *Laplacian matrix*) which is computed using the *degree matrix* and *adjacency matrix* of the graph. Although the Laplace matrix of a *directed graph* can also be defined, it is not unique because the degree matrix has two types, that is, the *indegree* and *outdegree matrices*.

N. Chino, *Structure and Dynamics of Asymmetric Interactions*,
https://doi.org/10.1007/978-981-97-8269-7_10

Sometimes, we solve the eigenvalue problem without computing the Laplace matrix. In this case, a special matrix called the *hyperlink matrix* corresponding to the adjacency matrix of a graph is computed, and the first eigenvector associated with the largest eigenvalue is computed by using a power method in numerical analysis. This means that nodes or vertices of a graph are embedded in a one-dimensional vector space.

We shall introduce some of the matrices including those discussed above, which play a fundamental role in the spectral graph theory to be discussed in Sect. 11.4 of Chap. 11 as well as the asymmetric MDS which includes symmetric MDS as a special case. The Laplace matrix is also utilized in the so-called *graph embedding* in machine learning and computer vision (e.g., Yan et al. [56]). We shall review them in Sect. 11.4.

In any case, in analyzing the structure of asymmetric interactions among elements, we embed these elements in a certain dimensional *vector space*. Thus, it is necessary and appropriate to examine the properties of the vector spaces mathematically. In this chapter, we shall begin with the basic concepts of the *group theory* relating to the vector space to understand the essential properties of the vector space, which will be introduced in Sect. 10.2. The group theory is also necessary to understand deeply the basic properties of psychological scales developed in psychometrics, as will be seen in Sect. 10.3.

In psychometrics, especially in psychological scaling, the literature which introduces or utilizes the concept of complex numbers has been minimal, although Stevens [53], for example, discusses the role of complex numbers in psychological scaling. However, in analyzing the structure of asymmetric interactions a complex number plays a fundamental role. For example, HFM in asymmetric MDS proposed by Chino and Shiraiwa [15] utilizes the complex number and embeds objects in a Hilbert space which is defined in Sect. 10.5. Another example is the Hermitian adjacency matrix introduced in analyzing mixed graphs in spectral graph theory, which has recently been discussed by several researchers (e.g., Guo & Mohar [24]; Liu & Li [37]; Mohar [42]; Yuan et al. [57]). A third example is the utilization of the *root of unity* in neural network models (e.g., Aizenberg et al. [1]; Aizenberg & Aizenberg [2]; Hirose [25]). Therefore, we shall present several basic definitions and theorems relating to the complex number in Sect. 10.4. In Sect. 10.5, we shall show how the different kinds of spaces can be induced, starting from an abstract vector space defined axiomatically in Sect. 10.2.

Finally, we shall refer to some of the notations which appear in this chapter. The first notation is a space R^n. As, for example, Boothby ([8], p. 2) points out, this space is used in several senses in the literature, and we must usually decide it from the context which one is intended. Sometimes it means merely R^n as *topological space*. Then, it serves as an n-dimensional *real vector space* or an n-dimensional *complex vector space*, and sometimes it is identified with *Euclidean space*. In this chapter we shall specify its sense just before the notation R^n like "the vector space R^n" whenever possible.

The second notation is V. It denotes a vector space in general and includes a real vector space and a complex vector space. The third notation is F. It means a field and includes real number R and complex number C. The fourth notation is E^n. It means a Euclidean space, while we sometimes denote an oriented Euclidean space by R^n. The fifth notation is W. It denotes a normed space. The sixth notation is a' set. We denote it using several alphabetical symbols like X, Y, Ω, Ψ.

10.2 Basic Concepts in Group Theory

Before proceeding, we must define basic classes of functions. A standard notation for a function from a set X to a set Y is $f : X \to Y, or y = f(x)$, which indicates the element $y \in Y$ that f associates with $x \in X$ (e.g., Bowen & Wang [9], p. 18). The set X is called the *domain* of the function f, while the set of all y for which there exists an x such that $y = f(x)$ is called the *codomain* (or sometimes called the *range*) of f.

According to Bowen and Wang (op. cit., p. 14, pp. 18, 19, pp. 23–25):

Definition 10.1 (*subset*) Let Ω and Ψ be sets. Then, Ψ is said to be a *subset* of Ω if every element of Ψ is also an element of Ω.

Remark Bowen and Wang (op. cit., p. 14) denote the above relationship as $\Psi \subset \Omega$, saying that it is a *customary notation.* On the other hand, Bloom ([7], p. 2) denotes it as $\Psi \subseteq \Omega$. Since the proper subset Ψ of Ω defined below is usually denoted as $\Psi \subset \Omega$ (e.g., Bloom, op. cit., p. 2), we shall denote the above relationship as $\Psi \subseteq \Omega$.

Definition 10.2 (*proper subset*) A nonempty subset Ψof Ω is called a proper subset of Ω if Ψ is not equal to Ω.

Example 10.1 The set of integers is a proper subset of the real number.

Definition 10.3 (*into, onto*) A function f is said to be *into* Y if $f(X)$ is a proper subset of Y, and it is said to be *onto* (or *surjective*) Y if $f(X) = Y$.

Definition 10.4 (*one-to-one*) A function f is said to be *one-to-one* (or *injective*) if for every $x_1, x_2 \in X$

$$f(x_1) = f(x_2) \Rightarrow x_1 = x_2. \tag{10.1}$$

Definition 10.5 (*one-to-one correspondence*) A function f is said to form a *one-to-one correspondence* (or to be *bijective*) from X to Y if $f(X) = Y$, and f is one-to-one.

Remark Any function is classified into one of the four classes by combining the notions of injection and surjection, that is, (1) injective and surjective (i.e., bijective), (2) injective and non-surjective, (3) non-injective and surjective, (4) noninjective and non-surjective.

Definition 10.6 (*semigroup*) A *semigroup* is a pair $(\Omega, *)$ consisting of a nonempty set Ω with an *associative* binary operation $*$:

$$(a * b) * c = a * (b * c) \text{ for all } a, b, c \in \Omega.$$

Example 10.2 The set of *positive integers* with addition constitutes a semigroup (e.g., Japanese Mathematical Society [28], p. 880).

In fact, it is easy to see that,

$$(a + b) + c = a + (b + c),$$

for all integers, a, b, c.

Remark Fountain ([21], p.117) shows an interesting example of a semigroup, which is composed of eight elements.

Definition 10.7 (*group*) A *group* is a pair $(\Omega, *)$ where Ω is a set and $*$ is a binary operation satisfying the following:

(a) $(a * b) * c = a * (b * c)$ for all $a, b, c \in \Omega$.
(b) There exists an element $e \in \Omega$ such that $a * e = e * a = a$ for all $a \in \Omega$. Such an element is called the *identity element*.
(c) For every $a \in \Omega$ there exists an element $a^{-1} \in \Omega$ such that $a * a^{-1} = a^{-1} * a = e$. Such an element is called the *inverse element*.

Remark 1 In (b), if $*$ is regarded as multiplicative, then the identity element is often called the *unity element*, while if it is additive, then it is called the *zero element*. Moreover, if the binary operation of the group is *commutative*, i.e., $a * b = b * a$, for all $a, b \in \Omega$, the group is said to be a commutative (or *Abelian*) *group*.

Example 10.3 The set $\{1, i, -i, -1\}$, where $i^2 = 1$, is a group with respect to the binary operation of multiplication (e.g., Bowen & Wang, op. cit., p. 25). It is easy to see that this set satisfies the three axioms of Definition 2.7 with respect to the binary operation of multiplication.

Example 10.4 The set Z of *all integers* with addition is an additive *abelian group*.

In this case, it is easy to see that the identity element is 0 and the inverse element of an integer n is $-n$. Similarly, it is easy to see that *rational numbers* Q, *real numbers* R, and *complex numbers* C are additive abelian group (e.g., Rotman [47], p. 129).

Remark 2 Later in this section, we shall define a more complicated group called *permutation group* which is included in a *symmetric group*. However, it is not so difficult to prove that it satisfies the three axioms in Definition 10.7.

Moreover, in Sect. 10.3 one can see that there correspond the mathematical group structures to the famous four scales of measurements in psychology, as Stevens [53] points out. These are the *permutation group*, the *isotonic group*, the *general linear group*, and the *similarity group*. We shall define them later in this section. As in the case of the permutation group, it is not so difficult to prove that these groups, except for the isotonic group, satisfy the three axioms in Definition 10.7. On the other hand, the isotonic group is a semigroup (e.g., Kim & Kozhuthov [27], p. 996).

Definition 10.8 (*subgroup*) A subset H of a group G is a subgroup if

(i) $1 \in H$; that is, an *identity element* is contained in H;
(ii) if $x, y \in H$, then $x * y \in H$; that is, H is *closed under* $*$;
(iii) if $x \in H$, then $x^{-1} \in H$; that is, an *inverse element* is contained in H.

Remark In general, a set S is called "*closed under* $*$" if $x * y \in S$ for all $x, y \in S$.

Next, we shall introduce several sets that form a *group* with respect to one binary operation and a *semigroup* with respect to another, according to Bowen and Wang ([9], pp. 31–35).

Definition 10.9 (*ring*) A *ring* is a triple $(\Omega, +, \cdot)$ consisting of a set Ω and two *binary operations* $+$ and $\cdot$ such that

(a) Ω with the operation $+$ is an *Abelian group.*
(b) The operation $\cdot$ is *associative.*
(c) Ω contains an *identity element*, denoted by 1, with respect to the operation $\cdot$, i.e.,

$$1 \cdot a = a \cdot 1 = a \text{ for all } a \in \Omega.$$

(d) The operations $+$ and $\cdot$ satisfy the *distributive axioms*

$$a \cdot (b + c) = a \cdot b + a \cdot c,$$
$$(b + c) \cdot a = b \cdot a + c \cdot a.$$

Remark The ring is called a *commutative ring* if the multiplication operator is also commutative.

Example 10.5 The sets of integers, rational numbers, and real numbers with the usual addition and multiplication operations are examples of rings.

Definition 10.10 (*integral domain*) A ring is an *integral domain* if it satisfies the following additional axioms.

(e) The operation $\cdot$ is *commutative.*
(f) If a, b, c are any elements of Ω with $c \neq 0$, then
$a \cdot c = b \cdot c \Rightarrow a = b$(cancellation law of multiplication).

Example 10.6 The sets of integers, rational numbers, and real numbers with the usual addition and multiplication operations which are examples of rings are also examples of integral domains.

Definition 10.11 (*field*) A *field* F is an integral domain, containing more than one element, and such that any element $a \in F\backslash\{0\}$ has an inverse with respect to multiplication.

Example 10.7 Real numbers, complex numbers, and rational numbers Q are examples of fields.

Definition 10.12 (*vector space*) A *set* V and a *field* F constitute a vector space if it satisfies the following rules:

(a) There exists a binary operation in V called *addition* and denoted by $+$ such that

  (1) $(u + v) + w = u + (v + w)$ for all $u, v, w \in V$.
  (2) $u + v = v + u$ for all $u, v \in V$.
  (3) There exists an element $0 \in V$ such that $u + 0 = u$ for all $u \in V$.
  (4) For every $u \in V$ there exists an element $-u \in V$ such that $u + (-u) = 0$.

(b) There exists an operation called *scalar multiplication* in which every scalar $\lambda \in F$ can be combined with every element $u \in V$ to give an element $\lambda u \in V$ such that

  (1) $\lambda(\mu u) = (\lambda\mu)u$,
  (2) $(\lambda + \mu)u = \lambda u + \mu u$,
  (3) $\lambda(u + v) = \lambda u + \lambda v$,
  (4) $1u = u$,

for all $\lambda, \mu \in F$ and all $u, v \in V$.

Figure 10.1 illustrates the dependence of the algebraic structures discussed above schematically, which was adapted from Figure 1 in Bowen and Wang [9].

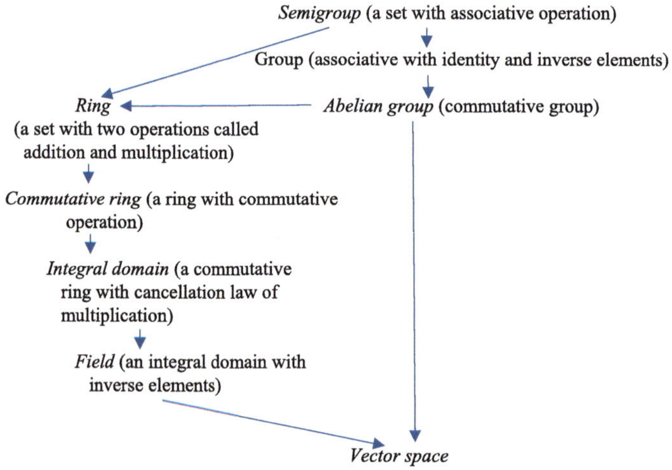

Fig. 10.1 Illustration of the algebraic structures introduced in this chapter

Remark 1 This figure indicates that the definition of ring, integral domain, and field are each a step more restrictive than the other. As Bowen and Wang (op. cit., p. 34) point out, any set that is a field is automatically an integral domain and a ring. Similarly, any set that is an integral domain is automatically a ring. In this sense, it can be said that a vector field inherits the properties of a ring and an Abelian group simultaneously.

Remark 2 Elements of V such as u, v, w in Definition 10.12 will be called *vectors*. Although Bowen and Wang (op. cit., p. 40) use boldface notation for vectors, we use lightface notation for them (e.g., Blank et al. [6]; Debnath & Mikusiński [19]), unless otherwise stated. If the field F is the field of *real numbers* R, V is called a *real vector space*, and if it is the field of *complex numbers* C, V is called a *complex vector space*. Vector space is sometimes called *linear space*.

In the following examples, we shall denote vectors using boldface notation in Examples 10.8 through 10.10 since we use vectors in the ordinary sense. It will not be so difficult to check whether the following Examples 10.8 through 10.11 satisfy all the rules listed in Definition 10.12.

Example 10.8 The vector space R^2 is the set of all 2-tuples of the form $\boldsymbol{x} = (x_1, x_2)$, and $x_1, x_2 \in \mathrm{R}$.

Example 10.9 The vector space R^n is the set of all n-tuples of the form $\boldsymbol{x} = (x_1, x_2, \cdots, x_n)$, and $x_1, x_2, \cdots, x_n \in R$.

Example 10.10 The vector space C^n is the set of all n-tuples of the form $\boldsymbol{z} = (z_1, z_2, \cdots, z_n)$, and $z_1, z_2, \cdots, z_n \in C$.

Example 10.11 The set V of all $m \times n$ complex matrices is a vector space with respect to the usual operation of matrix addition and multiplication by a scalar.

Example 10.12 (*Function spaces*). Let X be an arbitrary non-empty set and let V be a vector space. Denote by Γ the space of all functions from X to V. Then Γ becomes a vector space if the addition and multiplication by scalars are defined in the following natural way:

$$(f + g)(x) = f(x) + g(x), \quad (\lambda f)(x) = \lambda f(x).$$

The zero vector in Γ is the function which assigns the zero vector of V to every element of X (e.g., Debnath & Mikusiński, op., cit., p. 3).

Remark In *elementary mathematics*, vectors are treated as *concrete* ones which contrast with vectors in the *abstract* vector space defined above. For example, we associate with every point $P = (x_1, x_2, \cdots, x_n)$ a unique vector that has the *origin* as initial point and P as end point. This is the vector $\overrightarrow{OP} = (x_1, x_2, \cdots, x_n)$. Such a vector is called the *position vector* of P. The position vector corresponds to a vector shown in Example 10.9. By contrast, we usually define vectors as objects having *magnitude* and *direction* such as displacements, forces etc. Such a vector is called *a* free *vector*.

The following two definitions are concerned with the properties of a function which maps a group G to another group H. These two definitions are necessary, for example, to characterize a property of the general linear group which will be shown later.

Definition 10.13 (*homomorphism*) If G and H are two groups with the binary operations $*$ and $\cdot$, respectively, a function $f : G \to H$ is a homomorphism if

$$f(a * b) = f(a) \cdot f(b), \quad \text{for all } a, b \in G. \tag{10.2}$$

Definition 10.14 (*isomorphism*) A homomorphism $f : G \to H$ is an *isomorphism* if f is both one-to-one and onto.

Historically, it is said that the oldest part of group theory forms a *permutation group*, and its study dates back to the early nineteenth century (e.g., Dixon & Mortimer [20]). Here, a *permutation* is defined as follows (e.g., Bowen & Wang, op. cit., p. 108):

Definition 10.15 (*permutation*) A *permutation* is a one-to-one function from $\{\alpha_1, \alpha_2, \cdots, \alpha_n\}$ to $\{\alpha_1, \alpha_2, \cdots, \alpha_n\}$. If σ is a permutation, it is customary to write

$$\sigma = \begin{pmatrix} \alpha_1 & \alpha_2 & \cdots \alpha_n \\ \sigma(\alpha_1) & \sigma(\alpha_2) & \cdots \sigma(\alpha_n) \end{pmatrix}. \tag{10.3}$$

Example 10.13 Let n be five, and $\{\alpha_1\alpha_2\alpha_3\alpha_4\alpha_5\}$ be $\{12345\}$. An example of a permutations is

$$\sigma = \begin{pmatrix} 12345 \\ 41352 \end{pmatrix}.$$

In this permutation, $\{\sigma(1)\sigma(2)\cdots\sigma(5)\}$ is $\{41352\}$, and integers 1, 2, 3, 4, 5 are transformed to 4, 1, 3, 5, 2, respectively.

Definition 10.16 (*symmetric group*) The set of all permutations of an arbitrary nonempty set Ω forms a group, under composition of mappings, called the *symmetric group* on Ω, often written as $Sym\ (\Omega)$, which is written as S_n when n is a positive integer and $\Omega = \{1, 2, \cdots, n\}$.

Remark 1 To verify that all permutations of an arbitrary nonempty set Ω forms a group, we may check whether they satisfy the three axioms on a group in Definition 10.7. Axiom (a) which is an *associative law* may be almost trivial for all permutations. Axiom (b) which means the existence of the *identity element* for them may also be trivial because in this case we may choose σ which maps every element of the set to itself. Axiom (c) which means the existence of the *inverse element* for them may also be trivial because we may consider σ which exchanges the two rows in (10.3).

Remark 2 In this sense, the term *permutation group* means a subgroup of the symmetric group. It is apparent from the definition of S_n that its order (i.e., the number of elements in this group) is $n!$.

Definition 10.17 (*general linear group*) The nonsingular matrices in $M_n(F)$ form a group, the *general linear group*, often denoted GL (n, F).

Remark Here, $M_n(F)$ is an abbreviated form of $M_{n,n}(F)$ which means the set of all n-by-n matrices over the field F. Above definition is due to Horn and Johnson ([26], p. 14). Stevens ([53], Table 6) calls the mathematical group structure of the interval scale the general linear group. However, a more appropriate name for the group structure of this scale seems to be an *affine (transformation) group* defined below, although a general linear group is considered as a subgroup of an affine group, as can be seen from the following definition (e.g., Bloom [7], p. 264):

Definition 10.18 (**affine group**). The *affine (transformation) group* G_A^n is de-fined as

$$\left\{\begin{bmatrix} 1 & 0 \\ \boldsymbol{\alpha} & \boldsymbol{A} \end{bmatrix} | \boldsymbol{\alpha} \epsilon R^n, \boldsymbol{A} = \boldsymbol{GL}(n, R)\right\}, \tag{10.4}$$

where $\boldsymbol{A} \in M_n(R)$, $\boldsymbol{\alpha} = [\alpha_1, \alpha_2, \cdots, \alpha_n]^t$, and $\mathbf{0} = [0, 0, \cdots, 0]$.

Using the matrix $\begin{bmatrix} 1 & \mathbf{0} \\ \boldsymbol{\alpha} & \boldsymbol{A} \end{bmatrix}$, we can define an affine transformation as

$$\begin{bmatrix} 1 \\ \boldsymbol{y} \end{bmatrix} = \begin{bmatrix} 1 & \mathbf{0} \\ \boldsymbol{\alpha} & \boldsymbol{A} \end{bmatrix} \begin{bmatrix} 1 \\ \boldsymbol{x} \end{bmatrix}. \tag{10.5}$$

From (10.5), we have

$$\boldsymbol{y} = \boldsymbol{A}\boldsymbol{x} + \boldsymbol{\alpha}, \tag{10.6}$$

which transforms $\boldsymbol{x}$ to $\boldsymbol{y}$.

Or, equivalently, (10.5) can be rewritten as

$$\begin{bmatrix} 1 \\ y_1 \\ \vdots \\ y_n \end{bmatrix} = \begin{bmatrix} 1 & 0 & \dots 0 \\ \alpha_1 & a_{11} & \dots a_{1n} \\ \vdots & \vdots & \dots \vdots \\ \alpha_n & a_{n1} & \cdots a_{nn} \end{bmatrix} \begin{bmatrix} 1 \\ x_1 \\ \vdots \\ x_n \end{bmatrix}. \tag{10.7}$$

It is not difficult to see that $\begin{bmatrix} 1 & \mathbf{0} \\ \boldsymbol{\alpha} & \boldsymbol{A} \end{bmatrix}$ is nonsingular using a property of partitioned matrix which will be introduced in the next chapter.

Remark Equations (10.6) or (10.7) is a direct extension of the function described as the mathematical group of a general linear group, $x' = ax + b$, of the interval scale, which is stated as *general linear group* in Table 2.1 of Stevens ([52, 53]) in the next section.

A special case of the affine transformation described by (10.6), that is,

$$\boldsymbol{y} = \boldsymbol{A}\boldsymbol{x}, \tag{10.8}$$

is called a *linear transformation* T if it satisfies the following two conditions (e.g., Boom [7], p. 210; Lancaster & Tismenetsky, p. 117):

Definition 10.19 (*linear transformation*) Let V and V' be vector spaces over the field F. A mapping $T : V \to V'$ is called a linear transformation (abbreviated *L.T.*) if it preserves addition and scalar multiplication, that is, T preserves *addition* if

(1) $(\boldsymbol{u} + \boldsymbol{v})T = \boldsymbol{u}T + \boldsymbol{v}T$, for all $\boldsymbol{u}, \boldsymbol{v} \in V$.

Similarly, T preserves *scalar multiplication* if the image of cV is cv' for all $c \in F$ and all $v \in V$; that is, if

(2) $(c\boldsymbol{v})T = c(\boldsymbol{v}T)$, for all $\boldsymbol{v} \in V$ and $c \in F$.

In general, the transformation matrix $\boldsymbol{A}$ in linear transformation is *rectangular*.

$$T_A(\boldsymbol{x}) = \boldsymbol{A}\boldsymbol{x}, \quad \text{where } \boldsymbol{A} \in M_{m,n}(F). \tag{10.9}$$

However, in the *similarity transformations* and the similarity (transformation) group which we shall discuss below, we use a *special square matrix* $\boldsymbol{A}$ as the transformation matrix. This means that we consider a special case of the affine transformations, that is, (10.7). In addition, we consider a more special case in which this transformation is just equal to $\lambda\boldsymbol{x}$. As a result, we have

$$\boldsymbol{A}\boldsymbol{x} = \lambda\boldsymbol{x}. \tag{10.10}$$

This equation is well known as an *eigenvalue problem* which will be discussed in detail in Chap. 11. It is also well known that (10.9) can be rewritten as

$$\boldsymbol{A}\boldsymbol{X} = \boldsymbol{X} diag(\lambda_i) \boldsymbol{X}^{-1}\boldsymbol{A}\boldsymbol{X} = diag(\lambda_i), \tag{10.11}$$

If the eigenvalues of $\boldsymbol{A}$ are *distinct* (e.g., Wilkinson [58]). In this case we say that $\boldsymbol{A}$ and $\boldsymbol{X}^{-1}\boldsymbol{A}\boldsymbol{X}$ are *similar*. In other words, a *similarity transformation* reduces a matrix $\boldsymbol{A}$ to diagonal form, which is the theorem shown later in Chap. 11.

Definition 10.20 (*similarity group*) The similarity transformations in an n-dimensional Euclidean space form a *similarity (transformation) group*, G_E^n.

Remark The above definition is due to Kawata ([29], p. 64). According to Bloom [7], in the real Euclidean case, affine transformations are precisely those mappings which preserve *ratios of distances* along lines, which is the very property of the ratio scale in psychometrics (Stevens [52]). From the above discussion, we see that $G_E^n \subset GL(n,k) \subset G_A^n$.

Finally, we shall discuss the *isotonic group* which Stevens [53] refers to as the mathematical structure of ordinal scale in psychometrics, which will be discussed in the next section. This notion has rarely been used in group theory, although the notion of *isotonic transformation* is fairly well known in relation to the notion of *partial order*.

Before proceeding, it will be appropriate to introduce the notions of *relation* and *order relation* related to partial order. For example, Mahadev and Peled [39, p. 436] give the lists of these relations. The list of order relation might help us to distinguish between the partial order and the other order relations similar to it:

A. A list of relations

Let the operation $\neg$ denote negation and A be a set. A relation $R \subseteq A \times A$ is said to be

(1) *reflective* if $\neg aRa$ for each $a \in A$;
(2) *symmetric* if $aRb \Rightarrow bRa$ for all all $a, b \in A$;
(3) *asymmetric* if $aRb \Rightarrow \neg bRa$ for all $a, b \in A$;
(4) *antisymmetric* if for all $a \neq b$, $aRb \Rightarrow \neg bRa$;
(5) *transitive* if $(aRb, bRc) \Rightarrow aRc$ for all $a, b, c \in A$;
(6) *negatively transitive* if the complement of R (i.e., $(A \times A) - R$) is transitive;
(7) *complete* if for all distinct $a, b \in A$, aRb or bRa;

B. A list of order relations

Let R be a binary relation on A, i.e., $R \subseteq A \times A$. Then we have

(1) R is a *quasi order* if it is reflective and transitive;
(2) R is a *simple order* if it is transitive, antisymmetric and strongly complete;
(3) R is a *strict simple order* if it is transitive, antisymmetric and complete;
(4) R is a *strict weak order* if it is asymmetric and negatively transitive;
(5) R is a *partial order* if it is reflective, transitive and antisymmetric;
(6) R is a *strict partial order* if it is transitive and asymmetric;
(7) R is an *equivalence relation* if it is reflective, symmetric and transitive;

Remark 1 Partial order is sometimes called *weak partial order* or *natural partial order* (e.g., Dean & Keller [18]).

Now, we give the definition of isotonic transformation in the following, for example, according to Kim and Kozhukhov [30]:

Definition 10.21 (*isotonic transformation*) If X is a *partially ordered set*, then we denote by $O(X)$ the semigroup of all order-preserving transformations $\alpha : X \to X$ of X, that is, $x \leq y \Rightarrow x\alpha \leq y\alpha$ for any $x, y \in X$. We say that these transformations are *isotonic*.

Remark 2 In the above definition, Kim et al. use a special symbol in which a west-south line segment instead of a horizontal line segment is attached just under the symbol $<$, which is different from the symbol of *inequality*, that is, $\leq$. By contrast, for example, Clifford [16] uses this symbol, that is, the inequality. However, Molchanov [43] uses the same symbol as that of Kim et al. in defining *quasi order*.

Example 10.14 The equality relation is a trivial example of partial order.

Example 10.15 The inequality $\leq$ is another example of partial order (e.g., Bowen and Wang [9, p. 17]). It will not be difficult to verify, for example, that the inequality satisfies the three axioms of partial order, i.e., reflexivity, transitivity, and antisymmetry.

Remark 3 It should be noticed that the *strict inequality* $<$ does not satisfy the reflexivity and belongs to a *strict partial order*.

Example 10.16 The additive groups of integers, Z, rationals Q and real numbers R are all ordered groups under the natural ordering (e.g., Glass [22, p. 3]). The axiomatic definition of this group is given by the next definition, Definition 10.22.

Remark 4 We sometimes call an order relation $\geq$ the *dual* of $\leq$. As a primary notion of an ordered set A, if $x \leq a$ for every element x in a subset X, then we say that a is the *upper bound* of X. And if there exists such an a, then we say that X is *bounded from above*. On the other hand, we say that a is the *lower bound* of X if $x \geq a$ using the symbol of the dual of $\leq$. And if there exists such an a, then we say that X is *bounded from below* (e.g., Japanese Mathematical Society [28, p. 516]).

It should be noted that from Definition 10.21 we see that such a transformation forms a *semigroup*. At this point some natural questions arise concerning the isotonic transformation: Is the definition of *isotonic group* which Stevens refers to in the literature [52, 53] wrong? Are there any other definitions in which *partial order* forms a group? The answer to the first question is not wrong, but the answer to the second question is right.

As regards the first question, however, Stevens [[53], p. 24] uses verbose and equivocal expressions in explaining Fig. 10.1 in that he says that "The *increasing monotonic function*, which may have discontinuities of the type shown, belongs to the isotonic group and is a transformation applicable to ordinal scales". The reason for this is that a function is said to be *monotone* if it is either *decreasing* or *increasing* (e.g., Kubrusly [34, p. 10]).

As for the second question, there have been a few researchers who discuss the *group* of partial order (Clifford [16]; Dean & Keller [18]; Glass [22]). For example, according to Clifford:

Definition 10.22 (*partially ordered Abelian group*) A *partially ordered Abelian group* satisfies the following axioms:

I. G is an Abelian group under +.
II. G is a directed partially ordered set under $\leq$, that is:

(1) $a \leq a$,
(2) $a \leq b, b \leq a$ imply $a = b$,
(3) $a \leq b, b \leq c$ imply $a \leq c$,
(4) Given a and b, there exists c in G such that $a \leq c, b \leq c$.

III. $a \leq b$ implies $a + c \leq b + c$ for all c in G.

By contrast, there have recently been some researchers who discuss the semi*group* of partial order as Kim and Kozhukhov [30] whose definition of the semigroup was cited above (e.g., Mitsch [41]; Molchanov [43]; Nambooripad [45]). In particular, it is interesting to note that Molchanov gives a survey of the structure of semigroups of mappings and interrelations between properties of graphs and corresponding properties of semigroups associated with the graphs.

10.3 Fundamental Sales in Psychometrics

The problem of measurement is fundamental in any discipline of empirical sciences. Moreover, to measure some basic events to be observed in the discipline we need a scale on which these events are laid out. In general, measurement is defined as follows (e.g., Stevens [52, p. 677]):

Definition 10.23 (*measurement*) Measurement is defined as the *assignment of numerals to objects or events according to rules.*

In physics, seven basic scales, which are currently called (SI) *base units*, are time (second), length (meter), mass (kilogram), electric current (ampere), thermodynamic temperature (kelvin), amount of substance (mole), and luminous intensity (candela), and 22 *derived units* (Bureau international des poids et mesures [11]).

In psychology, however, it is generally very difficult to devise scales for perceptual events. For example, suppose we administer a questionnaire which is composed of a body of people's attitude items toward nuclear power using a typical five-level Likert item, i.e., strongly disagree, disagree, neither agree nor disagree, agree, strongly agree. What rules exist according to which we assign numerals to these levels of the Likert item?

According to Stevens ([52], p. 677), a committee of the British Association for the Advancement of Science started to debate the problem of measurement in 1932. This committee was composed of researchers to represent Section A (Mathematical and Physical Sciences) and Section J (Psychology), and was instructed to consider and report upon the possibility of "quantitative estimates of sensory events".

The conclusion reported by the committee after 8 years was that one must make explicit

(1) The various rules for the assignment of numerals,
(2) The mathematical properties (or group structure) of the resulting scales, and

(3) The statistical operations applicable to measurements made with each type of scale, in order to construct a sensory scale.

Moreover, it was contended that such a scale is possible only because there is a certain *isomorphism* between (a) what we can do with the aspects of objects or events, and (b) properties of a numerical series. Here, in dealing with the aspects of objects or events we administer empirical operations (1) for determining equality (classifying), (2) for rank-ordering, and (3) for determining when differences and when ratios between aspects of objects are equal. These operations are limited by the nature of the thing being scaled and by our choice of procedures.

Table 10.1 is a summary of the four scales (Stevens [53]), which is a revised version of the table which appeared in Stevens ([52], p. 678). At this point the reader might have realized that it is necessary to introduce the group concepts in set theory which have been discussed in the previous section in order to define the fundamental scales in psychology precisely.

Table 10.1 A summary table of scales of measurement

Scale	Basic empirical operations	Mathematical group structure	Permissible statistics (invariantive)	Typical examples
Nominal	Determination of equality	Permutation group x'=f(x) [f(x) means any one-to-one substitution]	Number of cases Mode Contingency corre-lation	"Numbering" of football players Assignment of type or model numbers to classes
Ordinal	Determination of Greater or less	Isotonic group x'=f(x) [f(x) means any increasing mono-tonic function]	Median Percentiles **Order correlation (type O)**	Hardness of mine-rals Quality of leather, Lumber, wool, etc. Pleasantness of odors
Interval	Determination of Equality of inter-vals or differences	General linear group x'=ax+b	Mean Standard deviation **Order correlation (type I)** Product-moment correlation	Temperature (Fahrenheit and centigrade) Energy Calendar dates "Standard scores" on achievement test (?)
Ratio	Determination of equality or ratios	Similarity group x'=ax	**Geometric mean** Coefficient of variation **Decibel transforma-tions**	Length, weight, density, resistance, etc. Pitch scale (mels) Loudness scale (sones)

(This table was reproduced from Table 6 of Stevens [53]. We denote words and phrases, which had been added to the contents of Table 1 of Stevens [52] by Stevens, by boldface letters.)

The followings are the concise but loose definitions of these four scales in psychometrics compared with the contents of Table 10.1.

Definition 10.24 (*nominal scale*) Nominal scale is a scale of which numerals assigned to the words or letters are used only as labels or type numbers.

Remark Stevens ([52], p. 678) distinguishes two types of nominal assignments, i.e., (a) the 'numbering' of football players for the identification of the individuals, and (b) the 'numbering' of types or classes, where each member of a class is assigned the same numeral.

Definition 10.25 (*ordinal scale*) Ordinal scale is a scale of which numerals assigned to the words or letters have only the order information.

Remark According to Stevens (op. cit., p. 679), this scale has the structure of what may be called the isotonic or order preserving group. He also says that most of the scales used widely and effectively by psychologists are *ordinal scales.*

Definition 10.26 (*interval scale*) Interval scale is a scale in which differences in the adjacent numerals assigned to the words or letters are all equal and the zero point on it is a matter of convenience.

Remark 1 Stevens (op. cit., p. 679) says that most psychological measurement aspires to create interval scales, and it sometimes succeeds. He also says that the problem usually is to devise operations for equalizing the units of the scales – a problem not always easy to solve, but one for which there are several possible modes of attack.

Definition 10.27 (*ratio scale*) Ratio scale is a scale in which differences in the adjacent numerals assigned to the words or letters are all equal and the absolute zero exists.

Remark 2 Stevens (op. cit., p. 679) says that ratio scales are *those most commonly encountered in physics* and are possible only when there exist operations for determining all four relations: equality, rank-order, equality of intervals, and equality of ratios.

Remark Stevens ([52, 53]) considered the fundamental scales and their properties in psychometrics in the middle of 1940s when *unidimensional scaling* still dominated psychometrics. However, the 1940s were also the dawn of *multidimensional scaling* (abbreviated as MDS) in psychometrics. In fact, as discussed in Chap. 7, Richardson [46] formed a basis of MDS and Torgerson ([54, 55]) established MDS (to be precise, classical MDS). In 1993, Chino and Shiraiwa proposed one of the *asymmetric MDS* called HFM in which the Euclidean space in MDS was extended to Hilbert space, as shown in Chap. 4. On the one hand, the similarity transformation in Euclidean space forms a *similarity group*, G_E^n, as shown in Definition 10.20. On the other hand, *unitary similarity transformation* in Hilbert space forms a *unitary group*, $U(n)$, which is an extension of the similarity group to complex space. In this sense, it will be necessary and appropriate to add a unitary group to the keyword "similarity group" in Table 10.1 (which is shown in Table 6 of Stevens [53]).

10.4 Complex Number and Its Basic Definitions and Theorems

A *familiar definition* of *complex number* is expressed in the form

$$z = x + iy, \tag{10.12}$$

where x is the *real part* of z, denoted as $\mathrm{Re}(z)$, and y is the *imaginary part* of z, denoted as $\mathrm{Im}(z)$, while the imaginary unit number i is defined, in Euler's notation, as $i^2 = -1$. We shall call $\overline{z}(= x - iy)$ the *(complex) conjugate* of z. It is evident that $z\overline{z}\left(= x^2 + y^2\right)$ is a real number, and z is a real number if $z = \overline{z}$.

Although there exist some variations in the *formal definition* of the complex number (e.g., Bak & Newman [5]; Lang [36]), we shall adopt Bak and Newman's definition.

Definition 10.28 (*complex field*) The *complex field C* is the set of ordered pairs of real numbers (a, b) with addition and multiplication defined by

$$(a, b) + (c, d) = (a + c, b + d),$$

$$(a, b)(c, d) = (ac - bd, ad + bc).$$

Remark The ordered pair of real numbers (a,b) means the complex number $z = a + ib$, and is an element of the complex field defined above. In other words, the complex numbers form a field defined in Definition 10.11.

It is apparent, from the formal definition of the complex field, that a complex number $z = x + iy$ can be interpreted as a two-dimensional vector (x,y) in the Cartesian plane. The real number x is associated with a point on the x-axis called the *real axis*, while the purely imaginary number iy corresponds to a point on the y-axis called as the *imaginary axis*. Such a plane is called the *complex plane*.

Complex numbers can be represented as follows using *polar coordinates* (r,θ) :

$$x = r\cos\theta, y = r\sin\theta, (r \geq 0). \tag{10.13}$$

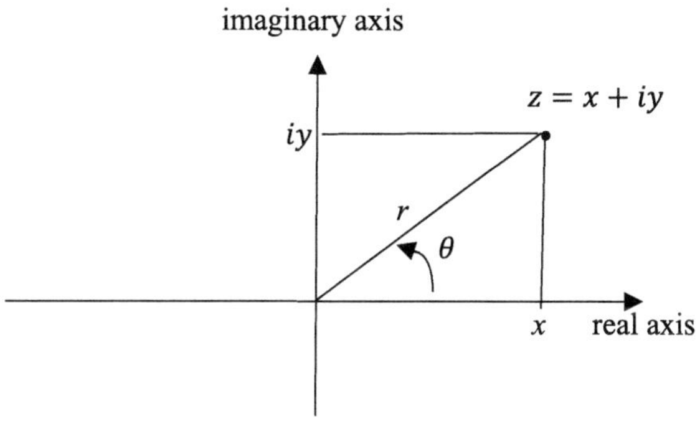

Fig. 10.2 Complex number z represented by polar coordinates in the complex plane.

Definition 10.29 (*modulus and argument*) The *modulus* or *absolute value* of the complex number z, $|z|$, is equal to $\sqrt{x^2+y^2}$, while the *argument* of z for $z \neq 0$, *Arg* z is the *angle* which the vector originating from 0 to z makes with the positive real x-axis.

Example 10.17 Let z be $1+i\sqrt{3}$. Then, the absolute value of z, $|z|$, is 2, and $Arg z$ is 60° which equals about 1.0472^r.

Remark In Fig. 10.2, the modulus and argument are denoted by r and θ of the polar coordinates, respectively. Apart from complex analysis, these are usually called the *radius* (or *radial coordinate*) and *polar angle* (or *angular coordinate*) in polar coordinate system, respectively.

It is apparent from (10.12) that a complex number $z = x + iy$ can be rewritten as

$$z = r(\cos\theta + i\sin\theta). \tag{10.14}$$

Moreover, if we use *Euler's formula*, that is,

$$e^{i\theta} = \cos\theta + i\sin\theta, \tag{10.15}$$

we have

$$z = re^{i\theta}, \tag{10.16}$$

which is called the *polar form* of the complex number $x + iy$.

Polar form has important properties such as given in the following theorems:

Theorem 10.1 *Let a, b be two real numbers. Then,*

$$e^{ia+ib} = e^{ia}e^{ib}. \tag{10.17}$$

For the proof of this theorem, see, for example, Lang ([29], p. 9).

Theorem 10.2 *Let u, v be two complex numbers. Then,*

$$e^{u+v} = e^u e^v. \tag{10.18}$$

For the proof of this theorem, see, for example, Lang (op. cit., p. 10).

We can also prove De Moivre's formula, that is,

$$(\cos x + i \sin x)^n = \cos nx + i \sin nx, \tag{10.19}$$

using Euler's formula.

In the next section, we shall introduce *Hilbert space*. In defining it, we need to consider a convergence problem of a sequence of complex numbers. Therefore, we shall present here two definitions and a proposition concerning this problem, following, for example, Bak and Newman ([5], p. 9):

Definition 10.30 (*convergence of a sequence*) The sequence of complex numbers $z_1, z_2, z_3, \cdots$ converges to z if the sequence of real numbers $|z_n - z|$ converges to 0. That is, $z_n \to z$ if $|z_n - z| \to 0$.

Definition 10.31 (*Cauchy sequence*) $\{z_n\}$ is called a *Cauchy sequence* if for each $\varepsilon > 0$ there exists an integer N such that $n, m > N$ implies $|z_n - z_m| < \varepsilon$.

Proposition 10.1 *$\{z_n\}$ converges if and only if $\{z_n\}$ is a* Cauchy sequence.

For the proof of this proposition, see, for example, Bak and Newman (op. cit., p. 10).

In Chap. 9, we have proposed a *complex difference equation model* which is a set of certain functions of complex numbers.

Here, we shall introduce a *polynomial function P of a complex number z*

$$P(z) = \alpha_n z^n + \alpha_{k-1} z^{k-1} + \cdots + \alpha_1 z + \alpha_0, \tag{10.20}$$

where α_k are (complex) constants.

Remark $P(z)$ defined above is a general form of each of the p elements $z_{j,n}^{(l)} l = 1, \cdots, p$, in (9.1). We have shown there several examples of (10.20) in relation to the complex difference equation model, like (9.9), (9.10), and (9.11). However, note that these examples include more than one complex number for each of the polynomials. For example, (9.9) includes two complex numbers, $z_{j,n}$ and $z_{k,n}$. The reason why it includes two complex numbers is that this model is concerned with *dyad*. Similarly, (9.10) has three complex numbers for each of the polynomials, because it is concerned with *triad*.

In considering such a complex-valued function, it is important to examine whether it is differentiable at z.

Definition 10.32 (*differentiability*) A complex-valued function f, defined in a neighborhood of z, is said to be *differentiable* at z if

$$\lim_{h \to 0} \frac{f(z+h) - f(z)}{h}$$

exists. Here, the limit is denoted $f'(z)$.

Remark Apparently, this definition is an extension of the differentiability of a real-valued function to the complex case. A complex-valued function in which the limit $f'(z)$ exists is called *holomorphic* (or *analytic*) (e.g., Rudin [48], p. 197).

Counterexample. As, for example, Bak and Newman (op. cit., pp. 21–1.2) show a complex function as a simple counterexample of a holomorphic function. In fact, the function $f(z) = \overline{z}$ is not differentiable at any point z since

$$(f(z+h) - f(z))/h = \overline{h}/h,$$

is equal to $+1$ if h is *real* and -1 if h is *purely imaginary*.

Finally, we shall define the (nth) *root of unity*, which is used in many branches of mathematics and applied mathematics such as in neural network theory and machine learning as will be introduced partly in the next chapter.

Definition 10.33 (*root of unity*) An (nth) root of unity is a number z satisfying the equation $z^n = 1$, where n is a positive integer.

Remark It is well known that the nth roots of unity are expressed as

$$\omega_k = e^{\left(\frac{2k\pi}{n}\right)i} = \cos\frac{2k\pi}{n} + i\sin\frac{2k\pi}{n}, k = 0, 1, \cdots, n-1, \tag{10.21}$$

Of course, as a special case when $k = 0$, $\omega_k = 1$. In *complex-valued neural networks* (CVNN), ω_k is denoted as $CSIGN\ (z)$ or more explicitly as $CSIGN_k(z)$ (e.g., Aizenberg et al. [1]; Aizenberg & Aizenberg [2]; Hirose [25]; Jankowski et al. [27]; Müezzinoğlu et al. [44]).

10.5 From Dot Product to Hilbert Space

In this book we shall deal with some spaces as hypothetical spaces in which we embed objects (or elements) with asymmetric interactions among them. Candidates for these spaces are Euclidean spaces, Minkowski spaces, Riemannian spaces, and Hilbert spaces, etc. Therefore, in this chapter we shall examine the properties of these spaces using various basic concepts relating to these properties. As defined in Sect. 10.2, elements of the *(abstract) vector space* have neither the notion of *length* of an element of a set nor that of *angle* between two elements in spite of the fact that these are *vectors*. However, we can generate various spaces equipped with the notions of length and/or angle by defining various concepts on the abstract vector space. Among these concepts, we shall first introduce the notions of *bilinear functional* (or *sesquilinear functional*) and *quadratic form* defined on a vector space. These two concepts do not require the structure of an *inner product space* (e.g., Debnath & Mikusiński [19], pp. 87, 88). We shall define the notions of a *dot product*, an *inner product* and a *norm* related to it soon:

Definition 10.34 (*bilinear functional*) By a bilinear functional ϕ on a *complex vector space* V over the filed F we mean a mapping $\phi{:}V \times V \to F$ such that

$$(a)\phi(\alpha x_1 + \beta x_2, y) = \alpha\, \phi(x_1, y) + \beta\, \phi(x_2, y),$$

$$(b)\ \phi(x, \alpha y_1 + \beta y_2) = \overline{\alpha}\, \phi(x, y_1) + \overline{\beta}\, \phi(x, y_2),$$

where $F{=}C$, α and β are contained in C, and $\overline{\alpha}$ and $\overline{\beta}$ are *complex conjugates* of α and β, respectively.

Remark 1 Debnath & Mikusiński (op. cit.) use the notation of the vector space as E instead of V in the above definition. The important point of this definition is that a bilinear functional is *linear* with respect to the first variable (or argument) and *antilinear* with respect to the second variable (or argument). However, it is not essential which of these variables (or arguments) is linear and which of these is antilinear. In any case, considering that a bilinear functional allows one of the arguments to be "twisted" in a semi-linear manner, the name "sesquilinear functional" might be thought of as more appropriate than the name "bilinear functional". In the special case when $\phi : V \times V \to R$, that is, the vector space V is *real* and $F{=}R$, then $\overline{\alpha} = \alpha$ and $\overline{\beta} = \beta$. As a result, the bilinear functional is *linear* with respect to *both variables (or arguments).*

Remark 2 Bilinear (or sesquilinear) functional is sometimes called *bilinear function* (or *sesquilinear function*) or *bilinear form* (or *sesquilinear form*). For example, Lancaster and Tismenetski ([35, p. 202]) call a bilinear functional *bilinear function,* and when it is expressed in terms of coordinates, they call it *bilinear form.* The term, *functional,* is customarily used in *functional analysis* (e.g., Maddox [38]). As for *sesquilinear form,* see, for example, Ball [4]. Horn and Johnson ([26], p. 169) distinguish between *real bilinear form* and *complex form,* saying that the latter is no longer bilinear.

The following are examples of bilinear functional shown in Debnath & Mikusiński (op. cit., p. 145):

Example 10.18 *Inner product* is a bilinear functional.

Note. Inner product will be defined soon in Definition 10.38. It should be noticed that in general inner product is defined on a complex vector space. The reader will be interested in the difference between a bilinear functional and an inner product.

Example 10.19 Let A and B be operators on an inner product space V. Then, $(Ax, y), (x, By), and\, (Ax, By)$ are bilinear functionals on V.

Example 10.20 Let f and g be linear functionals on a vector space V. Then, $\emptyset(x, y) = f(x)\overline{g(y)}$ is a bilinear functional on V.

Remark For further details of the bilinear functional shown in Example 10.20, see Definition 4.3.2 of Debnath and Mikusiński (op. cit., p. 145).

Next, we shall define a *quadratic form* according to Debnath and Mikusiński (op. cit., pp. 145, 146).

Definition 10.35 (*quadratic form*) Let ϕ be a bilinear functional on a vector space V. The functional $\Phi : V \to C$ defined by

$$\Phi(x) = \phi(x, x) \tag{10.22}$$

is called the *quadratic form* associated with ϕ.

At this point we shall digress for a while from the subjects of sesquilinear form and quadratic form, and define the notions of *dot product*, *inner product*, and *normed space*, according to Bloom ([7], pp. 173, 174).

Definition 10.36 (*dot product*) Let F be a field. If $\boldsymbol{u} = (x_1, x_2, \cdots, x_n)$ and $\boldsymbol{v} = (y_1, y_2, \cdots, y_n)$ are any elements of the vector space F, a *dot product* of $\boldsymbol{u}$ and $\boldsymbol{v}$ is defined as

$$\boldsymbol{u} \cdot \boldsymbol{v} = \sum_{i=1}^{n} x_i y_i \tag{10.23}$$

Theorem 10.3 *If F is any field, the following three statements hold for all $c \in F$ and all $\boldsymbol{u}, \boldsymbol{v}, \boldsymbol{w} \in F_n$:*

(a) $\boldsymbol{u} \cdot \boldsymbol{v} = \boldsymbol{v} \cdot \boldsymbol{u}$
(b) $\boldsymbol{u} \cdot (\boldsymbol{v} + \boldsymbol{w}) = (\boldsymbol{u} \cdot \boldsymbol{v}) + (\boldsymbol{u} \cdot \boldsymbol{w})$
(c) $c(\boldsymbol{u} \cdot \boldsymbol{v}) = (c\boldsymbol{u}) \cdot \boldsymbol{v} = \boldsymbol{u} \cdot (c\boldsymbol{v})$.

This theorem is an easy consequence of Definition 10.36 *shown above.*

Remark The dot product is defined in terms of coordinates and has some *algebraic uses*. However, it does not imply the notion of *length*. By contrast, for the dot product in R^n, which is called the *real dot product*, we need two further properties stated in the following theorem:

Theorem 10.4 *If $\boldsymbol{u}$ is any vector in R^n, then*

(a) $\boldsymbol{u} \cdot \boldsymbol{u} = |\boldsymbol{u}|^2$
(b) $\boldsymbol{u} \cdot \boldsymbol{u}=0$ if and only if $\boldsymbol{u} = 0$,

where

$$|\boldsymbol{u}| = \left(\sum_{i=1}^{n} x_i^2\right)^{1/2}, \tag{10.24}$$

is the length of $\boldsymbol{u}$.

Proof of this theorem may be trivial.

The notion of the real dot product can be generalized as follows (e.g., Bloom, op. cit., p. 174):

Definition 10.37 (*real inner product space*) Let V be any vector space over R. If there exists a function $f : V \times V \to R$ such that the laws

(a) $f(\boldsymbol{u}, \boldsymbol{v}) = f(\boldsymbol{v}, \boldsymbol{u})$;
(b) $f(\boldsymbol{u}, \boldsymbol{v} + \boldsymbol{w}) = f(\boldsymbol{u}, \boldsymbol{v}) + f(\boldsymbol{u}, \boldsymbol{w})$;
(c) $cf(\boldsymbol{u}, \boldsymbol{v}) = f(c\boldsymbol{u}, \boldsymbol{v}) = f(\boldsymbol{u}, c\boldsymbol{v})$;
(d) $f(\boldsymbol{u}, \boldsymbol{u}) \geq 0$;
(e) $f(\boldsymbol{u}, \boldsymbol{u}) = 0 \to \boldsymbol{u} = 0$;

hold for all $\boldsymbol{u}, \boldsymbol{v}, \boldsymbol{w} \in V$ and $c \in R$, then V is called a *real Euclidean space* (or *real inner product space*). In this case, we shall use E instead of V and the function f is called a *real inner product* on E.

The real inner product space can be generalized to the *complex inner product space*. To do so, we first define an inner product on a complex vector space V. In the following theorem we shall denote the inner product on the complex space simply as *inner product* according to Bowen & Wang ([10], p. 58).

Definition 10.38 (*inner product*) An *inner product* on a *complex vector space* V is a function $f : V \times V \to C$ with the following properties:

(1) $f(\boldsymbol{u}, \boldsymbol{v}) = \overline{f(\boldsymbol{v}, \boldsymbol{u})}$;
(2) $\lambda f(\boldsymbol{u}, \boldsymbol{v}) = f(\lambda \boldsymbol{u}, \boldsymbol{v})$;
(3) $f(\boldsymbol{u} + \boldsymbol{w}, \boldsymbol{v}) = f(\boldsymbol{u}, \boldsymbol{v}) + f(\boldsymbol{w}, \boldsymbol{v})$;
(4) $f(\boldsymbol{u}, \boldsymbol{u}) \geq 0 \, \text{and} \, f(\boldsymbol{u}, \boldsymbol{u}) = 0 \Leftrightarrow \boldsymbol{u} = 0$;

for all $\boldsymbol{u}, \boldsymbol{v}, \boldsymbol{w} \in V$ and $\lambda \in C$.

Remark It should be noted that according to the definition, the inner product of two vectors is generally a *complex number*.

Properties (2) and (3) of the above definition require that f be linear in its first argument, that is,

$$f(\lambda \boldsymbol{u} + \mu \boldsymbol{v}, \boldsymbol{w}) = \lambda f(\boldsymbol{u}, \boldsymbol{w}) + \mu f(\boldsymbol{v}, \boldsymbol{w}) \tag{10.25}$$

for all $\boldsymbol{u}, \boldsymbol{v}, \boldsymbol{w} \in V$ and $\lambda, \mu \in C$ (e.g., Bowen & Wang, op. cit., p. 58).

Property (1) and the linearity implied by Properties (2) and (3) ensure that f is conjugate linear in its second argument, that is,

$$f(\boldsymbol{u}, \lambda \boldsymbol{v} + \mu \boldsymbol{w}) = \overline{\lambda} f(\boldsymbol{u}, \boldsymbol{v}) + \overline{\mu} f(\boldsymbol{u}, \boldsymbol{w}) \tag{10.26}$$

for all $\boldsymbol{u}, \boldsymbol{v}, \boldsymbol{w} \in V$ and $\lambda, \mu \in C$ (e.g., Bowen & Wang, op. cit., p. 58; Debnath and Mikusiński [19], p. 88).

Remembering the two properties of Definition 10.34, we see that inner product is a bilinear functional (e.g., Debnath & Mikusiński, op. cit., p. 144).

If a function $f : V \times V \to C$ satisfies the conditions (1), (2), and (3) in Definition 10.38, we say that f is a *Hermitian function* or *Hermitian form*. Cristescu ([17], p. 48) calls it a *hermitian functional*. A hermitian functional is said to be *positive-definite* if $f(\boldsymbol{u}, \boldsymbol{u}) \geq 0$ for any $\boldsymbol{u}$ in V. In this case, $\sqrt{f(\boldsymbol{u}, \boldsymbol{u})}$ generates a *seminorm* on V. If $f(\boldsymbol{u}, \boldsymbol{u}) > 0$ for any $\boldsymbol{u} \neq 0$, f is called a scalar product (or inner product). In this case, $\sqrt{f(\boldsymbol{u}, \boldsymbol{u})}$ generates a *norm* on V.

The following definition of *inner product space* by Debnath and Mikusiński (op. cit., pp. 88, 89) includes an equivalent definition of an inner product to that presented above.

Definition 10.39 (*inner product space*) Let V be a complex vector space. A mapping $(\cdot, \cdot) : V \times V \to C$ is called an inner product in V if for any $x, y, z \in V$ and $\alpha, \beta \in C$ the following conditions are satisfied:

(a) $(x, y) = \overline{(x, y)}$;
(b) $(\alpha x + \beta y, z) = \alpha(x, z) + \beta(y, z)$;
(c) $(x, x) \geq 0$, and $(x, x) = 0$ implies $x = 0$.

A vector space with an inner product is called an *inner product space* or a *pre-Hilbert space* or a *unitary space*.

Remark There are many notations for the inner product. For example, Bowen and Wang ([10], p. 58) choose the "dot product" for it and write

$$f(\boldsymbol{u}, \boldsymbol{v}) = \boldsymbol{u} \cdot \boldsymbol{v}. \tag{10.27}$$

In the above definition, Debnath and Mikusiński (op. cit., p. 88) use the symbol, $\langle \cdot, \cdot \rangle$, instead of $(\cdot, \cdot)$. We will use $(\cdot, \cdot)$. Whether we shall use boldface notation or lightface notation for the elements in the vector space is not a matter of importance.

Example 10.21 The simplest example is the space of complex numbers C. The inner product is defined by $(u, v) = u\overline{v}$.

Example 10.22 Let us suppose that the ordered n-tuples $\boldsymbol{x} = (x_1, x_2, \cdots, x_n)$ and $\boldsymbol{y} = (y_1, y_2, \cdots, y_n)$ of real numbers are vectors in R^n. Then, the inner product is defined by

$$(\boldsymbol{x}, \boldsymbol{y}) = \sum_{i=1}^{n} x_i y_i. \tag{10.28}$$

Example 10.23 Let us suppose that the ordered n-tuples $\boldsymbol{u} = (u_1, u_2, \cdots, u_n)$ and $\boldsymbol{v} = (v_1, v_2, \cdots, v_n)$ of complex numbers are vectors in C^n. Then, the inner product is defined by

$$(\boldsymbol{u}, \boldsymbol{v}) = \sum_{i=1}^{n} u_i \overline{v}_i, \tag{10.29}$$

Example 10.24 The space $C([a, b])$ of all continuous complex valued functions on the interval $[a, b]$, with the inner product

$$(f, g) = \int_a^b f(x)\overline{g(x)}dx, \tag{10.30}$$

is an inner product space.

Once an inner product is introduced in an abstract vector space, we can induce the notion of length by defining

$$\|\boldsymbol{u}\| = \sqrt{(\boldsymbol{u}, \boldsymbol{u})}. \tag{10.31}$$

The quantity $\|\boldsymbol{u}\|$ is called a *norm*. The concept of norm in a vector space is an abstract generalization of the *length* of a vector in a Euclidean space, and a norm is generally defined on an abstract vector space as follows (e.g., Debnath and Mikusiński, op. cit., pp. 8, 9). At this point, we shall temporarily use lightface notation for a vector in this abstract space:

Definition 10.40 (*norm*) A real valued function $\|\cdot\|$ defined on a vector space V is called a (vector) *norm* if it satisfies the following axioms:

(1) $\|x\| \geq 0$ for all $x \in V$ and $\|x\| = 0$ if and only if $x = 0$;
(2) $\|\lambda x\| = |\lambda|\|x\|$ for all $x \in V$ and $\lambda \in C$;
(3) $\|x + y\| \leq \|x\| + \|y\|$ for all $x, y \in V$.

Let x be a typical vector of C^n (or R^n), $x = (x_1, x_2, \cdots, x_n)$. Then, we have the following:

Example 10.25 The function defined by

$$\|x\|_1 = \sum_{i}^{n} |x_i| \tag{10.32}$$

is a norm on C^n (or R^n).

Example 10.26 The function defined by

$$\|x\|_2 = \left(\sum_{i=1}^{n} |x_i|^2 \right)^{1/2}, \tag{10.33}$$

is a norm on C^n (or R^n), which is called the *Euclidean norm*.

Example 10.27 The following norm

$$\|x\|_p = \left(\sum_{i=1}^{n} |x_i|^p \right)^{1/p}, p \geq 1, \tag{10.34}$$

is called the *H ölder norm* on C^n (or R^n) (e.g., Lancaster & Tismenetsky [35], p. 351) or the L_p *norm* (e.g., Seber [49], pp. 65, 66). This norm is a generalization of the Euclidean norm shown above.

Example 10.28 The following norm

$$\|x\|_\infty = {}^{max}_{1 \leq i \leq n} |x_i|, \tag{10.35}$$

is called the *infinity norm* on C^n (or R^n). It is sometimes called the L_∞ *norm* (e.g., Seber, op. cit., p. 66).

Example 10.29 Let $z = \{z_n\} \in l^p$, for $p \geq 1$, where l^p is the space of all infinite sequences $\{z_n\}$ of complex numbers such that $\sum_{n=1}^{\infty} |z_n|^p < \infty$. Then, the function defined by

$$\|z\| = \left(\sum_{i=1}^{\infty} |z_i|^p \right)^{1/p} \tag{10.36}$$

is a norm in l^p, which is called the l^p *norm* (e.g., Debnath & Mikusińki [19], p. 9, p. 11).

As regards the notion of a vector norm, there exist two primary general notions relating to it, which are defined below (e.g., Horn and Johnson [26, p. 281, pp. 282, 283]):

Definition 10.41 (*ball of radius r around a point x*) Let $\|x\|$ be a vector norm on the real or complex vector space V, let x be a point of V, and let $r > 0$ be given. The *ball of radius r* around x is the set

$$B_{\|\cdot\|}(r; x) \equiv \{y \in V : \|y - x\| \leq r\}. \tag{10.37}$$

Definition 10.42 (*closed set and bounded set*) Let $\|x\|$ be a vector norm on the real or complex vector space V. A point $x \in S$ is said to be an *interior point* of S if there is some $\epsilon > 0$ such that $B(\epsilon; x) \subset S$. The set is said to be *open* if every point of S is an interior point; S is said to be *closed* if its complement is open. The set is bounded if there exists some $M > 0$ such that $S \subset B_{\|\cdot\|}(M; 0)$.

Definition 10.43 (*Normed Space*). A vector space with a norm is called a *normed space*.

Remark It is apparent from the definition of normed space that there exist various normed spaces. Moreover, if we add the restriction of *completeness* defined below to the normed space, then we have a *Banach space* (e.g., Debnath & Mikusiński, op. cit., p. 17):

Definition 10.44 (*Banach Space*) A normed space W is called *complete* if every Cauchy sequence in W converges to an element of W. A complete normed space is called a *Banach space*.

Let us now go back to an inner product space. Once a norm is induced by an inner product, the inner product space is considered also as a normed space. Then, we can prove the following two *fundamental inequalities* which hold for an inner product space.

Theorem 10.5 (Schwarz's Inequality) *For any two elements x and y of an inner product space we have*

$$|(x, y)| \leq \|x\| \|y\|. \tag{10.38}$$

The equality $|(x, y)| = \|x\| \|y\|$ *holds if and only if x and y are linearly dependent.*

For the proof of this theorem, see, for example, Debnath and Mikusiński (op. cit., pp. 90, 91).

Definition 10.45 (*Linear Independence*) A finite collection of vectors $\{x_1, x_2, \cdots, x_k\}$ is called *linearly independent* if $\alpha_1 x_1 + \alpha_2 x_2 + \cdots + \alpha_k x_k = 0$ only if $\alpha_1 = \alpha_2 = \cdots = \alpha_k = 0$. A collection of vectors which is not linearly independent is called *linearly dependent*.

Corollary 10.6 (Triangle Inequality). For *any two elements of* x and y of an *inner product space we have*

$$\|x + y\| \leq \|x\| + \|y\|. \tag{10.39}$$

For the proof of this theorem, see, for example, Debnath and Mikusiński (op. cit., pp. 90, 91).

The following theorem also holds for an inner product space:

Theorem 10.6 (Parallelogram Law) *For any two elements x and y of an inner product space we have*

$$\|x + y\|^2 + \|x - y\|^2 = 2\big(\|x\|^2 + \|y\|^2\big). \tag{10.40}$$

For the proof of this theorem, see, for example, Debnath and Mikusiński (op. cit., p. 91).

Definition 10.46 (*Law of Cosines*) Let $\boldsymbol{u}, \boldsymbol{v}$ be any nonzero vectors in a real Euclidian space E^n with inner product. Then, we have

$$\|\boldsymbol{u}-\boldsymbol{v}\|^2 = \|\boldsymbol{u}\|^2 + \|\boldsymbol{v}\|^2 - 2\|\boldsymbol{u}\|\|\boldsymbol{v}\|\cos(\boldsymbol{u},\boldsymbol{v}). \tag{10.41}$$

Remark Equation (10.41) can be rewritten, using an inner product, as

$$(\boldsymbol{u},\boldsymbol{v}) = \frac{1}{2}\left(\|\boldsymbol{u}\|^2 + \|\boldsymbol{v}\|^2 - \|\boldsymbol{u}-\boldsymbol{v}\|^2\right). \tag{10.42}$$

since, an inner product between the two vectors $\boldsymbol{u}$ and $\boldsymbol{v}$ are written as

$$(\boldsymbol{u},\boldsymbol{v}) = \|\boldsymbol{u}\|\|\boldsymbol{v}\|\cos(\boldsymbol{u},\boldsymbol{v}), \tag{10.43}$$

in a real Euclidian space E^n.

Let us now remember that an inner product space is also called a *pre-Hilbert space* or a *unitary space* (see, Definition 10.39). An important equality holds for the pre-Hilbert space:

Theorem 10.7a (Polarization Identity) **.** In any pre-Hilbert space, the *polarization identity holds:*

$$(\boldsymbol{u},\boldsymbol{v}) = \frac{1}{2}\left\{\|\boldsymbol{u}\|^2 + \|\boldsymbol{v}\|^2 - \|\boldsymbol{u}-\boldsymbol{v}\|^2 + i\left(\|\boldsymbol{u}\|^2 + \|i\boldsymbol{v}\|^2 - \|\boldsymbol{u}-i\boldsymbol{v}\|^2\right)\right\}, \tag{10.44}$$

where $\boldsymbol{u}$ *and* $\boldsymbol{v}$ *are vectors in a (complex) inner product space.*

For the proof of this theorem, see, for example, Cristescu ([17], pp. 49, 50).

Remark 1 Notice that the inner product, $(\boldsymbol{u},\boldsymbol{v})$, in (10.44) is a *complex number* as defined in Definition 10.35. Therefore, its right-hand side is composed of the *real part Re* $(\boldsymbol{u},\boldsymbol{v})$ and the *imaginary part Im* $(\boldsymbol{u},\boldsymbol{v})$, and thus (10.44) can be written as

$$(\boldsymbol{u},\boldsymbol{v}) = Re(\boldsymbol{u},\boldsymbol{v}) + iIm(\boldsymbol{u},\boldsymbol{v}), \tag{10.45}$$

as, for example, Bloom ([7], p. 526) points out. It is interesting to note that the real part of (10.44) is nothing but (10.42), which is a version of the famous *law of cosines* defined in a real Euclidean space. In other words, polarization identity can be said to be a generalization of the law of cosines to a complex space. Of course, the imaginary part of (10.44) is

$$Im(\boldsymbol{u},\boldsymbol{v}) = \frac{1}{2}\left(\|\boldsymbol{u}\|^2 + \|\boldsymbol{v}\|^2 - \|\boldsymbol{u}-i\boldsymbol{v}\|^2\right), \tag{10.46}$$

noticing that $\|i\boldsymbol{v}\|^2 = \|\boldsymbol{v}\|^2$. Therefore, (10.42) can also be written as

$$(\boldsymbol{u},\boldsymbol{v}) = \frac{1}{2}\left\{\|\boldsymbol{u}\|^2 + \|\boldsymbol{v}\|^2 - \|\boldsymbol{u}-\boldsymbol{v}\|^2 + i\left(\|\boldsymbol{u}\|^2 + \|\boldsymbol{v}\|^2 - \|\boldsymbol{u}-i\boldsymbol{v}\|^2\right)\right\}. \tag{10.47}$$

Remark 2 Polarization identity has some variant expressions. Bloom (op. cit., p. 526) uses (10.45) with $|\cdot|$ instead of $\|\cdot\|$, and calls it the polarization identity. Bowen and Wang ([9], p. 63) use a similar expression

$$(\boldsymbol{u}, \boldsymbol{v}) = \frac{1}{2}\left\{\|\boldsymbol{u}+\boldsymbol{v}\|^2 + i\|\boldsymbol{u}+i\boldsymbol{v}\|^2 - (1+i)\|\boldsymbol{u}\|^2 - (1+i)\|\boldsymbol{v}\|^2\right\}, \tag{10.48}$$

and call it a *polar identity*.

Cristescu (op. cit., p. 49) uses the following expression:

$$(\boldsymbol{u}, \boldsymbol{v}) = \frac{1}{4}\left\{\|\boldsymbol{u}+\boldsymbol{v}\|^2 - \|\boldsymbol{u}-\boldsymbol{v}\|^2 + i\|\boldsymbol{u}+i\boldsymbol{v}\|^2 - i\|\boldsymbol{u}-i\boldsymbol{v}\|^2\right\}, \tag{10.49}$$

with $\varphi(\boldsymbol{u}, \boldsymbol{v})$ instead of $(\boldsymbol{u}, \boldsymbol{v})$, which holds for a complex pre-Hilbert space. He calls it simply a *scalar product*. Debnath and Mikusiński ([19], p. 146) call it a polarization identity. Chino and Shiraiwa [15] follow this equation, and call it a polar identity. It is not difficult to show that in (10.49)

$$(\boldsymbol{u}, \boldsymbol{u}) = \|\boldsymbol{u}\|^2. \tag{10.50}$$

Horn and Johnson ([26], p. 263) call the following equation a polarization identity:

$$(\boldsymbol{u}, \boldsymbol{v}) = \frac{1}{4}\left\{\|\boldsymbol{u}+\boldsymbol{v}\|^2 - \|\boldsymbol{u}-\boldsymbol{v}\|^2\right\}, \tag{10.51}$$

and call the following equation

$$(\boldsymbol{u}, \boldsymbol{v}) = \frac{1}{2}\left\{\|\boldsymbol{u}+\boldsymbol{v}\|^2 - \|\boldsymbol{u}\|^2 - \|\boldsymbol{v}\|^2 + i\left(\|\boldsymbol{u}+i\boldsymbol{v}\|^2 - \|\boldsymbol{u}\|^2 - \|\boldsymbol{v}\|^2\right)\right\}, \tag{10.52}$$

an *inner product* for V as a vector space over C (Horn & Johnson [26], p. 264).

Equations (10.44), (10.47), (10.48), (10.49), and (10.52) can easily be proved to be the same, if we use the parallelogram law in Theorem 10.6 and if we notice that $Im\ (\boldsymbol{u}, \boldsymbol{v})$ in (10.45) can also be written as

$$Im(\boldsymbol{u}, \boldsymbol{v}) = \frac{1}{2}\left(\|\boldsymbol{u}+i\boldsymbol{v}\|^2 - \|\boldsymbol{u}\|^2 - \|\boldsymbol{v}\|^2\right). \tag{10.53}$$

By contrast, Debnath and Mikusiński (op. cit., p. 146) define a polarization identity of more general form:

Theorem 10.7b (Polarization Identity) *Let Ø be a bilinear functional on a complex vector space V and let Φ be the quadratic form associated with Ø. Then*

$$\text{Ø}(x, y) = \frac{1}{4}\{\Phi(x+y) - \Phi(x-y) + i\Phi(x+iy) - i\Phi(x-iy)\}, \tag{10.54}$$

for all $x, y \epsilon V$.

For the proof of this theorem, see, for example, Debnath and Mikusiński (op. cit., p. 146).

Polarization identity plays an important role in the Hermitian form model (abbreviated as HFM) for the analysis of asymmetry, as shown in Chap. 4.

We will now show that an inner product space is also a *metric space*. According to Bowen and Wang ([9], p. 61):

Definition 10.47a (*metric space*) A metric space is a nonempty set M equipped with a positive real-valued function $d : M \times M \to R$, called the *distance function*, that satisfies the following axioms:

(1) $d(\boldsymbol{u}, \boldsymbol{v}) \geq 0$ and $d(\boldsymbol{u}, \boldsymbol{v}) = 0 \Leftrightarrow \boldsymbol{u} = \boldsymbol{v}$;
(2) $d(\boldsymbol{u}, \boldsymbol{v}) = d(\boldsymbol{v}, \boldsymbol{u})$;
(3) $d(\boldsymbol{u}, \boldsymbol{w}) \leq d(\boldsymbol{u}, \boldsymbol{v}) + d(\boldsymbol{v}, \boldsymbol{w})$;

for all $\boldsymbol{u}, \boldsymbol{v}, \boldsymbol{w} \in M$.

Theorem 10.8 *An inner product space U is a metric space with the distance function given by*

$$d(\boldsymbol{u}, \boldsymbol{v}) = \|\boldsymbol{u} - \boldsymbol{v}\|. \tag{10.55}$$

The proof of this theorem is, for example, given by Bowen and Wang (op. cit., p. 62).

Blank, Exner, and Havlíček ([6], p. 5) use a basically equivalent definition of a metric space to that of Bowen and Wang (op. cit.):

Definition 10.47b (*metric space*) A metric on a set X is a map $\rho : X \times X \to [0, \infty)$, which is symmetric, $\rho(x, y) = \rho(x, y)$, $\rho(x, y) = 0$ if and only if $x = y$, and fulfils the triangle inequality,

$$\rho(x, z) \leq \rho(x, y) + \rho(y, z), \tag{10.56}$$

for any $x, y, z \in X$; the pair(X, ρ) is called a *metric space*. If X is a *normed space*, one can define a metric on X by $\rho(x, y) = \|x - y\|$. We say it is *induced by the norm*.

Remark It is apparent from the above definitions that a metric space does not necessarily require a normed space.

Finally, we shall define a Hilbert space, for example, according to Debnath and Mikusiński ([19], p. 93):

Definition 10.48 (*Hilbert space*) A *complete* inner product space is called a *Hilbert space*.

Remark The completeness of an inner product space U is the same as the completeness of a normed space W (see, Definition 10.41). Remembering Definition 10.39, we can also say that a *complete pre-Hilbert space* is called a Hilbert space.

Hilbert space plays an important role in asymmetric MDS, particularly in HFM developed by Chino and Shiraiwa [15] which is discussed in Chap. 4. Followings are some examples of Hilbert spaces (e.g., According to Debnath and Mikusiński, op. cit., pp. 92–94).

Example 10.31 $CandC^n$ are Hilbert spaces.

Example 10.32 l^2 is a Hilbert space. Here, l^2 is a special case of the l^p space shown in Example 10.29.

Remark There exists another Hilbert space using the space of all complex-valued functions on an interval $[a, b]$ which is called L^2. This space seems to be similar to l^2 in Example 10.32 but is different from it. However, the introduction of this space is beyond the scope of this book, and thus we shall merely refer to, for example, Debnath and Mikusiński (op. cit., pp. 67–77).

As discussed earlier, Hilbert space is an inner product space. Therefore, by Definition 10.38 it is characterized by the property $f(x, y) = \overline{f(y, x)}$, which is sometimes called *Hermitian symmetry*. Hermitian form also has this property, as introduced in defining an inner product on a complex vector space. This form is sometimes called a *symmetric sesquilinear form*. By contrast, a mapping s: $V \times V \to C$ which has the property such that

$$s(x, y) = -\overline{s(y, x)}, \tag{10.57}$$

is called an *antisymmetric sesquilinear form* or *skew-Hermitian form*, where V is a complex vector space. A special case of this form is a nondegenerate *skew-symmetric 2-form* given in R^{2n}, which is called a *symplectic structure* (e.g., Arnold [3], p. 219), as defined below:

Counterexample. Consider the space $C_0(R)$ of all complex-valued continuous functions that vanish outside some finite interval. This is an inner product space with the inner product

$$(f, g) = \int_{-\infty}^{\infty} f(x)\overline{g(x)}dx. \tag{10.58}$$

The space $C_0(R)$ is not a Hilbert space, since it is not *complete*. The proof of this result is shown in Debnath and Mikusiński (op. cit., pp. 95, 96).

Definition 10.49 (*symplectic structure*) A symplectic linear structure on R^{2n} is a nondegenerate *bilinear skew-symmetric 2-form* given in R^{2n}. This form is called the *skew-scalar product* and is denoted by $[\xi, \eta] = -[\eta, \xi]$. The space R^{2n} together with the *symplectic structure* $[\cdot, \cdot]$, is called a *symplectic vector space*. Here, a 2-form $[\cdot, \cdot]$ on R^{2n} is nondegenerate if $([\xi, \eta] = 0, \forall \eta) \Rightarrow (\xi = 0)$.

Remark 1 As discussed in Arnold (op. cit., p. 219), a *Euclidean structure* on a vector space is given by a symmetric bilinear form, and a *symplectic structure* by a skew-symmetric one. The geometry of a symplectic space is different from that of a Euclidian space, although there are many similarities.

Remark 2 In psychometrics, particularly in asymmetric MDS, Gower [23], Chino [12–14]; Chino and Shiraiwa [15]) utilize the symplectic space structure, as shown in Chaps. 4 and 7.

In order to understand further the notion of the skew-symmetric 2-form in Definition 10.54 introduced above, we need the notions of *k-forms* in *exterior algebra*, which is known to be one of the early precursors to the modern notion of a vector space. In the followings we shall introduce these notions according to Arnold ([3], pp. 163–166):

Definition 10.50 *(1-form) A form of degree 1* (or a *1-form*) is a linear function $\omega : R^n \to R$, i.e.,

$$\omega(\lambda_1 \boldsymbol{\xi}_1 + \lambda_2 \boldsymbol{\xi}_2) = \lambda_1 \omega(\boldsymbol{\xi}_1) + \lambda_2 \omega(\boldsymbol{\xi}_2), \forall \lambda_1, \lambda_2 \in R \text{ and } \boldsymbol{\xi}_1, \boldsymbol{\xi}_2 \in R^n, \quad (10.59)$$

where R^n means an n-dimensional real vector space.

Remark Schutz, B. ([50], pp. 49, 50) defines 1-forms in the context of tensor algebra. According to him, a *1-form* is a linear, real-valued function of vectors.

Definition 10.51 *(2-form) An exterior form of degree 2* (or a *2-form*) is a function on pairs of vectors $\omega^2 : R^n \times R^n \to R$, which is bilinear and skew symmetric:

$$\omega^2(\lambda_1 \boldsymbol{\xi}_1 + \lambda_2 \boldsymbol{\xi}_2, \boldsymbol{\xi}_3) = \lambda_1 \omega^2(\boldsymbol{\xi}_1, \boldsymbol{\xi}_3) + \lambda_2 \omega^2(\boldsymbol{\xi}_2, \boldsymbol{\xi}_3),$$

$$\omega^2(\boldsymbol{\xi}_1, \boldsymbol{\xi}_2) = -\omega^2(\boldsymbol{\xi}_2, \boldsymbol{\xi}_1), \forall \lambda_1, \lambda_2 \in R \text{ and } \boldsymbol{\xi}_1, \boldsymbol{\xi}_2, \boldsymbol{\xi}_3 \in R^n. \quad (10.60)$$

Example 10.33 Let $S(\boldsymbol{\xi}_1, \boldsymbol{\xi}_2)$ be the *oriented area of the parallelogram* spanned by the two vectors $\boldsymbol{\xi}_1$ and $\boldsymbol{\xi}_2$ of the oriented Euclidean plane R^2, i.e.,

$$S(\boldsymbol{\xi}_1, \boldsymbol{\xi}_2) = \begin{vmatrix} \xi_{11} & \xi_{12} \\ \xi_{21} & \xi_{1.2} \end{vmatrix}, \text{ where } \boldsymbol{\xi}_1 = \xi_{11} \boldsymbol{e}_1 + \xi_{12} \boldsymbol{e}_2, \ \boldsymbol{\xi}_2 = \xi_{21} \boldsymbol{e}_1 + \xi_{1.2} \boldsymbol{e}_2, \quad (10.61)$$

with $\boldsymbol{e}_1, \boldsymbol{e}_2$ a basis giving the orientation on R^2. $S(\boldsymbol{\xi}_1, \boldsymbol{\xi}_2)$ is a 2-form.

Remark The absolute value of the term $x_{j1}x_{k2} - x_{j2}x_{k1}$ in (2.4) of Chap. 2 is nothing but a 2-form, where the two vectors $\boldsymbol{x}_j = (x_{j1}, x_{j2})^t$ and $\boldsymbol{x}_k = (x_{k1}, x_{k2})^t$ in Chino's ASYMSCAL correspond to $\boldsymbol{\xi}_1$ and $\boldsymbol{\xi}_2$, respectively.

Now, we shall digress for a while from the subject of 2-forms to discuss the notions of orientations in real vector spaces.

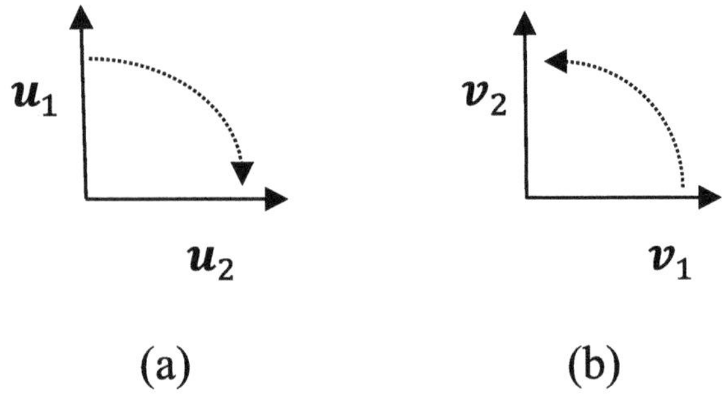

Fig. 10.3 Two pairs of vectors which are oriented opposedly

Remark By the oriented Euclidean plane R^2, we mean a real two-dimensional vector space R^2 with a Euclidean structure. As for the orientation in an n-dimensional vector space R^n, following cases will be distinguished (e.g., Bloom [7], pp. 41.2–426):

First, let $\boldsymbol{u}_1$ and $\boldsymbol{u}_2$ be *orthonormal* vectors in R^2, where orthonormal vectors are mutually *orthogonal* and each of them has norm 1, and $\boldsymbol{v}_1$ and $\boldsymbol{v}_2$ is also orthonormal in R^2. The 90 ° rotation which maps $\boldsymbol{u}_1 \rightarrow \boldsymbol{u}_2$ is *clockwise* (Fig. 10.3a), while the 90 ° rotation which maps $\boldsymbol{v}_1 \rightarrow \boldsymbol{v}_2$ is *counterclockwise* (Fig. 10.3b). An orthonormal pair such as $(\boldsymbol{u}_1, \boldsymbol{u}_2)$ is called *left-handed*, while the pair $(\boldsymbol{v}_1, \boldsymbol{v}_2)$ is called *right-handed*.

Second, the orthonormal triple $(\boldsymbol{u}_1, \boldsymbol{u}_2, \boldsymbol{u}_3)$ in three orthonormal vectors in R^3 is usually called a right-handed triple if an observer located at the terminal point of $\boldsymbol{u}_1$ sees the pair $(\boldsymbol{u}_2, \boldsymbol{u}_3)$ as right-handed, when we represent $\boldsymbol{u}_1, \boldsymbol{u}_2, \boldsymbol{u}_3$ as arrows with a common initial point P (Fig. 10.4a), while the triple $(\boldsymbol{v}_1, \boldsymbol{v}_2, \boldsymbol{v}_3)$ is called a left-handed triple, if an observer located at the terminal point of $\boldsymbol{v}_1$ sees the pair $(\boldsymbol{v}_2, \boldsymbol{v}_3)$ as left-handed, when we represent $\boldsymbol{v}_1, \boldsymbol{v}_2, \boldsymbol{v}_3$ as arrows with a common initial point P (Fig. 10.4b).

Third, let us definean *orientation in* R^n according to Bloom (op. cit., p. 426).

Theorem 10.9 (orientation in $\boldsymbol{R}^n$) *Two orthonormal n-tuples* $(\boldsymbol{u}_1, \cdots, \boldsymbol{u}_n)$ *and* $(\boldsymbol{v}_1, \cdots, \boldsymbol{v}_n)$ *in* R^n *have the same orientation if and only if the* $n \times n$ *matrices R* $(\boldsymbol{u}_1, \cdots, \boldsymbol{u}_n)$ *and R* $(\boldsymbol{v}_1, \cdots, \boldsymbol{v}_n)$ *have the same determinant.*

For the proof of this theorem, see, for example, Bloom (op. cit.).

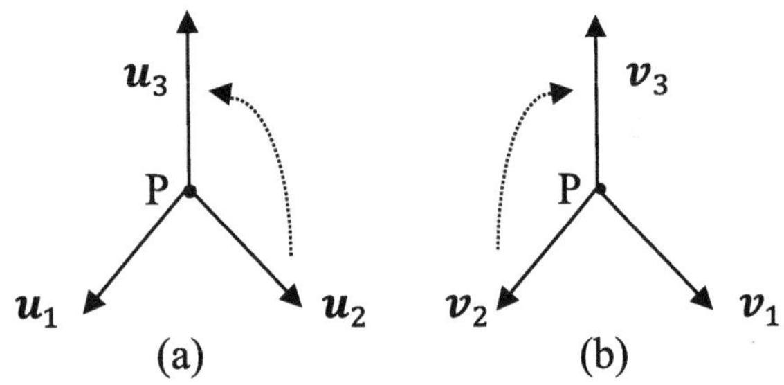

Fig. 10.4 Two orthogonal triples of vectors

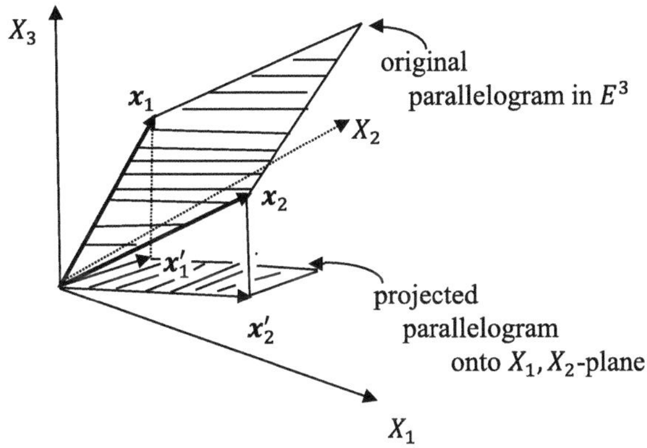

Fig. 10.5 A projection of the parallelogram spanned by two vectors $\boldsymbol{x}_1$ and $\boldsymbol{x}_2$ in E^3 onto X_1, X_2-plane

Remark In Bloom (op. cit., p. 364), R $(\boldsymbol{u}_1, \cdots, \boldsymbol{u}_m)$ generally denotes the $m \times n$ matrix whose *row vectors* in order are $\boldsymbol{u}_1, \cdots, \boldsymbol{u}_m$ in a vector space F_n of n-tuples over the field F. As regards the basic notions on matrix and determinant, see Chap. 11.

Now, we shall return to the subjects of 2-forms after a digression to the concept of orientation in some space. The next example of 2-forms is the oriented area of the *projection* of the parallelogram spanned by the two vectors in Euclidean R^3 (or sometimes denoted as E^3):

Example 10.34 The oriented area of the projection of the parallelogram spanned by two vectors $\boldsymbol{x}_1$ and $\boldsymbol{x}_2$ on the X_1, X_2-plane in E^3 is a 2-form.

Remark 1 Fig. 10.5 shows a projection of the parallelogram spanned by two vectors $\boldsymbol{x}_1$ and $\boldsymbol{x}_2$ in E^3.

Of course, we can project the parallelogram spanned by two vectors $\boldsymbol{x}_1$ and $\boldsymbol{x}_2$ in E^3 onto the other planes similarly, that is, X_2, X_3-plane and X_3, X_1-plane. The oriented areas of these projection are also 2-forms.

Remark 2 In an augmented inner product space model for asymmetric MDS by Chino [13], which is sometimes called Chino's ASYMSCAL, especially in the three-dimensional case, these forms are used as components of an asymmetric part (skew-symmetric part) of the model. Moreover, in this model a right-handed coordinate system is adopted as the orientation of the space E^3. The orthonormal basis vectors, say, $\boldsymbol{u}_1, \boldsymbol{u}_2, \boldsymbol{u}_3$ corresponding to the X_1, X_2, X_3 axes, respectively, are omitted in Fig. 10.5.

One-form and two-form introduced in Definitions 10.55 and 10.56, respectively, are generalized to *k-form* as follows (e.g., Arnold [3], p. 165):

Definition 10.52 (*k-form*) *An exterior form* of degree k, or a *k-form*, is a function of k vectors which is *k-linear* and *antisymmetric*:

$$\omega\left(\lambda_1 \xi_{1'} + \lambda_2 \xi_1'', \xi_2, \cdots, \xi_k\right) = \lambda_1 \omega(\xi_{1'}, \xi_2, \cdots, \xi_k) + \lambda_2 \omega\left(\xi_1'', \xi_2, \cdots, \xi_k\right),$$

$$\omega(\boldsymbol{\xi}_{i_1}, \boldsymbol{\xi}_{i_2}, \cdots, \boldsymbol{\xi}_{i_k}) = (-1)^{\nu}\omega(\boldsymbol{\xi}_1, \boldsymbol{\xi}_2, \cdots, \boldsymbol{\xi}_k), \tag{10.62}$$

where

$$\nu = \begin{cases} 0, \textit{ if the permutation } i_1, \cdots, i_k \textit{ is even}; \\ 1, \textit{ if the permutation} i_1, \cdots, i_k \textit{is odd}. \end{cases} \tag{10.63}$$

Remark In Chap. 4, we have referred to this form in equations (4.20) and (4.21).

Example 10.35 The oriented volume of the parallelepiped with edges $\boldsymbol{\xi}_1, \cdots, \boldsymbol{\xi}_n$ in oriented Euclidean space R^n is an n-form

$$\mathrm{V}(\boldsymbol{\xi}_1, \cdots, \boldsymbol{\xi}_n) = \begin{vmatrix} \xi_{11} & \cdots & \xi_{1n} \\ \vdots & \vdots & \vdots \\ \xi_{n1} & \cdots & \xi_{11} \end{vmatrix}. \tag{10.64}$$

At this point, the reader may want to know the relation between the 2-form and the *cross product* (or *vector product*) taught in the introductory course of analytic geometry. Let the two vectors $\boldsymbol{x}$ and $\boldsymbol{y}$ be defined in an oriented Euclidean space.

Then, the cross product $\boldsymbol{x} \times \boldsymbol{y}$ is a *vector* which is perpendicular to $\boldsymbol{x}$ and $\boldsymbol{y}$ and the *length* of $\boldsymbol{x} \times \boldsymbol{y}$ equals the area of the parallelogram spanned by $\boldsymbol{x}$ and $\boldsymbol{y}$, as shown in Fig. 10.6a. On the other hand, in Fig. 10.6b the cross product is shown

along with a Cartesian coordinate system as right-handed. It is evident from Fig. 10.6b that the *oriented area of the parallelogram* which appears in defining the cross product of two vectors is nothing but a *2-form* in Definition 10.49.

In the following, we shall refer to a formal definition of the cross product, for example, by Bloom ([7], p. 309):

Definition 10.53 (*cross product*) If $\boldsymbol{u} = (x_1, x_2, x_3)$ and $\boldsymbol{v} = (y_1, y_2, y_3)$ are vectors in a field F_3, a cross product of $\boldsymbol{u}$ and $\boldsymbol{v}$ is defined as

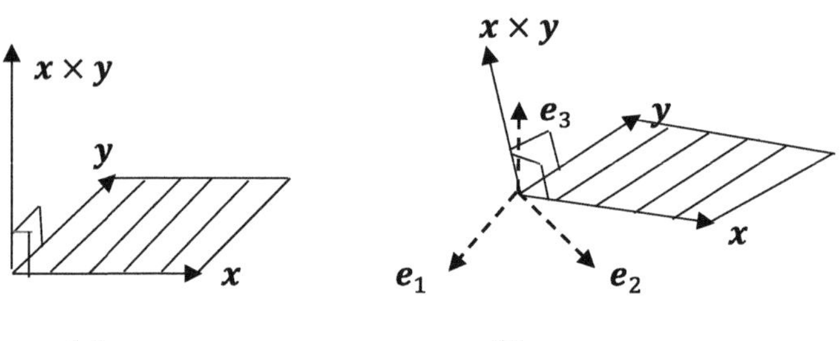

Fig. 10.6 The *cross product* $\boldsymbol{x} \times \boldsymbol{y}$ which is perpendicular to $\boldsymbol{x}$ and $\boldsymbol{y}$, and its length equals the area of the parallelogram spanned by the two vectors. In **a**, a Cartesian coordinate system is not exhibited, while in **b** it is exhibited using basis vectors, $\boldsymbol{e}_1, \boldsymbol{e}_1, \boldsymbol{e}_3$ as dotted arrows in a right-handed system.

$$u \times v = \left(det \begin{bmatrix} x_2 & x_3 \\ y_2 & y_3 \end{bmatrix}, -det \begin{bmatrix} x_1 & x_3 \\ y_1 & y_3 \end{bmatrix}, det \begin{bmatrix} x_1 & x_2 \\ y_1 & y_2 \end{bmatrix} \right), \tag{10.65}$$

where, by "det" we mean a *determinant* of a matrix. As regards a fundamental matrix theory, see Chap. 11.

Remark In the case when F is a three-dimensional Euclidean space, E^3, and x_3 and y_3 are both zeros in Definition 10.51, the cross product $\boldsymbol{u} \times \boldsymbol{v}$ in (10.65) becomes

$$u \times v = \left(0, 0, det \begin{bmatrix} x_1 & x_2 \\ y_1 & y_2 \end{bmatrix} \right) = (0, 0, x_1 y_2 - x_2 y_1). \tag{10.66}$$

Equation (10.64) clearly indicates that when the two vectors lie on the Euclidean plane formed by the basis vectors $\boldsymbol{e}_1$ and $\boldsymbol{e}_2$, the cross product lies on the line formed by the basis vector $\boldsymbol{e}_3$, and its length is $|x_1 y_2 - x_2 y_1|$. Particularly if we adopt a right-handed coordinate system as in Fig. 10.6b, the length is $x_1 y_2 - x_2 y_1$.

Let us now return to the k-form introduced in Definition 10.50. In general, k-forms can be generated by a product called *exterior product* of lower degrees of the forms, as will be introduced soon in Definition 10.52. The exterior product is sometimes called *wedge product*, and is denoted by the symbol, "$\wedge$".

We shall start with the exterior product of two 1-forms, according to, for example, Arnold (op. cit., pp. 166–170). The exterior product of 1-forms ω_1 and ω_2 on R^n generates a 2-form $\omega_1 \wedge \omega_2$ on R^n as follows:

Let $\boldsymbol{x}$ be a vector in R^n. Given two 1-forms ω_1 and ω_2, we can define a mapping of R^n to the plane $R \times R$ by associating to $\boldsymbol{x} \in R^n$ the vector $\omega(\boldsymbol{x})$ with components $\omega_1(\boldsymbol{x})$ and $\omega_2(\boldsymbol{x})$ in the plane with coordinates ω_1, ω_2. Then, we have

Definition 10.54 (*exterior product of two 1-forms*) The value of the exterior product $\omega_1 \wedge \omega_2$ on the pair of vectors $\boldsymbol{x}_1, \boldsymbol{x}_2 \in R^n$ is the *oriented area of the image of the parallelogram* with sides $\omega(\boldsymbol{x}_1)$ and $\omega(\boldsymbol{x}_2)$ on the ω_1, ω_2-plane:

$$(\omega_1 \wedge \omega_2)(\boldsymbol{x}_1, \boldsymbol{x}_2) = \begin{vmatrix} \omega_1(\boldsymbol{x}_1) & \omega_2(\boldsymbol{x}_1) \\ \omega_1(\boldsymbol{x}_2) & \omega_2(\boldsymbol{x}_2) \end{vmatrix}. \tag{10.67}$$

More generally, we have a following definition:

Definition 10.55 (*exterior multiplication*) The exterior product $\omega^k \wedge \omega^l$ of a k-form ω^k on R^n with an l-form ω^l on R^n is the k+l-*form* on R^n whose value on the k+l vectors $\boldsymbol{x}_1, \cdots, \boldsymbol{x}_k, \boldsymbol{x}_{k+1}, \cdots, \boldsymbol{x}_{k+l} \in R^n$ is equal to

$$\left(\omega^k \wedge \omega^l\right)(\boldsymbol{x}_1, \cdots, \boldsymbol{x}_{k+l}) = \sum (-1)^{\nu} \omega^k \left(\boldsymbol{x}_{i_1}, \cdots, \boldsymbol{x}_{i_k}\right) \omega^l \left(\boldsymbol{x}_{j_1}, \cdots, \boldsymbol{x}_{j_l}\right), \tag{10.68}$$

where $i_1 < \cdots < i_k$ and $j_1 < \cdots < j_l$; $(i_1, \cdots, i_k, j_1, \cdots, j_l)$ is a permutation of the number $(1, 2, \cdots, k + l)$; and

$$\nu = \begin{cases} 1 \ \textit{if this permutation is odd}; \\ 0 \ \textit{if this permutation is even}. \end{cases} \tag{10.69}$$

Example 10.36 If k=l=1, then

$$(\omega_1 \wedge \omega_2)(\boldsymbol{x}_1, \boldsymbol{x}_2) = \omega_1(\boldsymbol{x}_1)\omega_2(\boldsymbol{x}_2) - \omega_2(\boldsymbol{x}_1)\omega_1(\boldsymbol{x}_2). \tag{10.70}$$

This exterior form is equivalent to (10.67), which is the exterior form of two 1-forms.

Finally, we shall refer to the notion of *tensor* which is closely related to the notions of vectors and 1-forms. Before proceeding, we shall define a manifold. Let us denote by R^n the set of all n-tuples of real numbers $(x_1, x_2, \cdots, x_n)$. Then we have, for example, according to Schutz ([50], pp. 23–60),

Definition 10.56a (*manifold*) A set (of 'points') M is called a *manifold* if each point of M has an open neighborhood which has a continuous one-to-one map onto an open set of R^n for some n.

Remark The definition above means that M is *locally 'like'* R^n. A more explicit definition of manifold is the definition by Boothby ([8], p. 6).

Definition 10.56b (*manifold*) A *manifold M* of dimension n, or *n-manifold*, is a topological space with the following properties:

(i) M is *Hausdorff*,
(ii) M is *locally Euclidian* of dimension n, and
(iii) M has a countable basis of open sets.

Remark First, the topological space and the open sets in the above definition are defined, for example, according to Maddox ([38], p. 46), as

Definition 10.57 (*topological space*) A *topological space* (X, T) is a pair consisting of a non-empty set X and a class T of subsets of X satisfying the following axioms:
(T1) $\emptyset \in T$ and $X \in T$.
(T2) Any union (countable or not) of sets in T is in T.
(T3) The intersection of any finite number of sets in T is in T.
The sets of T which satisfy the above axioms are called *open sets* and T is called a *topology* for X.

Second, in (i) of Definition 10.54b, a Hausdorff space is a topological space defined as follows (e.g., Maddox, op. cit., p. 47):

There are many variations in the definitions of topological space in the literature (e.g., Blank et al.[6]; Boothby [8]; Krantz [31]; Maddox [38]; Rotman [47]). In these, the definitions by Maddox seems to be the most concise.

Definition 10.58 (*Hausdorff space*) A topological space (X, T) is called *Hausdorff* if and only if, for any x, y in X with $x \neq y$, there exist two disjoint open sets, one containing x and the other containing y.

Theorem 10.10 (Baire space) *A topological space X is a Baire space if and only if every countable point finite open cover of X is locally finite at a dense set of points.*
For the proof of this theorem, see McCoy [40].

Remark Theorem 10.10 contains some basic notions in the set theory. We shall introduce some of these notions according to Maddox (op. cit., p. 3, pp. 43, 44). First, let S be a class of sets A. Then the class S is called a *cover* for a set X if and only if $X = \cup\{A|A \in S\}$. Second, a subset Y of a metric space (X, d) is called *dense* in X if and only if $\overline{Y} = X$, where $\overline{Y}$ is the *closure* of Y. Here, the closure $\overline{Y}$ of Y is the smallest closed set containing Y: $\overline{Y} = \bigcap_{F \supset Y} F$.

After a short digression, we shall now define the notion of *tensor* as follows (e.g., Schutz, op. cit., p. 57), which we have already referred to in Chap. 4 as an important notion in interpreting HFM (i.e., Chino & Shiraiwa [15]):

Definition 10.59 (*tensor*) A *tensor* of type $\begin{pmatrix} N \\ N' \end{pmatrix}$ at P of a manifold M is a linear function which takes as arguments N *1-forms* and N' *vectors* and whose value is a real number.

Example 10.37 A scalar function on M is taken to be a tensor of type $\begin{pmatrix} 0 \\ 0 \end{pmatrix}$.

Example 10.38 Vectors are tensors of type $\begin{pmatrix} 1 \\ 0 \end{pmatrix}$.

Example 10.39 One-forms are tensors of type $\begin{pmatrix} 0 \\ 1 \end{pmatrix}$.

Example 10.40 All p-forms are tensors of type $\begin{pmatrix} 0 \\ p \end{pmatrix}$.

Section 10.6 Matrix norm

In Definition 10.40 we have already defined the notion of vector norm. In this section, we shall first define the notion of *matrix norm* of a square matrix axiomatically (e.g., Horn & Johnson [26, p. 290]).

Definition 10.60 A real valued function $\|\cdot\|$ defined on a vector space is called a *matrix norm* if it satisfies the following axioms:

(1) $\|\boldsymbol{A}\| \geq 0$ for all $\boldsymbol{A} \in M_n$ and $\|\boldsymbol{A}\| = 0$ if and only if $\boldsymbol{A} = \boldsymbol{O}$;
(2) $\|\lambda\boldsymbol{A}\| = |\lambda|\|\boldsymbol{A}\|$ for all $\boldsymbol{A} \in M_n$ and $\lambda \in C$;
(3) $\|\boldsymbol{A} + \boldsymbol{B}\| \leq \|\boldsymbol{A}\| + \|\boldsymbol{B}\|$ for all $\boldsymbol{A}, \boldsymbol{B} \in M_n$;
(4) $\|\boldsymbol{AB}\| \leq \|\boldsymbol{A}\|\|\boldsymbol{B}\|$ for all $\boldsymbol{A}, \boldsymbol{B} \in M_n$,

where M_n is a set of arbitrary square matrices of order n.

Remark 1 Horn and Johnson use the symbol composed of three vertical line segments instead of that composed of two vertical line segments, i.e. $\|\cdot\|$, in the above definition. Axioms (1–3) are identical to those for a vector norm in Definition 10.40. As for the notation M_n as well as other variants, we shall follow Horn and Johnson and we shall frequently use them in Chap. 11. It should be noticed that the set of all $m \times n$ matrices is considered as a vector space, as we have already pointed out in Example 10.11 in Section 10.2. Axiom (4) is called *submultiplicative.*

Basic matrix norms can be classified roughly into two. One is the matrix norm *subordinate to a vector norm* $\|\boldsymbol{x}\|$, and the other the *Euclidean norm.*

Definition 10.61 The following non-negative quantity with any matrix $\boldsymbol{A}$,

$$\|\boldsymbol{A}\| = sup_{\boldsymbol{x}\neq 0}\|\boldsymbol{A}\boldsymbol{x}\|/\|\boldsymbol{x}\|, \tag{10.71}$$

for all non-zero $\boldsymbol{x}$ is called the *matrix norm subordinate to the vector norm.*

Remark The matrix norm defined above is sometimes called the *matrix norm induced by the vector norm* (e.g., Lancaster & Tismenetsky [35, p. 363]; Seber [49, p. 68]).

The matrix norm subordinate to $\|\boldsymbol{x}\|_p$ is denoted by $\|\boldsymbol{A}\|_p$. The following are examples of $\|\boldsymbol{A}\|_p$ in the special cases when $p = 1, 2,$ and∞:

Example 10.41

$$\|\boldsymbol{A}\|_1 = max_j \sum_i |a_{ij}|, \tag{10.72}$$

where $\boldsymbol{A} = [a_{ij}]$.

Example 10.42

$$\|\boldsymbol{A}\|_2 = \max\left\{\sqrt{\lambda} : \lambda \text{ is an eigenvalue of} \boldsymbol{A}^*\boldsymbol{A}\right\}. \tag{10.73}$$

which is called the *spectral norm.* The $\sqrt{\lambda}$ is the *singular value* of $\boldsymbol{A}$, which will be discussed in more detail in Chap. 11. Moreover, $\boldsymbol{A}^*$ denotes the *conjugate transpose* of $\boldsymbol{A}$, which is called *Hermitian Matrix*. We shall define it more precisely in Definition 11.6 later in Chap. 11.

Example 10.43

$$\|\boldsymbol{A}\|_\infty = max_i \sum_j |a_{ij}|. \tag{10.74}$$

The other matrix norm which is basic and important is defined as follows:

Definition 10.62 The following norm

$$\|A\|_E = \left(\sum_{i,j=1}^{n} |a_{ij}|^2 \right)^{1/2}, A = [a_{ij}]_{i,j=1}^{n}, \tag{10.75}$$

defines a *matrix norm* on $M_n(C)$, which is called the *Euclidean* or *Schur norm* (e.g., Wilkinson [58]).

Remark 1 Although we have, from Definition 10.62, $\|I\| = 1$ for any subordinate norm, we have, from the above definition, $\|I\|_E = n^{1/2}$ for any Euclidean norm. Therefore, a Euclidean norm cannot be subordinate to any vector norm.

Remark 2 The Euclidean (matrix) norm defined above is a generalization of the Euclidean vector norm defined in the previous section. It is also called *Frobenius norm*, or *Hilbert-Schmidt norm* (e.g., Horn & Johnson [26], p. 291).

As discussed above, we usually confine our attention to norms of square matrices (e.g., Horn & Johnson, op. cit.; Lancaster & Tismenetsky [35]; Wilkinson, op. cit.). Of course, it is possible to define matrix norms of *rectangular matrices*. However, we shall not discuss them here because these matrix norms are beyond the scope of this book. As regards these norms, we shall refer to, for example, Stewart and Sun ([51, pp. 64–71]) and Seber ([49, pp. 67–73]).

References

1. Aizenberg, N. N., Ivaskiv, Y. L., & Pospelov, D. A. (1971). A certain generalization of threshold functions. *Dokrady Akademii Nauk SSSR, 196*, 1287–1290.
2. Aizenberg, N. N., & Aizenberg, I. N. (1992). CNN based on multi-valued neuron as a model of associative memory for grey scale images. In *CNNA '92 Proceedings of 2nd international workshop on cellular neural networks applications* (pp 36–41). Munich, Germany.
3. Arnold, V. I. (1978). *Mathematical methods of classical mechanics*. Springer-Verlag.
4. Ball, S. (2015). *Finite geometry and combinatorial applications*. Cambridge University Press.
5. Bak, J., & Newman, D. J. (1982). *Complex analysis*. Springer-Verlag.
6. Blank, J., Exner, P., & Havlíček, M. (1994). *Hilbert space operators in quantum physics*. AIP press.
7. Bloom, D. M. (1979). *Linear algebra and geometry*. Cambridge University Press.
8. Boothby, W. M. (1975). *An introduction to differentiable manifolds and Riemannian geometry*. Academic Press.
9. Bowen, R. M., & Wang, C.-C. (1976). *Introduction to vectors and tensors—Linear and multilinear algebra* (Vol. 1). Plenum Press.
10. Bowen, R. M., & Wang, C.-C. (1976). *Introduction to vectors and tensors—Vector and tensor analysis* (Vol. 2). Plenum Press.
11. Bureau international des poids et mesures. (2019). *Le Système International d'unités* [*The International System of Units*] (9th ed.). Sèvres Cedex.
12. Chino, N. (1977). N-ko no taisho-kan no hitaisho na kannkei o zusikika surutameno ichi-giho (A technique for representing asymmetric relationships between *N* objects. In: *Proceedings of the 5th annual meeting of the Behaviormetric Society of Japan* (pp. 146–149).
13. Chino, N. (1978). A graphical technique for representing the asymmetric relationships between *N* objects. *Behaviormetrika, 5*, 23–40.

14. Chino, N. (1990). A generalized inner product model for the analysis of asymmetry. *Behaviormetrika, 27*, 25–46.
15. Chino, N., & Shiraiwa, K. (1993). Geometrical structure of some non-distance models for asymmetric MDS. *Behaviormetrika, 20*, 35–47.
16. Clifford, A. H. (1940). Partially ordered Abelian group. *Annals of Mathematics, 41*, 465–473.
17. Cristescu, R. (1977). *Topological vector spaces*. Noordhoff International Publishing.
18. Dean, R. A., & Keller, G. (1966). Natural partial orders. *Canadian Journal of Mathematics, 20*, 535–554.
19. Debnath, L., & Mikusiński, P. (1999). *Introduction to Hilbert spaces with applications* (2nd ed.). Academic Press.
20. Dixson, J. D., & Mortimer, B. (1996). *Permutation groups*. Springer.
21. Fountain, J. (1979). Adequate semigroups. *Proceedings of the Edinburgh Mathematical Society, 22*, 113–125.
22. Glass, A. M. W. (1999). *Partially ordered groups*. World Scientific.
23. Gower, J. C. (1977). The analysis of asymmetry and orthogonality. In J. R. Barra, F. Brodeau, G. Romer, & B. van Cutsem (Eds.), *Recent development in statistics* (pp. 109–123). North Holland.
24. Guo, K., Mohar, B., & Suil, O. (2016). Hermitian adjacency matrix of digraphs and mixed graphs. *Journal of Graph Theory, 85*, 217–248.
25. Hirose, A. (2003). *Complex-valued neural networks—Theories and applications*. World Scientific.
26. Horn, R. A., & Johnson, C. A. (1985). *Matrix analysis*. Cambridge University Press.
27. Jankowski, S., Lozowski, A., & Zurada, J. M. (1996). Complex-valued multistate neural associative memory. *IEEE Transactions on Neural Networks, 7*, 1491–1496.
28. Japanese Mathematical Society. (2008). *Sugaku-Jiten* (Dictionary of Mathematics) (4th ed). Iwanami Book Company.
29. Kawada, Y. (1976). *Afin Kika • Syaei Kika [Affine Geometry and Projective Geometry]*. Iwanami.
30. Kim, V. I., & Kozhukhov, I. B. (2007). On semigroups of isotonic transformations of partially ordered sets. *Russian Mathematical Survey, 62*, 996–998.
31. Krantz, S. G. (1992). *Function theory of several complex variables* (2nd ed.). Wadsworth.
32. Kruskal, J. B. (1964). Multidimensional scaling by optimizing goodness of fit to a nonmetric hypothesis. *Psychometrika, 29*, 1–27.
33. Kruskal, J. B. (1964). Multidimensional scaling: A numerical method. *Psychometrika, 29*, 115–129.
34. Kubrusly, C. S. (2001). *Elements of operator theory*. Birkh ä user.
35. Lancaster, P., & Tismenetsky, M. (1985). *The theory of matrices* (2nd ed.). Academic Press.
36. Lang, S. (1993). *Complex analysis* (3rd ed.). Springer-Verlag.
37. Liu, J.-X., & Li, X.-L. (2015). Hermitian-adjacency matrices and Hermitian energies of mixed graphs. *Linear Algebra and Its Applications, 466*(181), 207.
38. Maddox, I. J. (1988). *Elements of functional analysis* (2nd ed.). Cambridge University Press.
39. Mahadev, N. V. R., & Peled, U. N. (1995). *Threshold graphs and related topics*. Elsevier Science B. V.
40. McCoy, R. A. (1972). A Baire space extension. *Proceedings of the American Mathematical Society, 33*, 199–202.
41. Mitsch, H. (1986). A natural partial order for semigroups. *Proceedings of the American Mathematical Society, 97*, 384–388.
42. Mohar, B. (2019). Hermitian adjacency spectrum and switching equivalence of mixed graphs. *Linear Algebra and Its Applications, 489*, 324–340.
43. Molchanov, V. A. (1983). Semigroups of mappings on graphs. *Semigroup Forum, 27*, 155–199.
44. Müezzinoğlu, M. K., Güzeliş, C., & Zurada, J. M. (2003). A new design method for the complex-valued multistate Hopfield associative memory. *IEEE Transactions on Neural Networks., 14*, 891–899.

45. Nambooripad, K. S. S. (1980). The natural partial order on a regular semigroup. *Proceedings of Edinburgh Mathematical Society, 23*, 249–260.
46. Richardson, M. W. (1938). Multidimensional psychophysics. *Psychological Bulletin, 35*, 659.
Rocci, R., & Bove, G. (2002). Rotation techniques in asymmetric multidimensional scaling. *Journal of Computational and Graphical Statistics, 11*, 405–419.
47. Rotman, J. J. (2006). *A first course in abstract algebra with applications* (3rd ed.). Pearson Prentice Hall.
48. Rudin, W. (1987). *Real and complex analysis* (3rd ed.). WCB/McGraw-Hill.
49. Seber, G. A. F. (2008). *A matrix handbook for statisticians*. John Wiley & Sons.
50. Schutz, B. (1980). *Geometrical methods of mathematical physics*. Cambridge University Press.
51. Stewart, G. W., & Sun, J. (1990). *Matrix perturbation theory*. Academic Press.
52. Stevens, S. S. (1946). On the theory of scales of measurement. *Science, 103*, 677–680.
53. Stevens, S. S. (1951). Mathematics, measurement, and psychophysics. In S. S. Stevens (Ed.) *Handbook of Experimental Psychology*. Wiley.
54. Torgerson, W. S. (1954). A law of categorical judgment. *The American Psychologist, 9*, 483.
55. Torgerson, W. S. (1958). *Theory and methods of scaling*. Wiley.
56. Yan, S., Zang, B., Zang, H.-J., Yang, Q., & Lin, S. (2007). Graph embedding and extensions: A general framework for dimensionality reduction. *IEEE Transactions on Pattern Analysis and Machine Intelligence, 29*, 40–51.
57. Yuan, B.-J., Wang, Y., Gong, S.-C., & Qiao, Y. (2020). On mixed graphs whose Hermitian spectral radii are at most 2. *Graphs and Combinatorics, 36*, 1573–1584.
58. Wilkinson, J. H. (1965). *The algebraic eigenvalue problem*. Clarendon Press.

Chapter 11
Definitions and Theories Useful for the Understanding of the Structures of Asymmetric Interactions

11.1 Introduction

In this chapter, we shall present various definitions and theories useful for the understanding of the contents of the structures of asymmetric interactions. As in the previous chapter, we shall omit all the proofs of the theorems listed in this chapter but give references to these proofs at the end of each theorem.

As pointed out in Sect. 1.1, theories and methods for analyzing the structure of asymmetric interactions have been scattered around many different branches of science. Among those, we shall deal with asymmetric MDS in psychometrics, spectral graph theory in mathematics, graph embedding in statistical and machine learning, chemical reaction networks, and neural networks. Some of the definitions and theories discussed in these branches are common to these theories and methods, but others are not. We shall first discuss those in matrix theory, which are common to all of these branches, in Sect. 11.2. Then, we shall discuss those which are specific to asymmetric MDS in Sect. 11.3, graph theory and spectral graph theory in Sect. 11.4, graph embedding in Sect. 11.5, chemical reaction networks and neural networks in Sect. 11.6.

11.2 Matrix Theory

Various important studies on matrix theory have been published since Wilkinson [139] wrote a comprehensive book titled *The Algebraic Eigenvalue Problem*. Some of them may be Horn and Johnson [58], Lancaster and Tismenetsky [69], and Stewart and Sun [117]. In this section, we shall pick up definitions and theories concerning the structures of asymmetric interactions, which are discussed in all these works. We assume throughout this book that matrix is defined in the field of *complex numbers* unless otherwise stated. We denote matrices by boldface capital letters, and the (i, j)

N. Chino, *Structure and Dynamics of Asymmetric Interactions*,
https://doi.org/10.1007/978-981-97-8269-7_11

element of a matrix $\boldsymbol{A}$ will be denoted by a_{ij}. Sometimes, we shall denote a matrix $\boldsymbol{A}$ briefly in terms of its elements as $\boldsymbol{A} = \{a_{ij}\}$. From now on, we shall use this notation frequently.

Definition 11.1 (*rectangular matrix*) An array of mn elements, which is written in the form

$$\boldsymbol{A} = \begin{bmatrix} a_{11} & a_{12} & \cdots & a_{1n} \\ a_{21} & a_{22} & \cdots & a_{2n} \\ \vdots & \vdots & \ddots & \vdots \\ a_{m1} & a_{m2} & & a_{mn} \end{bmatrix} \tag{11.1}$$

is called a *rectangular* $m \times n$ *matrix* from a *field* F.

Remark As regards a field, we shall refer to Sect. 10.2. Here, we shall assume F to be the set of all real or complex numbers, denoted by R and C, respectively. We shall denote the set of all $m \times n$ matrices with real elements simply as $R^{m\times n}$, and those with complex elements in the same manner as $C^{m\times n}$. To be precise, these sets are denoted by $M_{m,n}(F)$, and $M_{n,n}(F)$ is abbreviated to $M_n(F)$. In the case when $F = C$, $M_n(C)$ is sometimes abbreviated to M_n, and $M_{m,n}(C)$ to $M_{m,n}$ (e.g., Hohn and Johnson, [56], p. 4). We shall sometimes use this notation if necessary.

Example 11.1 The following is a rectangular 3×2 matrix from R.

$$\boldsymbol{A} = \begin{bmatrix} 1 & 2 \\ \sqrt{3} & 4 \\ 5 & \sqrt{6} \end{bmatrix},$$

which can also be denoted by $\boldsymbol{A} \in M_{3,2}(R)$ using the above notation.

If all the elements of the rectangular $m \times n$ matrix $\boldsymbol{A}$ are equal to zero, it is called a *zero-matrix*. If $m = n$, the matrix is said to be *square*. Most of the matrices introduced in Chap. 1 are square matrices with real elements. Two exceptions are shown in Tables 1.26 and 1.29, which are square matrices with complex elements.

Definition 11.2 (*transpose of a matrix*) Given a rectangular $m \times n$ matrix $\boldsymbol{A}$, the $n \times m$ matrix obtained by interchanging its rows and columns is called the *transpose of a matrix* and is denoted by $\boldsymbol{A}^t$:

$$\boldsymbol{A}^t = \begin{bmatrix} a_{11} & a_{21} & \cdots & a_{m1} \\ a_{12} & a_{22} & \cdots & a_{m2} \\ \vdots & \vdots & \ddots & \vdots \\ a_{1n} & a_{2n} & \cdots & a_{mn} \end{bmatrix}. \tag{11.2}$$

Example 11.2 A matrix $\boldsymbol{A}$ is the same as in Example 11.1. Then, the transpose of $\boldsymbol{A}$ is written by

$$A^t = \begin{bmatrix} 1 & \sqrt{3} & 5 \\ 2 & 4 & \sqrt{6} \end{bmatrix}.$$

Definition 11.3 (*conjugate transpose of a matrix*) Given a transposed matrix A^t of A, the following matrix is called the *conjugate transpose of a matrix*:

$$\overline{A}^t = \begin{bmatrix} \overline{a}_{11} & \overline{a}_{21} & \cdots & \overline{a}_{m1} \\ \overline{a}_{12} & \overline{a}_{22} & \cdots & \overline{a}_{m2} \\ \vdots & \vdots & \ddots & \vdots \\ \overline{a}_{1n} & \overline{a}_{2n} & \cdots & \overline{a}_{mn} \end{bmatrix}, \tag{11.3}$$

where $\overline{A}$ denotes the matrix whose elements are the complex conjugates of those of A.

Example 11.3 Let a 3×2 matrix A is a complex matrix defined by

$$A = \begin{bmatrix} 9 & 2+3i \\ -5+6i & 1-2i \\ 7+8i & 10 \end{bmatrix}.$$

Then, the conjugate transpose of A is

$$\overline{A}^t = \begin{bmatrix} 9 & -5-6i & 7-8i \\ 2-3i & 1+2i & 10 \end{bmatrix}.$$

Remark The conjugate transpose of a matrix, $\overline{A}^t$, is usually denoted by A^* (e.g., Seber [111], p. 54). $\overline{A}^t$ is sometimes called the *Hermitian adjoint* of $A \in M_{m,n}(C)$ (Horn and Johnson, op. cit., p. 6). If A is a *square matrix* then $\overline{A}^t$ is sometimes denoted by A^H and is referred to as the *Hermitian transposed matrix* (e.g., Wilkinson [139], p. 25). We shall define a *Hermitian matrix* soon.

Definition 11.4 (*diagonal matrix*) A special case of a square matrix for which all the elements lying outside the main diagonal are equal to zero is called a *diagonal matrix*:

$$A = \begin{bmatrix} a_{11} & 0 & \cdots & 0 \\ 0 & a_{22} & \cdots & 0 \\ \vdots & \vdots & \ddots & \vdots \\ 0 & 0 & \cdots & a_{nn} \end{bmatrix}, \tag{11.4}$$

which is sometimes written as $diag[a_{11}, a_{22}, \ldots a_{nn}]$.

Remark A degree matrix introduced in Sect. 1.3.1 is a typical example of diagonal matrix. In particular, the matrix $\boldsymbol{A}$ is called an *identity matrix* if $a_{11} = a_{22} = \ldots = a_{nn} = 1$. Hereafter, we shall denote it by $\boldsymbol{I}_n$.

Definition 11.5 (*symmetric matrix*) A square matrix $\boldsymbol{A}$ is said to be a *symmetric* if it is written in the form:

$$\boldsymbol{A} = \begin{bmatrix} a_{11} & a_{12} & \cdots & a_{1n} \\ a_{12} & a_{22} & \cdots & a_{2n} \\ \vdots & \vdots & \ddots & \vdots \\ a_{1n} & a_{2n} & \cdots & a_{nn} \end{bmatrix}, \tag{11.5}$$

so that $a_{ji} = a_{ij}$ for $i = 1, 2, \ldots, n, j = 1, 2, \ldots, n$.

Example 11.4 The following is a symmetric matrix:

$$\boldsymbol{A} = \begin{bmatrix} 3 & 2 & 1 \\ 2 & 4 & 6 \\ 1 & 6 & 5 \end{bmatrix}.$$

Remark Equivalently, a square matrix $\boldsymbol{A}$ is said to be *symmetric* if

$$\boldsymbol{A}^t = \boldsymbol{A}. \tag{11.6}$$

The adjacency matrix and the Laplacian matrix associated with a graph (but not digraph) introduced in Sect. 10.1 are symmetric matrices. In symmetric MDS which is introduced first in Sect. 11.3, special *symmetric matrices* such as the distance matrix shown in Tables 11.1 and 11.3 and the inner product matrix shown in Table 11.2 are analyzed. Diagonal elements of a distance matrix are all *zero*, as shown in Tables 11.1 and 11.3. Of course, a square matrix is generally an *asymmetric matrix*. *ASM* introduced in Sect. 1.1 is the very asymmetric matrix.

Definition 11.6 (*Hermitian matrix*) A square matrix $\boldsymbol{A}$ is said to be a *Hermitian (or self-adjoint)* if it is written in the form:

Table 11.1 The squared distance matrix $\boldsymbol{T}$ of four objects shown in Fig. 11.1

	v_1	v_2	v_3	v_4
v_1	0	1.25	1	1.25
v_2	1.25	0	1.25	4
v_3	1	1.25	0	1.25
v_4	1.25	4	1.25	0

Table 11.2 The doubly centered inner product matrix $\boldsymbol{B}_c$ of four objects shown in Fig. 11.1

	v_1	v_2	v_3	v_4
v_1	0.25	0	– 0.25	0
v_2	0	1.0	0	−1.0
v_3	– 0.25	0	0.25	0
v_4	0	−1.0	0	1.0

Table 11.3 Part of the distance matrix between nine Munsell colors obtained by Torgerson [131]

	v_1	v_2	v_3	...	v_9
v_1	0	1.23	3.48	...	5.83
v_2	1.23	0	2.59	...	4.38
v_3	3.48	2.59	0	...	6.22
⋮	⋮	⋮	⋮	...	⋮
v_9	5.83	4.38	6.22	...	0

$$\boldsymbol{A} = \begin{bmatrix} a_{11} & a_{12} & \cdots & a_{1n} \\ \overline{a}_{12} & a_{22} & \cdots & a_{2n} \\ \vdots & \vdots & \ddots & \vdots \\ \overline{a}_{1n} & \overline{a}_{2n} & \cdots & a_{nn} \end{bmatrix}, \tag{11.7}$$

so that $a_{ji} = \overline{a}_{ij}$ for $i = 1, 2, \ldots, n, j = 1, 2, \ldots, n$, where $\overline{a}_{ij}$ is the *complex conjugate* of a_{ij}.

Example 11.5 The following is a Hermitian matrix:

$$\boldsymbol{A} = \begin{bmatrix} 3 & 2+i & 1-7i \\ 2-i & 4 & -6+8i \\ 1+7i & -6-8i & 5 \end{bmatrix}.$$

Remark Equivalently a square matrix $\boldsymbol{A}$ is *Hermitian* if and only if

$$\boldsymbol{A}^* = \boldsymbol{A}, \tag{11.8}$$

where $\boldsymbol{A}^*$ is the *conjugate transpose* of $\boldsymbol{A}$, that is, $\boldsymbol{A}^* = \overline{\boldsymbol{A}}^t$. The matrix $\overline{\boldsymbol{A}}^t$ is frequently denoted by $\boldsymbol{A}^H$ (e.g., Wilkinson [140], p. 15), as pointed out in Remark on Definition 11.3. It is evident from the above definition that the main diagonal elements of a Hermitian matrix are *real*. Examples 1.23 and 1.24 in Sect. 1.2 are examples of Hermitian matrices. It is evident from Definition 11.6 that Hermitian matrix is *an extension* of the symmetric matrix defined in Definition 11.5. Hermitian matrix plays an important role in constructing HFM proposed by Chino and Shiraiwa [15], as shown in Chap. 4.

Definition 11.7 (*triangular matrix*) A matrix $\boldsymbol{A} = \left[a_{ij}\right] \in M_n$ is called *upper triangular* if $a_{ij} = 0$ whenever $j < i$. In contrast, $\boldsymbol{A}$ is called *lower triangular* if its transpose is upper triangular. Moreover, an upper triangular matrix is called *strictly upper triangular* if all of its diagonal elements are equal to zero. Similarly, a lower triangular matrix is called *strictly lower triangular* if all of its diagonal elements are equal to zero.

Example 11.6 The following matrices are an upper triangular matrix and a lower triangular matrix

$$\boldsymbol{A} = \begin{bmatrix} 3\,2\,1 \\ 0\,4\,6 \\ 0\,0\,5 \end{bmatrix} \text{ and } \boldsymbol{B} = \begin{bmatrix} 3\,0\,0 \\ 2\,4\,0 \\ 1\,6\,5 \end{bmatrix}, \text{ respectively.}$$

Definition 11.8 (*column vector and row vector*) If a rectangular matrix $\boldsymbol{A}$ defined by Definition 11.1 has only one column (i.e., $m \times 1$ matrix) or one row (i.e., $1 \times n$ matrix), we have

$$\boldsymbol{b} = \begin{bmatrix} b_1 \\ b_1 \\ \vdots \\ b_m \end{bmatrix}, \boldsymbol{c} = [c_1, c_2, \ldots, c_n]. \tag{11.9}$$

These matrices $\boldsymbol{b}$ and $\boldsymbol{c}$ are also referred to as *column vector* and *row vector*, respectively.

Definition 11.9 (*sum of the two matrices*) Let $\boldsymbol{A} = \left\{a_{ij}\right\}$ and $\boldsymbol{B} = \left\{b_{ij}\right\}$are two rectangular matrices of the same size, say, $m \times n$ matrices. Then, the sum of these two matrices is defined by

$$\boldsymbol{A} + \boldsymbol{B} = \left\{a_{ij} + b_{ij}\right\}. \tag{11.10}$$

Remark As regards the addition of two matrices, the *commutative* and *associative laws* hold for matrices of the same size,

$$\begin{gathered} \boldsymbol{A} + \boldsymbol{B} = \boldsymbol{B} + \boldsymbol{A}, \\ (\boldsymbol{A} + \boldsymbol{B}) + \boldsymbol{C} = \boldsymbol{A} + (\boldsymbol{B} + \boldsymbol{C}), \end{gathered} \tag{11.11}$$

respectively.

Definition 11.10 (*scalar multiplication*) The product of the matrix $\boldsymbol{A} = \left\{a_{ij}\right\}$ with a scalar $\alpha \in F$ is the matrix $\boldsymbol{C} = \left\{c_{ij}\right\}$ defined by

$$\boldsymbol{C} = \alpha \boldsymbol{A} = \left\{\alpha a_{ij}\right\}. \tag{11.12}$$

Remark The field F may be that of real number, R, or that of complex number, C. Moreover, If $\alpha = -1$ and $\boldsymbol{A} = \boldsymbol{B}$ in the above definition, then $\boldsymbol{C} = -\boldsymbol{B}$. Then, the sum of the two matrices, $\boldsymbol{A}$ and $\boldsymbol{C}$, becomes $\boldsymbol{A} - \boldsymbol{B}$ which defines the *difference of the two matrices*, $\boldsymbol{A}$ and $\boldsymbol{B}$.

The following properties of addition and scalar multiplication of matrices follow from the definitions and from the corresponding properties of numbers:

$$\begin{aligned} 0A &= \boldsymbol{O}, \\ (\alpha + \beta)\boldsymbol{A} &= \alpha\boldsymbol{A} + \beta\boldsymbol{A}, \\ \alpha(\boldsymbol{A} + \boldsymbol{B}) &= \alpha A + \alpha\boldsymbol{B}, \\ \alpha(\beta\boldsymbol{A}) &= (\alpha\beta)\boldsymbol{A}, \end{aligned} \tag{11.13}$$

where $\boldsymbol{O}$ is a zero-matrix.

Equations (11.9), (11.10), and (11.11) satisfy the axioms for the *abstract vector space* defined in Definition 10.12 of Chap. 10. In other words, matrices constitute a *vector space*. As also noted in that definition, matrices treated as vectors in abstract vector space are in marked contrast to vectors treated as *concrete ones* in elementary mathematics.

Definition 11.11 (*product of two matrices*) Let $\boldsymbol{A} = \{a_{ij}\}$ and $\boldsymbol{B} = \{b_{pq}\}$ be an $m \times n$ matrix and an $n \times l$ matrix, respectively. Then the product of these two matrices $\boldsymbol{C} = \{c_{jk}\} = \boldsymbol{AB}$ is defined in scalar notation by

$$c_{jk} = \sum_{p=1}^{n} a_{jp} b_{pk}, \tag{11.14}$$

where $\boldsymbol{C}$ is an $m \times l$ matrix.

Remark The following properties of matrix multiplication follow from the definition, which include the distributive and associative laws:

$$\begin{aligned} \boldsymbol{AI} &= \boldsymbol{IA}, \\ \boldsymbol{A}(\boldsymbol{B} + \boldsymbol{C}) &= \boldsymbol{AB} + \boldsymbol{AC},\ (\boldsymbol{B} + \boldsymbol{C})\boldsymbol{D} = \boldsymbol{BD} + \boldsymbol{CD},\ (distributivelaws). \\ \boldsymbol{A}(\boldsymbol{BC}) &= (\boldsymbol{AB})\boldsymbol{C},\ (associativelaw). \end{aligned} \tag{11.15}$$

However, matrix multiplication, in general, does not satisfy the *commutative law*, that is,

$$\boldsymbol{AB} \neq \boldsymbol{BA}. \tag{11.16}$$

Definition 11.12 (*determinant of a matrix*) Let $\boldsymbol{A}$ be a square matrix of order n. Then, the determinant of $\boldsymbol{A}$, denoted *det* $\boldsymbol{A}$ (or frequently denoted $|\boldsymbol{A}|$) is defined by

$$|\boldsymbol{A}| = \sum_{j} (-1)^{t(j)} a_{1j_1} a_{2j_2} \ldots a_{nj_n}, \tag{11.17}$$

where $t(j)$ is the number of *inversions* in the permutation $j = (j_1, j_2, \ldots, j_n)$ and j varies over all $n!$ permutations of $(1, 2, \ldots, n)$. Here, we say that a pair of numbers j_k and j_p in the permutation $(j_1, j_2, \ldots, j_n)$ form an inversion if $j_k > j_p$ while $k < p$, that is, if a large number in the permutation precedes a smaller one.

Remark This definition can be seen, for example, in Lancaster and Tismenetsky [68], pp. 25–26.

Another definition of the determinant of a matrix can be made using the notion of *minor*. Letting $\boldsymbol{A}$ be a square matrix of order n, a *minor* of order $n-1$ of $\boldsymbol{A}$ is defined to be the determinant of a submatrix of $\boldsymbol{A}$ obtained by deleting one row and one column from $\boldsymbol{A}$. We shall denote the minor obtained by deleting the ith row and jth column by m_{ij}. Moreover, we shall define a *cofactor* $\tilde{a}_{jk}$ of the element a_{jk} as

$$\tilde{a}_{jk} = (-1)^{j+k} m_{jk}. \tag{11.18}$$

A cofactor is a signed minor, which is a scalar (e.g., Lancaster and Tismenetsky [op. cit., pp. 32–33]).

Example 11.7 The following matrices are a minor m_{12} and a cofactor $\tilde{a}_{12}$ of a matrix $\boldsymbol{A}$ of order 3,

$$m_{12} = \begin{vmatrix} 5 & 6 \\ 3 & 5 \end{vmatrix} = (5 \times 5 - 3 \times 6) = 7,$$

and

$$\tilde{a}_{12} = (-1)^{1+2} m_{12} = (-1)^{1+2} \times 7 = -7,$$

respectively, where

$$\boldsymbol{A} = \begin{bmatrix} 3 & 2 & 1 \\ 5 & 4 & 6 \\ 3 & 7 & 5 \end{bmatrix}.$$

Definition11.13 (*adjoint of a matrix*) Let $\tilde{\boldsymbol{A}}$ be a square matrix of order n, composed of cofactors:

$$\tilde{\boldsymbol{A}} = \{\tilde{a}_{jk}\} = \begin{bmatrix} \tilde{a}_{11} & \tilde{a}_{12} & \cdots & \tilde{a}_{1n} \\ \tilde{a}_{21} & \tilde{a}_{22} & \cdots & \tilde{a}_{2n} \\ \vdots & \vdots & & \vdots \\ \tilde{a}_{n1} & \tilde{a}_{n2} & \cdots & \tilde{a}_{nn} \end{bmatrix}, \tag{11.19}$$

which is called a *matrix of cofactors* of $\boldsymbol{A}$. Then, the *adjoint of* $\boldsymbol{A}$, written as *adj* $\boldsymbol{A}$, is defined to be the transposed matrix of $\tilde{\boldsymbol{A}}$, that is,

$$adj\,\boldsymbol{A} = \tilde{\boldsymbol{A}}^t. \tag{11.20}$$

The adjoint of $\boldsymbol{A}$ is represented by

$$adj\boldsymbol{A} = \begin{bmatrix} \tilde{a}_{11} & \tilde{a}_{21} & \cdots & \tilde{a}_{\mathrm{n}1} \\ \tilde{a}_{12} & \tilde{a}_{22} & \cdots & \tilde{a}_{\mathrm{n}2} \\ \vdots & \vdots & \ddots & \vdots \\ \tilde{a}_{1n} & \tilde{a}_{2n} & \cdots & \tilde{a}_{nn} \end{bmatrix}.$$

Remark Horn and Johnson ([58], p. 20) call the adjoint of $\boldsymbol{A}$ *classical adjoint of* $\boldsymbol{A}$, or *adjugate of* $\boldsymbol{A}$ to avoid confusion with the Hermitian adjoint $\boldsymbol{A}^*$, which was noted in Definition 11.6.

Example 11.8 Let us first compute the matrix of cofactors of the matrix $\boldsymbol{A}$, in a similar way as shown in Example 11.7. Then, we have

$$\tilde{\boldsymbol{A}} = \begin{bmatrix} -22 & -7 & 23 \\ -3 & 12 & -15 \\ 8 & -13 & 2 \end{bmatrix}.$$

Then, by definition, we have

$$adj\,\boldsymbol{A} = \begin{bmatrix} -22 & -3 & 8 \\ -7 & 12 & -13 \\ 23 & -15 & 2 \end{bmatrix}.$$

Now, another definition of a determinant of a matrix is in order using the cofactor expansion:

Definition 11.14 (*cofactor expansion*) Let $\boldsymbol{A} = \{a_{jk}\}$be a square matrix of order n. Then, we have, for any j, k $(1 \le j, k \le n)$;

$$|\boldsymbol{A}| = \sum_{k=1}^{n} a_{jk}\tilde{a}_{jk}, \tag{11.21}$$

or

$$|\boldsymbol{A}| = \sum_{j=1}^{n} a_{jk}\tilde{a}_{jk}, \tag{11.22}$$

which are called the *row* and *column cofactor expansions* of determinant, respectively.

Example 11.9 If we adopt a row cofactor expansions of $\boldsymbol{A}$, we have, for example,

$$|A| = a_{11}\tilde{a}_{11} + a_{12}\tilde{a}_{12} + a_{13}\tilde{a}_{13} = 3 \times (-22) + 2 \times (-7) + 1 \times (23) = -57,$$

using a matrix of cofactors of $\boldsymbol{A}$ shown in Example 11.8.

Remark Cofactor expansion is sometimes called the *Laplace expansion* (e.g., Horn and Johnson [58], p. 7). Cofactor expansions can be generalized using a *minor of order p* $(1 \leq p \leq n)$, which enables us to define the other definition of determinant by *Laplace's theorem* (e.g., Lancaster and Tismenetsky [68], p. 37). However, we will not introduce the theorem, because this theorem is beyond the scope of this book.

Definition 11.15 (*singular matrix*)A square matrix $\boldsymbol{A}$ is called *singular* if $|\boldsymbol{A}|$ is zero, and otherwise *nonsingular*.

Definition 11.16 (*inverse matrix*) If a square matrix $\boldsymbol{A}$ is nonsingular, the *inverse of* $\boldsymbol{A}$ is computed as

$$\boldsymbol{A}^{-1} = adj\,\boldsymbol{A}/|\boldsymbol{A}|. \tag{11.23}$$

Next, we shall define the *rank* of a matrix. To do so, we must remember the notions of *linear dependence* and *independence* of a set of vectors in a vector space, which have already defined in Definition 2.48.

Remark The finite set of elements $\{\boldsymbol{x}_1, \boldsymbol{x}_2, \ldots, \boldsymbol{x}_n\}$ is said to be a (finite) *basis* of S if they are linearly independent and every element $\boldsymbol{x} \in S$ is a linear combination of the elements in $\{\boldsymbol{x}_1, \boldsymbol{x}_2, \ldots, \boldsymbol{x}_n\}$:

$$\boldsymbol{x} = \sum_{j=1}^{n} \alpha_j \boldsymbol{x}_j, \alpha_j \in F, j = 1, 2, \ldots, n. \tag{11.24}$$

Definition 11.17 (*rank of a matrix*) If $\boldsymbol{A}$ is an $m \times n$ rectangular matrix, $M_{m,n}(F)$, rank of a matrix $\boldsymbol{A}$ is defined to be the largest number of columns of $\boldsymbol{A}$ which constitute a linearly independent set. The rank of $\boldsymbol{A}$ is usually denoted by *rank* $\boldsymbol{A}$.

Remark It is also possible to define rank of a matrix in terms of *elementary row and column operations* (or *elementary transformation*) to an $m \times n$ matrix (e.g., Lancaster and Tismenetsky, op. cit., p. 51; Wilkinson [139], p. 45).

It is apparent from (11.23) that we cannot compute the "regular" inverse of a square matrix $\boldsymbol{A}$ if its rank is lower than its order. However, the corresponding problem in a more general case where $\boldsymbol{A}$ may be a rectangular or a square singular matrix has a solution which is called the *generalized inverse* (*g-inverse*). We shall first define a generalized inverse of a matrix $\boldsymbol{A}$ of order $m \times n$ below (e.g., Magnus and Neudecker [79], p. 38), and then show some of its variants next.

Definition 11.18 (*generalized inverse*) Any matrix $\boldsymbol{X}$ that satisfies $\boldsymbol{AXA} = \boldsymbol{A}$ is called a generalized inverse of $\boldsymbol{A}$ and denoted $\boldsymbol{A}^{-}$.

Definition 11.19 (*reflexive type generalized inverse*) Any matrix $\boldsymbol{X}$ is called a *reflexive generalized inverse* if

$$\boldsymbol{AXA} = \boldsymbol{A}, \tag{11.25}$$

$$\boldsymbol{XAX} = \boldsymbol{X}. \tag{11.26}$$

Remark Tanabe [123] discusses several generalized inverses and denotes the above generalized inverse by $\boldsymbol{A}_r^-$, while Lancaster and Tismenetsky [69] denote it by $\boldsymbol{A}^I$.

Definition 11.20 (*minimum norm type generalized inverse*) Any matrix $\boldsymbol{X}$ is called a *minimum norm type generalized inverse* if

$$\boldsymbol{AXA} = \boldsymbol{A}, \tag{11.27}$$

$$(\boldsymbol{XA})^* = \boldsymbol{XA}. \tag{11.28}$$

Remark Tanabe [op. cit.] denotes the above generalized inverse by A_m^-.

Definition 11.21 (*least square type generalized inverse*) Any matrix $\boldsymbol{X}$ is called a *least square type generalized inverse* if

$$\boldsymbol{AXA} = \boldsymbol{A}, \tag{11.29}$$

$$(\boldsymbol{AX})^* = \boldsymbol{AX}. \tag{11.30}$$

Remark Tanabe [op. cit.] denotes the above generalized inverse by A_m^-.

Definition 11.22 (*Moor-Penrose generalized inverse*) Any matrix $\boldsymbol{X}$ is called a *Moor-Penrose generalized inverse* if

$$\boldsymbol{AXA} = \boldsymbol{A}, \tag{11.31}$$

$$\boldsymbol{XAX} = \boldsymbol{X}, \tag{11.32}$$

$$(\boldsymbol{XA})^* = \boldsymbol{XA}, \tag{11.33}$$

$$(\boldsymbol{AX})^* = \boldsymbol{AX}. \tag{11.34}$$

Remark The Moor-Penrose generalized inverse has a unique solution. This inverse is originated by Moor [81] and the above definition is due to Penrose [101].

Example 11.10 The following matrix $\boldsymbol{A}$ is a rank 2 matrix,

$$
\boldsymbol{A} = \begin{bmatrix} 8.00 & 6.00 & -3.00 & -3.00 \\ 6.00 & 5.00 & -1.50 & -2.50 \\ -3.00 & -1.50 & 2.25 & 0.75 \\ -3.00 & -2.50 & 0.75 & 1.25 \end{bmatrix},
$$

computed in such a way that $\boldsymbol{A} = \boldsymbol{G}\boldsymbol{G}^t$, where

$$
\boldsymbol{G} = \begin{bmatrix} 2 & 2 \\ 2 & 1 \\ 0 & -1.5 \\ -1 & -0.5 \end{bmatrix}.
$$

This matrix appears in Gabriel [44] in computing a matrix for his biplot method.

The Moore Penrose inverse of $\boldsymbol{A}$, which we shall denote by $\boldsymbol{A}^-$, can be computed, for example, by a MATRAB function, pinv($\boldsymbol{A}$), as

$$
\boldsymbol{A}^- = \begin{bmatrix} 0.0455 & -0.0047 & -0.0753 & 0.0024 \\ -0.0047 & 1.1384 & 0.2147 & -0.0692 \\ -0.0753 & 0.2147 & 0.4350 & -0.1073 \\ 0.0024 & -0.0692 & -0.1073 & 0.0346 \end{bmatrix}.
$$

We can easily check that $\boldsymbol{G}^t\boldsymbol{A}^-\boldsymbol{G} = \boldsymbol{I}_2$, where $\boldsymbol{I}_2$ is the identity matrix of order 2.

Definition 11.23 (*similar matrix*) Square matrices $\boldsymbol{A}$ and $\boldsymbol{B}$ contained in $M_n(R)$ are said to be *similar*, if there is a nonsingular matrix $\boldsymbol{P}$ such that

$$
\boldsymbol{A} = \boldsymbol{P}\boldsymbol{B}\boldsymbol{P}^{-1}. \tag{11.35}
$$

A similar matrix can be extended to complex spaces. To do so, we must define a *unitary matrix* as follows:

Definition 11.24 (*unitary matrix*) A matrix $\mathbf{U} \in M_n(C)$ is called *unitary* if

$$
\boldsymbol{U}^*\boldsymbol{U} = \boldsymbol{I}. \tag{11.36}
$$

Remark It is apparent from (11.24) that $\boldsymbol{U}^{-1} = \boldsymbol{U}^*$.

Definition 11.25 (*unitary similar matrix*) Square matrices $\boldsymbol{A}$ and $\boldsymbol{B}$ contained in $M_n(C)$ are said to be *unitary similar*, if there is a unitary matrix $\boldsymbol{U}$ such that

$$
\boldsymbol{A} = \boldsymbol{U}\boldsymbol{B}\boldsymbol{U}^{-1} = \boldsymbol{U}\boldsymbol{B}\boldsymbol{U}^{*}. \tag{11.37}
$$

A similar matrix and the unitary similar matrix are deeply related to the *eigenvalue problem* to be discussed later. Therefore, we shall postpone this topic until the eigenproblem is defined.

Definition 11.21 (*circulant matrix*) A matrix $\mathbf{A} \in M_n(C)$ of the form

$$\boldsymbol{A} = \begin{bmatrix} a_1 & a_2 & \cdots & \cdots & a_n \\ a_n & a_1 & a_2 & \cdots & a_{n-1} \\ a_{n-1} & a_n & a_1 & \cdots & a_{n-2} \\ \vdots & \vdots & \ddots & \ddots & \vdots \\ a_2 & a_{11} & \cdots & a_n & a_1 \end{bmatrix}, \tag{11.38}$$

is called a *circulant matrix*.

Remark Examples 1.18 and 1.19 in Chap. 1 are examples of circulant matrices.

Definition 11.22 (*Toeplitz matrix*) A matrix $\boldsymbol{A} = \{a_{ij}\} \in M_{n+1}(C)$ of the form

$$\boldsymbol{A} = \begin{bmatrix} a_0 & a_1 & a_2 & \cdots & a_n \\ a_{-1} & a_0 & a_1 & \cdots & a_{n-1} \\ a_{-2} & a_{-1} & a_0 & \cdots & a_{n-2} \\ \vdots & \vdots & \ddots & \ddots & \vdots \\ a_{-\mathrm{n}} & a_{-\mathrm{n}+1} & \cdots & a_{-1} & a_0 \end{bmatrix}, \tag{11.39}$$

is called a *Toeplitz matrix*, where a_{ij} denotes a_{j-i} for some given sequence,

$$a_{-n}, a_{-n+1}, \ldots, a_{-1}, a_0, a_1, a_2, \ldots, a_{n-1}, a_n \in C. \tag{11.40}$$

Remark Example 1.20 in Chap. 1 is an example of Toeplitz matrices.

Definition 11.23 (*permutation matrix*) A matrix $\boldsymbol{P}$ is called a *permutation matrix* if exactly one entry in each row and column is equal to 1, and all other entries are 0.

Example 11.11 Let $\boldsymbol{P}$ be the following permutation matrix of order 4:

$$\boldsymbol{P} = \begin{bmatrix} 0 & 0 & 1 & 0 \\ 1 & 0 & 0 & 0 \\ 0 & 1 & 0 & 0 \\ 0 & 0 & 0 & 1 \end{bmatrix}. \tag{11.41}$$

Post-multiplication of $\boldsymbol{P}$ by the column vector $\boldsymbol{x} = [1, 2, 3, 4]^t$ results in

$$\boldsymbol{Px} = \begin{bmatrix} 3 \\ 1 \\ 2 \\ 4 \end{bmatrix}, \tag{11.42}$$

which is a permutation of the elements of the vector $\boldsymbol{x}$. On the other hand, pre-multiplication of $\boldsymbol{P}$ by the row vector $\boldsymbol{x} = [1, 2, 3, 4]$ results in

$$\boldsymbol{xP} = [2, 3, 1, 4], \tag{11.43}$$

which is another permutation.

Remark We have already introduced the notion of permutation in Definition 11.12. The permutation matrix discussed here provides us with a method for obtaining permutations $j = (j_1, j_2, \ldots, j_n)$ in which j varies over all permutations of $(1, 2, \ldots, n)$ discussed there. It is well known that every permutation matrix may be expressed as a product of matrices $\boldsymbol{I}_{ij}$ which are equal to the identity matrix except in rows i and j and columns i and j, which are of the form (e.g., Wilkinson [139], p. 44)

$$\begin{array}{c} \quad\quad col.i\,col.j \\ \boldsymbol{I}_{ij} = \begin{bmatrix} 0 & 1 \\ 1 & 0 \end{bmatrix} \begin{array}{l} row i \\ row j \end{array} . \end{array} \tag{11.44}$$

This matrix is known as one of the matrices of *elementary transformations* (or *operations*) which reduce a matrix of general form to one of the *canonical forms* (e.g., Horn and Johnson [58]; Lancaster and Tismenetsky [69]; Wilkinson, op. cit.). However, we shall not go into details of these transformations, because they are outside the scope of this book.

We shall now define the Gram matrix. Before doing so, we must define orthogonal and orthonormal sets or systems according to Lancaster and Tismenetsky [op. cit.].

Definition 11.24 (*orthogonal and orthonormal sets*) Any set of nonzero elements from a *unitary space* is called an *orthogonal set, or system*, if any two members of the set are orthogonal, or if the set consists of only one element. An orthogonal set or system in which each member is normalized is called as an *orthonormal set or system.*

Remark As regards the unitary space, we shall refer to Definition 10.39 of Chap. 10. In short, unitary space is an inner product space on a complex vector space C^n.

Definition 11.25 (*Gram matrix*) Let $\{\boldsymbol{x}_1, \boldsymbol{x}_2, \ldots, \boldsymbol{x}_r\} \subset C^n$ be any set of elements from a *unitary space*. Then, the $r \times r$ matrix $\boldsymbol{G} = \{(\boldsymbol{x}_j, \boldsymbol{x}_k)\}$ is called the *Gram matrix* of $\{\boldsymbol{x}_1, \boldsymbol{x}_2, \ldots, \boldsymbol{x}_r\}$, where $(\boldsymbol{x}_j, \boldsymbol{x}_k)$ is an inner product defined on the space.

Remark The determinant of the Gram matrix, $|\boldsymbol{G}|$, is called the *Gramian* or *Gram determinant* of $\{\boldsymbol{x}_1, \boldsymbol{x}_2, \ldots, \boldsymbol{x}_r\}$. The Gram matrix plays an important role in spectral graph theory. It should be noted from the definition that the Gram matrix is *Hermitian* and consequently, the Gramian is *real* (e.g., Lancaster and Tismenetsky [op. cit.], p. 110).

The Gram matrix can also be expressed in matrix form. Let $\boldsymbol{y}_1, \boldsymbol{y}_2, \ldots, \boldsymbol{y}_n$ be an orthonormal basis for the space and let $\boldsymbol{x}_j$ be such that

$$\boldsymbol{x}_j = \sum_{k=1}^{n} \alpha_{jk} \boldsymbol{y}_k, j = 1, 2, \ldots, r. \tag{11.45}$$

Now, let us define the $r \times n$ matrix $\boldsymbol{A} = \left[\alpha_{jk}\right]$. Here, it should be noticed that $\boldsymbol{A}$ has full rank if and only if the set $\{\boldsymbol{x}_1, \boldsymbol{x}_2, \ldots, \boldsymbol{x}_r\} \subset C^n$ is linearly independent. Now observe that

$$\left(\boldsymbol{x}_j, \boldsymbol{x}_k\right) = \sum_{u,v=1}^{n} \alpha_{ju}\overline{\alpha}_{kv}\left(\boldsymbol{y}_u, \boldsymbol{y}_v\right) = \sum_{u=1}^{n} \alpha_{ju}\overline{\alpha}_{ku}. \tag{11.46}$$

Here, it should be noted that $\left(\boldsymbol{y}_u, \boldsymbol{y}_v\right) = \delta_{uv}$, where δ_{uv} is the *Kronecker delta.* Then, we have

$$\boldsymbol{G} = \left[\left(\boldsymbol{x}_j, \boldsymbol{x}_k\right)\right] = \boldsymbol{A}\boldsymbol{A}^*. \tag{11.47}$$

(e.g., Lancaster and Tismenetsky, op. cit., p. 110).

The next theorem is an important one which characterizes the properties of the Gram matrix. Before proceeding to the theorem, let us revisit the remark on Definition 10.38 in which we introduced the notion of the positive definiteness. In that remark, we defined definiteness in terms of Hermitian functional. Here, Horn and Johnson ([58], p. 396) give a bit different definition of the positive definiteness from that of Cristescu ([21], p. 48) using matrix notation: An $n \times n$Hermitian matrix $\boldsymbol{A}$ is said to be *positive definite* if $\boldsymbol{x}^*\boldsymbol{A}\boldsymbol{x} > 0$ for all *nonzero* $\boldsymbol{x} \in C^n$. If the strict inequality required here is weakened to $\boldsymbol{x}^*\boldsymbol{A}\boldsymbol{x} \geq 0$, then matrix $\boldsymbol{A}$ is called *positive semidefinite* (abbreviated as *p.s.d*, unless otherwise stated).

Remark In matrix notation, a *Hermitian form* is defined by

$$\boldsymbol{x}^*\boldsymbol{A}\boldsymbol{x} = \sum_{i,j=1}^{n} a_{ij}x_i\overline{x}_j, \tag{11.48}$$

Then, we can also define the definiteness of a matrix in terms of its *eigenvalues* which will be discussed soon. In fact, a necessary and sufficient condition for $\boldsymbol{x}^*\boldsymbol{A}\boldsymbol{x}$ to be positive definite is that all the eigenvalues of $\boldsymbol{A}$ is *positive.* We shall refer, for example, to Wilkinson ([139], pp. 28–110) for this topic. A special case of the Hermitian form is the quadratic form denoted by $\boldsymbol{x}^t\boldsymbol{A}\boldsymbol{x}$ for $\boldsymbol{x} \in R^n$ and an $n \times n$ *symmetric matrix* $\boldsymbol{A}$.

Theorem 11.1 *Let $\boldsymbol{G} \in M_r$ be the Gram matrix of the vectors $\{\boldsymbol{x}_1, \boldsymbol{x}_2, \ldots, \boldsymbol{x}_r\} \subset C^n$ with respect to a given inner product $(\bullet, \bullet)$, and let $\boldsymbol{X} = [\boldsymbol{x}_1, \boldsymbol{x}_2, \ldots, \boldsymbol{x}_r] \in M_{n,r}$. Then,*

(1) *$\boldsymbol{G}$ is p.s.d;*
(2) *$\boldsymbol{G}$ is nonsingular if and only if the vectors $\boldsymbol{x}_1, \boldsymbol{x}_2, \ldots, \boldsymbol{x}_r$ are independent;*
(3) *There exists a positive definite matrix $\boldsymbol{P} \in M_n$ such that $\boldsymbol{G} = \boldsymbol{X}^*\boldsymbol{P}\boldsymbol{X}$, and*
(4) *rank $\boldsymbol{G}$ = rank $\boldsymbol{X}$ which is the maximum number of independent vectors in the set $\{\boldsymbol{x}_1, \boldsymbol{x}_2, \ldots, \boldsymbol{x}_r\}$.*

Proof of this theorem is, for example, in Horn and Johnson ([op. cit.], pp. 407–408).

Remark The Gram matrix appears in various fields of applied mathematics. For example, the Laplace matrix which plays a crucial role *in spectral graph theory* can be represented by a *Gram matrix* computed from the incidence matrix of a graph or a digraph, as will be discussed in the next section. A well-known *singular value decomposition* of an arbitrary matrix from $M_{m,n}(C)$ utilizes the two kinds of the Gram matrix, as will be discussed later in this section.

Theorem 11.2 (nomal matrix) *A square matrix* $\boldsymbol{A}$ *contained in* $M_n(C)$ *is normal if and only if*

$$\boldsymbol{AA}^* = \boldsymbol{A}^*\boldsymbol{A}. \tag{11.49}$$

For the proof of this theorem, see, for example, Lancaster and Tismenetsky ([69], pp. 176–177).

Remark It is apparent that a *Hermitian matrix* is a normal matrix from this theorem. It should be noticed that for any matrix $\boldsymbol{A} \in M_{m,n}(C)$ the matrices $\boldsymbol{AA}^*$ and $\boldsymbol{A}^*\boldsymbol{A}$ are *p.s.d* (e.g., Lancaster and Tismenetsky, op. cit., p. 180).

Let us now consider a *square root* $\boldsymbol{H}_0$ of a positive semidefinite matrix $\boldsymbol{H}$, which satisfies $\boldsymbol{H}_0^2 = \boldsymbol{H}$. Then, we have the following theorem:

Theorem 11.3 *A matrix* $\boldsymbol{H}$ *is positive definite (or semidefinite) if and only if it has a positive definite (respectively, semidefinite) square root* $\boldsymbol{H}_0$. *Moreover, rank* $\boldsymbol{H}_0 = rank\boldsymbol{H}$.

For the proof of this theorem, see, for example, Lancaster and Tismenetsky (op. cit., pp. 181).

Remark From the remark on Theorem 11.2, for any matrix $\boldsymbol{A} \in M_{m,n}(C)$ the matrix $\boldsymbol{A}^*\boldsymbol{A} \in M_n(C)$ is *p.s.d*. As a result, from Theorem 11.3 the matrix $\boldsymbol{A}^*\boldsymbol{A}$ has a positive semidefinite square root $(\boldsymbol{A}^*\boldsymbol{A})^{1/2}$ for any matrix $\boldsymbol{A} \in M_{m,n}(C)$.

Definition 11.26 (*irreducible matrix*) If $n \geq 2$, the matrix $\boldsymbol{A} \in M_n(C)$ is said to be *reducible* if there is an $n \times n$*permutation matrix* $\boldsymbol{P}$ such that

$$\boldsymbol{P}^t\boldsymbol{AP} = \begin{bmatrix} \boldsymbol{A}_{11} & \boldsymbol{A}_{12} \\ \boldsymbol{O} & \boldsymbol{A}_{22} \end{bmatrix}, \tag{11.50}$$

where $\boldsymbol{A}_{11}$ and $\boldsymbol{A}_{22}$ are square matrices of order less than n, and $\boldsymbol{O}$ is a zero-matrix. If no such $\boldsymbol{P}$ exists, then $\boldsymbol{A}$ is said to be *irreducible*.

Remark An important theorem on an irreducible matrix which plays a crucial role in spectral graph theory is the *Perron-Frobenius theorem* (e.g., Horn and Johnson [58], pp. 508–511; Lancaster and Tismenetsky, op. cit., p. 536). We shall discuss it in connection with an algebraic eigenvalue problem later.

We shall now introduce basic definitions and theorems on the algebraic eigenvalue problem. Although there is an immense literature on these definitions and theorems, we shall select a minimum of them which are necessary for the topics raised in this book.

Definition 11.27 (*eigenvalue problem*) Let $\boldsymbol{A} \in M_n(C)$. A nonzero vector $\boldsymbol{x} \in C^n$ is defined to be an *eigenvector* of $\boldsymbol{A}$ corresponding to the *eigenvalue* $\lambda \in C$ if

$$\boldsymbol{A}\boldsymbol{x} = \lambda\boldsymbol{x}, \boldsymbol{x} \neq 0. \tag{11.51}$$

Remark Let us consider the eigenproblem of the transposed matrix $\boldsymbol{A}^t$. By, definition,

$$\boldsymbol{A}^t\boldsymbol{y} = \lambda\boldsymbol{y}, \boldsymbol{y} \neq 0. \tag{11.52}$$

The eigenvalues of $\boldsymbol{A}^t$ are the same as those of $\boldsymbol{A}$, since the determinant of a matrix is equal to that of its transpose. In contrast, the eigenvectors of $\boldsymbol{A}^t$ are, in general, different from that of $\boldsymbol{A}$. Let us denote the eigenvectors of $\boldsymbol{A}^t$ associated with the eigenvalue λ_i by $\boldsymbol{y}_i$, so that we have

$$\boldsymbol{A}^t\boldsymbol{y}_i = \lambda_i\boldsymbol{y}_i, \tag{11.53}$$

which may be rewritten in the form

$$\boldsymbol{y}_i^t\boldsymbol{A} = \lambda_i\boldsymbol{y}_i^t. \tag{11.54}$$

The $\boldsymbol{y}_i^t$ is sometimes said to be a *left-hand eigenvector* of $\boldsymbol{A}$ associated with λ_i, while the vector $\boldsymbol{x}_i$ such that

$$\boldsymbol{A}\boldsymbol{x}_i = \lambda_i\boldsymbol{x}_i, \tag{11.55}$$

is said to be a *right-hand eigenvector* of $\boldsymbol{A}$ associated with λ_i. It is not difficult to show that
$\boldsymbol{x}_i^t\boldsymbol{y}_j = 0$ if

$$\lambda_i \neq \lambda_j. \tag{11.56}$$

It should be noticed that $\boldsymbol{x}_i^t\boldsymbol{y}_j$ is not an inner product as usually understood, since $\boldsymbol{x}_i$ and/or $\boldsymbol{y}_j$ may be complex vectors (see for example, Definition 10.39). In fact, we have $\boldsymbol{x}_i^t\boldsymbol{y}_j = \boldsymbol{y}_j^t\boldsymbol{x}_i$, and not $\boldsymbol{x}_i^t\boldsymbol{y}_j = \overline{\left(\boldsymbol{y}_j^t\boldsymbol{x}_i\right)}$.

Remark As regards the eigenproblem, the Hermitian matrix introduced in Definition 11.6 has an outstanding property. That is, the eigenvalues of a Hermitian matrix are *all real* in spite of the fact that its elements are generally *complex*. Researchers utilize this property in many fields of science. For example, in quantum mechanics the density matrix operator $\hat{\rho}$ is *Hermitian* (e.g., Stefanucci and Leeuwen [116], p. 101):

$$\hat{\rho} = \sum_n w_n \boldsymbol{x}_n \boldsymbol{x}_n^*, \tag{11.57}$$

where they use the notations of the *ket* and *bra* instead of the state vector $\boldsymbol{x}_n$ and $\boldsymbol{x}_n^*$, respectively. The $\boldsymbol{x}_n$ represents a physical state of a particle in the Hilbert space, and w_n is a probability of finding an isolated system of particles in the state $\boldsymbol{x}_n$ with $\sum_n w_n = 1$. In psychometrics any (real) ASM is converted into a Hermitian matrix $\boldsymbol{H}$ in HFM proposed by Chino and Shiraiwa [15], and objects are embedded in a Hilbert space using the eigenvectors of $\boldsymbol{H}$ if it is *p.s.d.* For further details, see Chaps. 4 and 5.

Definition 11.28 (*characteristic polynomial*) Let us consider the *determinant* of $\lambda \boldsymbol{I} - \boldsymbol{A}$. It is a monic (or univariate) polynomial of degree n in λ and is expressed by

$$det(\lambda \boldsymbol{I} - \boldsymbol{A}) = c_0 + c_1\lambda + \cdots + c_n\lambda^n,\ c_n = 1, \tag{11.58}$$

and is called the *characteristic polynomial of the matrix* $\boldsymbol{A}$.

Definition 11.29 (*spectrum and spectral radius*)The set of all distinct eigenvalues of $\boldsymbol{A} \in M_n$ is called the *spectrum* of $\boldsymbol{A}$ and is denoted by $\sigma(\boldsymbol{A})$, and the nonnegative real number $\rho(\boldsymbol{A}) = max\{|\lambda| : \lambda \in \sigma(\boldsymbol{A})\}$ is called the *spectral radius* of $\boldsymbol{A}$.

Definition 11.30 (*similarity transformation*) Let a matrix $\boldsymbol{H}$ be nonsingular. Then, transforms, $\boldsymbol{H}^{-1}\boldsymbol{AH}$, of a matrix $\boldsymbol{A}$ is called *similarity transforms* while $\boldsymbol{A}$ and $\boldsymbol{H}^{-1}\boldsymbol{AH}$ are said to be *similar*.

Theorem 11.4 *When the eigenvalues of* **A** *are distinct then there is a similarity transformation which reduces* **A** *to diagonal form.*

Proof of this theorem is, for example, in Wilkinson ([139], p. 6).

Example 11.12 Let **A** be the following sums of squares and products of four measurements within species (Fisher [37], p. 181):

$$\boldsymbol{A} = \begin{bmatrix} 19.1434 & 9.0356 & 9.7634 & 3.2394 \\ 9.0356 & 11.8658 & 4.6232 & 2.4746 \\ 9.7634 & 4.6232 & 12.2978 & 3.8794 \\ 3.2394 & 2.4746 & 3.8794 & 3.8794 \end{bmatrix}.$$

Then, the eigenvalues of **A** are 32.2801, 7.5674, 4.9317, and 0.9881. Since the eigenvalues are distinct, there is a similarity transformation which reduces **A** to diagonal form, as shown below:

$$\boldsymbol{H}^{-1}\boldsymbol{AH} = \boldsymbol{D},$$

where $\boldsymbol{H}$ is

$$\boldsymbol{H} = \begin{bmatrix} 0.7215 & -0.0314 & 0.6869 & 0.0820 \\ 0.4526 & -0.7305 & -0.4924 & -0.1379 \\ 0.4922 & 0.6688 & -0.4466 & -0.3331 \\ 0.1800 & 0.1341 & -0.2938 & 0.9291 \end{bmatrix},$$

and $\boldsymbol{D} = diag(32.2801, 7.5674, 4.9317, 0.9881)$.

In this way, this transformation matrix $\boldsymbol{H}$ reduces correlated scales to uncorrelated scales. If we rewrite the above equation as $\boldsymbol{AH} = \boldsymbol{HD}$, then we have $\boldsymbol{A}\ \boldsymbol{h}_j = \lambda_j \boldsymbol{h}_j$, where $\boldsymbol{h}_j$ is the eigenvector corresponding to the eigenvalue λ_j of $\boldsymbol{A}$. This is nothing but a similarity transformation in the multidimensional case compared with that in the unidimensional case of the *ratio scale* in psychological measurements (Stevens [118, 119]), which was discussed in Chap. 10.

Remark If a matrix $\boldsymbol{A}$ has *multiple eigenvalues*, things become complicated. Some matrices can be reduced to diagonal form by a similarity transformation, but others are not. In the former we may find independent eigenvectors associated with these multiple eigenvalues. However, in the latter we cannot reduce a matrix to diagonal form by the similarity transformation. In this case, we have only one eigenvector associated with multiple eigenvalues. The following matrix has such a property:

Definition 11.31 (*a simple Jordan submatrix of order r*) Let us define matrices $\boldsymbol{J}_r(a)$ of order r:

$$J_1(a) = [a], \boldsymbol{J}_r(a) = \begin{bmatrix} a & 1 & & & & \\ & a & 1 & & & \\ & & a & 1 & & \\ & & & \cdot & \cdot & \cdot \\ & & & & a & 1 \\ & & & & & a \end{bmatrix}, \tag{11.59}$$

It is called a *simple Jordan submatrix of order r*.

Remark 1 $\boldsymbol{J}_r(a)$ has an eigenvalue, a, of multiplicity r, but it has only one eigenvector $\boldsymbol{x} = \boldsymbol{e}_1 = [1, 0, 0, \ldots, 0]^t$.

Remark 2 Wilkinson (op. cit., p. 10) call (11.45) a *simple Jordan submatrix of order r*. Horn and Johnson ([58], p. 121) call it a *Jordan block*.

Definition 11.32a (*nilpotent matrix*) Let us consider a simple Jordan submatrix of order r. Then, in the special case when $a = 0$, that is,

$$\boldsymbol{J}_r(0) = \begin{bmatrix} 0 & 1 & & & & \\ & 0 & 1 & & & \\ & & 0 & 1 & & \\ & & & \cdots & \cdots & \\ & & & & 0 & 1 \\ & & & & & 0 \end{bmatrix}, \tag{11.60}$$

it is called a *nilpotent matrix*.

Remark 1 As defined above, nilpotent matrix is a matrix whose *superdiagonal* is composed of one and the other zero. However, sometimes it is defined as a transpose of $\boldsymbol{J}_r(0)$, which is a matrix whose *subdiagonal* is composed of one and the other zero.

Remark 2 For example, Horn and Johnson (op. cit., pp. 37, 139) call $\boldsymbol{J}_r(0)$ a *nilpotent matrix*, while Hirsch and Smale ([55], pp. 122–123) call it (to be precise, its transpose) an *elementary nilpotent block*, and sometimes call it *nilpotent operator*, or *nilpotent matrix*, depending on its context.

Horn and Johnson (op. cit., p. 37) define the nilpotent matrix as follows, which is another definition of a nilpotent matrix:

Definition 11.32b (*nilpotent matrix*) A matrix $\boldsymbol{A} \in M_n$ is called nilpotent if $\boldsymbol{A}^q = \boldsymbol{O}$ for some positive integer q, where $\boldsymbol{O}$ is a zero matrix.

Remark It is known that all eigenvalues of a nilpotent matrix are *0* (e.g., Horn and Johnson (op. cit., p. 37).

Example 11.13 The following is the nilpotent matrix of order 3.

$$N = \begin{bmatrix} 0 & 1 & 0 \\ 0 & 0 & 1 \\ 0 & 0 & 0 \end{bmatrix}.$$

The eigenvalues of $\boldsymbol{N}$ are all 0, and the eigenvectors are $[1, 0, 0]^t$, $[-1, 0, 0]^t$, and $[1, 0, 0]^t$.

In relation to the nilpotent matrix, there is an important matrix called *idempotent matrix*, which is defined below:

Definition 11.33 (*idempotent matrix*) A matrix $\boldsymbol{A} \in M_n$ is called *idempotent* if $\boldsymbol{A}^2 = \boldsymbol{A}$.

Remark It is known that each eigenvalue of an idempotent matrix is either 0 or 1 (e.g., Horn and Johnson (op. cit., p. 37).

Example 11.14 It is not difficult to see that the matrices

$$I_3 = \begin{bmatrix} 1 & 0 & 0 \\ 0 & 1 & 0 \\ 0 & 0 & 1 \end{bmatrix} \text{ and } P_3 = \begin{bmatrix} 0.6667 & -0.3333 & -0.3333 \\ -0.3333 & 0.6667 & -0.3333 \\ -0.3333 & -0.3333 & 0.6667 \end{bmatrix},$$

are idempotent. Here, I_3 is the identity matrix of order 3, and P_3 is an *orthogonal projector* defined by

$$P = A\left(A^*A\right)^{-1}A^*, \tag{11.61}$$

where A is a special $N \times k$ matrix. It is well known that an orthogonal projector is *idempotent* and *Hermitian* (e.g., Lancaster and Tismenetsky [69], p. 196). In this case, the matrix A is an 4×3 matrix

$$A = \begin{bmatrix} 2 & 2 & -4 \\ 2 & 1 & -3 \\ 0 & -\frac{3}{2} & \frac{3}{2} \\ -1 & -\frac{1}{2} & \frac{3}{2} \end{bmatrix},$$

which is used in Gabriel [41] in the context of his *biplot*. The biplot will be introduced in more detail soon in Remark 2 of Theorem 11.13.

Theorem 11.5 (Jordan canonical form) *Let $A \in M_n(C)$ be a given complex matrix. There is a nonsingular matrix $H \in M_n$ such that*

$$A = H \begin{bmatrix} J_{n_1}(\lambda_1) & & & O \\ & J_{n_2}(\lambda_2) & & \\ & & \ddots & \\ O & & & J_{n_k}(\lambda_k) \end{bmatrix} H^{-1} = HJH^{-1}, \tag{11.62}$$

and $n_1 + n_2 + \ldots, n_k = n$. The Jordan matrix J of A is unique up to permutations of the diagonal Jordan blocks. The eigenvalues λ_i, $i = 1, 2, \ldots, k$ are not necessarily distinct.

Proof of this theorem is, for example, in Horn and Johnson (op. cit., pp. 126–128).

Theorem 11.6 (real Jordan canonical form) *Each real matrix $A \in M_n(R)$ is similar to a block diagonal real matrix of the form, that is,*

$$
\mathrm{A} = \mathrm{H}\begin{bmatrix} \boldsymbol{C}_{n_1}(a_1, b_1) & & & & & & \\ & \boldsymbol{C}_{n_2}(a_2, b_2) & & & & \boldsymbol{O} & \\ & & \ddots & & & & \\ & & & \boldsymbol{C}_{n_p}(a_p, b_p) & & & \\ & & & & \boldsymbol{J}_{n_q}(\lambda_q) & & \\ & & \boldsymbol{O} & & & \ddots & \\ & & & & & & \boldsymbol{J}_{n_r}(\lambda_r) \end{bmatrix}
$$

$$
\boldsymbol{H}^{-1} = \boldsymbol{H}\boldsymbol{J}_R\boldsymbol{H}^{-1}, \tag{11.63}
$$

where $\lambda = a_k + ib_k$ is a nonreal eigenvalue of A for $k = 1, 2, \ldots, p$, a_k and a_k are real, and $\lambda_q, \lambda_{q+1}, \ldots, \lambda_r$ are the real eigenvalues of $\boldsymbol{A}$. Here, each real block triangular matrix $\boldsymbol{C}_{n_k}(a_k, b_k) \in M_{2n_k}$ is of the following form with the 2×2identity matrix $\boldsymbol{I}$ on the superdiagonal

$$
\boldsymbol{C}_{n_k}(a_k, b_k) = \begin{bmatrix} \boldsymbol{C}(a_k, b_k) & \boldsymbol{I} & & \boldsymbol{O} \\ & \boldsymbol{C}(a_k, b_k) & \ddots & \\ & & \ddots & \boldsymbol{I} \\ & \boldsymbol{O} & & \boldsymbol{C}(a_k, b_k) \end{bmatrix}, \tag{11.64}
$$

where

$$
\boldsymbol{C}(a_k, b_k) = \begin{bmatrix} a_k & b_k \\ b_k & a_k \end{bmatrix}, \tag{11.65}
$$

and the triangular matrix corresponds to a pair of conjugate Jordan blocks $\boldsymbol{J}_{n_k}(\lambda_k) \in M_{n_k}$ with nonreal λ_k in the Jordan canonical form (11.47) of $\boldsymbol{A}$. The real Jordan blocks $\boldsymbol{J}_{n_k}(\lambda_k)$ in (11.48) are exactly the Jordan blocks in (11.47).

A brief summary of the proof of this theorem is given, for example, in Horn and Johnson (op. cit., pp. 1511). Another proof is found in Hersch and Smale (op. cit., pp. 126–132).

Remark *Real Jordan canonical form* plays an important role in solving the *differential equation* with initial value (e.g., Hirsch and Smale, op. cit., pp. 133–141). We shall discuss this problem in Chap. 12. In asymmetric MDS, Chino [13] proposed a possibility of decomposing ASM using real Jordan canonical form. If we remain a real space, decomposition of any ASM using this form might be a most general method for obtaining configuration of objects.

We shall next introduce the eigenvalue problem of a special matrix called a *nonnegative matrices.*

Definition 11.33 (*nonnegative matrix*) A matrix $\boldsymbol{A} \in M_{m,n}(R)$ is said to be *nonnegative* if no element of $\boldsymbol{A}$ is negative.

Remark If all the elements of A are positive, then A is referred to as a positive matrix.

Definition 11.34 (*reducible matrix*) If $n \geq 2$, the matrix $\boldsymbol{A} \in M_n(C)$ is said to be *reducible* if there is an $n \times n$permutation matrix $\boldsymbol{P}$ such that

$$\boldsymbol{P}^t\boldsymbol{A}\boldsymbol{P} = \begin{bmatrix} \boldsymbol{A}_{11} & \boldsymbol{A}_{12} \\ O & \boldsymbol{A}_{22} \end{bmatrix}, \tag{11.66}$$

where $\boldsymbol{A}_{11}$ and $\boldsymbol{A}_{22}$ are square matrices of order less than *n*. If no such $\boldsymbol{P}$ exists then A is said to be *irreducible*.

Theorem 11.7 (Perron-Frobenius theorem) *If the matrix* $\boldsymbol{A} \in M_n(R)$ *is nonnegative and irreducible, then*

(a) *The matrix* **A** *has a positive eigenvalue, λ, equal to the spectral radius of* **A***;*
(b) *There is a positive (right) eigenvector associated with the eigenvalue λ;*
(c) *The eigenvalue λ has algebraic multiplicity 1.*

Proof of this theorem is, for example, in Lancaster and Tismenetsky ([69], pp. 536–537).

Remark 1 As introduced in Example 1.11 in Sect. 1.2 and discussed briefly in Sect. 1.3.1, *hyperlink matrices* are usually analyzed based on the above theorem in *spectral graph theory*.

Remark 2 It is interesting to note that the *Perron-Frobenius theorem*, Theorem 11.7, has been extended to the case where $\boldsymbol{A}$ has more than one eigenvalue with absolute value equal to λ, the spectral radius of $\boldsymbol{A}$ (e.g., Lancaster and Tismenetsky, op. cit., p. 540). However, we shall not introduce it here, because it is beyond the scope of this book.

Let us now remember that we have defined a *normal matrix* in Theorem 11.2 without referring to the eigenvalue problem. We shall now define a normal matrix in terms of the eigenvalue problem. Before proceeding, we shall define the notion of *unitary similarity*.

Definition 11.35 (*unitary similarity*) If the matrix $\boldsymbol{B}$ is *similar* to $\boldsymbol{A}$ and there exists a unitary transforming matrix $\boldsymbol{U}$ so that

$$\boldsymbol{B} = \boldsymbol{U}\boldsymbol{A}\boldsymbol{U}^* = \boldsymbol{U}\boldsymbol{A}\boldsymbol{U}^{-1}, \tag{11.67}$$

then we say that $\boldsymbol{B}$ is *unitarily similar* to $\boldsymbol{A}$.

Theorem 11.8 (normal matrix in terms of eigenvalue problem). A matrix $\boldsymbol{A} \in M_n(C)$ is *normal* if and only if it is *unitarily similar* to a diagonal matrix of its eigenvalues:

$$\boldsymbol{A} = \boldsymbol{U}\boldsymbol{D}\boldsymbol{U}^*, \tag{11.68}$$

where $\boldsymbol{D} = diag[\lambda_i]_{i=1}^n$, $\lambda_i \in \sigma(\boldsymbol{A})$, $i = 1, 2, \ldots, n$, and $\boldsymbol{U}$ is *unitary*.

Proof of this theorem can be seen, for example, in Horn and Johnson ([58], pp. 101–103). They prove the equivalence among the several statements on normal matrices, which include the content of the above theorem.

Remark In asymmetric MDS, Chino and Shiraiwa [15] utilize this theorem in decomposing the Hermitian matrix constructed from any ASM into a unitary matrix and a diagonal matrix in (4.9) of Chap. 4.

In general, Eq. (11.58) can be rewritten as $\boldsymbol{A} = \boldsymbol{U}\boldsymbol{D}\boldsymbol{U}^{-1}$ since $\boldsymbol{U}$ is unitary. Therefore, we have $\boldsymbol{A}\boldsymbol{U} = \boldsymbol{U}\boldsymbol{D}$. As a result, we have $\boldsymbol{A}\boldsymbol{u}_j = \mu_j \boldsymbol{u}_j$, where $\boldsymbol{u}_j$ is the eigenvector corresponding to the eigenvalue μ_j of $\boldsymbol{A}$. This transformation is called a *unitary similarity transformation*. This transformation is apparently an extension of a similarity transformation in Definition 11.30 to a complex case.

Example 11.15 In Chap. 5, we showed some applications of *HFM* to empirical and hypothetical data matrices. One of them is an international trade data among four nations in 2015, as shown in Table 5.3. We first administered a log transformation to each element of this matrix, as is sometimes done for economic data. Then, the *Hermitian matrix* computed from the transformed matrix $\boldsymbol{A}$ is

$$\boldsymbol{A} = \begin{bmatrix} 10.68 & 6.92+0.32i & 7.28-0.19i & 4.58-0.57i \\ 6.92-0.32i & 12.15 & 7.77-0.71i & 4.68-0.42i \\ 7.28+0.19i & 7.77+0.71i & 11.69 & 5.83+0.02i \\ 4.58+0.57i & 4.68+0.42i & 5.83-0.02i & 9.53 \end{bmatrix}$$

The eigenvalues of this matrix are 29.97, 6.11, 4.44, and 3.53. The unitary matrix $\boldsymbol{U}$ in the above remark is

$$\boldsymbol{U} = \begin{bmatrix} 0.50-0.03i & 0.22+0.11i & -0.70+0.14i & -0.32-0.27i \\ 0.54-0.06i & 0.45-0.11i & 0.54-0.35i & -0.27+0.05i \\ 0.55-0.01i & -0.03-0.05i & -0.03+0.21i & 0.77+0.23i \\ 0.39 & -0.85 & 0.15 & -0.31 \end{bmatrix}$$

The complex values which constitute each of the column vectors of $\boldsymbol{U}$ show the coordinates of the four nations in a unidimensional Hilbert space. Of course, the holistic complex space which is related to the symmetric part and skew-symmetric part of the observed real asymmetric data matrix is a four-dimensional Hilbert space, since the four eigenvalues associated with these column vectors are all positive.

It is easy to see that HFM begins with an *asymmetric data matrix*, and then it computes the *Hermitian matrix* corresponding to the real asymmetric matrix. Finally, HFM administers a *unitary similarity transformation* to this matrix.

The notion of unitary similarity can be generalized to that of *congruence* as follows:

Definition 11.36 (*congruent matrices*) Two square matrices $\boldsymbol{A}, \boldsymbol{B} \in M_n(C)$ are said to be *congruent* if there exists a *nonsingular matrix* $\boldsymbol{P} \in M_n(C)$ such that

$$A = PBP^*. \tag{11.69}$$

Theorem 11.9 (congruency of the square matrices). *A Hermitian matrix* $\boldsymbol{H} \in M_n(C)$ *is congruent to the matrix*

$$D_0 = \begin{bmatrix} I_s & O & O \\ O & -I_{r-s} & O \\ O & O & O_{n-r} \end{bmatrix}, \tag{11.70}$$

where $r = \text{rank}\boldsymbol{H}$ *and* s *is the number of positive eigenvalues of* $\boldsymbol{H}$ *counted according to multiplicities.*

Proof of this theorem is, for example, in Lancaster and Tismenetsky ([69], pp. 184–185).

Now we shall return to the discussion of the singular value decomposition mentioned in the Remark on Theorem 11.1 earlier. Before proceeding, we shall introduce the following theorem:

Theorem 11.10 (positive definite and positive semidefinite matrices). *A matrix* $\boldsymbol{A}$ *is positive definite (or semidefinite) if and only if it has a positive definite (respectively, semidefinite) square root* $\mathbf{A}_0$ *such that* $\boldsymbol{A}_0^2 = \boldsymbol{A}$. *Moreover, rank* $\mathbf{A}_0$ *is equal to rank* $\mathbf{A}$.

Proof of this theorem is, for example, in Lancaster and Tismenetsky (op. cit., p. 181).

Remembering here that a square matrix $\boldsymbol{A}^*\boldsymbol{A}$ is p.s.d. (see the remark on Theorem 11.2), it is apparent from the above theorem that given an arbitrary matrix $\boldsymbol{A} \in M_{m,n}(C), \boldsymbol{A}^*\boldsymbol{A}$ has a positive semidefinite square root $(\boldsymbol{A}^*\boldsymbol{A})^{1/2}$. The eigenvalues $\lambda_1, \lambda_2, \ldots, \lambda_n$ of $(\mathbf{A}^*\mathbf{A})^{1/2}$ are said to be the *singular values* $s_1, s_2, \ldots, s_n$ of $\mathbf{A}$.

Theorem 11.11 *The nonzero eigenvalues of the matrices* $(\boldsymbol{A}^*\mathbf{A})^{1/2}$ *and* $(\boldsymbol{A}\boldsymbol{A}^*)^{1/2}$ *coincide.*

Proof of this theorem is, for example, in Lancaster and Tismenetsky (op. cit., pp. 182–183).

Theorem 11.12 (singular-value decomposition) *Let an arbitrary matrix* $\mathbf{A}$ *be in* $M_{m,n}$, *and let* $\{s_i\}_{i=1}^r$ *be the nonzero singular values of* $\mathbf{A}$. *Then* $\mathbf{A}$ *can be decomposed as*

$$A = UDV^*, \tag{11.71}$$

where $\boldsymbol{U} \in M_m$ *and* $\boldsymbol{V} \in M_n$ *are unitary and* $\boldsymbol{D} \in M_{m,n}$ *has* s_i *in the (i, i) position* $(1 \leq i \leq r)$ *and zeros elsewhere. Moreover, the numbers* $\{s_i\}$ *are nonnegative square roots of the eigenvalues of* $\boldsymbol{A}^*\boldsymbol{A}$ *(and is also those of* $\boldsymbol{A}\boldsymbol{A}^*$*). The columns of* $\boldsymbol{U}$ *are eigenvectors of* $\boldsymbol{A}\boldsymbol{A}^*$ *and those of V are eigenvectors of* $\boldsymbol{A}^*\boldsymbol{A}$ *arranged in the same order as the corresponding eigenvalues* s_i^2.

Proof of this theorem is, for example, in Horn and Johnson ([58], p. 415).

Remark Singular value decomposition is usually abbreviated as *SVD*. The SVD defined by the above theorem is an *extension* of the SVD defined on the real field to the complex field. In fact, the following theorem is an SVD defined on the real field (e.g., Lawson and Hanson [71], p. 18):

Theorem 11.13 (singular-value decomposition defined on the real field) *Let a matrix $\boldsymbol{A}$ of rank k be in $M_{m,n}(R)$. Then there is an orthogonal matrix $\boldsymbol{U} \in M_m(R)$, an orthogonal matrix $\boldsymbol{V} \in M_n(R)$, and a diagonal matrix $\boldsymbol{S} \in M_{m,n}$ such that*

$$\boldsymbol{A} = \boldsymbol{USV}^t, \tag{11.72}$$

Here the diagonal entries of $\boldsymbol{S}$ can be arranged to be nonincreasing: all these entries are nonnegative and exactly k of them are strictly positive.

We shall omit the proof of this theorem.

Remark 1 Although Lawson and Hanson (op. cit.) call $\boldsymbol{S}$ a diagonal matrix in the above theorem, it is, of course, a rectangular matrix since $\boldsymbol{S} \in M_{m,n}$ in the strict sense. In other words, the submatrix of rank k in $\boldsymbol{S}$ corresponding to the nonzero singular values is diagonal in the strict sense.

Remark 2 In the special case when m equals n, the matrix $\boldsymbol{A}$ in the above theorem is an arbitrary real asymmetric matrix, that is, an ASM. Some researchers have proposed to utilize SVD as a model for analyzing ASM in asymmetric MDS (e.g., Gower [47]; Constantine and Gower [18]; Okada [97, 98]). For example, Constantine and Gower apply SVD to the *skew-symmetric part* of ASM, $\boldsymbol{A}_{sk}$. Here, in general we can decompose any ASM uniquely into the symmetric part $\boldsymbol{A}_s$ and skew-symmetric part $\boldsymbol{A}_{sk}$ such that

$$\boldsymbol{A} = \left(\boldsymbol{A} + \boldsymbol{A}^t\right)/2 + \left(\boldsymbol{A} - \boldsymbol{A}^t\right)/2 = \boldsymbol{A}_s + \boldsymbol{A}_{sk}. \tag{11.73}$$

According to Constantine and Gower, $\boldsymbol{A}_{sk}$ can be decomposed as

$$\boldsymbol{A}_{sk} = \boldsymbol{X\Lambda KX}^t. \tag{11.74}$$

Here $\boldsymbol{\Lambda}$ is a special diagonal matrix composed of the *singular values* of $\boldsymbol{A}_{sk}$. They call (11.64)*canonical analysis of skew symmetry*. We saw in more detail in Chap. 7. In any case, the configuration of objects obtained by Eq. (11.64) has not a *metric property* as Constantine and Gower (op. cit.) point out. In fact, Chino and Shiraiwa [15] point out that the $\boldsymbol{A}_{sk}$ has a *symplectic structure* which is not considered as a metric structure. We have already gone into details of this notion in Sect. 10.5 of Chap. 10.

Okada [97, 98] propose an asymmetric MDS based on SVD. He applies SVD directly to an ASM, obtain two kinds of orthogonal matrices $\boldsymbol{X}$ and $\boldsymbol{Y}$, which are composed of left singular vectors and right singular vectors, respectively, such that

$$A = XDY^t, \tag{11.75}$$

and interpret the element a_{jk} in $A \in M_n(R)$ geometrically. However, the *simultaneous interpretation of the configurations* obtained by the X and Y needs care. The reason for this is that a_{jj} in (11.75) cannot be defined uniquely, although X and Y can be embedded in Euclidean spaces, or stated another way, induce two kinds of metrics, because they are composed of eigenvectors of AA^t and A^tA, respectively.

Okada's method is considered as a special case of *biplot* originally proposed by Gabriel [43, 44] and developed further by Gower and Hand [48]. In particular, Gabriel [43] begins with a *rank factorization* (e.g., Rao [102]) of any rectangular matrix.

Finally, SVD has been extended to higher order SVD (abbreviated as *HOSVD* (e.g., De Lathauwer et al. [25]; Kroonenberg [68]). However, we shall not discuss it, because it is beyond the scope of this book.

Another important theorem on SVD is concerned with the *approximation* of one matrix by another of lower rank known as the Eckart-Young theorem (Eckart and Young [30]). Since the original theorem by Eckart and Young is rather lengthy, we shall summarize it according to Lawson and Hanson ([71], p. 26):

Theorem 11.14 (Eckart and Young) *Given a matrix $A \in M_{m,n}(R)$ of rank k, and a nonnegative integer $r < k$, the matrix $B \in M_{m,n}(R)$ of rank r that minimizes $B - A_F$ is $\tilde{B} = U\tilde{S}V^t$, where $U \in M_{m,m}(R)$ and $V \in M_{n,n}(R)$ are orthogonal matrix and $\tilde{S} \in M_{m,n}(R)$ is constructed from $S \in M_{m,n}(R)$ by replacing s_{r+1} through s_k by zeros. Here, $A = USV^t$ is a singular value decomposition of $A \in M_{m,n}(R)$ with ordered singular values $s_1 \geq s_2 \geq \ldots s_k > 0$.*

Proof of this theorem is, for example, in Eckart and Young (op. cit.) and Lawson and Hanson (op. cit.).

Remark There exists the *Schmidt-Mirsky theorem*, which is a natural generalization of the Eckart-Young theorem to the *complex field* (e.g., Stewart and Sun [117], p. 208). However, we shall not introduce it because it is beyond the scope of this book.

Definition 11.37 (*Kronecker product*) If $A = \{a_{ij}\} \in M_m(F)$ and $B = \{b_{ij}\} \in M_n(F)$, then the *right Kronecker product* of A and B, $A \otimes B \in M_{mn,mn}(F)$, is defined by the partitioned matrix

$$A \otimes B = \begin{bmatrix} a_{11}B & a_{12}B & \cdots & a_{1m}B \\ a_{21}B & a_{22}B & \cdots & a_{2m}B \\ \vdots & \vdots & & \vdots \\ a_{m1}B & a_{m2}B & \cdots & a_{mm}B \end{bmatrix}, \tag{11.76}$$

where F is a field.

Remark Although the *left* Kronecker product of A and B can be defined similarly to be

$$\begin{bmatrix} Ab_{11} & Ab_{12} & \cdots & Ab_{1n} \\ Ab_{21} & Ab_{22} & \cdots & Ab_{2n} \\ \vdots & \vdots & & \vdots \\ Ab_{n1} & Ab_{n2} & \cdots & Ab_{nn} \end{bmatrix},$$

we shall call the right Kronecker product simply the Kronecker product, following, for example, Lancaster and Tismenetsky [69].

Definition 11.38 (*vec function*) A vector-valued function associated with a matrix $\boldsymbol{A} = \{a_{ij}\} \in M_{m,n}(F)$ defined by

$$\begin{bmatrix} \boldsymbol{a}_1 \\ \boldsymbol{a}_2 \\ \vdots \\ \boldsymbol{a}_n \end{bmatrix} \in F^{mn},$$

is called the *vec-function* of $\boldsymbol{A}$, where $\boldsymbol{a}_1, \boldsymbol{a}_2, \ldots, \boldsymbol{a}_n$ are columns of $\boldsymbol{A}$.

Remark There is a close relationship between the vec-function and the Kronecker product. For example, we have

$$vec(\boldsymbol{AX}) = (\boldsymbol{I}_n \otimes \boldsymbol{A})vec\boldsymbol{X}, \tag{11.77}$$

where $\boldsymbol{I}_n$ is an identity matrix of order n. For further interesting relations between the vec-function and the Kronecker product, see, for example, Lancaster and Tismenetsky (op. cit., pp. 409–411).

11.3 Asymmetric MDS

In asymmetric MDS, we embed objects in a certain dimensional metric space, given an ASM. However, there is a *necessary and sufficient condition* for an ASM to be embedded in a metric space.

Before proceeding, we shall first introduce a basic theorem on symmetric MDS since asymmetric MDS is a natural generalization of symmetric MDS. In symmetric MDS, we assume that the similarity matrix is *symmetric*. We shall hereafter abbreviate such a symmetric similarity matrix as *SSM*. As regards the SSM, the Young and Householder theorem is well known (Young and Householder [143]).

Theorem 11.15 (Young and Householder) *A necessary and sufficient condition for a set of numbers $d_{jk} = d_{kj}$ to be the mutual distances of a real set of points in Euclidean space is that the matrix $\boldsymbol{B} = \{b_{jk}\}$,*

$$b_{jk} = \left(d_{j0}^2 + d_{k0}^2 - d_{jk}^2\right)/2, \tag{11.78}$$

be p.s.d; and in this case the set of points is unique apart from a Euclidean transformation. Here, 0 in (11.68) denotes the origin.

Remark 1 The b_{jk} in (11.68) is nothing but an *inner product* of any nonzero vectors in a real Euclidean space, which can be derived from the familiar *law of cosines* (see Definition 10.46 in Chap. 10). It is interesting to note that (11.68) is a special case of the *polarization identity* in any *pre-Hilbert space* (see Theorem 10.7a in Chap. 10).

Remark 2 Although Theorem 11.15 by Young and Householder is well known in psychometrics, there are some predecessors who proved essentially the same theorem as Theorem 11.15 (Schoenberg [109]) or conducted a preliminary investigation of Euclidean spaces (Menger [82]).

In psychometrics it is usual that the Euclidean distances d_{jk} between objects (or nodes) are estimated from the data. As a result, the inner products b_{jk} between objects must be estimated from the data. In fact, Torgerson [130, 131], who proposed a classical MDS, estimated d_{jk} between nine Munsell colors using the *method of triadic combinations*. We will not go into details of this method (Torgerson [131], pp. 262–277), but will discuss a bit soon. Once we have estimated distances between objects, we can compute the estimated inner product between objects. Let us denote it by $\hat{b}_{jk}$. Here, we denote the corresponding matrix by $\hat{\boldsymbol{B}} = \left\{\hat{b}_{jk}\right\}$. According to the Young-Householder theorem, we can embed objects in an Euclidean space if $\hat{\boldsymbol{B}}$ is *p.s.d.*. Configuration of objects is obtained by solving the eigenvalue problem of $\hat{\boldsymbol{B}}$. However, it is usually the case that we shall not compute $\hat{\boldsymbol{B}}$ by (11.68) but compute a special inner product matrix $\boldsymbol{B}_c$ by utilizing the following *centering matrix* $\boldsymbol{C}$ (Horst [59])

$$\boldsymbol{B}_c = -\frac{1}{2}\boldsymbol{C}\boldsymbol{T}\boldsymbol{C}^t, \tag{11.79}$$

where $\boldsymbol{T}$ is the *estimated squared distance matrix* between objects, and

$$\boldsymbol{C} = \boldsymbol{I} - \frac{1}{N}1_N 1_N^t. \tag{11.80}$$

Here, N is the number of objects, $\boldsymbol{I}$ is an identity matrix of order N, and 1_N is the vector whose elements are all 1. $\boldsymbol{B}_c$ is said to be *doubly centered* by the centering matrix $\boldsymbol{C}$.

Example. 11.16 Let original hypothetical objects be located in a Euclidean plane, as depicted in Fig. 11.1. The squared distance matrix $\boldsymbol{T}$ computed from this figure is shown in Table 11.1, and the doubly centered inner product matrix $\boldsymbol{B}_c$ is shown in Table 11.2.

The eigenvalues of $\boldsymbol{B}_c$ are 2, 0.5, 0, and 0. It is evident that $\boldsymbol{B}_c$ is p.s.d. and is also positive definite in the two-dimensional case corresponding to the positive

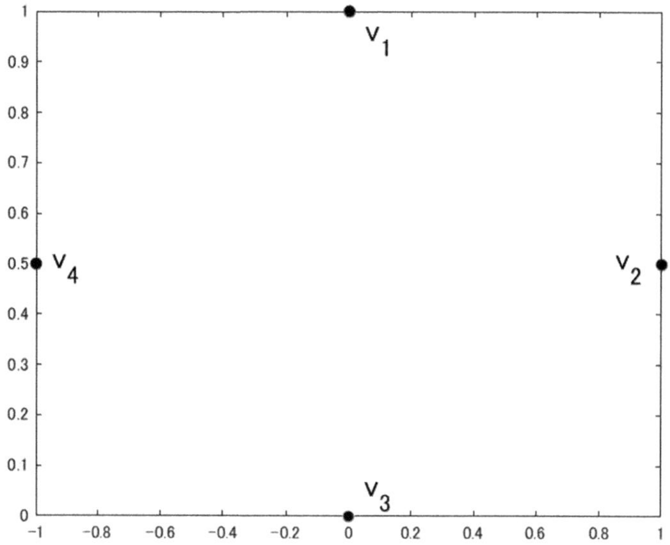

Fig. 11.1 Original hypothetical configuration of four objects in a Euclidean plane

eigenvalues. However, it is not necessarily p.s.d. if the squared distance matrix is an *empirically observed one,* as will be discussed soon. In any case, the eigenvectors corresponding to these eigenvalues can be said to be a solution of the *classical MDS* by Torgerson, although, for example, Torgerson [131] factored $\boldsymbol{B}_c$ by the *centroid method* developed by Thurstone [126, 127]. The eigenvectors corresponding to the largest and second largest eigenvalues are (0, – 0.7071, 0, 0.7071) and (– 0.7071, 0, 0.7071, 0), respectively. Here, it should be noted that these eigenvectors are normalized to 1 (in MATLAB). Therefore, it will be necessary to adjust these eigenvectors, according to the sizes of the corresponding eigenvalues. Figure 11.2 shows the recovered configuration of four objects, which is rotated by π radian in a counterclockwise direction from Fig. 11.1. Note that in this figure the centroid of the configuration is the *origin.*

In contrast, if the distance or squared distance matrix is obtained *empirically*, the matrix $\boldsymbol{B}_c$ is not necessarily p.s.d. as pointed above. A historical example of such distance matrices is the distance matrix between nine Munsell colors obtained by Torgerson (op. cit.).

Example 11.17 Table 11.3 shows part of the estimated distance matrix by Torgerson (op. cit., p. 287, Table 7).

The eigenvalues of the doubly centered inner product matrix $\boldsymbol{B}_c$ through the squared distance matrix $\boldsymbol{T}$ obtained by the distance matrix shown in Table 11.3 are 30.0649, 24.5244, 3.2891, 2.0671, 0.7007, 0.0, – 0.2833, – 0.5304, and – 1.6391. It is evident that $\boldsymbol{B}_c$ is not p.s.d. at least mathematically. Therefore, according to the Young-Householder theorem, we cannot embed nine Munsell colors in a Euclidean space. However, if we restrict the nine-dimensional holistic space of Munsell colors

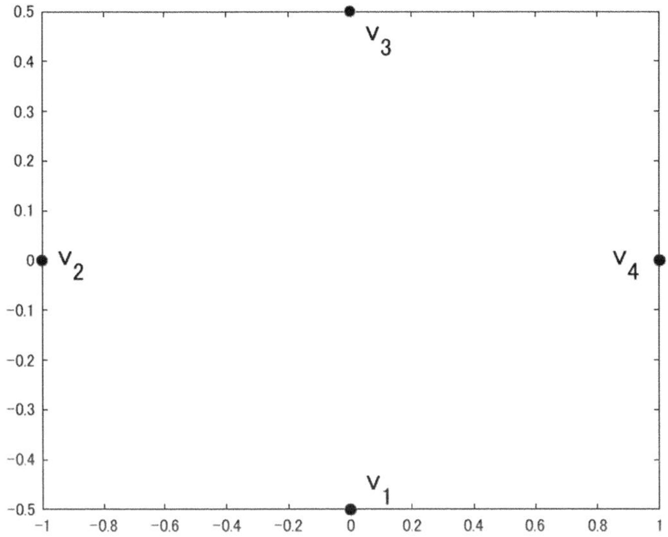

Fig. 11.2 The recovered configurations of four objects by the classical MDS

to two dimensions corresponding to the largest and second largest eigenvalues, then we can approximate the holistic space by a two-dimensional Euclidean space.

In asymmetric MDS, there have been a number of models since Young [143] proposed an asymmetric MDS, as reviewed, for example, in Chino [14]. As can be seen in that review, most of them *augment a Euclidean distance* between objects by adding a certain asymmetric term to it. However, researchers are rare who refer to a *holistic metric space* in which objects are embedded in their asymmetric MDS models. By a holistic metric space, we mean a well-established *mathematical space* which includes information about both the symmetric part and the asymmetric part of an ASM simultaneously. In contrast, Chino and Shiraiwa [15] proved that given an ASM measured at a ratio scale level, objects can be embedded in a finite-dimensional *Hilbert space* or an *indefinite metric space* under a mild condition. It should be noticed that Hilbert space is a natural extension of Euclidean space, as discussed earlier in Chap. 10. The Chino-Shiraiwa theorem is clearly a generalization of the Young-Householder theorem to complex spaces. Although this theorem in their original paper includes the proof of the theorem, we shall state the theorem more simply and formally by dropping the proof as follows:

Before proceeding, a few general remarks are in order, according to Chino and Shiraiwa (op. cit., pp. 36–37). Let us first construct the *Hermitian matrix* $\boldsymbol{H} \in M_{N,N}(C)$ uniquely from an observed ASM, say $\boldsymbol{S} = \left\{s_{jk}\right\} \in M_{N,N}(R)$, in such a way that

$$\boldsymbol{H} = \boldsymbol{S}_s + i\boldsymbol{S}_{sk}, \tag{11.81}$$

where $\boldsymbol{S}_s = \{(s_{jk} + s_{kj})/2\}$ and $\boldsymbol{S}_{sk} = \{(s_{jk} - s_{kj})/2\}$. Here, we assume that $\boldsymbol{S}$ is measured at a ratio scale level.

Then, we consider the eigenvalue problem of $\boldsymbol{H}$, namely,

$$\boldsymbol{H}\boldsymbol{u}_i = \lambda_i \boldsymbol{u}_i,\ 1 \le i \le n, \tag{11.82}$$

where λ_i is the ith eigenvalue of $\boldsymbol{H}$, and $\boldsymbol{u}_i$ is the eigenvector corresponding to λ_i.

Here, we have $\lambda_i \neq 0 (1 \le i \le n)$, and $\lambda_{n+1} = \lambda_{n+2} = \cdots = \lambda_N = 0$, without loss of generality. If we let

$$\boldsymbol{U} = \{\boldsymbol{u}_1, \boldsymbol{u}_2, \ldots, \boldsymbol{u}_n, \boldsymbol{u}_{n+1}, \boldsymbol{u}_{n+2}, \ldots, \boldsymbol{u}_N\} = (\boldsymbol{U}_1, \boldsymbol{U}_2),$$

then (11.68) can also be written as

$$\boldsymbol{H}\boldsymbol{U} = (\boldsymbol{U}_1, \boldsymbol{U}_2)\begin{pmatrix} \boldsymbol{\Lambda} & \mathbf{O} \\ \mathbf{O} & \mathbf{O} \end{pmatrix} = \boldsymbol{U}\tilde{\boldsymbol{\Lambda}},$$

where $\boldsymbol{\Lambda} = diag(\lambda_1, \lambda_2, \ldots, \lambda_n)$. Then we have

$$\boldsymbol{H} = \boldsymbol{U}_1 \boldsymbol{\Lambda} \boldsymbol{U}_1^*, \tag{11.83}$$

since $\boldsymbol{U}$ is *unitary*. Above equation enables us to define a *Hermitian form*, that is,

$$\varphi(\boldsymbol{\zeta}, \boldsymbol{\tau}) = \boldsymbol{\zeta} \boldsymbol{\Lambda} \boldsymbol{\tau}^*, \tag{11.84}$$

in an complex vector space, where $\boldsymbol{\tau}$ and $\boldsymbol{\zeta}$ are row vectors in this space. Moreover, a Hermitian form is called *(Hermitian) inner product* (or scalar product) if $\varphi(\boldsymbol{\zeta}, \boldsymbol{\zeta}) > 0$ for any $\boldsymbol{\zeta} \neq 0$. The inner product is sometimes called a *pre-Hilbert space* or a *unitary space*, as was discussed in Sect. 10.5. Furthermore, it should be noticed that a *complete* inner product space is called a *Hilbert space* by Definition 10.48 in Sect. 10.5.

Theorem 11.16 (Chino and Shiraiwa) *A necessary and sufficient condition for a set of numbers $d_{jk} = d_{kj}$ to be the mutual distances of a real set of points in a finite-dimensional Hilbert space is that the matrix $\boldsymbol{H} = \{h_{jk}\}$,*

$$h_{jk} = \frac{1}{2}\left(d_{jo}^2 + d_{ko}^2 - d_{jk}^2\right) + \frac{1}{2}i\left(d_{jo}^2 + d_{ko}^2 - \overline{d}_{jk}^2\right),\ 1 \le j, k \le N. \tag{11.85}$$

be p.s.d; and in this case the set of points is unique apart from an arbitrary unitary transformation. Here,

$$d_{jk} = \left\|\boldsymbol{v}_j - \boldsymbol{v}_k\right\|,\ 1 \le j \le N, \tag{11.86}$$

$$d_{jo} = \left\|\boldsymbol{v}_j\right\|,\ 1 \le j \le N, \tag{11.87}$$

$$\overline{d}_{jk} = \|\boldsymbol{v}_j - i\boldsymbol{v}_k\|, 1 \leq j, k \leq N. \tag{11.88}$$

and $\boldsymbol{v}_j(1 \leq j \leq N)$ *be the row vector of order n, which corresponds to the j-th row of* $\boldsymbol{U}_1$ *in (11.73).*

Remark 1 Equation (11.85) is nothing but a variation of the *polarization identity*, which we showed in Theorem 10.7a. In the original paper by Chino and Shiraiwa (op. cit.), it is called the *polar identity*, which is shown in Eq. (40). Moreover, it is evident that (11.85) is an extension of (11.78) in the Young-Householder theorem.

Remark 2 Takane [122] gives another expression for this theorem in introducing our theorem.

Example 11.18 Here, we shall use Example 1.24 in Chap. 1 again as an example of an application of Theorem 11.16 to empirical data. Table 1.27 in Example 1.24 is an international trade data among four nations as an observed ASM, whose elements are assumed to be measured at a ratio scale level. As a preprocessing operation, we administered a log-transformation to the original data. The transformed matrix is shown in Table 1.28. Using this matrix as an observed ASM, we computed a Hermitian matrix by (11.73), and obtained the matrix shown in Table 1.29. If this matrix is p.s.d., we can embed four nations in a Hilbert space. As is shown in Sect. 5.3, we can embed these nations in a four-dimensional Hilbert space since eigenvalues of this matrix are all positive and they are 29.9714, 6.1145, 4.4377, and 3.5309. We shall refer to this section as regards the interpretation of the configuration of four nations in this Hilbert space.

Suppose now that we are interested in the problem of whether these eigenvalues are statistically meaningful or not. Here, we shall assume in this book that the reader is familiar with the basic concepts and theorems in mathematical statistics (e.g., Kendall and Stuart [65, 66]. However we shall recapitulate some of them if we think it necessary to do so. We shall show a theorem which is related to an answer to such a question below, although we will not go into details of such a problem as it is beyond the scope of this book. Before proceeding, we shall recapitulate a *random matrix* (e.g., Anderson [3], p. 17) and a *multivariate normal distribution* (e.g., Muirhead [87], pp. 5–9).

Definition 11.39 (*random matrix*) A random matrix $\boldsymbol{Z} \in M_{m,n}(F)$ is a matrix

$$\boldsymbol{Z} = \left\{z_{ij}\right\}, \tag{11.89}$$

of random variables $z_{11}, \ldots, z_{mn}$.

Remark A *random (scalar) variable* and a *random vector* are special cases of the random matrix. Here, the field F in Definition11.37 may be *real* or *complex*. Sometimes, F may be *quaternion* (e.g., Dyson [29]).

Definition 11.40 (*multivariate normal distribution*) The $m \times 1$ random vector $\boldsymbol{x}$ is said to have a *multivariate normal distribution* if, for every $\boldsymbol{\alpha} \in R^m$, the distribution of $\boldsymbol{\alpha}^t\boldsymbol{x}$ is *univariate normal.*

Remark Normal distribution is sometimes called *Gaussian distribution.*

Theorem 11.17 (existence of mean μ and covariance matrix Σ) *If $m \times 1$ random vector $\boldsymbol{x}$ has a multivariate normal distribution then both a mean vector $\boldsymbol{\mu} = E(\boldsymbol{x})$ and a covariance matrix $\boldsymbol{\Sigma} = Cov(\boldsymbol{x})$ exist and the distribution of $\boldsymbol{x}$ is determined by $\boldsymbol{\mu}$ and $\boldsymbol{\Sigma}$.*

Proof of this theorem is, for example, in Muirhead (op. cit., p. 5).

Remark A multivariate normal distribution is usually abbreviated as $N_m(\boldsymbol{\mu}, \boldsymbol{\Sigma})$.

Theorem 11.18 (density function of the multivariate normal distribution) *If $m \times 1$ random vector $\boldsymbol{x}$ is $N_m(\boldsymbol{\mu}, \boldsymbol{\Sigma})$ and $\boldsymbol{\Sigma} \in M_{\boldsymbol{m}}(R)$ is positive definite then the density function of $\boldsymbol{x}$ is denoted by*

$$f_x(\boldsymbol{x}) = (2\pi)^{-m/2}(det\boldsymbol{\Sigma})^{-1/2}exp\left[-\frac{1}{2}(\boldsymbol{x}-\boldsymbol{\mu})^t\boldsymbol{\Sigma}^{-1}(\boldsymbol{x}-\boldsymbol{\mu})\right]. \tag{11.90}$$

Proof of this theorem is, for example, in Muirhead (op. cit., p. 9).

We are now ready to state the main theorem regarding the distribution of the eigenvalues of a Hermitian matrix.

Theorem 11.19 (joint probability density function for the eigenvalues of matrices from a Gaussian ensemble) *The joint probability density function for the eigenvalues of matrices from a Gaussian orthgonal, Gaussian unitary, or Gaussian symplectic ensemble is given by*

$$P_{N\beta}(\lambda_1, \ldots, \lambda_N) = C_{N\beta}exp\left(-\frac{1}{2}\beta\sum_{j=1}^{N}\lambda_j^2\right)\prod_{j<k}|\lambda_j - \lambda_k|^{\beta}, \tag{11.91}$$

where $\beta = 1$ if the ensemble is orthogonal, $\beta = 2$ if it is unitary, and $\beta = 4$ if it is symplectic. The constant $C_{N\beta}$ is chosen in such a way that the $P_{N\beta}$ is normalized to 1, and it is given by

$$C_{N\beta}^{-1} = (2\pi)^{N/2}\beta^{-N/2-\beta N(N-1)/4}[\Gamma(1+\beta/2)]^{-N}\prod_{j=1}^{N}\Gamma(1+\beta j/2). \tag{11.92}$$

Proof of this theorem is, for example, in Mehta ([80], pp. 54–56).

Remark Dyson [29] introduces three types of ensembles of *Hermitian matrices*. Here, an *ensemble* is a set, which is frequently used in statistical mechanics. It is sometimes used as a probability distribution (e.g., Johnstone [63]). This concept is said to be coined by Gibbs [46]. Dyson (op. cit.) considers a square matrix $\boldsymbol{M} \in M_n(F)$ whose elements are of the form

$$m_{ij} = \sum_{\alpha=0}^{\beta-1} m_{ij\alpha}e_{\alpha}, i, j = 1, \ldots, n, \tag{11.93}$$

where $m_{ij\alpha}$ are real coefficients, and the e_α $(\alpha = 0, 1, \ldots, \beta - 1)$ are units of a certain algebra Φ. The three possible choices for Φ are

(i) $\Phi = R$, the algebra of *real numbers*, for which $\beta = 1$ and $e_0 = 1$;
(ii) $\Phi = C$, the algebra of *complex numbers*, for which $\beta = 2$ and $e_0 = 1, e_1 = i$;
(iii) $\Phi = Q$, the algebra of *real quaternions*, for which $\beta = 4$ and $e_0 = 1, e_1 = i$, $e_2 = j, e_{11} = k$,

where $i^2 = j^2 = k^2 = ijk = -1$. Moreover, in the special case when $\beta = 1$, the distribution function (11.81) is known as the *Wishart distribution* (Wishart [141]).

As for the various definitions useful for the understanding of the contents of asymmetric MDS we shall omit them in this section because we will discuss them in Chap. 7 in reviewing asymmetric MDS in detail.

11.4 Graph Theory and Spectral Graph Theory

In this section, we shall deal with basic definitions and theorems concerning the *graph theory* as well as the *spectral graph theory*, many of which include matrix theory. We shall pick up these definitions and theorems mainly from Brouwer and Haemers [11], Christofides [21], Chung [16], and Harary [51]. Details are referred to these books.

Graph theory is said to have a long history. Moreover, it is also said to have been independently discovered many times, as, for example, Harary (op. cit., p. 1) points out. It almost certainly began when, in 1735, Leonhard Euler solved the so-called *K önigsberg* bridge problem (e.g., Alexanderson [2], p. 567). As such, there exist a formidable number of definitions in order to make available the basic concepts and terminology of graph theory. However, in this book we shall mainly state those definitions and theorems related to asymmetric matrices associated with various *digraphs*. Thus, we shall state those related to undirected graph, i.e., *graph*, only if we think it necessary to discuss them.

As Wilson and Beineke [140] point out, not all the terminology in graph theory is completely standardized. Therefore, we shall refer to the literature on the definitions concerning graphs if necessary. We shall begin with the definition of graph, according to them:

Definition 11.41 (*graph*) A *graph* G is a pair $(V(G), E(G))$, where $V(G)$ is a finite nonempty set of *vertices*, and $E(G)$ is a finite family of (not necessarily distinct) *unordered pairs* of $V(G)$ called *edges*.

Remark 1 Vertex is sometimes called *node*, *point*, *terminal* etc., while edge is sometimes called *arc*, *line*, *branch* etc. (Harary, op. cit., p. 9). The number of vertices of a graph is called the *order* of G. Let v and w are vertices of G. Then, we shall denote the edge $\{v, w\}$ by vw. If $e = vw$ is an edge of G, then e is said to *join* the vertices

v and w, and these vertices are said to be *adjacent*. As regards the relation between edge and vertex, we say that e is *incident* to v and w. If two or more edges join the same pair of vertices, then we shall call them *multiple edges*.

Remark 2 Although Harary (op. cit.) lists the synonyms of vertex and edge as above, Cristofides for example distinguishes between *lines* and *arcs*. According to him, if the lines of a graph have a *direction* which is usually be an arrow he calls them *arcs*, and the resulting graph is a directed graph, i.e., digraph. Wilson and Beineke (op. cit., pp. 1–2) also distinguish between them, and call the loops with arrow arcs. Brouwer and Haemers [11] do not distinguish between them.

Definition 11.42 (*order of a graph*) The number of vertices of a graph G is called the *order* of G.

Definition 11.43 (*simple graph*) A *simple graph* is an unweighted, undirected graph containing no graph loops or multiple edges.

Definition 11.44 (*complete graph*) A *complete graph* is a simple graph in which every two vertices are adjacent.

Definition 11.45 (*subgraph*) A *subgraph* of a graph $G = (V(G), E(G))$ is a graph $S = (V(S), E(S))$ such that $V(S) \subseteq V(G)$ and $E(S) \subseteq E(G)$.

Definition 11.46 (*clique*) If a subgraph S of a graph G is complete, then S is called a *clique* of G.

Definition 11.47 (*connected graph*) A graph G is connected if every pair of vertices are joined by an edge.

Remark Christofides ([21], p. 12), for example, subdivides a connected graph further into three parts: *strongly* connected graph, *unilaterally* connected graph, and *weakly* connected graph. According to him, a graph is said to be strongly connected or *strong* if for any two distinct vertices i and j, there is at least one path going from i to j. The above definition implies that any two vertices of a strong graph are mutually reachable. A graph is said to be unilaterally connected or *unilateral* if for any two distinct vertices i and j, there is at least one path going from either i to j or from i to j, or both. Finally, a graph is said to be weakly connected or *weak* if there is at least one chain joining every pair of distinct vertices. If for a pair of vertices such a chain does not exist, the graph is said to be *disconnected*. Harary et al. ([52], pp. 65–68) raise some theorems on the connectedness criteria.

Definition 11.48 (*loop*) A *loop* of G is an arc whose initial and final vertices are the same.

Remark This definition is for example made by Cristofides (op. cit., p. 5). By a graph which includes the term, *arc*, he implicitly assumes that G is a *digraph*. Wilson and Beineke ([140], pp. 1–3) raise both examples of the *nondirected loop* in Fig. 1 and the *directed loop* in Fig. 6, and call the latter loop as an arc. In contrast, Harary ([51], p. 10) indirectly defines a loop by a nondirected graph, i.e., a graph, in which one line joins a point to itself.

Definition 11.49 (*arc-weighted versus vertex-weighted*) If a number is associated with an arc of G, these numbers are called *weights*, *lengths* or *costs*, and the graph is called *arc-weighted*, while if it is associated with a vertex of G, then the graph is called *vertex-weighted*. If a graph is both arc and vertex weighted, then it is simply called *weighted*.

Remark Cristofides (op. cit., p. 5) distinguishes between arc-weighed and vertex-weighted. In contrast, Wilson and Beineke (op. cit., p. 2) do not distinguish between them, and define a *weighted graph* as a graph to each edge of which is assigned a number (usually positive) called its weight. We shall call the matrix composed of these weights the *weight matrix*.

Definition 11.50 (*digraph*) A digraph D is a pair $(V(G), E(G))$, where $V(G)$ is a finite nonempty set of *vertices*, and $E(G)$ is a finite family of (not necessarily distinct) *ordered pairs* of $V(G)$ called *edges*.

Remark 1 Most of the graphs introduced in Chap. 1 are digraphs with loops and multiple edges, because the corresponding matrix which is thought of as a weight matrix is *asymmetric*.

Remark 2 Classically, a digraph is defined to have no loops or multiple edges. For example, Harary ([51], p. 10) defines a digraph as a directed graph which 'consists of a finite nonempty set of points together with a prescribed collection of ordered pairs of distinct points'. Moreover, there are some definitions for digraphs regarding the *connectivity* which are similar to those for graphs introduced previously in Definition 11.47. In fact, a digraph D is called *strongly connected* if, for each pair of vertices v and w, there is a directed path in D from v to w, *unilaterally connected* if there is either a directed path in D from v to w or a directed path from w to v, and *connected* if there is a path from v to w in the underlying graph of D (e.g., Wilson and Beineke [140], p. 8).

Definition 11.51 (*degrees of a vertex*) The number of edges which have a vertex v_i as their *initial* vertex is called the *outdegree* of vertex v_i, while the number of edges which have v_i as their *final* vertex is called the *indegree* of vertex v_i.

Remark 1 Let us denote the *indegree* and the *outdegree* of vertex v_i of G by $d_{in}(v_i)$ and $d_{out}(v_i)$, respectively. Then, it is obvious that

$$\sum_{i=1}^{n} d_{in}(v_i) = \sum_{i=1}^{n} d_{out}(v_i) = m, \tag{11.94}$$

where n is the total number of vertices and m is the total number of edges of G.

Remark 2 Definition 11.51 is by Cristofides (op. cit., p. 7), in which he implicitly assume a *digraph*. Of course, he refers to the definition of degrees of a *graph*, i.e., nondirected graph, as a special case of the degree of a digraph. In contrast, Harary (op. cit., p. 14) defines the degree of a vertex in a graph by the number of lines (edges) incident with vertex v_i, and defines *indegree* and *outdegree* in digraph in Chap. 16. In other words, the indegree of a vertex v_i of a digraph is the number of vertices adjacent to it, and the outdegree is the number of vertices adjacent from it. Degree of a graph is sometimes called *valency* (e.g., Wilson and Beineke [140]). As a result, indegree and outdegree in a digraph are also called *in-valency* and *out-valency*, respectively.

From a statistical viewpoint, researchers have long been interested in *random graphs* and distributions of degrees (or *degree distributions*) in graph and digraph (e.g., Barabási and Albert [4]; Bollobás [9]; Ciotti [17]; Erdös and Rényi [32]; Newman et al. [91]; Saberi et al. [108]; Solomonoff and Rapaport [115]). As regards the degree distributions, there exists an important notion called *scale-free network* whose degree distribution follows a power law (e.g., Barabási and Albert, op. cit., Newman et al. [91]; Price [103]). However, we will not review them, because the history of the statistical development of these notions is beyond the scope of this book.

Definition 11.52 (*regular graph*) Let us consider a graph G with p vertices and q edges. If the minimum degree among the vertices and the maximum degree among them are both $r \leq p - 1$, then such a graph is called *regular* of degree r.

Definition 11.53 (*walk*) A walk of a graph G is an alternating sequence of vertices and edges $v_0, e_1, v_1, e_2, v_2, v_3, \ldots, v_{n-1}, e_n, v_n$, beginning and ending with vertices, in which each edge is incident with the two vertices immediately preceding and following it. Here, v_0 is the initial vertex, and v_n the terminal vertex. This walk joins v_0 and v_n.

Remark The above definition is made by Harary ([51], p. 13), and a bit modified by the author based on the definition by Wilson and Beineke (op. cit., p. 7).

Definition 11.54 (*path in a digraph*) A *path* in a digraph is any sequence of edges where the final vertex of one is the initial vertex of the next one.

Remark A *path in* a *graph* is defined in terms of *walk*, for example, by Harary ([op. cit.], p. 13) and Wilson and Beineke (op. cit., p. 7).

Definition 11.55 (*circuit in a digraph*). A *circuit* is a path in which the initial vertex of the initial path coincides with the final vertex of the final path.

Remark 1 It should be noticed that the 'circuit' in electrical circuit discussed in Chap. 1 does not necessarily mean the circuit defined above.

Remark 2 A *circuit in a graph* is also defined in terms of *walk*, for example, by Wilson and Beineke (op. cit., p. 7).

Definition 11.56 (*chain*) A *chain* is the nondirected counterpart of the path in a digraph and applies to graphs with the direction of its edges disregarded.

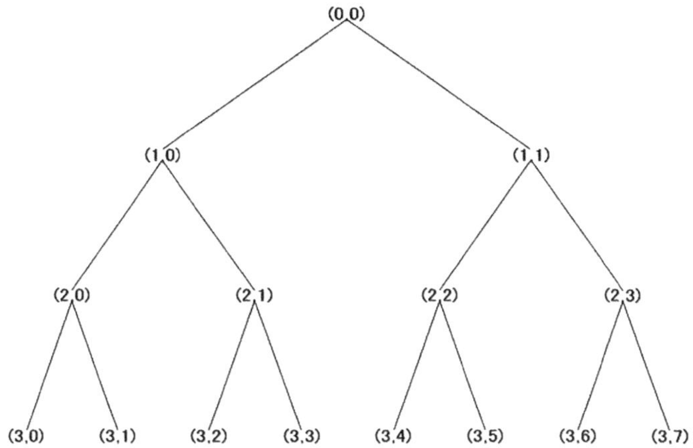

Fig. 11.3 An example of a binary tree, which was depicted by the *ntree* function of MATLAB

Definition 11.57 (cycle) A *cycle* is the nondirected counterpart of the circuit in a digraph, and it is a chain in which the beginning and end vertices are the same.

Remark 1 A graph is called *acyclic* if it has no cycles.

Definition 11.58 (*tree and forest*) A *tree* is a connected acyclic graph, and any graph without cycles is called a *forest*. Therefore, the components of a forest are trees.

Definition 11.59 (*a binary tree*) A *binary tree* is a tree in which each vertex has at most two *children*, which are referred to as the *left child* and the *right child*.

Example 11.19 (*Binary Tree*). Figure 11.3 shows an example of a binary tree, which was depicted by the *ntree* function of MATLAB.

Definition 11.60 (*distance in a graph*) The *distance* $d(u, v)$ between two vertices u and v in a graph G is the length of a *shortest path* joining them if any, and otherwise $d(u, v) = \infty$.

Example 11.20 (*Distance Matrix in a Graph*) Figure 11.4 is an example of a graph G, and Table 11.4 shows the distance matrix of G whose (i, j) element denotes the *shortest path* between vertex i and vertex j, which was computed by the *distances* function of MATLAB.

Remark 1 Computation of the shortest distances between vertices is easy for such a graph in Fig. 11.4 whose number of vertices is small. However, if the number of vertices is large, such a computation becomes laborious. In such a case, it is convenient to use a computer program such as the *distances* function of MATLAB. MATLAB also has a function called *shortestpath* which enables us to compute the shortest path and to show it using a special color, as shown in Fig. 11.5.

In a connected graph, *distance* defined above is a *metric* which satisfies the three axioms in Definition 10.47a(*metric space*) in Chap. 10. A shortest path is often

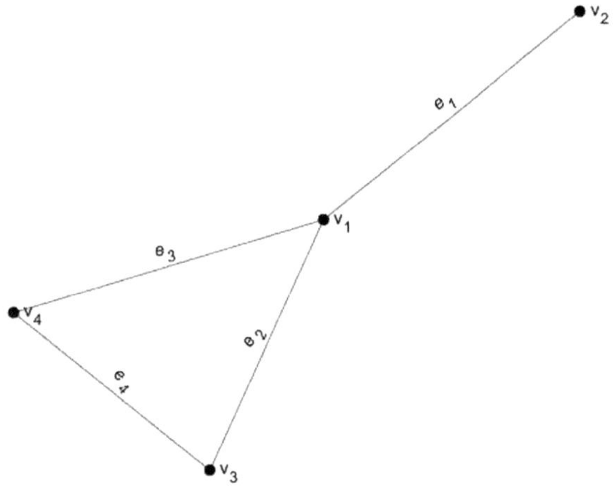

Fig. 11.4 A graph G without loops, which is composed of 4 vertices and 4 edges

Table 11.4 The distance matrix of a graph G shown in Fig. 11.4 whose (i, j) element denotes the *shorted path* between vertex i and vertex j

	v_1	v_2	v_3	v_4
v_1	0	1	1	1
v_2	1	0	2	2
v_3	1	2	0	1
v_4	1	2	1	0

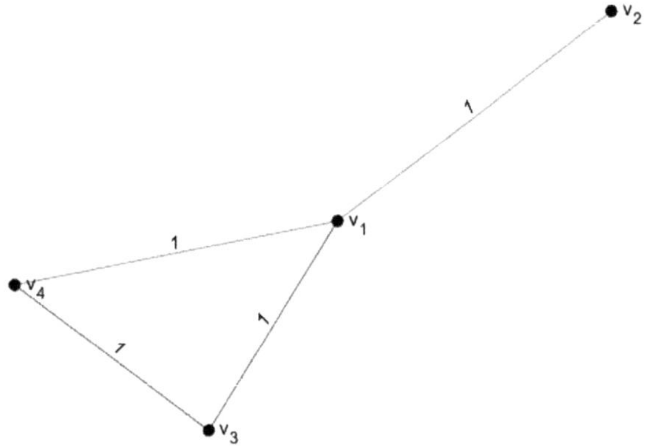

Fig. 11.5 Indication of a specified shortest path in a graph G by red edges

called a *geodesic*, and the *diameter* of a connected graph G is the length of any longest geodesic.

Remark 2 Definition 11.60 is by Harary [51]. In contrast, the notion of the distance of a graph G which is accompanied by weights was introduced by Dijkstra [26]. The distance between two vertices u and v in a *weighted graph* defined by him is shown in Definition 11.61. His algorithm for finding the shortest path between vertices in a weighted graph is called the *Dijkstra algorithm.* As regards a *weighted digraph*, we have an algorithm called the *Bellman-Ford algorithm* or the *Bellman-Ford-Moore algorithm* (e.g., Bellman [7]; Ford [38]; Moore [85]; Shimbel [112]). We shall introduce the definition of the distance in the weighted digraph by Bellman (op. cit.) in Definition 11.62. All these algorithms are implemented, for example, in the *distances* function in the recent versions of MATLAB. As regards the shortest path problem, there has been a body of literature which generalizes it to the *k shortest paths problem* (e.g., Eppstein [31]). However, we shall not discuss them, since a complete treatment is beyond the scope of this book.

Definition 11.61 (*distance in a weighted graph*) Let n points (vertices), some or all pairs of which be connected by an edge and let us assume that the length of each edge is given. Moreover, let us restrict ourselves to the case where at least one path exists between any two vertices. The distance $d(u, v)$ between two given vertices u and v is the path of minimum total length between them.

Example 11.21 (*Distance Matrix of a Weighted Graph*) Figure 11.6 is an example of a weighted graph G, and Table 11.5 shows the distance matrix of G whose (i, j) element denotes the shortest path between vertex i and vertex j, which was computed by the *distances* function of MATLAB. The default of the algorithm of this function requires the weights to be nonnegative, which corresponds to choose the Dijkstra algorithm in MATLAB. Moreover, the shortest path, for example, between vertex 2 and vertex 4 is computed as (v_2-v_1-v_4), and the path length is calculated as 10 by the *shortestpath* function of MATLAB.

The following definition by Bellman [7] is slightly adapted to suit more general notations of a digraph than those which he used.

Definition 11.62 (*distance in a weighted digraph*) Let a set of N points (vertices) are connected in a weighted digraph. Given an asymmetric weight matrix $\boldsymbol{W} = \{w_{ij}\}$,

Fig. 11.6 A weighted graph with 5 vertices and 8 edges

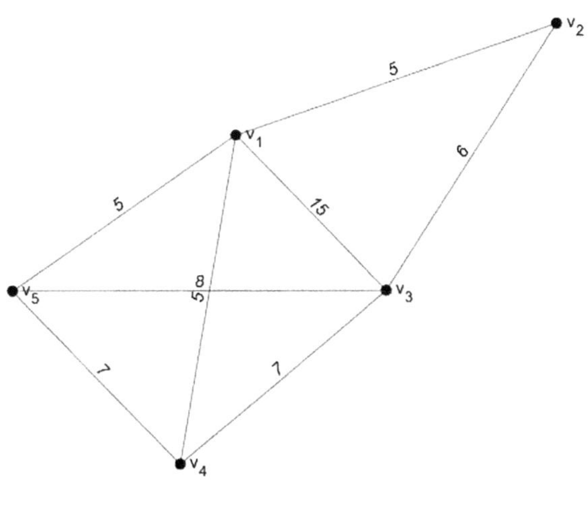

Table 11.5 The distance matrix of a weighted graph G shown in Fig. 11.6 whose (i, j) element denotes the *shorted path* between vertex *i* and vertex *j*

	v_1	v_2	v_3	v_4	v_5
v_1	0	5	11	5	5
v_2	5	0	6	10	10
v_3	11	6	0	7	8
v_4	5	10	7	0	7
v_5	5	10	8	7	0

where w_{ij} is the time required to travel from point *i* to point *j*, the path which consume minimum time is the distance between these two points.

Remark In the Bellman-Ford-Moore algorithm, a weighted digraph is assumed to have possibly negative weight and no *negative cycle*. Here, a negative cycle is a path that leads from a node back to itself and the sum of the edge weights on the path is negative.

Example 11.22 (*distance matrix of a weighted digraph*) Figure 11.7 is an example of a weighted digraph *G*, and Table 11.6 shows the distance matrix of *G* whose (i, j) element denotes the shortest path between vertex *i* and vertex *j*, which was computed by the *distances* function of MATLAB. Moreover, the shortest path, for example, between vertex 3 and vertex 4 is computed as (v_{11}–v_5–v_4), and the path length is calculated as 12 by the *shortestpath* function of MATLAB.

It is evident from Table 11.6 that the distance matrix between vertices is *no longer symmetric*. As a result, distances in this case do not satisfy the symmetric axiom required for the metric space.

Recently, a *fixed-point theorem* in mathematics has received increasing attention in graph theory in connection with a shortest path in digraph (e.g., Espínola and Kirk

Fig. 11.7 A weighted digraph G of order 5

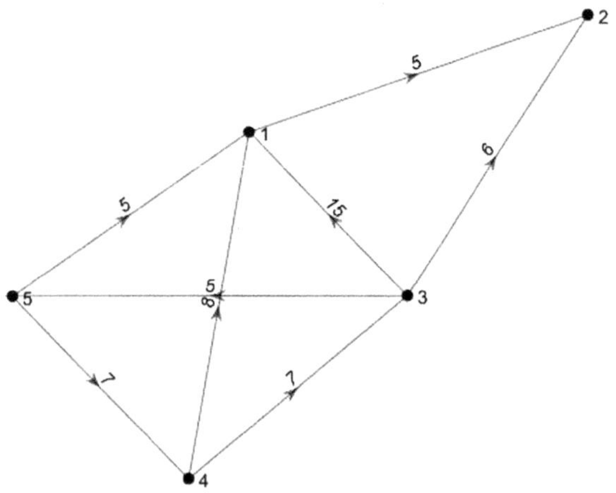

Table 11.6 A distance matrix of G of a digraph

	v_1	v_2	v_3	v_4	v_5
v_1	0	5	∞	∞	∞
v_2	∞	0	∞	∞	∞
v_3	10	6	0	12	5
v_4	8	13	7	0	12
v_5	5	10	14	7	0

[34]; Jachymski [61]; Neog and Debnath [89]). For example, Neog and Debnath (op. cit.) established several interesting theorems on the fixed point of a digraph. We shall introduce some of them below. Before proceeding, we shall introduce three definitions necessary for understanding these theorems according to Neog and Debnath (op. cit.):

Definition 11.63 (*nonempty closed and bounded subsets of X in a complete metric space*) Let (X, d) be a complete metric space and $CB(X)$ be the class of all nonempty closed and bounded subsets of X. For $A, B \in CB(X)$, let

$$D(A, B) = max\{sup_{b\in B}d(b, A), sup_{a\in A}d(a, B)\}, \tag{11.95}$$

where $d(a, B) = inf_{b\in B}d(a, b)$. Then $(CB(X), D)$ is a metric space and D is said to be a *Hausdorff metric* induced by d.

Remark 1 The Hausdorff metric defined above is a distance function, and the distance is defined between two subsets in the same metric space. However, it should be noted that a Hausdorff space introduced in Definition 10.58 in Chap. 10 includes no such distance function. It is apparent that the definition of the Hausdorff space is merely based on an abstract group theory.

Remark 2 We can generalize the *Hausdorff distance* defined above to the *Gromov-Hausdorff distance* which permits us to compare the distance of two compact metric spaces (e.g., Mémoli [81]). However, we will not discuss the Gromov-Hausdorff distance since this notion is beyond the scope of this book.

Definition 11.64 (*existence of a path in a digraph*) Let $CB(X)$be the class of all nonempty closed and bounded subsets of X. For each $x \in V(G)$, the notation $[Tx]_G$ denotes a class of nonempty closed and bounded subsets of G such that $[Tx]_G = \{A \in CB(X) : thereexistsapathfromutoxforsomeu \in A\}$.

Remark As regards the closed and bounded sets, see Definitions 10.41 and 10.42 in Chap. 10.

Definition 11.65 (*existence of a self-path map in a digraph*) We say that the set valued map $T : X \to CB(X)$ is a *self-path map*, whenever, for each $x \in V(G)$, there is a path from u to x for some $u \in Tx$, we denote this by $Tx \in [Tx]_G$.

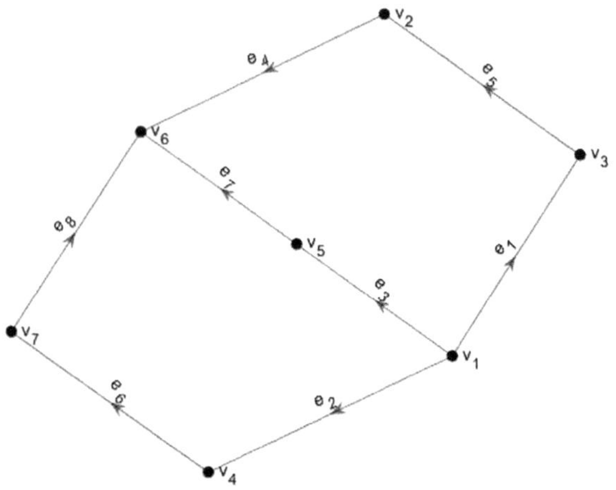

Fig. 11.8 Existence of a path with a start point in a subgraph S of a digraph G (This graph was adapted from that of Fig. 1 in Neog and Debnath (op. cit., p. 11)

When $x \neq y$, by the notation $Tx \in [Tx]_G$, we mean that there is a path from x to Ty for some $u \in Ty$. Also, we define $[y]_{G^l}$ as $[y]_{G^l} = \{x \in G : thereexistsapathfromytox\}$. Furthermore, the point $x \in V(G)$ is said to be a *fixed point* of the set valued map $T : X \to CB(X)$ if $x \in Tx$.

Theorem 11.20 (*existence of start point or cycle in a digraph*) Let G be a digraph such that every path in G has a lower bound within itself. Then, there exists a path in G, considered as a subgraph of G, which has a *start point* or a *cycle*.

Proof of this theorem is given by Neog and Debnath (op. cit., p. 3).

Example 11.24 Let G be a digraph and V a vertex set $\{v_1, v_2, v_3, \ldots\}$. Suppose that S is a subgraph of G with vertices $V(S) = \{v_1, v_2, v_3, v_4, v_5, v_6, v_7\}$ and edges $E(S) = \{e_1, e_2, e_3, e_4, e_5, e_6, e_7, e_8\}$ which are shown in Fig. 11.8. It is easy to see that S has no cycle and every path in S has a lower bound within itself, and $v_1 \to v_5 \to v_6$ is the shortest path with a start point v_1.

Theorem 11.21 (existence of fixed point in a digraph) *Let* (X, d) *be a complete metric space,* $\emptyset : X \to (-\infty, \infty)$ *a map bounded from above and upper semi-continuous and* $T : X \to CB(X)$ *a self path map satisfying the condition* $d(u, x) \leq \emptyset(u) - \emptyset(x)$, *for all* $x \in X$ *and* u $\in Tx$. *Then, T has a fixed point.*

Proof of this theorem is given by Neog and Debnath ([89], p. 6).

Remark According to Maddox ([77], p. 51), the condition for continuity of a real f at x_0 on the real line can be split into two: (i) $f(x_0) - \varepsilon < f(x)$ if $|x - x_0| < \delta$ and (ii) $f(x) < f(x_0) + \varepsilon$ if $|x - x_0| < \delta$. We thus define f as *lower semicontinuous* at x_0 if and only if for every $\varepsilon > 0$ there exists $\delta > 0$ such that $f(x_0) - \varepsilon < f(x)$ if $|x - x_0| < \delta$. Similarly for *upper semicontinuous* function. In general, a *complete metric space* is an example of the *Baire space* (e.g., De Groot [24]).

According to Neog and Debnath (op. cit., p. 6), the following example verifies Theorem 11.20.

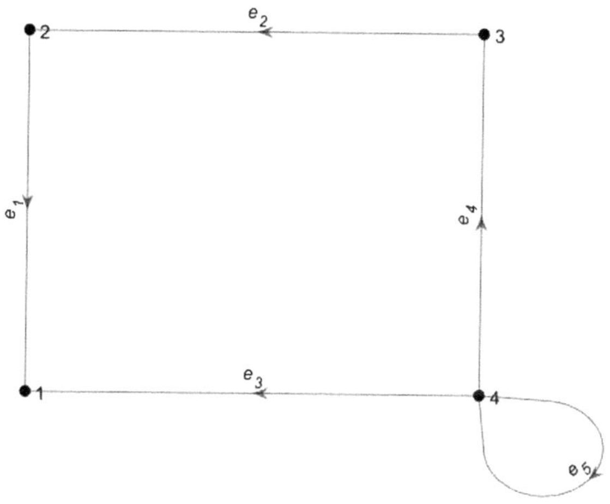

Fig. 11.9 A digraph with a fixed point, which assumes a metric space with edges satisfying a metric condition (This figure is adapted from that of Fig. 3 in Neog and Debnath (op. cit., p. 6)

Example 11.25 Let G be a digraph, and $X = \{1, 2, 3, 4\} = V(G)$ and $E(G) = \{(2, 1), (3, 2), (4, 1), (4, 3), (4, 4)\}$, which is shown in Fig. 11.9. In this figure, let us consider the closed and bounded subsets of X as $\{4\}$ and $\{3, 4\}$. Moreover, let us define $T : X \to CB(X)$ as follows:

$$Tx = \begin{cases} \{4\}, x \in \{11, 4\} \\ \{11, 4\}, x \notin \{11, 4\} \end{cases}. \tag{11.96}$$

Then, it is easy to verify that T is a *self path map*. Let $V(G)$ be endowed with metric $d : X \times X \to (0, \infty)$ by

$$d(x, y) = |x - y|, \text{ for all } x, y \in X. \tag{11.97}$$

Moreover, let us define $\emptyset : X \to (-\infty, \infty)$ by $\emptyset(t) = \frac{5t}{2}$. Then, the conditions of Theorem 11.18 are satisfied. Thus, T has a fixed point 4.

It is interesting to note that the distance between two vertices in a graph satisfies the metric axioms, as was discussed in Example 11.20 of Definition 11.60. However, in Definition 11.60 vertices are treated as *qualitative labels*. In contrast, in Example 11.25 of Theorem 11.21 vertices are treated as *quantitative numbers*, and the distance function between two vertices is defined by using these numbers, as shown in (11.87).

Definition 11.66 (*mean distance in a graph*) Let G be a connected graph with vertex set $V(G)$, and let d_{ij} denote the distance between vertices i and j (that is, the length of the *shortest path* between i and j). Moreover, let n be the number of vertices. Then, the *mean distance* in G is given by

$$\mu(G) = \frac{1}{\binom{n}{2}} \sum_{i<j} d_{ij}. \tag{11.98}$$

Remark 1 The mean distance varies depending on the types of graphs. In some cases, it varies depending on whether the number of vertices is even or odd.

Remark 2 The sum of distances between all pair of vertices of a graph, which is usually denoted by

$$W(G) = \sum_{\{i,j\} \subseteq V(G)} d_{ij}, \tag{11.99}$$

is called the *Wiener index* and was initiated by Wiener [138]. $W(G)$ is equivalent to $\sum_{i<j} d_{ij}$ in (11.98).

Example 11.26 Figure 11.10 is a *circuit graph* of order 5, and Table 11.7 is the distance matrix of this figure.

It is evident from Table 11.7 that $\sum_{i<j} d_{ij} = 15$ and $\mu(G) = 1.5$ since $n = 5$.

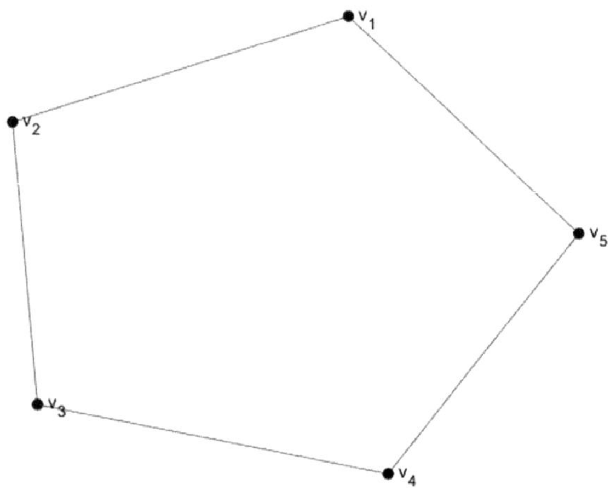

Fig. 11.10 A circuit graph of order 5

Table 11.7 A distance matrix of the circuit graph shown in Fig. 11.10

	v_1	v_2	v_3	v_4	v_5
v_1	0	1	2	2	1
v_2	1	0	1	2	2
v_3	2	1	0	1	2
v_4	2	2	1	0	1
v_5	1	2	2	1	0

Definition 11.67 (*mean distance in a weighted graph*). Let $G = (V, E)$be a connected graph with n vertices and m edges, and let $f : E \rightarrow R^+$be a weight function. Moreover, let $d_G(i, j; f)$ be the minimum length of a path between vertex i and vertex j according to the weight function f. Then, the *mean distance* in G is defined by

$$\mu_f(G) = \frac{1}{\binom{n}{2}} \sum_{\{i,j\} \subseteq V} d_G(i, j; f). \tag{11.100}$$

Remark 1 It should be noted that weights to be attached to edges are *nonnegative* in this definition. To be precise, R^+ is the set of all positive real numbers.

Remark 2 This definition is due to Djelloul and Kouider [27]. An alternative definition of mean distance in a weighted graph is made by Dankelmann [22], in which a nonnegative weight function is assumed to vertices. Recently an interesting extension of connected unweighted graphs to continuous domain has been done by Garijo et al. [45] which is called the *continuous mean distance* of a weighted graph.

Definition 11.68 (*mean distance in a digraph*) Let $G = (V, E)$ be a *strongly connected digraph* with n vertices. Moreover, let $d_G(i, j)$ be the distance *from* vertex i *to* vertex j in G. The *mean (or average) distance* in G is defined by

$$\mu(G) = \frac{1}{n(n-1)} \sum_{\{i,j\} \subseteq V} d_G(i, j). \tag{11.101}$$

Remark This definition is due to Dankelmann et al. [23]. Mean distance in a digraph is said to be initiated by Harary [50], who was motivated to develop a formula to measure a positional aspect of the *status of a person* in an organization or a group. As a result, his digraph was restricted to tree structures. However, Harary et al. ([52], p. 185) extend it to more general cases, in which the distance sum (i.e., the Wiener index) is defined as the sum a_i of the *finite* distances $d(i, j)$ for all j in G or the sum b_j of the *finite* distances $d(i, j)$ for all i in G. That is, a_i is the sum of the finite entries in the ith row of the distance matrix of G, while b_j is the sum of the finite entries in the jth column of the distance matrix. As regards the definition of the *strongly connected digraph*, see Remark 2 of Definition 11.50 (digraph).

Example 11.27 Figure 11.11 shows a digraph with tree structure, which was adapted from Fig. 1 in Harary (op. cit.), and Table 11.8 is the distance matrix associated with it. Figure 11.11 is a digraph version of a binary tree shown in Example 11.19 before. However, it is apparent that this digraph is not strongly connected because some two points are not necessarily mutually reachable. In this case, the number of edges with finite distance is 34, and the sum of distances is 62 ($= 1 \times 14 + 2 \times 12 + 3 \times 8$), according to the formula by Harary et al. (op. cit.). Therefore, the mean distance of this digraph is 62/34.

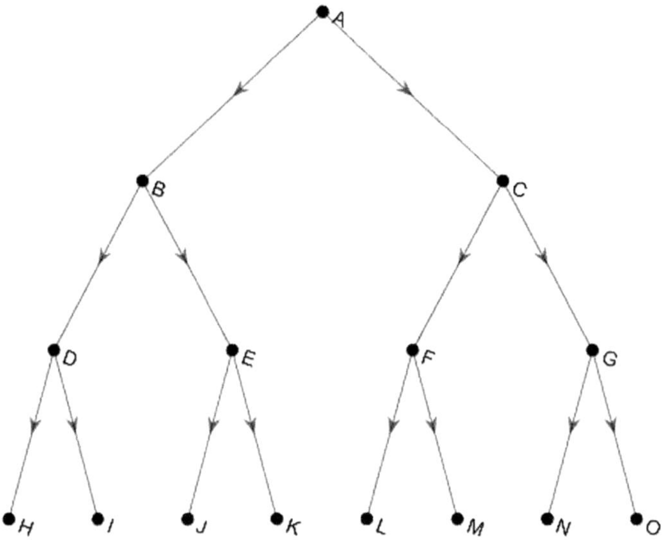

Fig. 11.11 A digraph version of a binary tree shown in Example 11.19, which was adapted from Harary (op. cit.). In terms of *positional status* in an organization, person (vertex) A is, for example, an *official* and the others (i.e., persons B to O) are *subordinates*

Table 11.8 A distance matrix associated with the digraph shown in Fig. 11.11

	A	B	C	D	E	F	G	H	I	J	K	L	M	N	O
A	0	1	1	2	2	2	2	3	3	3	3	3	3	3	3
B	∞	0	∞	1	1	∞	∞	2	2	2	2	∞	∞	∞	∞
C	∞	∞	0	∞	∞	1	1	∞	∞	∞	∞	1	1	1	1
D	∞	∞	∞	0	∞	∞	∞	1	1	∞	∞	∞	∞	∞	∞
E	∞	∞	∞	∞	0	∞	∞	∞	∞	1	1	∞	∞	∞	∞
F	∞	∞	∞	∞	∞	0	∞	∞	∞	∞	∞	1	1	∞	∞
G	∞	∞	∞	∞	∞	∞	0	∞	∞	∞	∞	∞	∞	1	1
H	∞	∞	∞	∞	∞	∞	∞	0	∞	∞	∞	∞	∞	∞	∞
I	∞	∞	∞	∞	∞	∞	∞	∞	0	∞	∞	∞	∞	∞	∞
J	∞	∞	∞	∞	∞	∞	∞	∞	∞	0	∞	∞	∞	∞	∞
K	∞	∞	∞	∞	∞	∞	∞	∞	∞	∞	0	∞	∞	∞	∞
L	∞	∞	∞	∞	∞	∞	∞	∞	∞	∞	∞	0	∞	∞	∞
M	∞	∞	∞	∞	∞	∞	∞	∞	∞	∞	∞	∞	0	∞	∞
N	∞	∞	∞	∞	∞	∞	∞	∞	∞	∞	∞	∞	∞	0	∞
O	∞	∞	∞	∞	∞	∞	∞	∞	∞	∞	∞	∞	∞	∞	0

Remark 1 As is apparent from the above example, the distance matrix is generally *asymmetric* in digraphs except for special cases. This means that in digraphs distances do not necessarily satisfy the metric axioms which hold for graphs.

Remark 2 Harary et al. ([52], p. 186) show a few interesting digraphs with *symmetric* distance matrix as well as *asymmetric* distance matrix and compute their Wiener indices. They also raise a fundamental theorem on the strongly connected digraph (or strong graph) and its corollaries. The next theorem shows this.

Theorem 11.22 (Bounds of the row distance sum and column distance sum of a strongly connected digraph) *Let G be a strong digraph with n vertices, a_i is the sum of the finite entries in the ith row of the distance matrix of G, and b_i is the sum of the finite entries in the jth column of the distance matrix. Then,*

$$n - 1 \leq a_i \leq n(n-1)/2, \tag{11.102}$$

and

$$n - 1 \leq b_i \leq n(n-1)/2. \tag{11.103}$$

Remark 1 It is apparent that the following inequality holds for a strong digraph, which is the Corollary 6.16b by Harary [51].

$$n(n-1) \leq \sum d_{ij} \leq n^2(n-1)/2. \tag{11.104}$$

Remark 2 As for the bounds of graphs and digraphs, there exists some important literature other than Harary (op. cit.). For example, Ng and Teh [93] raise some theorems on the Wiener index. Plesník [102] summarizes various bounds for the Wiener indices of graphs and digraphs. We will not go into detail about such literature because such a topic is beyond the scope of this book.

Definition 11.69 (*clustering coefficients*) Let $G = (V, E)$ be a graph with n vertices and m edges. A *local clustering coefficient* is defined for each vertex i by

$$C_i = \frac{2m_i}{k_i(k_i - 1)}, \tag{11.105}$$

where k_i is the number of vertices which are located at the nearest neighbors of a vertex i, while m_i is the number of edges between these nearest neighbors whereas a *global clustering coefficient* or *overall clustering coefficient* is defined by

$$\overline{C} = \frac{1}{n}\sum_{i=1}^{n} C_i. \tag{11.106}$$

Remark 1 The above definition is adapted from the definition made by Watts and Strogatz [137]. The m_i can be extended to a more general case in which the edges are the shortest paths of length l. The shortest path in the above definition is a special case when $l = 1$.

Remark 2 Watts and Strogatz [op. cit.] introduced these measures to determine whether a graph is a *small-world network* by analogy with the small world phenomenon (e.g., Milgram [83]). They found that many real-world networks are highly clustered (namely, have a high clustering coefficient), yet have small characteristic path length (namely, have a small average shortest path length).

We shall now introduce a concept which was derived from the degrees of vertex of a graph. This concept is known as *centrality measure*, and has long been discussed in the literature (e.g., Bavelas [5]; Bonacich [8]; Freeman [40]; Katz [64]; Landau [70]; Negre et al. [80]; Newman et al. [91]; Shaw [112]). Historically, centrality measures started with the question of what the most important or central person in the social network is. However, we shall not go into details of these measures and merely introduce some of them in the following. For further details, see the above literature.

Definition 11.70 (*degree centrality*) *Degree centrality* of a graph G is defined by the degree of G.

Remark 1 In the case of digraph, we must use two measures of degree centrality, that is, the indegree or outdegree.

Remark 2 According to Freeman (op. cit.), Shaw (op.cit.) introduced the degree centrality.

Definition 11.71 (*eigenvector centrality*) In general, *eigenvector centrality* is defined by the elements of an *eigenvector* of a matrix associated with a graph or digraph.

Remark 1 An example of the matrix associated with a digraph defined above is a *hyperlink matrix* shown in Example 1.11 of Chap. 1. The first eigenvector of this matrix is obtained by applying the Perron-Frobenius theorem of Theorem 11.7 in Sect. 11.2. Another example is a choice matrix shown in Katz (op. cit.), which is a sociometric data gathered by Forsyth and Katz [39]. This matrix is considered as an *adjacency matrix* of an unweighted digraph. Although we have shown several examples of adjacency matrix in Chap. 1, we shall define an adjacency matrix formally later in this section. As a third example, we shall propose to use a *Hermitian matrix* which is computed from any (real) ASM by HFM (Chino and Shiraiwa [15]). If we identify the ASM with a *weighted matrix of a digraph with loops*, we can compute eigenvectors of the Hermitian matrix, as is shown in Sect. 1.3.1.

Remark 2 The notion of eigenvector centrality is said to have been coined by Landau [70] (e.g., Volchenkov [135]).

Definition 11.72 (*closeness centrality*) *Closeness centrality* of a vertex, say, i, is defined by the mean geodesic distance from vertex i to every other vertex.

Remark Closeness centrality is said to go back to Leavitt [72, 73]. For further details, we shall refer to Freeman (op. cit.).

Definition 11.73 (*betweenness centrality*) Let g_{ij}be the number of geodesics linking vertices v_i and v_j in a connected graph, and let $g_{ij}(v_k)$ be the number of geodesics linking vertices v_i and v_j that contain a third vertex v_k. Then, the *partial betweenness*, $b_{ij}(v_k)$, is defined by

$$b_{ij}(v_k) = g_{ij}(v_k)/g_{ij}. \tag{11.107}$$

And the *betweenness centrality*, $C_B(v_k)$, is defined by

$$C_B(v_k) = \sum_{i<j}^{n} b_{ij}(v_k), \tag{11.108}$$

where n is the number of vertices in the graph.

Remark 1 Betweenness centrality is a centrality measure in *communication networks*, which regards the role of the *relayer* between two sites as important. For further details, see Freeman [40].

Remark 2 Betweenness centrality was introduced by Freeman (op. cit.).

Now we shall introduce some key concepts and terminology in graph theory. First, we shall define an incidence matrix. There are a few versions of this concept. The first two assume *graphs without loops*, and these are defined by Brouwer and Haemers ([11], p. 1).

Definition 11.74 (*incidence matrix of a graph without loops*) Let G be an undirected graph without loops. The (vertex-edge) *incidence matrix* of G is the 0–1 matrix $\boldsymbol{B} = \{b_{ij}\} \in M_{n,m}$, with rows indexed by the vertices and columns indexed by the edges, where $b_{ij} = 1$ when vertex i is an endpoint of edge.

Example 11.28 Suppose that we have a *graph G* without loops shown earlier in Fig. 11.14. Since both n and m are four, we have a 4×4 incidence matrix of G shown in Table 11.9. In this table v_i denotes vertices, while e_j edges.

Table 11.9 The incidence matrix associated with a graph shown in Fig. 11.13

	e_1	e_2	e_{11}	e_4
v_1	1	1	1	0
v_2	1	0	0	0
v_3	0	1	0	1
v_4	0	0	1	1

Definition 11.75 (*incidence matrix of a digraph without loops*) Let G be a directed graph, i.e., a digraph, without loops. The *directed incidence matrix* of G is the matrix $\boldsymbol{B} = \{b_{ij}\} \in M_{n,m}$, with rows indexed by the vertices and columns by the edges, where $b_{ij} = -1, 1, 0$ when vertex i is the head of edge j, the tail of j, or not on j, respectively.

Example 11.29 Suppose that we have a *digraph* G without loops shown in Fig. 11.12. Since both n and m are four, we have a 4×4incidence matrix of G shown in Table 11.10. In this table v_i denotes vertices, while $e_j$edges.

Cristofides ([21], p. 14) gives a more general definition of the incidence matrix than Brouwer and Holmes (op. cit.), in which a digraph has *loops*:

Definition 11.76 (*incidence matrix of a digraph with loops*). Let G be a directed graph, i.e., a digraph, with loops. The *directed incidence matrix* of G is the matrix $\boldsymbol{B} = \{b_{ij}\} \in M_{n,m}$, with rows indexed by the vertices and columns by the arcs, where $b_{ij} = -1, 1$ if vertex i is the head of arc j, the tail of j, respectively, and $b_{ij} = 0$ if vertex i is not a terminal vertex of arc j or if the arc is a loop.

Remark It should be remembered that as pointed out earlier Cristofides (op. cit.) calls a directed edge *an arc*.

Example 11.30 Suppose that we have a *digraph* G with a loop shown in Fig. 11.13.

Since n equals 4 and m equals 5, we have a 4×5incidence matrix of G shown in Table 11.11. In this table v_i denotes vertices, while e_j edges.

Table 11.10 The incidence matrix associated with a digraph shown in Fig. 11.12

	e_1	e_2	e_3	e_4
v_1	1	– 1	– 1	0
v_2	– 1	0	0	0
v_3	0	1	0	1
v_4	0	0	1	– 1

Fig. 11.12 A digraph G without loops, which is composed of 4 vertices and 4 edges

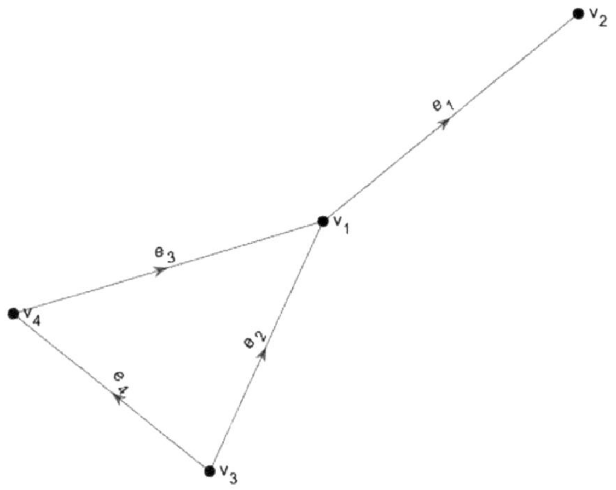

Fig. 11.13 A digraph G with a loop, which is composed of 4 vertices and 5 edges

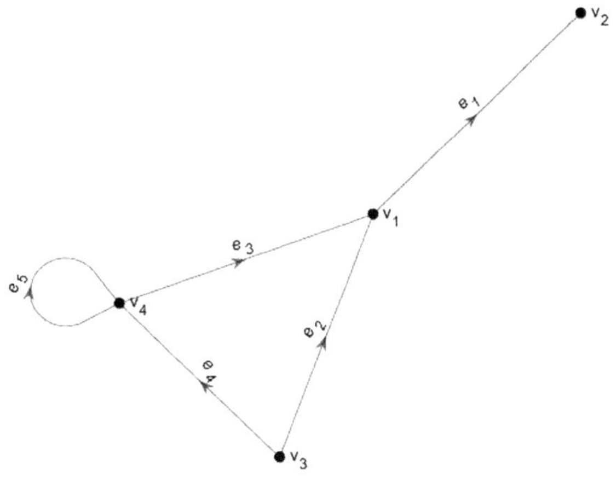

Fig. 11.14 A graph with a loop

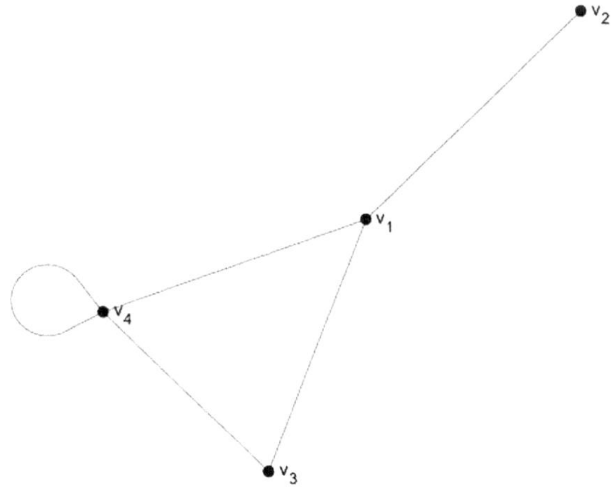

Table 11.11 The incidence matrix associated with a digraph shown in Fig. 11.13

	e_1	e_2	e_3	e_4	e_5
v_1	1	− 1	− 1	0	0
v_2	− 1	0	0	0	0
v_3	0	1	0	1	0
v_4	0	0	1	− 1	0

Second, we shall define an adjacency matrix. In Sect. 1.2 we have already shown some empirical examples of the adjacency matrix. In Sect. 1.3.1 we have defined it briefly. Here, we shall define it more precisely. Although there have been various definitions of the adjacency matrix in the literature, the definition by Brouwer and Haemers ([11], p. 1) seems to be the most precise:

Definition 11.77 (*adjacency matrix of a graph*) Let G be a graph without multiple edges. The *adjacency matrix* of G is the 0–1 matrix $\boldsymbol{A} = \{a_{ij}\}$ indexed by the vertex set $V(G)$ of G, where $a_{ij} = 1$ when there is an edge from i to j and $a_{ij} = 0$ otherwise. Occasionally, we consider multigraphs (possibly with loops), in which case a_{ij} equals the number of edges from i to j.

Remark Apparently, this definition enables us to include the case of *digraph*, since the phrase "an edge from i to j" implies the direction in the edge. Table 1.30 in Sect. 1.3.1 is an adjacency matrix associated with the undirected graph without multiple edges and loops, which is depicted in Fig. 1.28. It is apparent from the definition that the adjacency matrix of an undirected graph (i.e., a graph) is *symmetric*, while the adjacency matrix of a digraph is *asymmetric*. All the adjacency matrices shown in Sect. 1.1 are asymmetric.

Example 11.31 Figure 11.14 is a *graph with a loop*, and Table 11.12 is the adjacency matrix associated with Fig. 11.14.

Remark Figure 11.14 is the same as Fig. 11.13 except that Fig. 11.13 has directed edges. It should be noted in Table 11.9 that the *loop* which is attached to vertex v_4 *adds 2* to the corresponding cell in the adjacency matrix (e.g., Shum and Blake [114]).

Example 11.32 Figure 11.12 shown in Example 11.29 is a *digraph without loops*, and Table 11.13 is an adjacency matrix corresponding to the digraph.

Example 11.33 Figure 11.13 shown in Example 11.30 is a *digraph with a loop*, and Table 11.14 is the corresponding adjacency matrix.

Third, we shall define a Laplacian matrix of a graph. In Sect. 1.3.1 we have defined it briefly in introducing spectral graph theory. In this section, we will define

Table 11.12 The adjacency matrix corresponding to the graph shown in Fig. 11.14

	v_1	v_2	v_3	v_4
v_1	0	1	1	1
v_2	1	0	0	0
v_3	1	0	0	1
v_4	1	0	1	2

Table 11.13 The adjacency matrix corresponding to the digraph shown in Fig. 11.12

	v_1	v_2	v_3	v_4
v_1	0	1	0	0
v_2	0	0	0	0
v_3	1	0	0	1
v_4	1	0	0	0

Table 11.14 The adjacency matrix associated with a digraph shown in Fig. 11.13

	v_1	v_2	v_3	v_4
v_1	0	1	0	0
v_2	0	0	0	0
v_3	1	0	0	1
v_4	1	0	0	1

it precisely, according to Brouwer and Haemers ([11], p. 1) and Chung ([16], p. 12). Brouwer and Haemers call the Laplacian matrix *Laplace matrix*

Definition 11.78 (*Laplacian matrix of an undirected graph without multiple edges and loops*) Let $\boldsymbol{A}_u = \{a_{ij}\}$be the adjacency matrix of *G*. The *Laplacian matrix* $\boldsymbol{L}_u = \{l_{ij}\}$ of an *undirected graph G without multiple edges and loops* is the matrix indexed by the vertex set of *G*, with zero row sums, where $l_{ij} = -a_{ij}$ for $i \neq j$. If $\boldsymbol{D}_u = \{d_{ij}\}$ is the diagonal matrix, indexed by the vertex set of *G* such that d_{ii} is the degree of the vertex *i*, then the Laplacian matrix is written as

$$\boldsymbol{L}_u = \boldsymbol{D}_u - \boldsymbol{A}_u, \tag{11.109}$$

and

$$\boldsymbol{Q}_u = \boldsymbol{D}_u + \boldsymbol{A}_u, \tag{11.110}$$

is called the *signless Laplacian matrix* of G.

Remark As Brouwer and Haemers (op. cit.) point out, $\boldsymbol{L}_u$ and $\boldsymbol{Q}_u$ are both *positive semidefinite*. In fact, we have

$$\boldsymbol{L}_u = \boldsymbol{N}\boldsymbol{N}^t, \boldsymbol{Q}_u = \boldsymbol{M}\boldsymbol{M}^t, \tag{11.111}$$

where $\boldsymbol{M}$ is the *incidence matrix* of *G* and $\boldsymbol{N}$ is the *directed incidence matrix* of the *directed graph* obtained by orienting the edges of *G* in an arbitrary way.

Example 11.34 Let *G* be the *undirected graph without loops* shown in Fig. 11.4 in Example 11.20 (also shown in Fig. 1.28 in Sect. 1.3.1). The adjacency matrix $\boldsymbol{A}_u$ of *G* has already been shown in Table 1.3.1 in Sect. 1.30, while the degree matrix of G is written as $\boldsymbol{D}_u = \text{diag}\,(3, 1, 2, 2)$. Moreover, the *incidence matrix* $\boldsymbol{M}$ of *G* is shown in Table 11.9 in Example 11.28. Finally, let the *directed graph* obtained by orienting the edges of *G* in an arbitrary way be the digraph shown in Fig. 11.12 in Example 11.29. Then, the *directed incident matrix* $\boldsymbol{N}$ is equal to the matrix shown in Table 11.10. Then we have

$$\boldsymbol{L}_u = \boldsymbol{D}_u - \boldsymbol{A}_u = \boldsymbol{N}\boldsymbol{N}^t = \begin{bmatrix} 3 & -1 & -1 & -1 \\ -1 & 1 & 0 & 0 \\ -1 & 0 & 2 & -1 \\ -1 & 0 & -1 & 2 \end{bmatrix}. \tag{11.112}$$

and

$$\boldsymbol{Q}_u = \boldsymbol{D}_u + \boldsymbol{A}_u = \boldsymbol{M}\boldsymbol{M}^t = \begin{bmatrix} 3 & 1 & 1 & 1 \\ 1 & 1 & 0 & 0 \\ 1 & 0 & 2 & 1 \\ 1 & 0 & 1 & 2 \end{bmatrix}. \tag{11.113}$$

Remark Two kinds of incidence matrices $\boldsymbol{N}$ and $\boldsymbol{M}$ defined in (11.101) can be viewed as special cases of a more general *Gram matrix* $\boldsymbol{G} = \boldsymbol{A}\boldsymbol{A}^*$ in (11.38), where $\boldsymbol{A}^*$ is the *conjugate transpose* of a *complex* rectangular matrix $\boldsymbol{A} \in M_{m,n}(C)$. Moreover, it is known that for any such matrix $\boldsymbol{A}$, the matrices $\boldsymbol{A}\boldsymbol{A}^*$ and $\boldsymbol{A}^*\boldsymbol{A}$ are *positive semidefinite* (Lancaster and Tismenetsky, [69], p. 180). As for an important property of Gram matrix, we have already introduced it in Theorem 11.1. By the way, eigenvalues of $\boldsymbol{L}_u$ in (11.102) are 4.0, 3.0,1.0, and 0, while those of $\boldsymbol{Q}_u$ in (11.103) are 4.5616, 2.0, 1.0, 0.4384 (to be precise, positive definite).

Chung ([16], p. 12) shows an extension of the Laplacian matrix in Definition 11.78 to that of a weighted undirected graph possibly with loops. He calls the Laplacian matrix *Laplacian for graphs*:

Definition 11.79 (*Laplacian matrix of a weighted undirected graph possibly with loops*). Let G be a *weighed undirected graph possibly with loops*, and each edge (i, j) formed by vertices i and j be associated with a symmetric weight matrix $\boldsymbol{W} = \{w_{ij}\}$ with nonnegative values. Then, the Laplacian matrix $\boldsymbol{L} = \{l_{ij}\}$ is defined by

$$l_{ij} = \begin{cases} d_{jj}^{(w)} - w_{jj}, & \textit{if } i = j, \\ -w_{ij}, & \textit{if } i \textit{ and } j \textit{ are adjacent}, \\ 0, & \textit{otherwise}, \end{cases} \tag{11.114}$$

where $d_{jj}^{(w)} = \sum_i w_{ij}$. If we normalize $\boldsymbol{L}$ using the diagonal matrix $\boldsymbol{D}_w = diag\left\{d_{11}^{(w)}, \ldots, d_{nn}^{(w)}\right\}$, then we have $\boldsymbol{D}_w^{-1/2}\boldsymbol{L}\boldsymbol{D}_w^{-1/2}$. Here, n is the number of vertices.

Example 11.35 Figure 11.15 shows a *weighted undirected graph without loops*. If the weights in this graph are $w_{12} = 6, w_{13} = 4, w_{14} = 5, and w_{34} = 3$, then we have $\boldsymbol{D}_w = diag\{15, 6, 7, 8\}$,

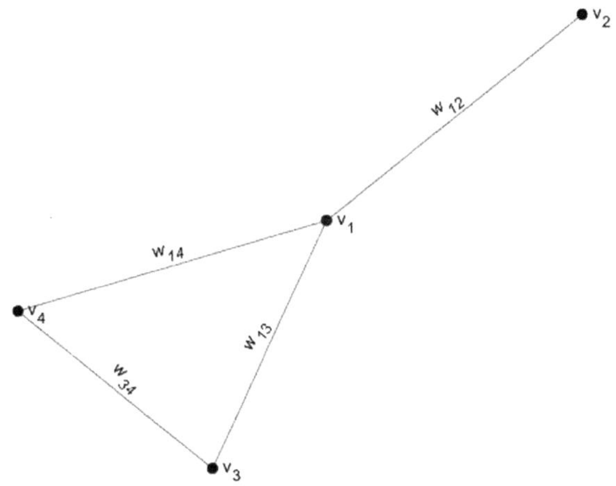

Fig. 11.15 A special weighted undirected graph without loops

$$W = \begin{bmatrix} 0 & 6 & 4 & 5 \\ 6 & 0 & 0 & 0 \\ 4 & 0 & 0 & 3 \\ 5 & 0 & 3 & 0 \end{bmatrix}, \text{ and } L = \begin{bmatrix} 15 & -6 & -4 & -5 \\ -6 & 6 & 0 & 0 \\ -4 & 0 & 7 & -3 \\ -5 & 0 & -3 & 8 \end{bmatrix}. \quad (11.115)$$

Remark 1 In this case, the eigenvalues of $\boldsymbol{L}$ in (11.105) are 20.2002, 10.4626, 5.3372, and 0.0, which shows that $\boldsymbol{L}$ is also *positive semidefinite*. It is evident that the Laplacian matrix $\boldsymbol{L}$ in Definition 11.79 is a generalization of that in Definition 11.78.

Remark 2 Belkin and Niyogi [6] utilize the same Laplacian matrix as the matrix defined in (11.104). Yan et al. [142] define a similar Laplacian matrix in the context of dimensionality reduction problem and machine learning problem. They also assume an undirected weighted graph, but $d_{jj}^{(w)}$ in Definition 11.79 by Chung (op. cit., p. 12) is defined as $\sum_{i \neq j} w_{ij}$ by Yan et al. (op. cit.). This means that weights of the loops in this case are not considered. Stated another way, Yan et al. assume a graph without loops.

Laplacian matrices defined up to now assume that the graphs corresponding to them are *undirected*. In contrast, in any *directed graph* (i.e., *digraph*) the degree matrix corresponding to it cannot be defined uniquely. Moreover, both adjacency matrix and weight matrix are in general *asymmetric*. Therefore, for example Chung (op. cit., p. 22) focuses his attention on the Laplacian of a graph and does not deal with that of a digraph. Recently, however, there has been an increasing attention paid to the study of asymmetric Laplacian matrices. For example, Agaev and Chebotarev [1] examine the spectra of *asymmetric Laplacian matrices*. Veerman and Lyons [134] discuss *Laplacian dynamics* in digraph. The following definition of an asymmetric Laplacian matrix is due to Agaev and Chebotarev (op. cit.):

Definition 11.80 (*Laplacian matrix of a weighted digraph*) Let the matrices $\boldsymbol{L} = \{l_{ij}\} \in M_n(R)$ such that

$$l_{ij} \le 0, j \ne i, \tag{11.116}$$

$$\sum_{j=1}^{n} l_{ij} = 0, i = 1, \ldots, n. \tag{11.117}$$

If $\boldsymbol{L}$ is *symmetric* and $l_{ij} \in \{0, -1\}, i \ne j$, then $\boldsymbol{L}$ is the *Laplacian matrix of an undirected graph*: $l_{ij} = -1$ if and only if the graph has an edge $(i, j), i \ne j$; l_{ii} is the degree of vertex i.

In contrast, if every matrix $\boldsymbol{W} = \{w_{ij}\}$ of the form of (11.106) and (11.107) is associated with a *weighed digraph* $\boldsymbol{G}$ whose arc weights are *strictly positive*:

$$l_{ij} = \begin{cases} -\sum_{k \ne i} w_{ik}, & \text{if } i = j, \\ -w_{ij} & \text{if } i \ne j \text{ and } G \text{ has an } (i, j) \text{ arc}, \\ 0 & \text{if } i \ne j \text{ and } G \text{ has no } (i, j) \text{ arc}. \end{cases} \tag{11.118}$$

then $\boldsymbol{L} = \{l_{ij}\}$ is the *Laplacian matrix of a weighted digraph.*

Remark It should be noted that the weight matrix $\boldsymbol{W} = \{w_{ij}\}$ in (11.118) is generally *asymmetric*. This means that eigenvalues of this matrix are in general *complex*. Of course, the Laplacian matrix defined above is the same as that defined in Definition 11.79 in the special case when $\boldsymbol{W}$ is symmetric, except that the diagonal elements are row sum of the weights.

Example 11.36 Figure 11.16 shows a general weighted digraph with multiple edges and *loops*. If, for example, the weight matrix $\boldsymbol{W}$ is such that

$$\boldsymbol{W} = \begin{bmatrix} 10 & 6 & 4 & 5 \\ 4 & 8 & 0 & 0 \\ 8 & 2 & 9 & 3 \\ 3 & 3 & 5 & 7 \end{bmatrix}, \tag{11.119}$$

The eigenvalues of $\boldsymbol{W}$ are 19.6901, 8.2055, 3.0522 + 1.9571i. and 3.0522 − 1.9571i.

If we analyze this matrix via HFM (Chino and Shiraiwa [15]), the eigenvalues of the Hermitian matrix constructed from $\boldsymbol{W}$ are 20.5122, 7.5222, 5.4190, and 0.5466, which indicate that the four vertices can be embedded in a four-dimensional *Hilbert space*.

On the other hand, $\boldsymbol{D}_w = diag\{15, 4, 13, 11\}$, and the Laplacian matrix is as follows:

$$\boldsymbol{L} = \begin{bmatrix} 15 & -6 & -4 & -5 \\ -4 & 4 & 0 & 0 \\ -8 & -2 & 13 & -3 \\ -3 & -3 & -5 & 11 \end{bmatrix}. \tag{11.120}$$

Fig. 11.16 A weighted digraph with multiple edges and loops

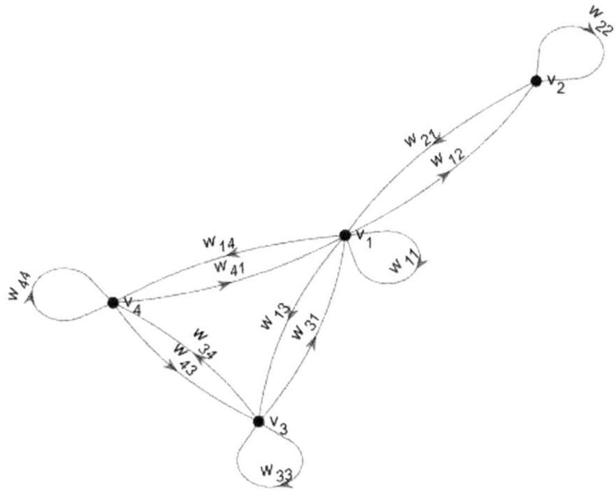

The eigenvalues of $\boldsymbol{L}$ are 19.3667, 16.7494, 6.8839, and 0.0, which indicate that the four vertices can be embedded in a three-dimensional *Euclidean space*. Ordinarily,the Laplacian matrix is known to be p.s.d. (e.g., Brouwer and Haemers [11]; Chung [16]).

Remark Agaev and Chebotarev (op. cit.) examine the spectrum of a *standardized Laplacian matrix* $\tilde{\boldsymbol{L}} = (nb)^{-1}\boldsymbol{L}(G)$,where n is the order of a weighted digraph G and b is positive arc weights exceeding $b > 0$, and obtain a region that contains the Laplacian spectra in a complex region (see Theorem 5, p. 163). In quantum physics, *magnetic Laplacian* for a digraph which utilizes *Hermitian matrix* is known (e.g., Furutani et al. [43]; Shubin [113]).

In any case, there has been an immense literature on *spectral graph theory* which deals with the eigenstructures of adjacency matrices and Laplacian matrices in an undirected graph (or simply graph) (e.g., Brouwer and Haemers [11]; Chung [16]). For example, Brouwer and Haemers discuss various propositions and theorems on the largest and second largest eigenvalues of graphs. Chung examines eigenvalues of various types of graphs and subgraphs. However, we will not introduce these propositions and theorems, since they are not concerned with the structure and dynamics of *asymmetric interactions* between objects which are major topics in this book.

At this point a natural question arises concerning Laplacian matrices in spectral graph theory. Why are the Laplacian matrices so frequently utilized in the theory? One reason may be that Laplacian matrices have very desirable properties in that they are all *positive semidefinite*. For example, Laplacian matrices of undirected unweighted graphs are expressed as *Gram matrices* as discussed earlier, and they induce *inner product* on a vector space. Moreover, the Laplacian matrix of an undirected weighted graph is denoted by a *quadratic form* with the weight matrix, which means that the Laplacian matrix is positive semidefinite, as, for example, Belkin and Niyogi ([6], p. 588) point out.

However, there exists a special restriction on the Laplacian matrix $\boldsymbol{L} = \{l_{ij}\}$ of a graph, since it is related to the following *heat equation* in physics

$$\frac{\partial u}{\partial t} = \nabla^2 u, \tag{11.121}$$

where ∇^2 is called *Laplacian*, or *Laplacian operator*. The Laplacian is associated with the rate of *dissipation of heat*, as for example, Chung (op. cit., p. 149) points out. The Laplacian matrix is known as a *discrete Laplacian operator*. Tittelfitz [129], for example, discusses the heat equation on discrete domains. Stated another way, the Laplacian matrix of a graph and digraph can be said to be restrictive in that the dynamics among nodes with a Laplacian matrix are related to the dynamics of the cooling processes of nodes. If such a dynamic is not congruent with the dynamics of an observed graph or digraph, it seems to be more appropriate and natural to choose alternative matrices in examining the spectral properties of the graph or digraph. The following is an alternative. It is based on the Chino-Shiraiwa theorem (i.e., Theorem 11.16) discussed in the previous section as well as in Chaps. 4 and 5:

Definition 11.81 Let an observed weight matrix $\boldsymbol{W} \in M_n(R)$ among nodes, whose elements are not necessarily positive, be associated with a *digraph G*. If we compute a *Hermitian matrix* $\boldsymbol{H}_w \in M_n(C)$ such that

$$\boldsymbol{H}_w = \left(\boldsymbol{W} + \boldsymbol{W}^t\right)/2 + i\left(\boldsymbol{W} - \boldsymbol{W}^t\right)/2, \tag{11.122}$$

then, nodes can be embedded in a finite-dimensional *Hilbert space* if $\boldsymbol{H}_w$ is *positive semidefinite*. In contrast, if $\boldsymbol{H}_w$ is *indefinite*, nodes can be embedded in an *indefinite metric space*. In all cases, the spectra of $\boldsymbol{H}_w$ are *real*.

Remark 1 It is apparent that we *need not impose any restrictions* on the elements of $\boldsymbol{W}$, except that $\boldsymbol{H}_w$ constructed from an observed weight matrix is positive semidefinite. This property of $\boldsymbol{H}_w$ should be contrasted with that of the Laplacian matrices. Moreover, even if $\boldsymbol{H}_w$ is indefinite, we can consider any dimension corresponding to a negative eigenvalue as a one-dimensional Hilbert space by changing the sign of the eigenvalue.

Remark 2 The construction of a Hermitian matrix given a general *ASM*, say, $\boldsymbol{S} \in M_n(R)$, instead of $\boldsymbol{W}$, goes back to Escoufier and Grorud [33]. However, they did not examine the space property of this type of Hermitian matrix. As discussed in Example 1.23 in Chap. 1, there has been an increasing attention paid to the study of Hermitian adjacency matrices (e.g., Greaves et al. [49]; Liu and Li [75]; Moher [84]; Yuan et al. [144]). As pointed out there, such complex matrices can neither be observed in vivo nor in daily life, although these matrices are meaningful from the viewpoint of pure mathematics. This property of these Hermitian adjacency matrices is also contrasted with the Hermitian matrix $\boldsymbol{H}_w \in M_n(C)$ based on the Chino-Shiraiwa theorem.

Example 11.37 Let the weight matrix associated with a digraph be the matrix shown in Table 1.28 in Chap. 1. The original data is an international trade data shown in Table 1.27 in Chap. 1 as well as Table 5.3 in Chap. 5, and Table 1.28 is the log-transformed trade data among four nations. If we identify this transformed ASM (asymmetric similarity matrix) with the observed weight matrix of a digraph, then $\boldsymbol{H}_w \in M_n(C)$ is computed as in Table 1.29. Moreover, the eigenvalues of $\boldsymbol{H}_w$ are 29.9714, 6.1145, 4.4377, and 3.5309, as shown in Sect. 5.3. This means that we can embed the four nations in a *four-dimensional Hilbert space*. Figure 5.5 in Sect. 5.3 depicts these configurations of nations separately in each of the *complex planes* corresponding to the four dimensions. In general, each of the configuration of nodes (in this case, nations) in which nodes are located in a complex plane can be easily interpreted based on the Hermitian form model (HFM) by Chino and Shiraiwa [15]. For further details of how to interpret the configuration of objects, we shall refer to Sect. 5.2 in Chap. 5.

11.5 Graph Embeddings

As discussed in Sect. 1.3.1, the scope of graph embedding is very wide. In this section we shall supplement some of the major methods for graph embedding discussed there in relation to MDS, the graph theory and spectral graph theory introduced in the previous section. These methods include *Isometric Feature mapping (ISOMAP)* by Tenenbaum et al. [125], *Locally Linear Embedding (LLE)* by Roweis and Saul [107], *Laplacian Eigenmap (LE)* by Belkin and Niyogi [6], and Marginal Fisher Analysis (MFA) by Yan et al. [142].

(1) ISOMAP

ISOMAP as a method for nonlinear dimensionality reduction enables us to recover the nonlinearly embedded manifold such as a *Swiss roll* (e.g., Martinez and Martinez [79]), given a body of high-dimensional observations. Figure 11.17 shows a Swiss roll, which is a manifold of dimension 2 in a subspace of E^3. Let $\boldsymbol{x}_i$ be the p-dimensional coordinate vector of the ith observation. For example, 32×32 pixel grey-scale face images can be thought of as points in a 1,024-dimensional observation space. Input data of the ISOMAP thus can be summarized in a matrix $\boldsymbol{X} \in M_{n,p}(R)$, where n is the number of observations. Before proceeding, we must compute the *distances* $d_X(i,j)$ between all pairs i, j from n data points.

According to Tenenbaum et al. (op. cit.), ISOMAP algorithm has three steps. The first step is to construct a *neighborhood graph* G over all data points by connecting points i and j if (as measured by $d_X(i,j)$) they are closer than ε (ε-ISOMAP) or if i is one of the *K nearest neighbors* of j (K-ISOMAP). Here, edge lengths are set equal to $d_X(i,j)$. The second step is to compute *shortest paths distances*. First, we initialize $d_G(i,j) = d_X(i,j)$ if i, j are linked by an edge; and $d_G(i,j) = \infty$ otherwise. Then for each value of n observations in turn, we replace all entries $d_G(i,j)$ by $min\{d_G(i,j), d_G(i,k) + d_G(k,j)\}$. The matrix of final values $\boldsymbol{D}_G = \{d_G(i,j)\} \in M_n(R)$ contains the shortest path distances between all pairs of points in G. The third step is to construct d-dimensional embedding. First, we compute the squared distance $\boldsymbol{T} \in M_n(R)$ using $\boldsymbol{D}_G$. Then, we compute $\boldsymbol{B}_c \in M_n(R)$ applying (11.69)

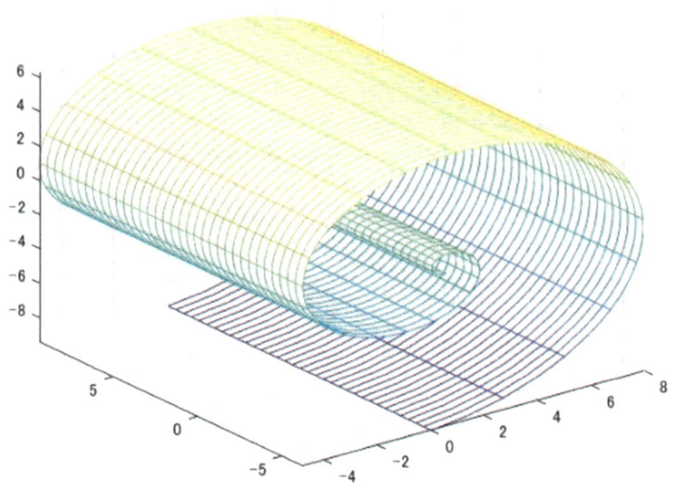

Fig. 11.17 A *Swiss roll*, which is a manifold of dimension 2 in a subspace of E^3

to $\boldsymbol{T}$ of ISOMAP. Tanenbaum et al. denote it by $\tau(\boldsymbol{D}_G)$. Finally, we may solve the eigenvalue problem of $\boldsymbol{B}_c$ by the classical MDS.

Remark An excellent feature of ISOMAP is that it may be possible to capture a locally linear structure such as a Swiss roll in the configuration space. In this sense, ISOMAP is a desirable generalization of (symmetric) MDS developed in psychometrics. MATLAB has a function *isomap.*

(2) **LLE**

LLE proposed by Roweis and Saul [107] is also a method for nonlinear dimensionality reduction by locally linear embedding. LLE obtains the final configuration of data points in three steps, given the p-dimensional coordinate vector $\boldsymbol{x}_i$ of the ith observation, which is thought of as an *unsupervised learning algorithm.* They expect that each data point and its neighbor lie on or close to a locally linear patch of the manifold. The first step is to select neighbors to each data point. We assign neighbors to each data point $\boldsymbol{x}_i$, for example, by using the *K nearest neighbors.* The second step is to compute the weights W_{ij} that best linearly reconstruct $\boldsymbol{x}_i$ from its neighbors, solving the constraint least-squares problem in the following equation,

$$\varepsilon(W) = \sum_i \left| \boldsymbol{x}_i - \sum_j W_{ij}\boldsymbol{x}_j \right|^2, \tag{11.123}$$

which adds up the squared distance between all the data points and their reconstruction. The third step is to compute the low-dimensional embedding vectors $\boldsymbol{y}_i$ best reconstruct by W_{ij}, minimizing the following equation,

$$\Phi(y) = \sum_i \left| y_i - \sum_j W_{ij} y_j \right|^2, \tag{11.124}$$

by finding the smallest *eigenmodes* (e.g., Königsberg [68]) of the sparse symmetric matrix $\boldsymbol{M} = \{M_{ij}\} \in M_n(R)$, where

$$\boldsymbol{M} = (\boldsymbol{I} - \boldsymbol{W})^t (\boldsymbol{I} - \boldsymbol{W}). \tag{11.125}$$

and $\boldsymbol{W} = \{W_{ij}\} \in M_n(R)$. For further details, see Roweis and Saul (op. cit.).

Remark It is apparent that $\boldsymbol{M}$ in (11.115) is p.s.d. since this matrix is a special form of $\boldsymbol{A}^*\boldsymbol{A}$ where $\boldsymbol{A}$ is generally a square matrix in $M_n(C)$, as we pointed out in the remark of Theorem 11.2 in Sect. 11.2. This property guarantees that nodes can be embedded in a lower dimensional Euclidean space.

(3) **LE**

LE proposed by Belkin and Niyogi (op. cit.) is a third method for nonlinear dimensionality reduction by locally linear embedding. LE also obtains the final configuration of n data points in three steps, given the p-dimensional coordinate vector $\boldsymbol{x}_i$ of the ith observation, which is thought of as a *manifold learning algorithm.* The first step is to construct a neighborhood graph which they call an *adjacency graph*, as is the same as ISOMAP. The second step is to compute the weights to be attached to the graph. They have two variations for weighting the edges. One is to use the *heat kernel.* If nodes i and j are connected, they compute the weight such that

$$W_{ij} = exp\left(-\|\boldsymbol{x}_i - \boldsymbol{x}_j\|^2 / t\right), \tag{11.126}$$

where $t \in R$ is a parameter. The other is to use the *simple-minded*, in which $W_{ij} = 1$, if and only if nodes i and j are connected by an edge. In this case, the weight matrix $\boldsymbol{W} = \{W_{ij}\} \in M_n(R)$ is nothing but a usual *adjacency matrix* of a graph. The third step is to compute the *Laplacian matrix* of a weighted undirected graph, that is, they compute the Laplacian matrix $\boldsymbol{L} \in M_n(R)$, according to (11.104) in Definition 11.79 by Chung ([16], p. 12), and they solve its eigenvalue problem.

Remark He and Niyogi [53] proposed a linear approximation of LE discussed above and called it *Locally preserving projection (LPP).* They consider a linear transformation that maps p-dimensional data point $\boldsymbol{x}_i$ to an l-dimensional point $\boldsymbol{y}_i$ first, such that $\boldsymbol{y}_i = \boldsymbol{A}^t \boldsymbol{x}_i$, where $l \leq p$ and $\boldsymbol{A} \in M_{p,l}(R)$. Then, they administer LE to these transformed points, and obtain the following eigenvalue problem

$$\boldsymbol{XLX}^{\mathrm{t}}\boldsymbol{a} = \lambda \boldsymbol{XLX}^{\mathrm{t}}\boldsymbol{a}, \tag{11.127}$$

where $\boldsymbol{X} = \{\boldsymbol{x}_1, \boldsymbol{x}_2, \ldots, \boldsymbol{x}_n\} \in M_{p,n}(R)$, $\boldsymbol{L} \in M_n(R)$ is a Laplacian matrix, $\boldsymbol{a}$ is a p-dimensional eigenvector of the matrix $\boldsymbol{XLX}^{\mathrm{t}} \in M_p(R)$, and n is the number of observations. Let the column eigenvectors in the above equation be $\boldsymbol{a}_0, \boldsymbol{a}_1, \ldots, \boldsymbol{a}_{l-1}$, which are associated with the eigenvalues, $\lambda_0 < \lambda_1 < \cdots < \lambda_{l-1}$. Then, $\boldsymbol{A} = (\boldsymbol{a}_0, \boldsymbol{a}_1, \ldots, \boldsymbol{a}_{l-1})$.

ISOMAP, LLE, and LE for nonlinear dimensionality reduction construct a neighborhood graph in common, given a high-dimensional data points. However, these methods adopt different strategies for obtaining a low-dimensional configuration of nodes of a graph. ISOMAP utilizes the classical MDS by computing the distances between nodes. LLE solves an eigenvalue problem by computing a special matrix using the weights associated with a neighborhood graph which are estimated from the data points by a least-squares method. LE computes the weights of the graph directly using the heat kernel etc. and obtains a configuration of nodes by solving an eigenvalue problem of the usual Laplacian matrix constructed from these weights. Yan et al. [142] summarizes the features of these methods including PCA and LDA (linear discriminant analysis) from somewhat different viewpoint.

(4) **MFA**

MFA proposed by Yan et al. (op. cit.) is an algorithm which overcomes a limitation of LDA, using the graph embedding framework. They construct two kinds of graphs called an *intrinsic graph* and a *penalty graph*. The algorithm of MFA is composed of four steps. The first step is to project the data into the PCA subspace by retaining $N - N_c$ dimensions, where N is the number of data points and N_c is the number of classes of these data points. Here, let $\boldsymbol{W}_{PCA}$ be the transformation matrix of PCA (e.g., Turk and Pentland [132]). The second step is to construct the *intraclass compactness* and *interclass separability graphs*. In the former graph, set the adjacency matrix $\boldsymbol{W} = \{W_{ij}\}$ such that $W_{ij} = W_{ji} = 1$ if sample $\boldsymbol{x}_i$ is among the k_1-nearest neighbors of $\boldsymbol{x}_j$ in the same class. In the latter graph, set the similarity matrix $\boldsymbol{W}^{(p)} = \left\{W_{ij}^{(p)}\right\}$ such that $W_{ij}^{(p)} = 1$ if the pair (i, j) is among the k_2 shortest pairs among the set $\{(i, j), i \in \pi_c, j \notin \pi_c\}$. The third step is to compute the Marginal Fisher Criterion

$$\boldsymbol{w}^* = \arg\min_{\boldsymbol{w}} \frac{\boldsymbol{w}^t \boldsymbol{X}(\boldsymbol{D} - \boldsymbol{W})\boldsymbol{X}^t \boldsymbol{w}}{\boldsymbol{w}^t \boldsymbol{X}(\boldsymbol{D}^p - \boldsymbol{W}^p)\boldsymbol{X}^t \boldsymbol{w}}, \tag{11.128}$$

which is a special linearization of the graph embedding framework with $\boldsymbol{B} = \boldsymbol{D}^p - \boldsymbol{W}^p$. The fourth step is compute the final linear projection direction as $\boldsymbol{w} = \boldsymbol{W}_{PCA}\boldsymbol{w}^*$.

The dimensionality reduction methods discussed up to now are all concerned with the symmetric weight matrix of an undirected graph (or simply, graph). Recently, however, Iwai et al. [60] proposed an *asymmetric LPP* (ALPP), which is an extension of LPP proposed by He and Niyogi [53]. In ALPP, the Laplacian matrix $\boldsymbol{L} = \boldsymbol{D} - \boldsymbol{W}$ in LPP is replaced by $\boldsymbol{L}_H = \boldsymbol{D}_H - \boldsymbol{H}_{\alpha}$, where $\boldsymbol{H}_{\alpha}$ is a revised version of (11.71) parametrized by α such that

$$\boldsymbol{H}_{\alpha} = (1 - \alpha)\boldsymbol{S}_s + \alpha \boldsymbol{S}_{sk}, \tag{11.129}$$

and $\boldsymbol{D}_H$ is a diagonal matrix whose (i, j) element is $\sum_{j=1}^{n} S_{ij}^{(s)}$. Here, $S_{ij}^{(s)}$ is the (i, j) element of $\boldsymbol{S}_s$.

11.6 Chemical Reaction Networks and Neural Networks

In this section, we shall define several basic matrices peculiar to chemical reaction networks and (artificial) neural networks, although we have already introduced them briefly in Chap. 1. These matrices play fundamental roles in describing not only the *structures* but also the *dynamics* of asymmetric interactions in chemical reaction networks and neural networks. As a result, we shall raise these issues again later in Chap. 12 from the viewpoint of dynamical system theory.

11.6.1 Chemical Reaction Networks

Since chemical reaction networks obey the laws of chemical reactions, we need some matrices peculiar to them. In particular, we need some special matrices, since chemical reaction networks obey the *law of mass action*. Before proceeding we shall define the notions of *mole* and *mol*, the latter of which is one of the *International System of Units* (abbreviated as *SI*) unit.

Definition 11.83 (*mole and mol*) One mole of a substance is defined as containing 6.02214076 elementary entities of that substance. *Mol* is the symbol of the mole, which is the unit of amount of substance in the SI unit.

Remark 1 The value 6.02214076 is called *Avogadro's number*.

Remark 2 A mole had been defined as the number of atoms present in 12.0 g of ^{12}C isotope until 2018. The above definition of the mole by National Institute of Standards and Technology (NIST [95]) has been used since 2019.

Definition 11.84 (*Law of Mass Action*) The velocity of a chemical reaction is proportional to the product of the molecular concentrations of the reacting substances, each concentration being raised to the power which is numerically equal to the coefficient of that substance in the equation representing the reaction.

Remark 1 The Law of Mass Action is sometimes called the *law of molecular concentration.*

Remark 2 The above definition is due to Brinkley ([10], p. 160). The Law of Mass Action is said to be initiated by Guldberg and Waage in 1867. According to Brinkley (op. cit.), the relation of reaction velocity to concentration was stated quantitatively in 1863 by Guldberg and Waage. However, as usual, the origins of the Law of Mass Action have a longer history. For their details, see for example Ferner and Aronson [36].

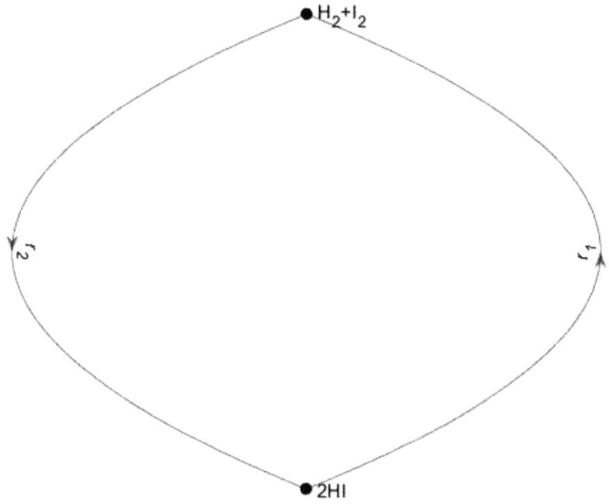

Fig. 11.18 A simple chemical reaction network composed of chemical complexes, *HI* and $H_2 + I_2$

Example 11.38 Figure 11.18 is a weighted digraph which denotes a simple chemical reaction network (e.g., Timm [128], pp. 240–241). In this figure vertices are composed of two chemical complexes, *HI* (hydrogen iodide) and $H_2 + I_2$ (hydrogen and iodine), and edge labels, r_1 and r_2, denote the *reaction rates* of these chemical reactions. According to the Law of Mass Action, these rates are denoted by

$$r_1 = k_1[HI]^2, r_2 = k_2[H_2][I_2], \tag{11.130}$$

where k_1 and k_2, denote the *rate constants*, while the brackets indicate *molecular concentration.* At *equilibrium*, these two rates are equal and
we have

$$\frac{[H_2][I_2]}{[HI]^2} = k_2/k_1 = K, \tag{11.131}$$

where K is called the *equilibrium constant* of the reaction.

One of the special matrices peculiar to chemical reactions is a stoichiometric matrix, an example of which has already been shown in Table 1.31 in Chap. 1.

Definition 11.85 (*stoichiometric matrix*) A *stoichiometric matrix* is defined by an $m \times r$ matrix $\boldsymbol{S} = \{s_{jk}\} \in M_{m,r}(R)$,

$$\boldsymbol{S} = \begin{bmatrix} s_{11} & s_{12} & \cdots & s_{1r} \\ s_{21} & s_{22} & \cdots & s_{2r} \\ \vdots & \vdots & \ddots & \vdots \\ s_{m1} & s_{m2} & \cdots & s_{mr} \end{bmatrix}, \tag{11.132}$$

where m is the number of chemical species (or substrates) and r is the number of chemical reactions in the network, while s_{jk} is the number obtained by subtracting the number of moles of jth substrate from that of the kth product.

Remark The term *stoichiometry* is first used by Richter [106]. The word stoichiometry comes from the Greek *stoikheion* which means *element* plus *o* plus *metry* (Sykes [121]).

Example 11.39 Here, we shall again show Example 1.12 in Chap. 1. Fig. 11.19 is the copy of Fig. 1.12 in that chapter. The stoichiometric matrix of this chemical reaction network is shown in Table 11.15, which is a copy of Table 1.31 in Chap. 1.

Stoichiometric matrix defined by (11.122) is not sufficient to describe the structure of chemical reaction networks such as a digraph shown in Fig. 11.19. In order to describe it, it is convenient to define the following matrix called the *complex (stoichiometric) matrix* (Rao et al. [105]).

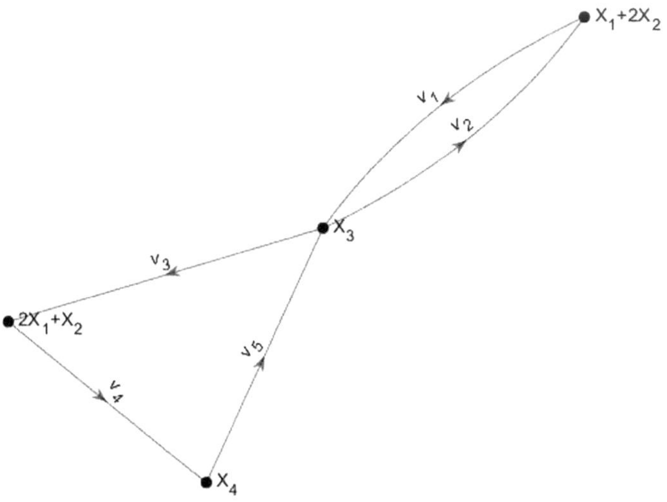

Fig. 11.19 A digraph composed of the four chemical complexes, $X_1 + 2X_2$, X_3, $2X_1 + X_2$, and X_4, which are composed of the four chemical substrates X_1, X_2, X_3 and X_4, which was adapted from Fig. 2. of Rao et al. (op. cit.)

Table 11.15 The stoichiometric matrix of the chemical reaction network shown in Fig. 11.18

Substrates/reactions	v_1	v_2	v_3	v_4	v_5
X_1	− 1	1	2	− 2	0
X_2	− 2	2	1	− 1	0
X_3	1	− 1	− 1	0	1
X_4	0	0	0	1	− 1

Definition 11.86 (*chemical complexes*) The set of chemical complexes of a chemical reaction network is defined as the union of all the different left- and right-hand sides (i.e., substrates and products) of the reactions in the network.

Definition 11.87 (*complex (stoichiometric) matrix*) A *complex (stoichiometric) matrix* is defined by an $m \times c$ matrix $\boldsymbol{Z} = \{z_{jk}\} \in M_{m,c}(R)$,

$$\boldsymbol{Z} = \begin{bmatrix} z_{11} & z_{12} & \cdots & z_{1c} \\ z_{21} & z_{22} & \cdots & z_{2c} \\ \vdots & \vdots & \ddots & \vdots \\ z_{m1} & z_{m2} & \cdots & z_{mc} \end{bmatrix}, \tag{11.133}$$

where m is the number of chemical species (or substrates) and c is the number of *chemical complexes* in the network, while z_{jk} is the number of moles of the jth chemical species (or substrates) in the kth complex.

Example 11.40 In the case of the chemical reaction networks whose digraph is shown in Fig. 11.18, the complex (stoichiometric) matrix $\boldsymbol{Z}$ is shown in Table 11.16. This matrix has already appeared in (1.24) in Chap. 1.

Remark 1 The complex (stoichiometric) matrix $\boldsymbol{Z}$ can be linked to the stoichiometric matrix $\boldsymbol{S}$ shown in Table 11.15 via the incidence matrix $\boldsymbol{B}$ of this network which is shown in Table 11.17. In fact, we can write $\boldsymbol{S} = -\boldsymbol{Z}\boldsymbol{B}$, as has already been pointed out in Example 1.12 in Chap. 1. Here, Table 11.17 is based on the definition of the incidence matrix due to Cristofides [21]. If we define it due to Rao et al. (op. cit.) and denote it $\boldsymbol{B}_R$, then $\boldsymbol{B}_R = -\boldsymbol{B}$. As a result, if we use $\boldsymbol{B}_R$, then we can write the stoichiometric matrix $\boldsymbol{S} = \boldsymbol{Z}\boldsymbol{B}_R$.

Remark 2 Complex (stoichiometric) matrix is coined by Horn and Jackson [57], in which they call it *complex matrix*. Rao et al. (op. cit.) call it the complex stoichiometric matrix.

So far we have introduced a few basic definitions and a law on chemical reaction networks. However, there is an immense literature on them and has been currently growing at a very fast rate. Therefore, it is impossible to introduce these definitions, and such a topic is beyond the scope of this book. Instead, we shall refer to some

Table 11.16 A complex (stoichiometric) matrix of the chemical reaction network shown in Fig. 11.18

substrate/complex	$X_1 + 2X_2$	X_3	$2\,X_1 + X_2$	X_4
X_1	1	0	2	0
X_2	2	0	1	0
X_3	0	1	0	0
X_4	0	0	0	1

Table 11.17 The incidence matrix B defined by Cristofides (op. cit.) of the chemical reaction network shown in Fig. 11.18

Substrates/reactions	v_1	v_2	v_3	v_4	v_5
X_1	1	− 1	0	0	0
X_2	− 1	1	1	0	− 1
X_3	0	0	− 1	1	0
X_4	0	0	0	− 1	1

of the literature here. As for examples of classic textbooks about general chemistry, we shall refer to Brinkley [10] and Timm [128]. We shall also refer to a famous textbook by Nicolis and Prigogine [94] whose main theme is the self-organization in nonequilibrium systems. Especially in Chap. 13, they introduce an interesting chemical reaction known as the *Belousov-Zhabotinski reaction* (Zaikin and Zhabotinski [146]). Recently, Feinberg [35] published a textbook about foundations of chemical reaction network theory. As for examples of research papers, Craciun and Pantea [19] discuss the dynamics of chemical reaction networks under the assumption of mass-action kinetics. They introduce several definitions and theorems around the chemical reaction networks and discuss the *identifiability* of rate constants given a chemical reaction network. Van der Schaft et al. [133] characterize the notions of complex and detailed balancing for mass action kinetics chemical reaction networks from the perspective of algebraic graph theory.

11.6.2 Neural Networks

In neural networks, basic matrices appear in interactions between neurons. As regards the connection between two neurons, we have a famous postulate called *Hebb's rule*, *Hebb's postulate*, or *cell assembly theory* introduced by Hebb [54].

Definition 11.88 (*Hebb's postulate*) When an axon of cell i is near enough to excite a cell j and repeatedly or persistently takes part in firing it, some metabolic change takes place in one or both cells such that i's efficiency, as one of the cells firing j, is increased.

Remark 1 This postulate implies two aspects, i.e., *connection* and *causality*. The connection means that "neurons wire together if they fire together", as Löwel and Singer [76] point out. The causality indicates the direction of synaptic modification, as Caporale and Dan point out [12]. In other words, connection implies *symmetry*, while causality *asymmetry*. Cell's efficiency between two neural cells is usually denoted by *connection weight*, w_{ij}, which means, to be precise, the *synaptic strength* between these cells.

Remark 2 Hebb's postulate has been gaining strong experimental support with the finding of *long-term potentiation* (*LTP*) of synaptic transmission, as for example

Caporale and Dan (op. cit.) point out. LTP was discovered in the rabbit hippocampus by Lømo [74]. According to him, a single pulse of electrical stimulation to fibers of the *perforant pathway* caused excitatory postsynaptic potentials in cells of the *dentate gyrus*. Moreover, the postsynaptic cells' response to these single-pulse stimuli could be enhanced for a long period of time if a high-frequency train of stimuli to the presynaptic fibers is delivered to the presynaptic fibers. There has been a large literature which is concerned with LTP and related topics (e.g., Park et al. [100]; Tazerart et al. [124]; Volk et al. [136]) since Lømo (op. cit.) found LTP. However, we shall not give an overview of this topic, since it is beyond the scope of this book.

The Hebbian postulate has been written in several forms. For example, Hopfield [56] defines it by

$$\Delta w_{ij} = \left[x_i(t)x_j(t)\right]_{average}. \tag{11.134}$$

Here, w_{ij} is the strength of connection between neurons i and j, and Δw_{ij} is the *modification of* w_{ij}. The average is some appropriate calculation over past history.

Parisi [99] defines the Hebbian postulate by

$$w_{ij}^{new} = f\left(w_{ij}^{old} + cx_ix_j\right), \tag{11.135}$$

where w_{ij}^{new} is the new synaptic strength of the ith neuron on the jth neuron after having stored a pattern $\{x_1, \ldots, x_n\}$, each of which has only the values ± 1, while w_{ij}^{old} is the old value of the synaptic strength. The value c is a normalization constant, which for convenience he takes equal to $n^{-1/2}$.

Kosko [67] defines the Hebbian postulate by

$$\frac{dw_{ij}}{dt} = x_ix_j, \tag{11.136}$$

where w_{ij} is the (i, j) element of a connection matrix between two nodes i and j, and x_i is the state of node i. Mathematically, dw_{ij} in (11.126) is the limit of Δw_{ij} in (11.124) when the increment of time t, Δt, approaches zero.

Kosko (op. cit.) also defines an alternative to (11.126) defined by

$$\frac{dw_{ij}}{dt} = \frac{dx_i}{dt}\frac{dx_j}{dt}. \tag{11.137}$$

which he calls the *differential Hebbian law.*

Jankowski et al. [62] define the Hebbian postulate by

$$w_{ij} = \frac{1}{N}\sum_{\mu=1}^{p} \varepsilon_i^{\mu}\overline{\varepsilon}_j^{\mu}, i, j = 1, \ldots n, \tag{11.138}$$

where

$$\varepsilon_i = z^{\xi_i}, i = 1, \ldots, n, \tag{11.139}$$

$\xi_i \in [0, 1, \ldots, k - 1]$, and z is the kth root of one. Moreover, ε_i^{μ} is called *phasor i* in the kth of p learned patterns (e.g., Noest [96], p. 2196). It is apparent from (11.128) that w_{ij} is *Hermitian*. We have gone into details of this kind of *complex-valued neural network models* in (10.21) in Chap. 10.

Remark Recently, Chino has been developing a nonlinear complex difference system model called Dynamical (or dynamic) weighted Digraph (abbreviated as DWD) since 2014, which we have discussed in detail in Chap. 9. Using DWD, we can define an alternative to the Hermitian connection weight matrix for Hebbian postulate. Let us imagine an empirical neural network composed of n neurons. Such a network can be expressed as a weighted digraph of order n. In general, the observed weight matrix W associated with the digraph has a one-to-one correspondence to Hermitian matrix $\boldsymbol{H}_w$ defined by (11.112) and nodes of the digraph can be embedded in a finite-dimensional Hilbert space if $\boldsymbol{H}_w$ is p.s.d., as in Definition 11.81. In DWD, the configuration of the embedded nodes in a Hilbert space is viewed as a constituent of a state space in which nodes change dynamically through their asymmetric interactions, and these changes are assumed to be described by a set of complex nonlinear difference equations. A general complex asymmetric interaction matrix appears as part of the parameters in this set of equations. This matrix can be an alternative to the Hermitian matrix defined by (11.112). Of course, this matrix is no longer a Hermitian matrix, and a general complex asymmetric matrix. We have seen this in more detail in Chap. 9.

Appendix

In this appendix, we shall show a few examples of the MATLAB codes which depict digraphs and a configuration shown in Example 11.2 of this chapter.

Example A.1 (Codes for Classical MDS and Its Output Shown in Fig. 11.1) The following MATLAB commands are written to draw Fig. 11.1 in Sect. 11.3 which is a configuration of a hypothetical objects obtained by applying the classical MDS to the squared distance matrix shown in Table 11.1.

```
N=4;
disp('The order of a squared distance matrix');
disp(N)
dn=1-1/N;
in=-1/N;
figure(1);
plot(0,1, 1,0.5, 0,0, -1,0.5,'Marker','o','MarkerFaceColor','k')
text(0.05,0.95,'v_1','FontSize',16);
text(0.85,0.5,'v_2','FontSize',16);
text(0.05,0.05,'v_3','FontSize',16);
text(-0.95,0.5,'v_4','FontSize',16);
T=[0,1.25,1,1.25; 1.25,0,1.25,4; 1,1.25,0,1.25; 1.25,4,1.25,0];
disp('2. Input squared distance matrix');
disp(T)
C=[dn,in,in,in; in,dn,in,in; in,in,dn,in; in,in,in,dn];
disp('3. The centering matrix of order N');
disp(C)
B=-C*T*C'/2;
disp('4.  A double-centered inner product matrix');
disp(B)
[V,D]=eig(B);
disp('5. Eigenvalues of B');
disp(D)
disp('6. Eigenvectors of B');
disp(V)
figure(2);
plot(0,-0.5, -1,0, 0,0.5, 1,0,'Marker','o','MarkerFaceColor','k')
text(0.05,-0.45,'v_1','FontSize',16);
text(-0.95,0,'v_2','FontSize',16);
text(0.05,0.45,'v_3','FontSize',16);
text(0.85,0,'v_4','FontSize',16);
```

Example A.2 (Codes for Depicting a Binary Tree in Example 11.19 Shown in Fig. 11.3) The following MATLAB commands are written to draw a binary tree shown in Fig. 11.3 in Sect. 11.4. To do this job, MATLAB has an *ntree* function.

```
set(gca,'Visible','off')
ax=gca;
treeEx1=ntree(2,3);
h=plot(treeEx1);
whitebg(h)
h1=findobj(h,'Type','Text');
set(h1,'Color',[0,0,0]);
print(h,'-deps','eps-files/Example1-unweighted-tree.eps')
```

Example A.3 (Codes for Computing the Distance Matrix in a Weighted Digraph) The following MATLAB commands are written to compute the distance matrix and the shortest path between two vertices of the digraph shown in Fig. 11.7 in Sect. 11.4. To do this job, MATLAB has a *distances* function and a *shortestpath* function. Table 11.6 is the output of the distances function for this digraph, and the result of the *shortestpath* function output the shortest path in this case is the vertices set (3 5 4), which means that the shortest path from vertex 3 to vertex 4 is the rote from vertex 3 to vertex 4 via vertex 5.

```
set(gca,'Visible','off')
ax=gca;
s=[1 3 3 3 4 4 5 5];
t=[2 1 2 5 1 3 1 4];
weights=[5 15 6 5 8 7 5 7]';
code={'5' '15' '6' '5' '8' '7' '5' '7'}';
EdgeTable=table([s' t'],weights, code,...
   'VariableNames',{'EndNodes' 'Weight' 'Code'})
names={'Node1' 'Node2' 'Node3' 'Node4' 'Node5'}';
country_code={'1' '2' '3' '4' '5'}';
NodeTable=table(names, country_code,'VariableNames',{'Name' 'Country'})
G=digraph(EdgeTable,NodeTable);
plot(G,'k','NodeLabel',G.Nodes.Country,'EdgeLabel',G.Edges.Code)
D=distances(G);
[path1,d]=shortestpath(G,3,4)
disp=('1. The distance matrix of G')
D
P=shortestpath(G,3,4)
```

References

1. Agaev, R., & Chebotarev, P. (2005). On the spectra of nonsymmetric Laplacian matrices. *Linear Algebra and its Applications, 399*, 157–168.

2. Alexanderson, G. L. (2006). About the cover: Euler and K önigsberg bridges: A historical view. *Bulletin of The American Mathematical Society, 43*, 567–573.
3. Anderson, T. W. (1984). *An introduction to multivariate statistical analysis* (2nd ed.). Wiley.
4. Barabási, A.-L., & Albert, R. (1999). Emergence in scaling in random networks. *Science, 286*, 509–512.
5. Bavelas, A. (1948). A mathematical model for group structures. *Applied Anthropology, 7*, 16–30.
6. Belkin, M., & Niyoghi, P. (2001). Laplacian eigenmaps and spectral techniques for embedding and clustering. *Advances in Neural Information Processing System, 14*, 585–591.
7. Bellman, R. (1958). On a routing problem. *Quarterly of Applied Mathematics, 16*, 87–90.
8. Bonacich, P. (1987). Power and centrality: A family of measures. *American Journal of Sociology, 92*, 1170–1182.
9. Bollob á s, B. (1985). *Random graphs*. Academic Press.
10. Brinkley, S. R. (1951). *Principles of general chemistry*. Macmillan.
11. Brouwer, A. E., & Haemers, W. H. (2012). *Spectra of graphs*. Springer.
12. Caporale, N., & Dan, Y. (2008). Spike timing-dependent plasticity: A Hebbian learning rule. *The Annual Review of Neuroscience, 31*, 25–46.
13. Chino, N. (1980). A unified geometrical interpretation of the MDS techniques for the analysis of asymmetry and related techniques. In *Paper presented at the symposium on "Asymmetric multidimensional scaling" at the Spring Meeting of the Psychometric Society*, Iowa City, Iowa, USA.
14. Chino, N. (2012). A brief survey of asymmetric MDS and some open problems. *Behaviormetrika, 39*, 127–165.
15. Chino, N., & Shiraiwa, K. (1993). Geometrical structures of some non-distance models for asymmetric MDS. *Behaviormetrika, 20*, 35–47.
16. Chung, F. R. K. (1997). *Spectral graph theory*. The American Mathematical Society.
17. Ciotti, V. (2015). Degree correlations in signed social networks. *Physica A: Statistical Mechanics and its Applications, 422*, 25–39.
18. Constantine, A. G., & Gower, J. C. (1978). Graphical representation of asymmetric matrices. *Applied Statistics, 27*, 297–304.
19. Cracium, G., & Pantea, C. (2008). Identifiability of chemical reaction networks. *Journal of Mathematical Chemistry, 44*, 244–259.
20. Cristescu, R. (1977). *Topological vector spaces*. Noordhoff International Publishing.
21. Cristofides, N. (1975). *Graph theory, an algorithmic approach*. Academic Press.
22. Dankelmann, P. (2012). Average distance in weighted graphs. *Discrete Mathematics, 312*, 12–20.
23. Dankelmann, P., Oellermann, O. R., & Wu, J.-L. (2004). Minimum average distance of strong orientations of graphs. *Discrete Applied Mathematics, 143*, 204–212.
24. De Groot, J. (1963). Subcompactness and the Baire category theorem. *Indagationes Mathematicae, 25*, 761–767.
25. De Lathauwer, L., De Moor, B., & Vandewalle, J. (2000). A muti-linear singular value decomposition. *SIAM Journal on Matrix Analysis and Applications, 21*, 1253–1278.
26. Dijkstra, E. W. (1959). A note on two problems in connexion with graphs. *Numerische Mathematik, 1*, 269–271.
27. Djelloul, S., & Kouider, M. (2001). On weighted mean distance. *Australasian Journal of Combinatorics, 23*, 181–195.
28. Dyson, F. J. (1962). A Brownian-motion model for the eigenvalues of a random matrix. *Journal of Mathematical Physics., 3*(6), 1191–1198.
29. Dyson, F. J. (1962b). Statistical theory of the energy levels of complex systems. I. *Journal of Mathematical Physics, 3*(No.1), 140–156.
30. Eckart, C., & Young, G. (1936). The approximation of one matrix by another of lower rank. *Psychometrika, 1*, 211–218.
31. Eppstein, D. (1994). Finding the k shortest paths. *ICS Technical Reports* 94-26. University of California, Irvine, USA.

32. Erdös, P., & Rényi, A. (1959). On random graphs I. *Publicationes Mathematicae, 6*, 290–297.
33. Escoufier, Y., & Grorud, A., et al. (1980). Analyse factorielle des matrices carrees non symetriques [Factor analysis of square asymmetric matrices]. In E. Diday (Ed.), *Data analysis and informatics* (pp. 263–276). North Holland.
34. Espínola, R., & Kirk, W. A. (2006). Fixed point theorems in R-trees with applications to graph theory. *Topology and its Applications, 153*, 1046–1055.
35. Feinberg, M. (2019). *Foundations of chemical reaction network theory*. Springer.
36. Ferner, R. E., & Aronson, J. K. (2015). Cato Guldberg and Peter Waage, the history of the Law of Mass Action, and its relevance to clinical pharmacology. *British Journal of Clinical Pharmachology, 81*, 52–55.
37. Fisher, R. A. (1936). The use of multiple measurements in taxonomic problems. *Annals of Human Genetics, 7*, 179–188.
38. Ford, L. R. (1956). Network flow theory. *The Rand Corporation, P.923*, 1–12.
39. Forsyth, E., & Katz, L. (1946). A matrix approach to the analysis of sociometric data: Preliminary report. *Sociometry, 9*, 340–347.
40. Freeman, L. C. (1977). A set of measures of centrality based on betweenness. *Sociometry, 40*, 35–41.
41. Freeman, L. C. (1978/1979). Centrality in social networks conceptual clarification. *Social Networks, 1*, 215–239.
42. Furutani, S., Shibahara, T., Akiyama, M., Hato, K., & Aida, M. (2020). In B. Ulf, E. Fromont, A. Hotho, A. Knobbe, M. Maathus, & C. Robardet (Eds.), *Machine Learning and Knowledge Discovery in Databases, European Conference, ECML PKDD 2019, Würzburg, Germany, September 16–20, Proceedings, Part I* (pp. 447–463). Springer International Publishing.
43. Gabriel, K. R. (1971). The biplot graphic display of matrices with application to principal component analysis. *Biometrika, 58*, 453–467.
44. Gabriel, K. R. (1972). Analysis of meteorological data by means of canonical decomposition and biplots. *Journal of Applied Meteorology, 11*, 1071–1077.
45. Garijo, D., Alberto, M., & Rodrigo I. S. (2022). https://arxiv.org/pdf/2103.11676.pdf
46. Gibbs, J. W. (1902). *Elementary principles in statistical mechanics*. Charles Scribner's Sons.
47. Gower, J. C. (1977). The analysis of asymmetry and orthogonality. In J. R. Barra, F. Brodeau, G. Romer, & B. van Cutsem (Eds.), *Recent developments in statistics* (pp. 109–123). North Holland.
48. Gower, J. C., & Hand, D. J. (1996). *Biplots*. Chapman & Hall.
49. Greaves, G., Mohar, B., et al. (2019). Interlacing families and the Hermitian spectral norm of digraphs. *Linear Algebra and its Applications, 564*, 201–208.
50. Guldberg, C. M., & Waage, P. (1867). *Études sur les affinités chimiques* [Studies of chemical affinities]. Brøgger & Christíe.
51. Harary, F. (1969). *Graph theory*. Addison-Weley.
52. Harary, F., Norman, R. Z., & Cartwright, D. (1965). *Structural models: An introduction to the theory of directed graphs*. Wiley.
53. He, X., & Niyogi, P. (2003). Locality preserving projections. In *Proceedings of the Conference on Advances in Neural Information Processing Systems*.
54. Hssebb, D. O. (1949). *The organization of behavior*. Wiley.
55. Hirsch, M. W., & Smale, S. (1974). *Differential equations, dynamical systems, and linear Algebra*. Academic Press.
56. Hopfield, J. J. (1982). Neural networks and physical systems with emergent collective computational abilities. *Proceedings of the National Academy of Sciences of the United States of America, 79*, 2554–2558.
57. Horn, F., & Jackson, R. (1972). General mass action kinetics. *Archive for Rational Mechanics and Analysis, 47*, 81–116.
58. Horn, R. A., & Johnson, C. A. (1985). *Matrix analysis*. Cambridge University Press.
59. Horst, P. (1965). *Factor analysis of data matrices*. Holt, Rine-hart and Winston Inc.
60. Iwai, Y., Nishiyama, M., & Yoshimura, H. (2017). Asymmetric locality preserving projection and its application to k-nearest neighbor method. In *Fifteenth IAPR International Conference on Machine Vision Applications* (pp. 42–45). Nagoya University.

61. Jachymski, J. (2008). The contraction principle for mappings on a metric space with a graph. *Proceedings of the American Mathematical Society, 136*, 1359–1373.
62. Jankowski, S., Lozowski, A., & Zurada, J. M. (1996). Complex-valued multistate neural associative memory. *IEEE Transactions on Neural Networks, 7*, 1491–1496.
63. Johnstone, I. M. (2008). Multivariate analysis and Jacobi ensembles: Largest eigenvalue, Tracy-Widom limits and rates of convergence. *The Annals of Statistics, 36*, 2638–2716.
64. Katz, L. (1953). A new status index derived from sociometric analysis. *Psychometrika, 18*, 39–43.
65. Kendall, M., & Stuart, A. (1977). *The advanced theory of statistics* (Vol. 1). Charles Griffin & Company Limited.
66. Kendall, M., & Stuart, A. (1973). *The advanced theory of statistics* (Vol. 2). Charles Griffin & Company Limited.
67. Kosko, B. (1986). Differential Hebbian learning. In *AIP Conference Proceedings* (pp. 277–282).
68. Königsberg, Z. R. (2009). A generalized eigenmode algorithm for reducible regular matrices over the max-plus algebra. *International Mathematical Forum, 4*, 1157–1171.
69. Kroonenberg, P. M. (2020). Multiway extension of the SVD. In T. Imaizumi, A. Nakayama, & S. Yokoyama, *Advanced studies in behaviormetrics and data science* (pp. 141–175). Springer Nature Singerpore Pte Ltd.
70. Lancaster, P., & Tismenetsky, M. (1985). *The theory of matrices* (2nd ed.). Academic Press.
71. Landau, E. (1895). Zur relativen Wertbemessung der Turnierresultate. *Deutsches Wochenschach, 11*, 366–369.
72. Lawson, C. L., & Hanson, R. J. (1974). *Solving least squares problems*. Prentice-Hall Inc.
73. Leavitt, H. J. (1949). *Some effects of certain communication patterns on group performance*. Unpublished Ph.D. dissertation. Massachusetts Institute of Technology.
74. Leavitt, H. J. (1951). Some effects of communication patterns on group performance. *Journal of Abnormal and Social Psychology, 46*, 35–50.
75. Lømo, T. (1966). Frequency potentiation of excitatory synaptic activity in the dentate area of the hippocampal formation. *Acta Physiologica Scandinavia, 68*, 128–130.
76. Liu, J., & Li, X. (2015). Hermitian-adjacency matrices and Hermitian energies of mixed graphs. *Linear Algebra and its Applications, 466*, 182–207.
77. Löwel, S., & Singer, W. (1992). Selection of intrinsic horizontal connections in the visual cortex by correlated neuronal activity. *Science, 255*, 209–212.
78. Maddox, I. J. (1988). *Elements of functional analysis* (2nd ed.). Cambridge University Press.
79. Magnus, J. R., & Neudecker, H. (1988). *Matrix differential calculus with applications in statistics and econometrics*. Wiley.
80. Martinez, W. L., & Martinez, A. R. (2008). *Computational statistics handbook with MATLAB* (2nd ed.). Chapman & Hall/CRC.
81. Mehta, M. L. (2004). *Random matrices* (3rd ed.). Elsevier.
82. Mémoli, F. (2012). Some properties of Gromov-Hausdorff distances. *Discrete & Computational Geometry, 48*, 416–440.
83. Menger, K. (1931). New foundations of Euclidean geometry. *American Journal of Mathematics, 53*, 721–745.
84. Milgram, S. (1967). The small-world problem. *Psychology Today, 1*, 61–67.
85. Mohar, B. (2016). Hermitian adjacency spectrum and switching equivalence of mixed graph. *Linear Algebra and its Applications, 489*, 324–340.
86. Moore, E. F. (1959). The shortest path through a maze. In *Proceedings of the International Symposium on Switching Theory, Part II* (pp. 285–292). Harvard University Press.
87. Moore, E. H. (1920). *Bulletin of the American Mathematical Society, 26*, 394–395.
88. Muirhead, R. J. (1982). *Aspects of multivariate statistical theory*. Wiley.
89. Negre, C. F. A., Morzan, U. N., Hendrickson, H. P., Pal, R., Lisi, G. P., Loria, J. P., Rivalta, I., Ho, J., & Batista, V. S. (2018). *Proceeding of the National Academy of Sciences of the United States of America, 115*, E12201–E12208.

90. Neog, M., & Debnath, P. (2017). Fixed points of set valued mappings in terms of start point on a metric space endowed with a directed graph. *Mathematics, 5*, 1–7.
91. Newman, M. E. J. (2006). *The mathematics of networks*. http://www-personal.umich.edu/~mejn/papers/palgrave.pdf
92. Newman, M. E. J., Barabási, A.-L., & Watts, D. J. (Eds.). (2006). *The structure and dynamics of networks*. Princeton University Press.
93. Newman, M. E. J., Strogatz, S. H., & Watts, D. J. (2001). Random graphs with arbitrary degree distributions and their applications. *Physical Review E, 64*, 026118-1–026118-17.
94. Ng, C. P. & Teh, H. H. (1966/67). On finite graphs of diameter 2. *Nanta Mathematica, 1*, 72–75.
95. Nicolis, G., & Prigogine, I. (1977). *Self-organization in nonequilibrium systems from dissipative structures to order through fluctuations*. Wiley.
96. NIST. (2018). *Redefining the mole*. https://www.nist.gov/si-redefinition/redefining-mole
97. Noest, A. J. (1988). Discrete-state phasor neural networks. *Physical Review A, 38*, 2196–2199.
98. Okada, A. (2010). Analysis of car switching data by asymmetric multidimensional scaling based on singular value decomposition. In *Proceedings of the 34th annual conference of the Gesellschaft für Klassification* (pp. 211–212). Karlsruhe, Germany.
99. Okada, A. (2012). Analysis of car switching data by asymmetric multidimensional scaling based on singular value decomposition. In W. Gaul, A. Geyer-Schulz, L. Schmidt-Thieme, & J. Kunze (Eds.), *Challenges at the interface of data analysis, computerscience, and optimization* (pp. 143–150). Springer.
100. Parisi, G. (1986). A memory which forgets. *Journal of Physics A: Mathematical and General, 19*, L617–L620.
101. Park, J. M., Jung, S.-C., & Eun, S.-Y. (2014). Long-term synaptic plasticity: Circuit perturbation and stabilization. *Korean Journal of Physiology & Pharmacology, 18*, 457–460.
102. Penrose, R. (1955). A generalized inverse for matrices. *Proceedings of the Cambridge Philosophical Society, 51*, 406–413.
103. Plesnik, J. (1984). On the sum of all distances in a graph or digraph. *Journal of Graph Theory, 8*, 1–21.
104. De Solla Price, D. (1976). A general theory of bibliometric and other cumulative advantage processes. *Journal of the American Society for Information Science, 27*, 292–306.
105. Rao, C. R. (1973). *Linear statistical inference and its applications* (2nd ed.). Wiley.
106. Rao, S., van der Schaft, A., & Jayawardhana, B. (2031). A graph-theoretical approach for the analysis of model reduction of complex-balanced chemical reaction networks. *Journal of Mathematical Chemistry, 51*, 2401–2422.
107. Richter, J. B. (1792). *Anfangsgründe der Stöchyometrie oder Meßkunst chymischer Elemente* (Vol. 2). Georg Olms Verlag.
108. Rowels, S. T., & Saul, L. K. (2000). Nonlinear dimensionality reduction by locally linear embedding. *Science, 290*, 2323–2326.
109. Saberi, M., Khosrowabadi, R., Khatibi, A., Misic, B., & Jafari, G. (2021). Topological impact of negative links on the stability of resting-state brain network. *Scientific Reports, 11*, 2176.
110. Schoenberg, I. J. (1935). Remarks to Maurice Fréchet's article "Sur la definition axiomatique d'une classe d'espace distanciés vector-iellement applicable sur l'espace de Hilbert." *Journal of Mathematics, 36*, 724–732.
111. Seber, G. A. F. (2008). *A matrix handbook for statisticians*. Wiley.
112. Shaw, M. E. (1954). Group structure and the behavior of individuals in small group. *Journal of Psychology, 38*, 139–149.
113. Shimbel, A. (1953). Structural parameters of communication networks. *The Bulletin of Mathematical Biophysics, 15*, 501–507.
114. Shubin, M. A. (1994). Discrete magnetic Laplacian. *Communications in Mathematical Physics, 164*, 259–275.
115. Shum, K., & Blake, L. (2005). Expander graphs and codes. In A. Ashikhmin & A. Barg (Eds.), *Algebraic coding theory and infirmation theory: DIMACS workshop Algebraic coding theory and information theory* (pp. 57–68). The American Mathematical Society.

116. Solomonoff, R., & Rapoport, A. (1951). Connectivity of random nets. The *Bulletin of Mathematical Biophysics, 13*, 107–117.
117. Stefanucci, G., & van Leeuwen, R. (2013). *Nonequilibrium many-body theory of quantum systems. A modern introduction*. Cambridge University Press.
118. Stewart, G. W., & Sun, J. (1990). *Matrix perturbation theory*. Academic Press.
119. Stevens, S. S. (1946). On the theory of scales of measurement. *Science, 103*, 677–680.
120. Stevens, S. S. (1951). Mathematics, measurement, and psychophysics. In S. S. Stevens (Ed.), *Handbook of experimental psychology*. Wiley.
121. Sykes, J. B. (Ed.). (1976). *The concise oxford dictionary of current English*. Oxford University Press.
122. Takane, Y. (2004). Matrices with special reference to applications in psychometrics. *Linear Algebra and Its Applications, 388*, 341–361.
123. Tanabe, K. (1975). Neumann-Type expansion of reflexive generalized inverses of a matrix and the hyperpower iterative method. *Linear Algebra and Its Applications, 10*, 163–175.
124. Tazerart, S., Mitchell, D. E., Miranda-Rottmann, S., & Araya, R. (2020). *Nature Communications, 11*, 1–16.
125. Tenenbaum, J. B., de Silva, V., & Langford, J. C. (2000). A global geometric framework for nonlinear dimensionality reduction. *Science, 290*, 2319–2323.
126. Thurstone, L. L. (1931). Multiple factor analysis. *Psychological Review, 38*, 406–427.
127. Thurstone, L. L. (1947). *Multiple factor analysis*. University of Chicago Press.
128. Timm, J. A. J. (1956). *General chemistry*. McGraw-Hill.
129. Tittelfitz, J. (2006). *The heat equation on discrete domains*. https://sites.math.washington.edu/~reu/papers1/2006/justin/heateqn.pdf
130. Torgerson, W. S. (1954). A law of categorical judgement. *The American Psychologist, 9*, 483.
131. Torgerson, W. S. (1958). *Theory and methods of scaling*. Wiley.
132. Turk, M., & Pentland, A. (1991). Eigenfaces for recognition. *Journal of Cognitive Neuroscience, 3*, 71–86.
133. Van der Schaft, A., Rao, S., & Jayawardhana, B. (2015). Complex and detailed balancing of chemical reaction networks revisited. *Journal of Mathematical Chemistry, 53*, 1445–1458.
134. Veerman, J. J. P., & Lyons, R. (2020). A primer on Laplacian dynamics in directed graphs. *Nonlinear Phenomena in Complex Systems, 23*, 196–206.
135. Volchenkov, D. (2021). Infinite ergodic walks in finite connected undirected graphs. *Entropy, 23*, 205. https://doi.org/10.3390/e23020205.
136. Volk, L. J., Bachman, J. L., Johnson, R., Yu, Y., & Huganir, R. L. (2013). PKM-ζ is not required for hippocampal synaptic plasticity, learning and memory. *Nature, 493*, 7432.
137. Watts, D. J., & Strogatz, S. H. (1998). Collective dynamics of 'small-world' networks. *Nature, 393*, 440–442.
138. Wiener, H. (1947). Structural determination of paraffin boiling points. *Journal of the American Chemical Society, 69*, 17–20.
139. Wilkinson, J. H. (1965). *The Algebraic Eigenvalue problem*. Clarendo Press.
140. Wilson, R. J., & Beineke, L. W. (1979). *Applications of graph theory*. Academic Press.
141. Wishart, J. (1928). The generalized product moment distribution in samples from a normal multivariate population. *Biometrika A, 20*, 32–43.
142. Yan, S., Xu, D., Zhang, B., Zhang, H.-J., Yang, Q., & Lin, S. (2007). Graph embedding and extensions: A general framework for dimensionality reduction. *IEEE Transactions on Pattern Analysis and Machine Intelligence, 29*, 40–51.
143. Young, F. W. (1975). An asymmetric Euclidean model for multi-process asymmetric data. In *Paper presented at U.S.-Japan Seminar on MDS*, San Diego, U.S.A.
144. Young, G., & Householder, A. S. (1938). Discussion of a set of points in terms of their mutual distances. *Psychometrika, 3*, 1–3.
145. Yuan, B.-J., Wang, Y., Gong, S.-C., & Qiao, Y. (2020). On mixed graphs Whose Hermitian Spectral Radii are at most 2. *Graphs and Combinatorics, 36*, 1573–1584.
146. Zaikin, A. N., & Zhabotinski, A. M. (1970). Concentration wave propagation in two-dimensional liquid-phase self-oscillating system. *Nature, 225*, 535–537.

Chapter 12
Definitions and Theorems Useful for the Understanding the Dynamics of Asymmetric Interactions

12.1 Introduction

In this chapter, we shall present various definitions and theorems useful for the understanding of the dynamics of the asymmetric interaction among objects (or nodes) which is observed in many fields of science as introduced in Chap. 1. These dynamics can usually be described by a set of *differential equations* or *difference equations*, as already suggested in Sect. 1.4.

In Sect. 12.2 we shall introduce basic concepts in differential equation which describes the phenomena in various fields of science. Differential equation modelling may be natural and appropriate in the case when objects are considered to interact with each other in a continuous time.

In Sect. 12.3 we shall introduce the notion of *chaos*, and several *routes to chaos*. We shall also introduce various definitions related to chaos and show examples of applications to some famous differential and difference equations.

In Sect. 12.4 we shall introduce basic concepts in difference equation which describes the phenomena in many fields of science. Difference equation modelling may be natural and appropriate in the case when objects are considered to interact with each other in a discrete time.

In both cases of differential and difference equations we must assume a *state space* in which nodes or objects are embedded at a given time. For example, in the physical system of particles the state space is the Cartesian product $R^3 \times R^3$ of pairs $(\boldsymbol{x}, \boldsymbol{v})$, $\boldsymbol{x}$, $\boldsymbol{v}$ in R^3, where $\boldsymbol{x}$ is the position, while $\boldsymbol{v}$ the velocity that a particle might have at a given moment (e.g., Hirsch and Smale [34]). The R^3 is sometimes identified with a three-dimensional *Euclidean space*. In a chemical system the state of a chemical species (substrate) is usually its concentration. In these systems we can measure these quantities which constitute the state space empirically.

However, there exist a variety of system in which we cannot observe the state space directly. In such cases we propose to choose a *Hilbert space* or *indefinite metric space* as a state space, based on the Chino-Siraiwa theorem on asymmetric

N. Chino, *Structure and Dynamics of Asymmetric Interactions*,
https://doi.org/10.1007/978-981-97-8269-7_12

similarity matrix between nodes or objects. Hilbert space is a natural extension of the Euclidean space to a complex space. Here, it should be noticed that in these cases the position of each node or object in a Hilbert space or indefinite metric space which constitutes a state space cannot be observed empirically. However, it should also be noticed that the position vector of nodes or objects is associated with an empirical asymmetric similarity matrix.

In Sect. 12.5, we shall summarize basic concepts and theorems in complex dynamical system. Complex dynamical system is described by a polynomial difference equation in complex variables. Even the polynomial difference equation in one complex variable exhibits various interesting and fascinating solution curves called fractals. In Sect. 12.5.1 we shall introduce important notions on fixed point and periodic point in complex differential equations. In Sect. 12.5.2 we shall introduce *Mandelbrot sets*, *Julia set*, and *Fatou set* in complex dynamical systems. The reader will be surprised by the fact that even in a dyadic relation in DWD fractals appear under a special condition, which we have pointed out in (9.37) in Chap. 9. We shall show MATLAB codes for some of the figures in the appendix of this chapter. We use *Sunday ChaosTimes* developed by R&D Team [48] in depicting some of the figures related to chaos.

12.2 Basic Concepts in Differential Equation

12.2.1 State Space, Vector Field, and Solution Curves

Roughly speaking, a differential equation is a kind of equation which describes change in a quantity, which specifies the instantaneous state of a system, over time. Such a quantity x is called the *state variable* for the system, which constitutes a *state space*. If the state space is p-dimensional, state variables constitute a *state vector*, $x = \left(x_1, x_2, \ldots, x_p\right)^t$, whose components are coordinates of a point in the space. Since velocity is a rate at which the state variable x_i changes in time, any velocity vector is defined in terms of $\boldsymbol{x}$ as

$$\dot{x} = \frac{dx}{dt} = \left(\frac{dx_1}{dt}, \frac{dx_2}{dt}, \ldots, \frac{dx_p}{dt}\right)^t. \tag{12.1}$$

Mathematically, each component of $\dot{\boldsymbol{x}}$, i.e., $\frac{dx_i}{dt}$, is the *derivative* of x_i with respect to time t. If $\dot{\boldsymbol{x}}$ is any function of $\boldsymbol{x}$, we can write it generally as

$$\dot{x} = f(x) \tag{12.2}$$

The above equation is called *autonomous*, which means that the function f does not depend on time. The map f: $R^p \rightarrow R^p$ assigns $\dot{\boldsymbol{x}}$ to $\boldsymbol{x}$. On the other hand, if the function depends on time, then we have

$$\dot{x} = f(x, t) \tag{12.3}$$

The above equation is called *nonautonomous*. In this case, the map $f: R^p \times R \to R^p$ assigns $\dot{x}$ to $(\boldsymbol{x}, t)$. In other words, we assign a vector $\dot{\boldsymbol{x}} \in R^p$ based at each point $\boldsymbol{x} \in R^p$ at time t. Both these maps introduced above can be considered a vector field on R^p. Figure 12.1 is an example of a vector field on R^2, which is a copy of Fig. 8.5 in Chap. 8.

If the function described by (12.2) is linear in $\boldsymbol{x}$, we have the following differential equation:

$$\dot{x} = Ax, \tag{12.4}$$

where

$$A = \begin{bmatrix} a_{11} & \cdots & a_{1p} \\ \vdots & \cdots & \vdots \\ a_{p1} & \cdots & a_{pp} \end{bmatrix}. \tag{12.5}$$

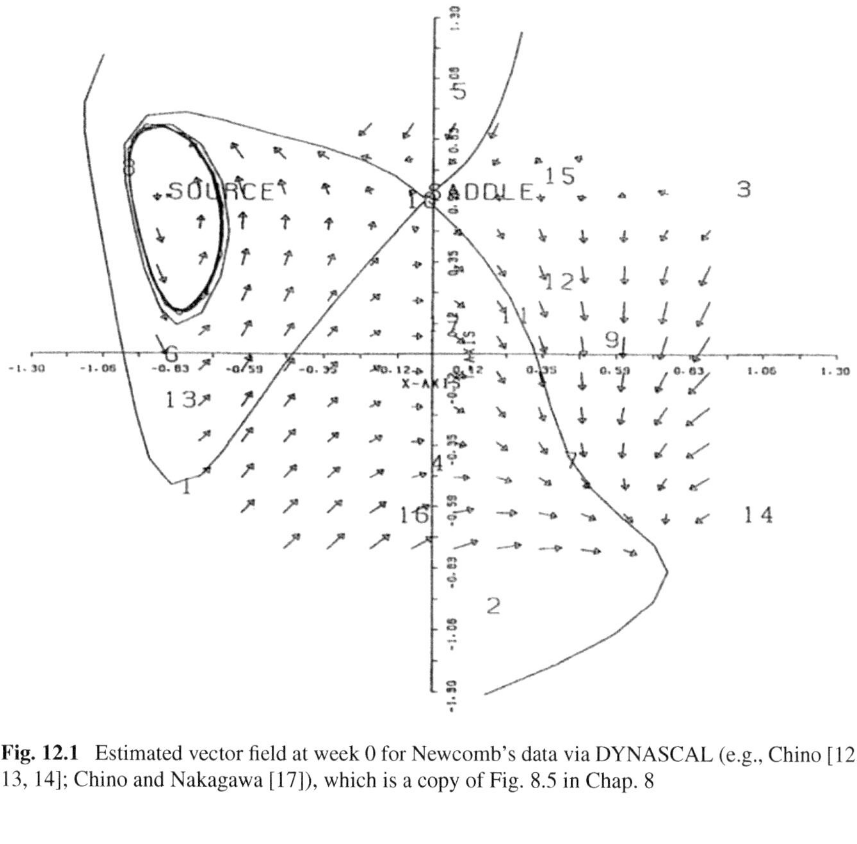

Fig. 12.1 Estimated vector field at week 0 for Newcomb's data via DYNASCAL (e.g., Chino [12, 13, 14]; Chino and Nakagawa [17]), which is a copy of Fig. 8.5 in Chap. 8

Equation (12.4) assigns the vector based at $\boldsymbol{x}$ which is a translate of $\boldsymbol{Ax}$. Therefore, the matrix $\boldsymbol{A}$ can be considered as a map $\boldsymbol{A} : R^p \rightarrow R^p$ (e.g., Hirsch and Smale [34], p. 11).

Example 12.1 For example, if $p = 2$, then (12.4) is written as

$$\begin{bmatrix} dx_1/dt \\ dx_2/dt \end{bmatrix} = A\begin{bmatrix} x_1 \\ x_2 \end{bmatrix} = \begin{bmatrix} a_{11} & a_{12} \\ a_{21} & a_{22} \end{bmatrix}\begin{bmatrix} x_1 \\ x_2 \end{bmatrix}, \tag{12.6}$$

whose function f in (12.2) is linear.

Example 12.2 The following is an estimated vector field whose function f in (12.3) is nonlinear and $p = 2$. This vector field is a copy of the vector field which has already been shown in Chap. 8.

It is common to restate (12.2) in the form of an *initial value problem*:

$$\dot{x} = f(x), x(0) = k. \tag{12.7}$$

where $\boldsymbol{k}$ is a prescribed initial value in R^p at time *0*. Similarly, we can restate (12.3) in the form of another initial value problem:

$$\dot{x} = f(x, t), x(0) = k. \tag{12.8}$$

As regards the initial value problem of the *autonomous system* described by (12.7), there exists a *fundamental local theorem* of ordinary differential equations. The reason why it is called the local theorem is that it deals with the nature of the vector field near some point $\boldsymbol{x}(0)$ (e.g., Hirsch and Smale [34]). Let us now denote E by a normed vector space, W is an open set in E, and the map f in (12.7) is a continuous map $W \rightarrow E$. Then, we have the following theorem (e.g., Hirsch and Smale, op. cit., pp. 162–163):

Theorem 12.1 (fundamental local theorem) *Let $W \subset E$ be an open subset of a normed vector space, $f : W \rightarrow E$ a C^1 (continuously differentiable) map, and $x_0 \in W$. Then there is some $a > 0$ and a unique solution*

$$x : (-a, a) \rightarrow W$$

of the differential equation

$$\dot{x} = f(x)$$

satisfying the initial condition

$$x(0) = x_0.$$

For the proof of this theorem, see, for example, Hirsch and Smale (op. cit., pp. 163–167).

As for the initial value problem of the *nonautonomous system* described by (12.8), various types of existence proofs are known (Hartman [32], pp. 8–15). According to him, one of the simplest and most useful is the following.

Theorem 12.2 (Picard-Lindel ö f) *Let* $x, f \in R^p$; $f(x,t)$ *continuous on a parallelepiped R:* $t_0 \leq t \leq t_0 + a$, $|x - x_0| \leq b$ *and uniformly Lipschiz continuous with respect to x. Let M be a bound for* $|f(x,t)|$ *on R;* $\alpha = \min(a, b/M)$. *Then*

$$\dot{x} = f(x,t), x(t_0) = x_0 \tag{12.9}$$

has a unique solution $x = x(t)$ *on* $[t_0, t_0 + \alpha]$.

For the proof of 0.2, see, for example, Hartman (op. cit., pp. 8–10).

In any case, we can find a curve in R^p *that satisfies (12.7) or (12.8) and passes through the point* $u = (u_1, \ldots, u_p)$ *when* $t = 0$. *The reader can find some curves in* R^2 *in Fig. 12.1. In dynamical system, solution curves are spoken of as trajectories or orbits.*

Let us now return to (12.4). As for its solution with initial values, we have the following theorem.

Theorem 12.3 *Let* **A** *be an operator on* R^p *having n distinct, real eigenvalues. Then for all* $x_0 \in R^p$, *the linear differential equation*

$$\dot{x} = Ax; x(0) = x_0, \tag{12.10}$$

has a unique solution.

For the proof of 12.3, see, for example, Hirsch and Smale ([34], pp. 47–48).

Remark The map defined by (12.4) is also called an *operator* in the case where the domain and range are the same.

Solution curves of (12.10) are given by the following theorem:

Theorem 12.4 *Let the p × p matrix* **A** *have n distinct real eigenvalues* $\lambda_1, \ldots, \lambda_p$. *Then every solution to the differential equation*

$$\dot{x} = Ax, x(0) = u, \tag{12.11}$$

is of the form

$$x_i(t) = c_{i1} \exp(t\lambda_1) + \cdots + c_{ip} \exp(t\lambda_p); .i = 1, \cdots, p, \tag{12.12}$$

for unique constants $c_{i1}, \ldots, c_{ip}$ *depending on* $\boldsymbol{u}$.

According to Hirsch and Smale (op. cit., p. 51), this theorem is an immediate consequence of the proof of 12.3.

Remark It is known that (12.12) has another expression $e^{tA}\boldsymbol{u}$ (e.g., Hirsch and Smale, op. cit., p. 90; Guckenheimer and Holmes [28], p. 8). Here, in general, an *exponential of a matrix* $\boldsymbol{X} \in M_p(C)$ is defined as

$$\exp(X) = e^X = \sum_{k=0}^{\infty} \frac{X^k}{k!}. \tag{12.13}$$

Here, $k!$ is *k factorial*, the product of the first k positive integers if $k > 0$, and $0! = 1$ by definition.

12.2.2 Singular Points of an Autonomous System

In the previous section we have introduced some basic concepts and theorems on differential equation. In this section we introduce an important notion called a *singular point* of the solution curves in autonomous system. Here let us consider a differential equation defined by (12.7) in the previous section.

Definition 12.1 Let the function f in (12.7) be C^1 or continuously differentiable, as in 0.1. Then, a point $\boldsymbol{x}_s \in W \subset R^p$ of (12.7) is a *singular point* if $f(x_s) = 0$.

Remark The singular point is sometimes called *singularity* (e.g., Birkhoff and Rota [9]; Hartman [33]), *equilibrium point* (e.g., Guckenheimer and Holmes [28]; Hirsch and Smale [34]; Seydel [51]), or critical point (e.g., Rosen [48]; Seydel, op. cit.). It is occasionally called *rest point* or *zero point*. Hereafter we will adopt the singular point among these synonyms, unless otherwise stated.

Particularly in the two-dimensional *linear* autonomous system, the detailed classification of the singular point is well known. Figure 12.2 shows the basic singular points in the classification, which is depicted in Fig. 8.1 of Chap. 8.

Six singular points shown in Fig. 12.2 are determined by the eigenvalues of $\boldsymbol{A}$, which are obtained by a similarity transformation of $\boldsymbol{A}$ in (12.6) such that

$$B = PAP^{-1}, \tag{12.14}$$

where $\boldsymbol{P}$ is a nonsingular matrix. As regards the similarity transformation, see Definition 11.30 in Chap. 11. It is known that $\boldsymbol{B}$ has one of the following forms:

$$(1)\begin{bmatrix} \lambda & 0 \\ 0 & \mu \end{bmatrix};\ (2)\begin{bmatrix} a & -b \\ b & a \end{bmatrix};\ (3)\begin{bmatrix} \lambda & 0 \\ 1 & \lambda \end{bmatrix},$$

(e.g., Hirsch and Smale [34], p. 86). Here, it should be noticed that these matrices are composed of the basic elements in the real Jordan canonical form which was introduced in Theorem 11.6 in Chap. 11.

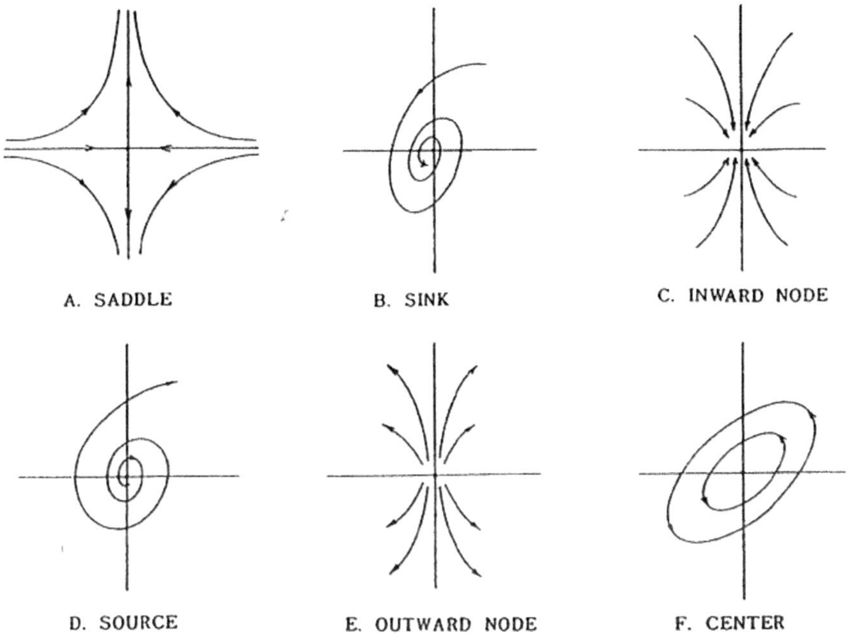

Fig. 12.2 Basic singular points in a two-dimensional autonomous system (adopted from Chino and Nakagawa [17])

In any case, using (12.14) we can compute the exponential of the matrix $\boldsymbol{A}$, that is, e^A such that

$$e^A = e^{P^{-1}BP} = P^{-1}e^B P. \tag{12.15}$$

As a result, we can compute the solution to (12.11). This is the reason why we compute the transformation of $\boldsymbol{A}$ using (12.14).

The qualitative features of the solution curves near the origin become the basic six patterns shown in Fig. 12.2, depending on the eigenvalues of $\boldsymbol{B}$ as follows:

(1) If the eigenvalues are both real with opposite signs, we have a *saddle* shown in Fig. 12.2a.
(2) If the eigenvalues are complex with negative real parts, we have a *sink* shown in Fig. 12.2b.
(3) If the eigenvalues are both real with negative signs, we have an *inward node* depicted in Fig. 12.2c.
(4) If the eigenvalues are complex with positive real parts, we have a *source* shown in Fig. 12.2d
(5) If the eigenvalues are both real with positive signs, an outward node shown in Fig. 12.2e.
(6) If the eigenvalues are pure imaginary, we have a *center* shown in Fig. 12.2f.

Example 12.3 In Fig. 12.1 there are two singular points, a source and a saddle, in this vector field. It should be noted, however, that the system described by this vector field is assumed to be nonlinear and is estimated from a set of empirical data. Moreover, we consider the estimated vector field at each of the fifteen weeks as an autonomous system with a certain parameter value, although we assume a set of nonlinear nonautonomous differential equations at the outset in estimating the holistic vector fields over a week. This means that we interpret the nonautonomous system as changes in the vector field described by an autonomous system over a certain parameter value, as was discussed in Chap. 8.

In the second quadrant of Fig. 12.1 we can identify a curious closed curve which is accompanied by a source. This closed curve is not the singular point, *center*, in Fig. 12.2f, but is peculiar to a nonlinear system and is called a *limit cycle*. We shall introduce it later.

Returning now to a general nonlinear autonomous system described by (12.7), how can we find its singular points? The answer to this question is to *linearize* this equation at a singular point $\boldsymbol{x}_s$ by studying the linear system

$$\dot{\xi} = Df(x_s)\xi, ..., \xi \in R^p, \tag{12.16}$$

where $\boldsymbol{Df} = \left[\partial f_i/\partial x_j\right]$ is the *Jacobian matrix* of first partial derivatives of the function

$$f = \left(f_1\left(x_1, \ldots, x_p\right), f_2\left(x_1, \ldots, x_p\right), \ldots, f_p\left(x_1, \ldots, x_p\right)\right)^t, \tag{12.17}$$

namely,

$$Df = \begin{bmatrix} \partial f_1/\partial x_1 & \cdots & \partial f_1/\partial x_p \\ \vdots & \vdots & \vdots \\ \partial f_p/\partial x_1 & \cdots & \partial f_p/\partial x_p \end{bmatrix}, \tag{12.18}$$

and

$$x = x_s + \xi, \ldots \xi \ll 1, \tag{12.19}$$

(e.g., Guckenheimer and Holmes [28], p. 12).

It is apparent that (12.16) is a *linear autonomous system.*

Example 12.4 (*a famous nonlinear autonomous system*) The following set of differential equations is a famous two-dimensional nonlinear autonomous system in which a *limit cycle* appears under a certain condition in an electrical circuit. Such a closed curve is peculiar to the so-called *Hopf bifurcation* (e.g., Hirsch and Smale [34], pp. 227–228), which will be discussed in detail later:

$$\begin{bmatrix} dx/dt \\ dy/dt \end{bmatrix} = \begin{bmatrix} y - f_\mu(x) \\ -x \end{bmatrix}, \tag{12.20}$$

where

$$f_\mu(x) = x^3 - \mu x, -1 < \mu < 1. \tag{12.21}$$

Sometimes parameter μ is called a bifurcation parameter. In this case, it may be the temperature of the resistor.

The Jacobian matrix of this set of differential equations can be computed as

$$Df = \begin{bmatrix} -3x^2 + \mu & 1 \\ -1 & 0 \end{bmatrix}. \tag{12.22}$$

In order to examine the qualitative features of the system near the singular point, we may compute the eigenvalues of this matrix. If we put $K = 3x^2 - \mu$, these *eigenvalues* of $\boldsymbol{Df}$ are

$$\lambda = \frac{-K \pm \sqrt{K^2 - 4}}{2}. \tag{12.23}$$

It is not difficult to see that we have a sink if $-1 \leq \mu \leq 0$, and a source if $0 < \mu \leq 1$, where $|x| < \sqrt{3\mu}/3$. According to Hirsch and Smale (op. cit., p. 228), E. Hopf proved that for fairly general one-parameter families of equations $dx/dt = f_\mu(x)$, there must be a closed orbit for $\mu > \mu_0$ if the eigenvalue character of a singular point changes suddenly at μ_0 from a sink to a source. In general, there are two types of limit cycles, as depicted in Fig. 12.3. The limit cycle discussed in this example is the type depicted in Fig. 12.3b.

Example 12.5 The following is a typical list of examples of limit cycles:

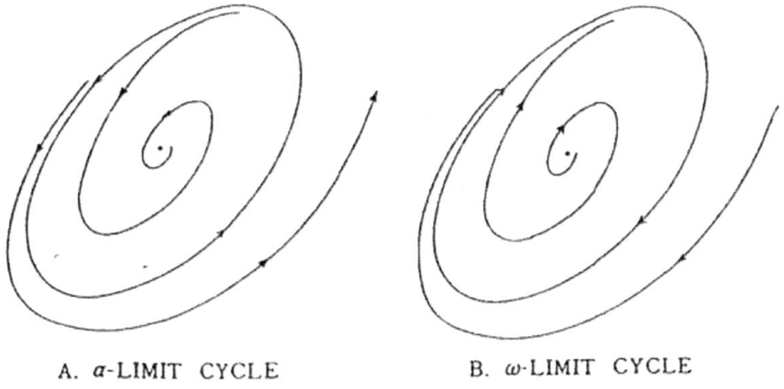

Fig. 12.3 Two types of limit cycles, which are adopted from Chino and Nakagawa [17]

12.2.3 Stability Notions of a Dynamical System

In this section, we shall introduce the two kinds of notions of *stability*. One is the *local stability* of singular points, and the other is the *structural stability* of dynamics of a system. These notions are very important in discussing differential equations in empirical sciences such as physical, biological, chemical sciences. We shall give some definitions and theorems concerning these two notions, according to Hirsch and Smale ([34], pp. 185–188) and Guckenheimer and Holmes [28]. Moreover, we shall introduce the notion of *attractor* which is related to the asymptotic stability which will be discussed soon.

Before proceeding, we shall define a concept on differentiation below.

Definition 12.2 (*r times differentiable function*) A function is said to be C^r if it is *r times differentiable* and each derivative is *continuous*, if $r = 0$ then the function is said to be continuous.)

Definition 12.3 (*stable singular point*) Suppose $\boldsymbol{x}_s \in W$ is a singular point of the differential equation

$$\dot{x} = f(x), \tag{12.24}$$

where $f : W \to E$ is a C^1 map from an open set W of the vector spave E into E. Then $\boldsymbol{x}_s$ is a *stable* singular point if for every neighborhood U of $\boldsymbol{x}_s$ in W there is a neighborhood U_1 of $\boldsymbol{x}_s$ in U such that every solution $\boldsymbol{x}(t)$ with $\boldsymbol{x}(0)$ in U_1 is defined and in U for all $t > 0$.

Definition 12.4 (*asymptotically stable singular point*) If U_1 can be chosen so that in addition to the properties described in Definition 12.2, $\lim_{t\to\infty} x(t) = x_s$, then $\boldsymbol{x}_s$ is asymptotically stable.

Definition 12.5 (*unstable singular point*) A singular point $\boldsymbol{x}_s$ that is not stable is called *unstable*. This means that there is a neighborhood U of $\boldsymbol{x}_s$ such that for every neighborhood U_1 of $\boldsymbol{x}_s$ in U, there is at least one solution $\boldsymbol{x}(t)$ starting at $x(0) \in U_1$, which does not lie entirely in U.

Remark A *sink* in Fig. 12.2b is asymptotically stable and therefore stable. A *center* in Fig. 12.2f is stable but not asymptotically stable. Qualitative feature of the center is destroyed by the small nonlinear perturbation to it. Such a property is concerned with the notion of *structural stability*, which will be discussed later. A *source* is unstable.

Definition 12.6 (*hyperbolic singular point*) A singular point $\boldsymbol{x}_s$is *hyperbolic* if the derivatives $\boldsymbol{D}f(\boldsymbol{x}_s)$ has no eigenvalue with real part zero.

Corollary *A hyperbolic singular point is either unstable or asymptotically stable.*

Theorem 12.5 (Lyapunov' stability) *Let $\boldsymbol{x}_s \in W$ be a singular point for (12.24). Let $V : U \to R$ be a continuous function defined on a neighborhood $U \subset W$ of $\boldsymbol{x}$, differentiable on $U - x_s$, such that*

(a) $V(\boldsymbol{x}_s) = 0$ and $V(\boldsymbol{x}) > 0$ if $x \neq x_s$;
(b) $\dot{V} \leq 0$ in $U - x_s$.

Then $\boldsymbol{x}_s$ is stable, Furthermore, if also

(c) $\dot{V} < 0$ in $U - x_s$,
then $\boldsymbol{x}_s$ is asymptotically stable.

For the proof of this theorem, see, for example, Hirsch and Smale ([34], pp. 194–195).

It will be appropriate to introduce here the notion of *attractor* which is closely related to the notion of asymptotical stability. Roughly speaking, asymptotically stable singular points are called attractors (e.g., Seydel [50]). Although there is no universal agreement about what the best definition is (e.g., Eckmann [20]), we shall introduce the following definition of attractor by Seydel ([51], p. 389) which is adapted from Eckmann (op. cit., p. 644).

Definition 12.7 (*attractor*) A compact region A is an *attractor* of a flow or a solution curve φ of (12.24) or more generally $\dot{\boldsymbol{x}} = f(\boldsymbol{x}, \mu)$ with a parameter μ, if the following four hypotheses hold:

(1) A is invariant under φ.
(2) A has a shrinking neighborhood under φ.
(3) The flow is recurrent—that is, trajectories starting from any (open) subset of A hit that subset again and again for arbitrarily large values of t; the flow is nowhere transient.
(4) The flow cannot be decomposed—that is, A cannot be split into two nontrivial invariant parts.

Remark The largest neighborhood of (2) is called the *basin of attraction.*

Definition 12.8 (*strange attractor*) An attractor is called a *strange attractor* if flow lines depend sensitively on the initial values.

Remark *Asymptotically stable equilibria (singular points)* are attractors (e.g., Seydel, op. cit., p. 8). Another simple attractor is a *limit cycle*. If a system settles down to a *periodic orbit*, then the system is said to have a periodic attractor, while if the system behaves chaotically with an *aperiodic orbit*, then the system is said to have a chaotic attractor which is commonly referred to as a strange attractor (e.g., Addison [2], p. 93). *K-torus* is also an attractor. The logistic map to be discussed later in Sect. 12.3, for example, has period 1 attractor, period 2 attractor, period 4 attractors etc.

Next, we shall introduce the notions and theorems related to *structural stability*. An important theory related to it is the *Hartman-Grobman theorem*. According to

Wiggins ([56], p. 234), this theorem is originally proved independently by Hartman [31] and Grobman [27]. The following is due to Guckenheimer and Holmes ([28], p. 13).

Theorem 12.9 (Hartman-Grobman) *If $Df(\mathbf{x}_s)$ has no zero or purely imaginary eigenvalues then there is a homeomorphism h defined on some neighborhood U of $\mathbf{x}_s$ in R^p locally taking orbits of the nonlinear flow of (12.7), to those of the linear flow of (12.16). The homeomorphism preserves the sense of orbits and can also be chosen to preserve parametrization by time.*

For the proof of this theorem, see, for example, Hartman ([32], Chap. IX).

Next, the following is a definition of the structural stability by Guckenheimer and Holmes (op. cit.):

Definition 12.9 (*structural stability*) A map $F \in C^r(R^p)$ (respectively a C^r vector field f) is *structurally stable* if there is an $\varepsilon > 0$ such that all C^1, ϵ perturbations of F (respectively of f) are topologically equivalent to F (respectively f).

Remark An alternative definition of structural stability is, for example, given by Wiggins ([56], p. 99).

In the case of two-dimensional vector fields, the following result is known (e.g., Wiggins, op. cit., p. 100), which is given by Peixoto [42]:

Theorem 12.10 (Peixoto) *A C^r vector field on a compact boundaryless two-dimensional manifold M is structurally stable if and only if*

(1) *The number of singular points and periodic orbits is finite and each is hyperbolic;*
(2) *There are no orbits connecting saddle points.*
(3) *The nonwandering set consists of singular points and periodic orbits.*

Moreover, if M is orientable, then the set of such vector fields is open and dense in $C^r(M, M)$ (note: this is stronger than generic).

Remark 1 A point p is called nonwandering for the flow φ_t of a vector filed if, for any neighborhood U of p, there exists arbitrarily large t such that $\varphi_t(U) \cap U \neq \emptyset$ (e.g., Guckenheimer and Holmes [28], p. 33). $C^r(M, M)$ in the above theorem denotes the space of C^r maps of M into M.

Remark 2 Studies of the Hartman-Grobman theorem is still an active area of research. For example, Palmer [40] proposes a generalization of this theorem. Pinto and Robledo [44] discuss this theorem for differential equations with piecewise constant arguments of mixed type. Zhang, Liu, and Lei [59] discuss Hö*lder*dependence of the parametrized Hartman-Grobman theorem.

12.2.4 Bifurcations of Singular Points of a Vector Field

In this section we shall deal with the system in which qualitative behaviors of a vector field change according to certain parameter values. We have already shown an example of such a problem in Example 12.2. These changes are called *bifurcations* and the parameter values are said to be *bifurcation values*. As was discussed in Chap. 8, we are specifically interested in a phenomenon in which certain factors influence a vector field and the values of these factors change continuously by the complicated interactions among individuals or some influence from outside the system.

Definition 12.10 (*bifurcation problem*) A p-dimensional autonomous differential equation with bifurcation parameter values can be written as

$$\dot{x} = f_{\mu}(x), ...\mu = \left(\mu_1, \ldots, \mu_q\right)^t, \tag{12.25}$$

where $\boldsymbol{\mu}$ is a q-dimensional bifurcation parameter vector, as this equation indicates, while the state vector $\boldsymbol{x}$ is in general p-dimensional. Moreover, singular points of this equation are the points where $f_{\mu}(x) = 0$.

Remark As this parameter vector $\boldsymbol{\mu}$ varies, the *implicit function theorem* implies that these singular points are described by smooth functions of $\boldsymbol{\mu}$ away from those points at which the Jacobian derivative of $f_{\boldsymbol{\mu}}(\boldsymbol{x})$ with respect to $\boldsymbol{x}$, $\boldsymbol{D_x} f_{\boldsymbol{\mu}}$ has a zero eigenvalue (e.g., Guckenheimer and Holmes ([28], p. 118). As for the implicit function theorem and its proof, see, for example, Hirsch and Smale ([34], Appendix IV). It is usual that we deal with a special case when $q = 1$. The singular points at which all eigenvalues of the linearized field are different from zero are called *nondegenerate*. The bifurcation parameter and the bifurcation point in the above theorem can be characterized by the following definition (e.g., Arnold [5], p. 258).

Definition 12.11 (*degenerate singular point*) A value of the parameter to which a degenerate singular point corresponds is called bifurcation value of the parameter, and the degenerate singular point itself in the direct product of the phase space with the axis of the values of the parameter is called a bifurcation point.

In relation to the nondegenerate singular point of a vector field, the following theorem is important (Arnold, op. cit.).

Theorem 12.11 *Any singular point of a vector field depending smoothly on a parameter depends smoothly on the parameter itself, provided that all eigenvalues of the linear part of the field at the singular point are different from zero.*

For the proof of this theorem, see, for example, Arnold, op. cit., p. 258).

In any case, as Guckenheimer and Holmes (op. cit., p. 117) point out, the detailed changes in the topological equivalence class of a flow can be exceedingly complicated, since the bifurcation problem defined above is related to a degenerate singular point. Therefore, in this section we shall deal with the simple bifurcations of individual singular points and periodic orbits, as they did.

In bifurcation theory two types of bifurcations are known. One is the *local bifurcation* and the other the *global bifurcation*. The former deals with the local qualitative behaviors of the vector filed described by (12.25) near singular points, while the latter deals with the global qualitative behaviors of the system. As was discussed in Chap. 8, four kinds of bifurcations are the best known bifurcations which depend on a single parameter in a two-dimensional autonomous differential equation (e.g., Guckenheimer and Holmes [28]). These are *saddle-node*, *pitchfork*, *transcritical*, and the *Hopf bifurcations*. Figure 12.4 shows examples of two types of bifurcations.

Example 12.6 The following is a list of examples of bifurcations:

As discussed in Chap. 8, the saddle-node bifurcation is the process in which there exists no singular point at the outset and then a *degenerate singular point* is born to bifurcate into a saddle and a sink. The reverse process is also a saddle-node bifurcation, and in this case a saddle and a sink approach gradually and are suddenly unified into a degenerate singular point. Figure 12.4a draws the former case. Degenerate singular points cannot be observed in principle since they are structurally unstable. Therefore, we might observe a saddle and a node as if they suddenly appeared or disappeared.

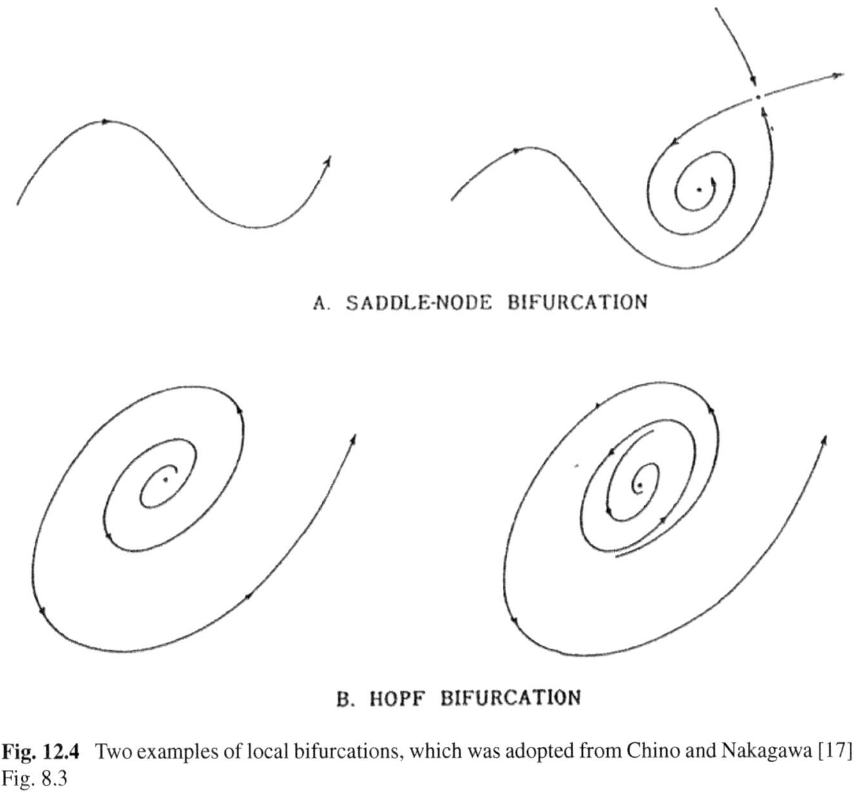

Fig. 12.4 Two examples of local bifurcations, which was adopted from Chino and Nakagawa [17], Fig. 8.3

On the other hand, the Hopf bifurcation is the case where eigenvalues of the Jacobian at singular point are pure imaginary. In this case either an *α-limit cycle* or an *ω-limit cycle* ω-limit cycle appears before or after the bifurcation. Figure 12.4b shows the case in which the the α-limit cycle appears. We have shown in Chap. 8 that these two local bifurcations can be interpreted psychologically or sociologically.

Example 12.7 The following two types of bifurcations are well-known examples of global bifurcations:

We shall now introduce the *global bifurcations*. They are relatively poorly understood in mathematics compared with the local ones and the global bifurcation theory is still an active area of research. In such a situation *saddle loop* and *saddle connection* are a few well known examples. Figure 12.5 illustrate them. In the former case one of the two orbits which go out of a saddle enters the saddle. On the other hand, in the latter case two orbits, one of which goes out of a saddle and the other enters the other saddle, connect to form an orbit. Psychological or sociological interpretations of these global bifurcations are also possible (see, Chino [15]).

Remark For further details of bifurcation problem, see, for example, Arnold [5], Guckenheimer and Holmes [28], Seydel [50], Thiel et al. [55], Wiggins [56].

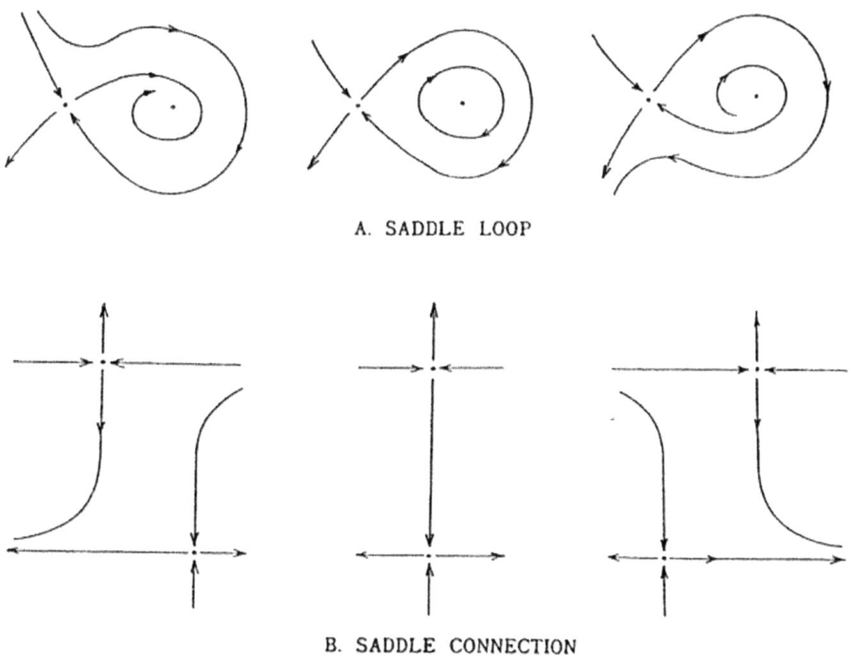

Fig. 12.5 Examples of global bifurcation, which is adopted from Chino and Nakagawa [17]

12.3 Chaos and the Routes to Chaos

In this section we shall introduce the notion of chaos, and several routes to chaos. As Seydel (op. cit.) points out, bifurcations discussed in the previous section play a crucial role in initiating chaos, although there is no unique way in which chaos arises. First, we shall define chaos. Second, we shall define the *Lyapunov exponent* whose sign signifies chaos and whose value measures how chaotic, and show examples of its applications to some famous difference and differential equations. Finally, we shall show various examples of difference equation models for the analysis of asymmetric structure and dynamics.

The word "chaos" is derived from the the Greek "$\chi\alpha o\varsigma$" (*khaos* in the Roman alphabet). According to the Oxford Dictionary of English (3rd ed.) (Stevenson, [54]), "chaos" is the property of a complex system whose behavior is so unpredictable as to appear random, owing to great sensitivity to small changes in conditions. Although there seems to be no universally accepted strict definition of chaos, we shall adopt the definition of chaos by Sprott [53] in the following:

Definition 12.12 (*chaos*) *Chaos* is the *aperiodic*, long-term behavior of a *bounded, deterministic* system that exhibits *sensitive dependence on initial conditions.*

Remark 1 The deterministic system in the above definition is defined in either *differential* or *difference equations*. Schuster ([49], p. 15) present four possible criteria for chaotic motion.

(1) The time dependence of the signal "looks chaotic",
(2) The power spectrum exhibits broadband noise,
(3) The autocorrelation function decays rapidly,
(4) The Poincaré map shows space-filling points.

Remark 2 Chaotic motion was discovered by Poincaré [45] (e.g., Schuster, op. cit., p. 1), but it took many years until the meteorologist Lorenz [35] found that even a simple set of three coupled, first order, nonlinear differential equations can lead to completely chaotic trajectories.

As for the basic concepts in time series analysis such as power spectrum and autocorrelation function in the above remark, see, for example, Anderson [4]. The Poincaré map is defined as follows (e.g., Guckenheimer and Holmes [28], p. 23):

Definition 12.13 (*Poincaré map*) Let γ be a periodic orbit of some flow φ_t, in R^n arising from a nonlinear vector field $f(x)$. Take a *local cross section* $\Sigma \subset R^p$, of dimension $p - 1$, first. The hypersurface Σ need not be planar, but must be chosen so that the flow is everywhere *transverse* to it. Denote the (unique) point where γ intersects Σ by p, and let $U \subseteq \Sigma$ be some neighborhood of p. Then the *first return* or *Poincaré map* $P : U \to \Sigma$ is defined for a point $q \in U$ by

$$P(q) = \varphi_t(q), \tag{12.26}$$

where $\tau = \tau(q)$ is the time taken for the orbit $\varphi_t(q)$ based at q to first return to Σ.

Remark For further details, see, for example, Guckenheimer and Holmes (op. cit., pp. 22–27).

We shall now define the Lyapunov exponent which plays a decisive role in measuring how chaotic the system under consideration is. We can consider at least three cases in defining the Lyapunov exponent. The first case to consider is the Lyapunov exponent for one-dimensional map (e.g., Seydel [51], p. 410).

Definition 12.14 (*Lyapunov exponent for one-dimensional map*) Let us consider a one-dimensional map described by the difference equation

$$x_{n+1} = f(x_n). \tag{12.27}$$

Then, the Lyapunov exponent is defined as follows:

$$L(x) = \lim_{N\to\infty} \frac{1}{N} \sum_{n=0}^{N-1} \ln f'(x_n), \tag{12.28}$$

where $f'(x_n)$ is the first derivative with respect to x_n.

Remark The Lyapunov exponent L serves as a measure for exponential divergence ($L > 0$) or contraction ($L < 0$) of the orbit of the system.

The second case to consider is the Lyapunov exponent for p-dimensional map (e.g., Aihara [3]; Sprot [53]).

Definition 12.15 (*Lyapunov exponent for p-dimensional map*) Let us consider a p-dimensional map described by the difference equation

$$\boldsymbol{x}_{n+1} = \boldsymbol{f}(\boldsymbol{x}_n). \tag{12.29}$$

where $\boldsymbol{x}_n$ is a p-dimensional vector. By the linear approximation of (12.29) we have the following map

$$\Delta\boldsymbol{x}_{n+1} = \boldsymbol{J}_n \Delta\boldsymbol{x}_n, \tag{12.30}$$

where $\boldsymbol{J}_n$ is the Jacobian matrix of first derivatives of $\boldsymbol{f}$, as defined by (12.18).

Moreover, let us define the following positive definite matrix

$$\boldsymbol{\Gamma}(\boldsymbol{x}_0, N) = \left[\{\boldsymbol{M}(\boldsymbol{x}_0, N)\}^t \boldsymbol{M}(\boldsymbol{x}_0, N)\right]^{1/2N}, \tag{12.31}$$

where

$$\boldsymbol{M}(x_0, N) = \prod_{n=0}^{N-1} \boldsymbol{J}_n. \tag{12.32}$$

Then, the Lyapunov exponent (Lyapunov spectrum) is defined as

$$\lambda_i = \lim_{N \to \infty} \frac{1}{N} \ln \sigma_i(N), \tag{12.33}$$

where $\sigma_i(N)$ is the ith eigenvalue of $\mathbf{\Gamma}\ (x_0, N)$.

Remark According to Aihara [3], we have a difficulty in computing (12.31), and researchers have proposed methods for coping with this problem (e.g., Shimada and Nagashima [52]). For further details, see, for example, Aihara, op. cit.). Sprott ([53], p. 110) discusses a special case in which p equals to 2 in the above definition.

The final case to consider is the Lyapunov exponent for ordinary differential equations (e.g., Seydel [51], pp. 410–411).

Definition 12.16 (*Lyapunov exponent for ordinary differential equations*) Let us consider ordinary differential equations

$$\dot{\boldsymbol{x}} = \boldsymbol{f}(\boldsymbol{x}, \mu), \tag{12.34}$$

where $\boldsymbol{x}$ is a state vector in R^p, and μ is a parameter which is not essential in this context.

By linearizing the above equation, we have

$$\dot{\boldsymbol{h}} = \boldsymbol{f}_x(\boldsymbol{x}, \mu)\boldsymbol{h}, \tag{12.35}$$

where the Jacobian $\boldsymbol{f}_x$ is evaluated along the particular flow line $\boldsymbol{x}$ that is investigated. The solution $\boldsymbol{h}(t)$ can be expressed as

$$\boldsymbol{h}(t) = \mathbf{\Phi}(t, \boldsymbol{x})\boldsymbol{h}(0), \tag{12.36}$$

where $\boldsymbol{h}(0)$ is any initial vector of the system. Then, the Lyapunov exponents L_k of the trajectory $\boldsymbol{x}(t)$ are real numbers defined by

$$L_k = \lim_{t \to \infty} \frac{1}{t} \ln|\mu_k(t)| \tag{12.37}$$

where $\mu_k(t), k = 1, \ldots, p$ are the eigenvalues of $\boldsymbol{\Phi}(t, \boldsymbol{x})$.

The Lyapunov exponent discussed up to now is a measure of the *dynamic properties* of the system described by differential equations or difference equations. On the other hand, there are a lot of measures of the *geometric properties* of the system, which are called *fractal dimensions*. For example, Addison ([2], Sect. 7.5) lists, as these dimensions, Box counting dimension, the information dimension, correlation dimension, the pairwise and averaged pointwise dimension, and the Lyapunov dimension. Sprott (op. cit., this chapter) makes a list from a different viewpoint, namely, similarity dimension, capacity dimension, correlation dimension, entropy, BDS (Brock, Dechert, Scheinkman) statistic, and minimum mutual information. For the definitions and details of these concepts, see, for example, these books. In this chapter, we shall only introduce some of them.

Before proceeding, we must at least refer to the notions of *fractal* and *self-similarity*. The term 'fractal' was originated by Mandelbrot [36]. According to the Oxford Dictionary of English (3rd ed.) (Stevenson [54]), this word is derived from the Latin *fract* 'broken', from the verb *frangere*, and means that a curve or geometrical figure, each part of which has the same statistical character as the whole. On the other hand, Mandelbrot (op. cit., p. 15) defines a fractal as follows.

Definition 12.17 (*Mandelbrot*) A *fractal* is a set for which the *Hausdorff Besicovitch dimension D* strictly exceeds the topological dimension D_T.

Remark As regards the topological dimension, we shall define it soon. The Hausdorf Besicovitch dimension (or *Hausdorff dimension*) is often called *fractional dimension.* Moreover, every set with a noninteger D is a *fractal* (e.g., Mandelbrot, op. cit., p. 15). Hausdorff dimension is defined below.

Definition 12.18 (*Hausdorff dimension*) Let X be a compact subset of a metric space and let $d > 0$. The d-dimensional outer measure $m_d(X)$ is obtained from

$$m_d(X, \varepsilon) = \inf\left\{\sum_{i \in I}(diamS_i)^d\right\}, \tag{12.38}$$

where the *inf* is over all finite coverings of X by sets S_i with diameter less than $\varepsilon > 0$ in such a way that

$$m_d(X) = \lim_{\varepsilon \to 0} m_d(X, \varepsilon). \tag{12.39}$$

Here, $m_d(X)$, depending on the choice of d, may be finite or infinite. Then, we have the *Hausdorff dimension*

$$h(X) = \sup\{d \in R_+ : m_d(X) = \infty\}. \tag{12.40}$$

Remark Although there exist various definitions in the literature, the above definition by Peitgen and Richter ([41], p. 12.2) seems to be very succinct and easy to understand. Hereafter, we shall denote the Hausdorff dimension by D_H instead of $h(X)$, following Mandelbrot (op. cit., p. 15).

Although Mandelbrot's Definition 12.17 introduced above is simple, it is criticized, for example, by Falconer ([24], p. xxv). Falconer contends that this definition is unsatisfactory in that it excluded a number of sets that clearly ought to be regarded as fractals. For this reason, he recommends to give a list of properties which characterizes the fractal. These are listed below:

(1) A set F has a fine structure, i.e. detail on arbitrarily small scales.
(2) F is too irregular to be described in traditional geometrical language, both locally and globally.
(3) Often F has some form of self-similarity, perhaps approximate or statistical.

(4) Usually, the 'fractal dimension' of F (defined in some way) is greater than its topological dimension.
(5) In most cases of interest F is defined in very simple way, perhaps recursively.

As regards the following four dimensions, we shall refer to Sprott [53].

Definition 12.19 (*similarity dimension*) Let us form a set by successively cutting a uniform string, the number of pieces N at each step is related to the length reduction R at that step by $N \propto R^{D_s}$ or

$$D_s = \frac{\log N}{\log R}, \tag{12.41}$$

where D_s is called the similarity dimension.

Definition 12.20 (*topological dimension*) Let a system described by difference or differential equations has multiple Lyapunov exponents, λ_j. The largest D for which $\sum_{j=1}^{D} \lambda_j \geq 0$ is called the *topological dimension.*

Definition 12.21 (*Kaplan-Yorke dimension*) Let us consider the largest D Lyapunov exponents for which $\sum_{j=1}^{D} \lambda_j \geq 0$, and the Lyapunov exponent λ_{D+1} corresponding to the next higher dimension $D+1$. Then the Kaplan-York (or Lyapunov) dimension D_{KY} is defined as

$$D_{KY} = D + \frac{1}{|\lambda_{D+1}|} \sum_{j=1}^{D} \lambda_j. \tag{12.42}$$

Definition 12.22 (*correlation dimension*) Suppose that the object is formed from N points in some embedding space. Let $C(r)$ be the number of points within all the circles of radius r, normalized so that $C(r)$ is 10 when r is sufficiently large that it includes all the points without double counting

$$C(r) = \frac{2}{N(N-1)} \sum_{i=1}^{N} \sum_{j=1}^{N} \Theta\big(r - r_{ij}\big), \tag{12.43}$$

where Θ is the Heaviside function

$$\Theta(x) = \begin{cases} 0 \text{ for } & x < 0 \\ 1 \text{ for } & x \geq 0 \end{cases} \tag{12.44}$$

and r_{ij} is the spatial separation between two points labeled i and j, usually given in an m-dimensional time-delay embedding by the Euclidean norm

$$r_{ij} = \sqrt{\sum_{k=0}^{m-1} \big(X_{i-k} - X_{j-k}\big)^2}. \tag{12.45}$$

$C(r)$ is called the correlation sum, and it converges to the correlation integral for $N \to \infty$.

Then, the *correlation dimension* is defined as

$$D_2 = \lim_{r\to 0}\lim_{N\to\infty} \frac{d \log C(r)}{d \log r}. \tag{12.46}$$

Remark For example, Sprott ([53], Chaps. 12 and 13) lists the definitions of the various fractal dimensions including the above definitions. Moreover, he collects various common chaotic systems described by differential equations and difference equations, and shows the values of Lyapunov exponent, Kaplan-Yorke dimension, and correlation dimension with their references. Furthermore, he shows inequalities on the values of several fractal dimensions (Sprott, op. cit., p. 339), including the *Euclidean dimension* (the integer dimension of the space in which the set is embedded), capacity dimension, Hausdorff dimension, information dimension, correlation dimension, and topological dimension. Berry [8] shows the relationship between Hausdorff dimension and power spectrum in that $D_H = (5-\alpha)/2$, where α is related to the power spectrum $p(f) \propto f^{-\alpha}$. Higuchi [33] proposes a method for estimating D_H even in the case of a small number of data.

We shall now show two examples of chaos. First, we shall introduce the famous *logistic map* described by a one-dimensional difference equation which simulates growth of a population. It is a quadratic map proposed by May [38] and is expressed as

$$x_{n+1} = ax_n(1-x_n), \quad (0 \le a \le 4), \tag{12.47}$$

where a is the growth rate of a population, while x_n is the population size in year n. We shall introduce it as a simple example which exhibits a chaotic motion depending on a, although it is not a differential equation model to be treated in this section. It is known that (1) if a equals 0.9 then the population size decays to zero as year proceeds, (2) if a equals 2.6 then it exhibits a steady state of period 1, and (3) if a equals 4.0 then it exhibits a chaotic motion (e.g., Addison [2], p. 90). Figure 12.6 shows iterates of the logistic map defined by (12.47) starting from $x_1 = 0.1$, $a = 4.0$ for 1200 iterations with the discrete delay increment 5. An inspection reveals that a chaotic behavior seems to take place. In order to ascertain empirically whether these iterates really show a chaotic behavior or not, we computed the Lyapunov exponent corresponding to them. Figure 12.7 shows this. This figure was drawn by Sunday Chaos Times. Estimated value of this Lyapunov exponent was nearly equal to the theoretical value $\ln 2 = 0.69314\ldots$ (e.g., Sprott [53], p. 417).

Second, we shall introduce a famous three-dimensional differential equation model called *Lorenz equation*, which was proposed by meteorologist Lorenz [35].

$$\begin{bmatrix} dx/dt \\ dy/dt \\ dz/dt \end{bmatrix} = \begin{bmatrix} \sigma(y-x) \\ -xz + rx - y \\ xy - bz \end{bmatrix}, \tag{12.48}$$

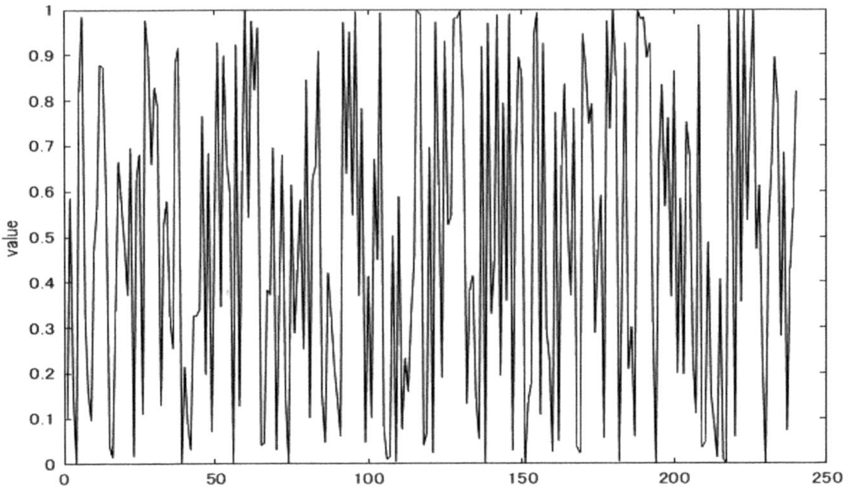

Fig. 12.6 Iterates of the logistic map with $a = 4$. The original data generated by (12.27) for 1000 iterations was thinned out with discrete delay increment 5

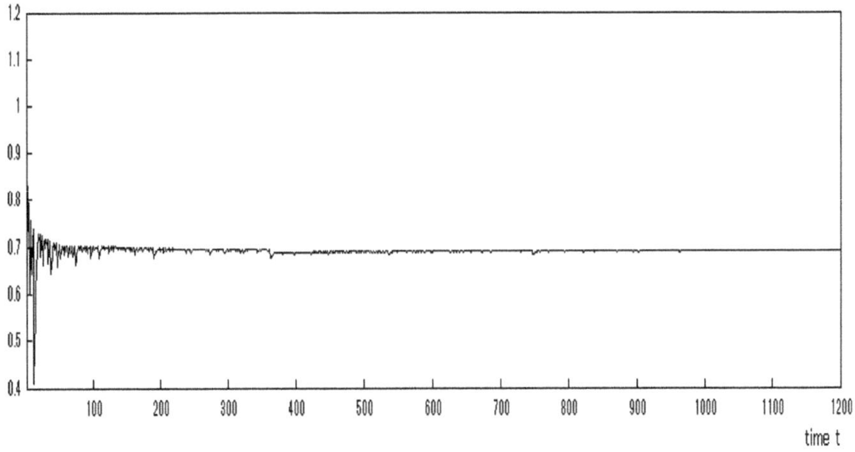

Fig. 12.7 Lyapunov exponent for the iterates of the logistic map shown in Fig. 12.6

where σ, r, b are real constants. It is known that we have a chaos in the case when $\sigma = 10, r = 28$, and$b = 8/3$ The chaotic trajectory of this model called *Lorenz attractor* shows sensitive dependence on initial conditions. In this case the initial condition was set equal to $(0, -0.01, 9)$, which is equal to that shown in Sprott (2003, p. 431). Figure 12.8 shows the time series obtained by solving this equation using MATLAB.

Figures 12.9, 12.10, and 12.11 show the attractor projected onto the xy plane, the xz plane, and yz plane of this attractor, respectively.

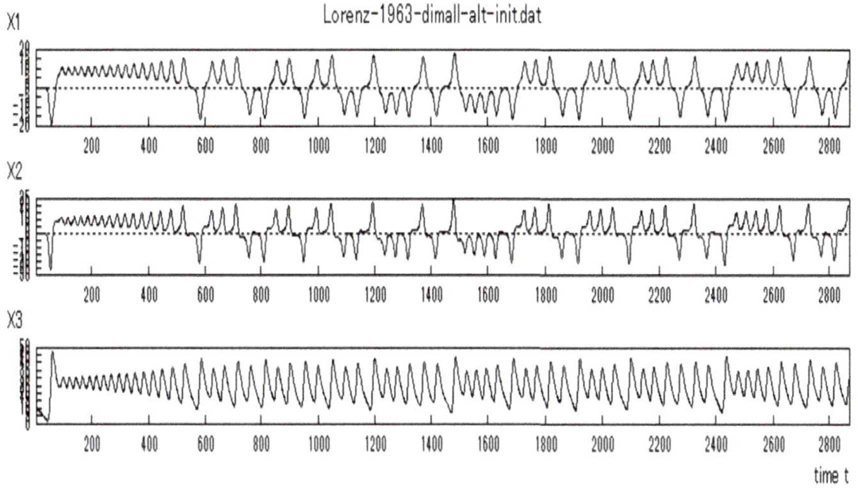

Fig. 12.8 Time series for the Lorenz attractor in the chaotic mode specified by the parameter values $\sigma = 10,\ r = 28,$ and $b = 8/3$

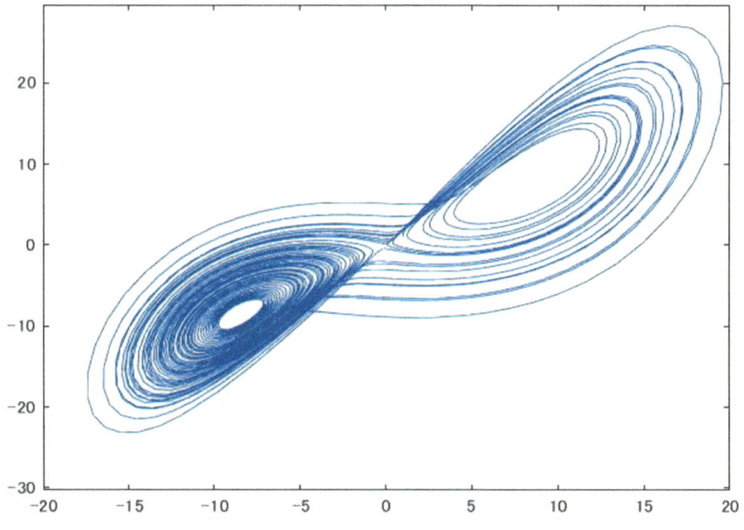

Fig. 12.9 Lorenz attractor projected onto the *xy* plane

The Lyapunov exponents for the three-dimensional attractor shown in Fig. 12.8 are depicted in Fig. 12.12. The largest Lyapunov exponent, the second exponent, and the third exponent were computed as 0.0186, − 0.0012, and − 0.1246, respectively. These values are not equal to those shown in Sprott ([53], p. 431), although both have the largest exponent with a small positive value, zero or nearly equal to zero, and a negative value. The reason why they are not the same may be due to the difference

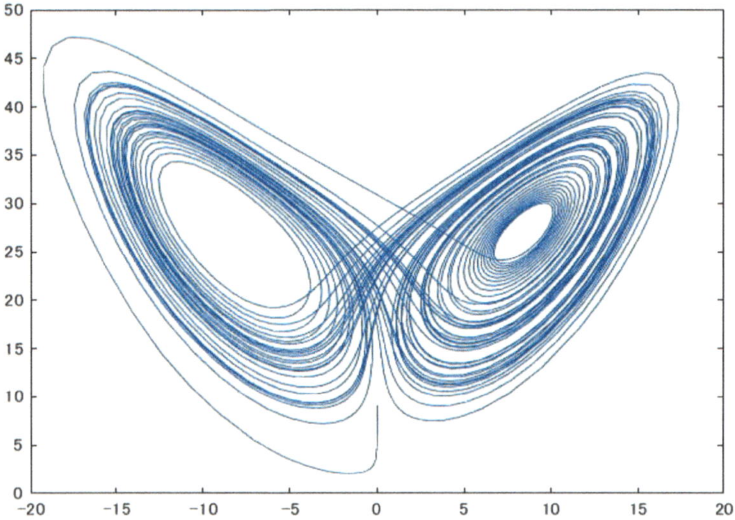

Fig. 12.10 Lorenz attractor projected onto the *xz* plane

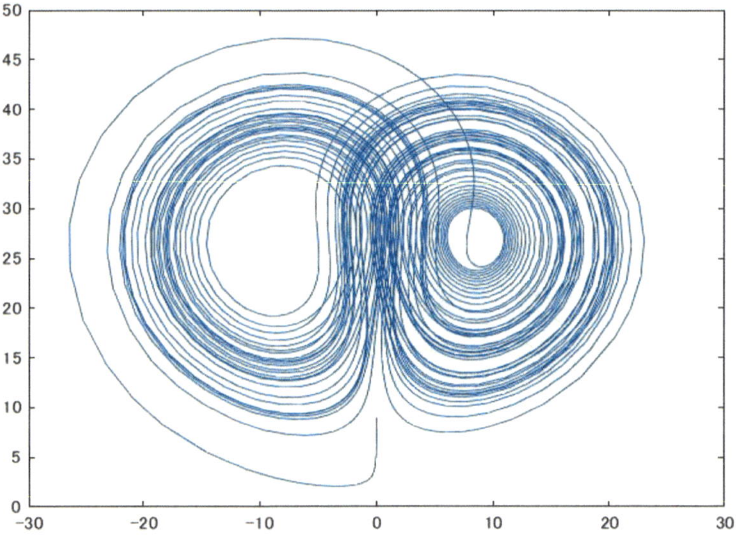

Fig. 12.11 Lorenz attractor projected onto the *yz* plane

of the specification of parameter values or the difference of the method in estimating the Lyapunov exponent.

Finally, we shall introduce various routs to chaos. It is well known that at least three routs to chaos exist (e.g., Addison [2]; Seydel [51]; Sprott [53]). These are *periodic doubling*, *quasiperiodicity*, and *intermittency*. Periodic doubling route to chaos has

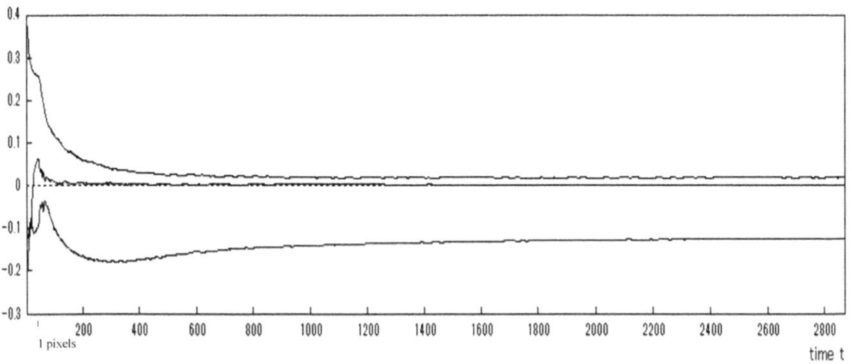

Fig. 12.12 Lyapunov exponents for the three-dimensional Lorenz attractor

been described in connection with the logistic map. If we change the growth rate parameter in (12.27) within an appropriate range, the qualitative behavior of this system changes drastically. Such a phenomenon is considered as a bifurcation of singular points of a vector field described by a difference or differential equation, the latter of which has already been introduced in 10.2.4.

Figure 12.13 shows a bifurcation diagram of the *logistic map* described earlier by (12.47). Although the logistic map starts with the fixed point $x = 0$ when the parameter value a equals 1, we zoomed into the logistic diagram over the range $2.8 \leq a \leq 4.0$. It is apparent from Fig. 12.13 that, as this parameter changes from 2.8 to about 3.5, there appear periodic attractors, period 1, period 2, period 4 attractors (e.g., Addison [2], p. 93) seriatim. It is curious that we have a period 3 attractor too in this diagram. This attractor can be more visible if we zoom into it over the range $3.825 \leq a \leq 3.86$. Figure 12.14 shows this.

Quasiperiodicity route to chaos is the route in which a singular point, spiral, becomes unstable and gives birth to a limit cycle in a Hopf bifurcation and then becomes a torus by a secondary Hopf bifurcation (which is sometimes called a Neimark-Sacker bifurcation) after which the torus usually develops and becomes chaotic (e.g., Sprott [53]). As an example of this route, we shall introduce a numerical experiment conducted by Harlim and Langford [30]. They used the following nonlinear differential equation:

$$\begin{bmatrix} dx/dt \\ dy/dt \\ dz/dt \end{bmatrix} = \begin{bmatrix} (z+\mu_1)x - \omega y \\ \omega x + (z+\mu_1)y \\ \mu_2 + \mu_3 z - z^3 + k\left(x^2+y^2\right) \end{bmatrix}, \tag{12.49}$$

where $\mu_1, \mu_2, \mu_3, \omega, andk$ are parameters. Specifically, if we change μ_1 as a bifurcation parameter, then we have a variety of solution curves of this system, as described below. Although they show only the solution curves corresponding to the parameter values $\mu_1 = -0.6, -0.5,$ and -0.2, we show $\mu_1 = -1.5, -1.1, -1.0, -0.7, -0.6, -0.5,$ and -0.4, corresponding to *sink*, *limit cycle*, a variant of

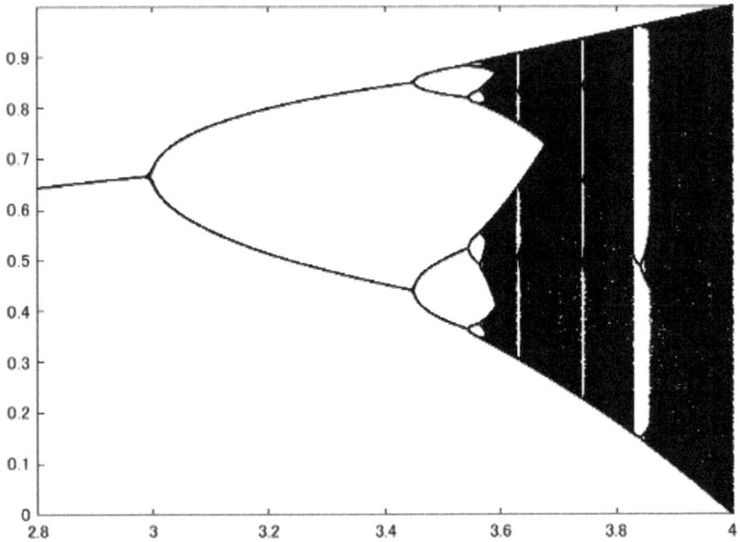

Fig. 12.13 A bifurcation diagram for the logistic map in the range $2.8 \leq a \leq 4.0$. The MATLAB codes for this map was adapted from Hanchak [29]

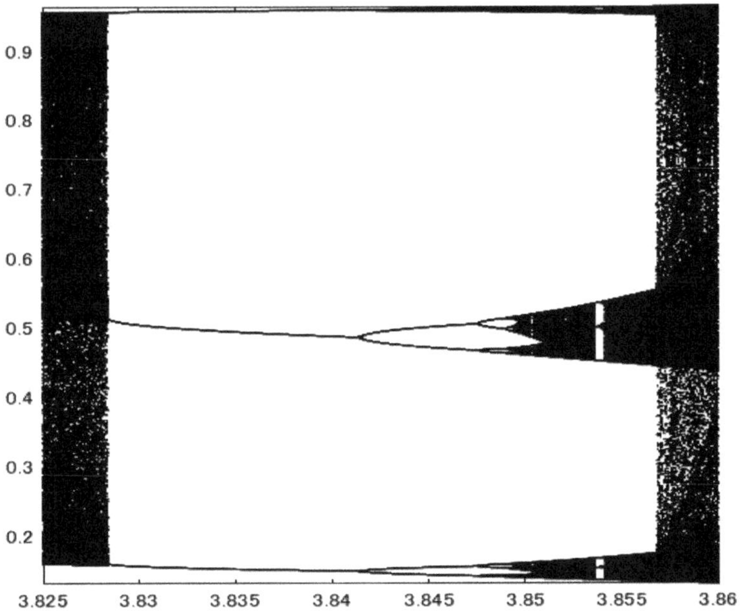

Fig. 12.14 A bifurcation diagram for the logistic map in the range $3.825 \leq a \leq 3.86$

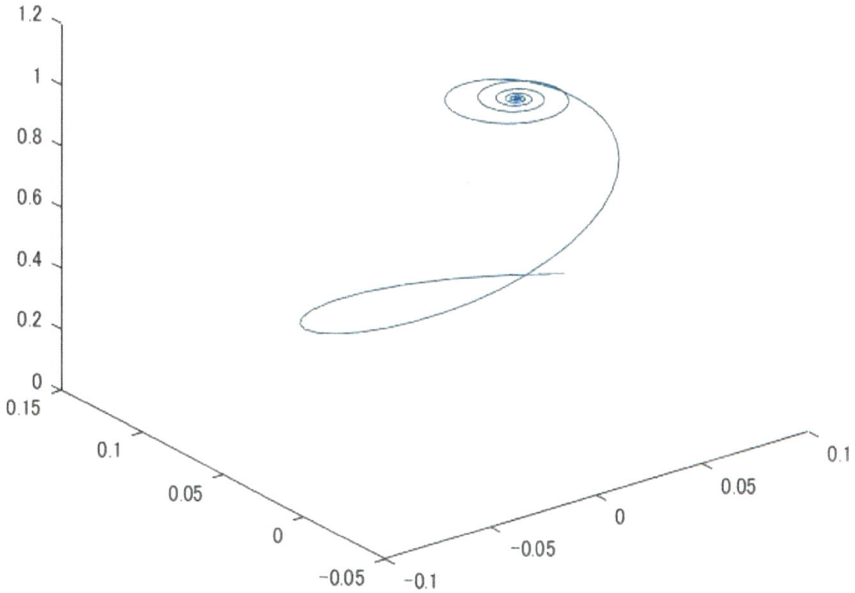

Fig. 12.15 The solution curve corresponding to (12.49) in the case when $\mu_1 = -1.5$

limit cycle, *deforming limit cycle*, *torus*, *transient chaos*, respectively. In all of these cases, the other parameters of (12.49) were set equal to, $\mu_2 = 0.4$, $\mu_3 = 1.0$, $\omega = 3.5$, $k = -1.0$, and the initial values of this equation were set equal to $(0.1, 0.1, 0.1)$, which are the same as those in Harlim and Langford (op. cit.).

Figure 12.15 shows the solution curve of (12.49) in the case when $\mu_1 = -1.5$. This curve indicates a sink.

Figure 12.16 shows the solution curve of (12.49) in the case when $\mu_1 = -1.1$. This curve apparently indicates a limit cycle. The coordinates of this curve on the vertical axis become almost the same after about 600 iterations.

A Hopf bifurcation might have given rise to the limit cycle in the interval $(-1.5, -1.1)$of the bifurcation parameter μ_1.

As the parameter value increases a bit, a variant of the limit cycle appears. Figure 12.17 shows the solution curve of (12.49) in the case when $\mu_1 = -1.0$. This curve indicates a variant of the limit cycle in Fig. 12.16.

In this figure, the coordinates on the vertical axis become almost the same after about 500 iterations, which indicate that the limit cycle is on a plane after 500 iterations.

Figure 12.18 shows the solution curve of (12.49) in the case when $\mu_1 = -0.7$. This curve still indicates a limit cycle. The coordinates on the vertical axis are almost the same after about 500 iterations. This means that the limit cycle shown above is on a plane after a few hundred iterations.

Figure 12.19 shows the solution curve of (12.49) in the case when $\mu_1 = -0.6$, which is essentially equivalent to Fig. 11a of Harlim and Langford ([30], p. 2566).

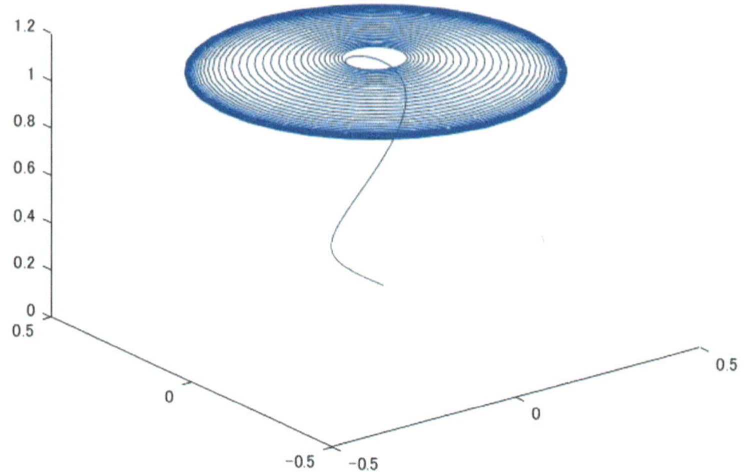

Fig. 12.16 The solution curve corresponding to (12.49) in the case when $\mu_1 = -1.1$

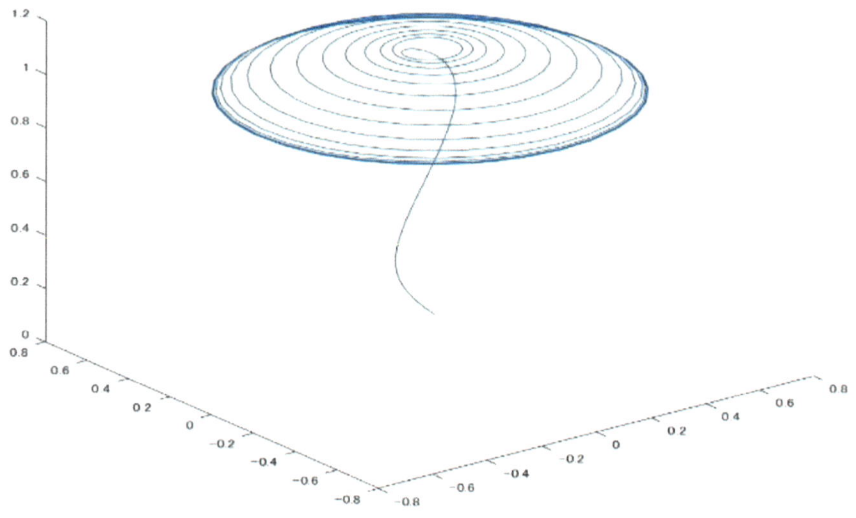

Fig. 12.17 The solution curve corresponding to (12.49) in the case when $\mu_1 = -1.0$

This curve still indicates a limit cycle. In this case, the coordinates on the vertical axis are almost the same after about 3000 iterations. This may be one reason why the variations of the solution curve in the vertical direction are greater than those in Figs. 12.17 and 12.18.

Figure 12.20 shows the solution curve of (12.49) in the case when $\mu_1 = -0.5$, which is essentially equivalent to Fig. 11b of Harlim and Langford [30].

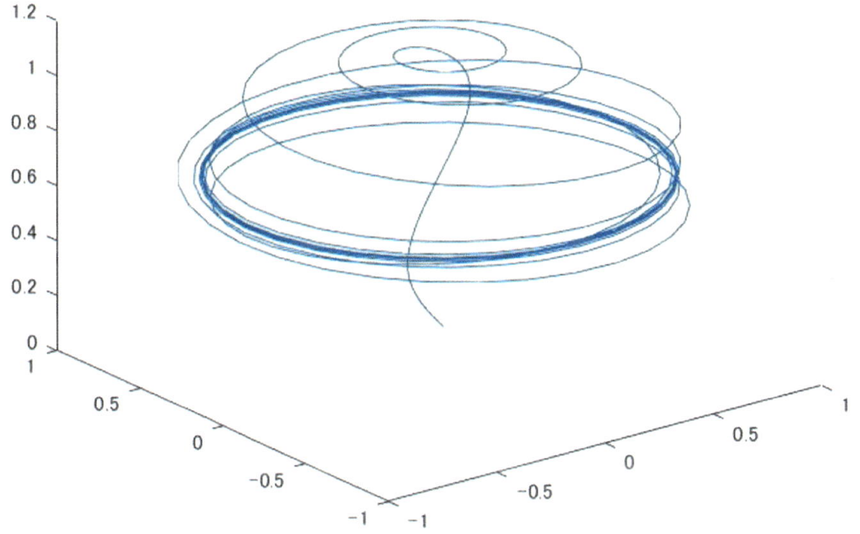

Fig. 12.18 The solution curve corresponding to (12.49) in the case when $\mu_1 = -0.70$

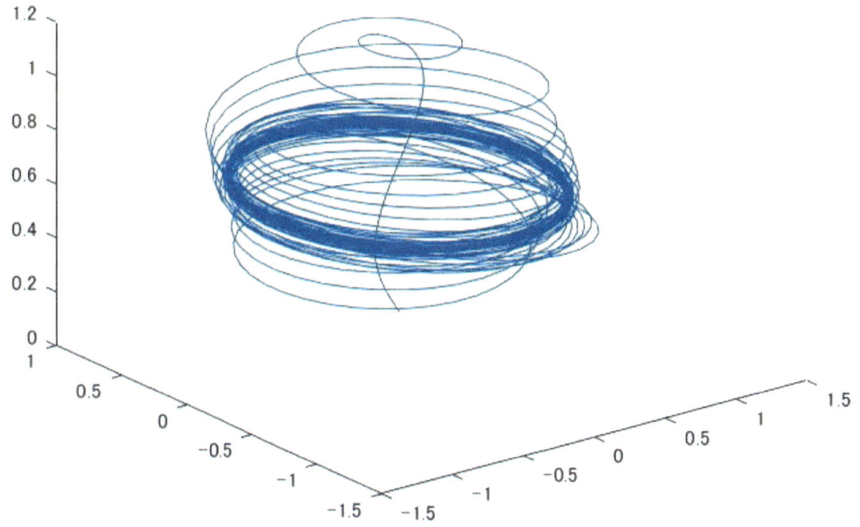

Fig. 12.19 The solution curve corresponding to (12.49) in the case when $\mu_1 = -0.60$

This curve indicates an invariant torus, as Harlim and Langford (op. cit., p. 2566) point out.

Figure 12.21 shows the solution curve of (12.49) in the case when $\mu_1 = -0.4$, which is a similar curve to that in Fig. 11b of Harlim and Langford ([30], p. 2568). They call it a *bursting oscillation*.

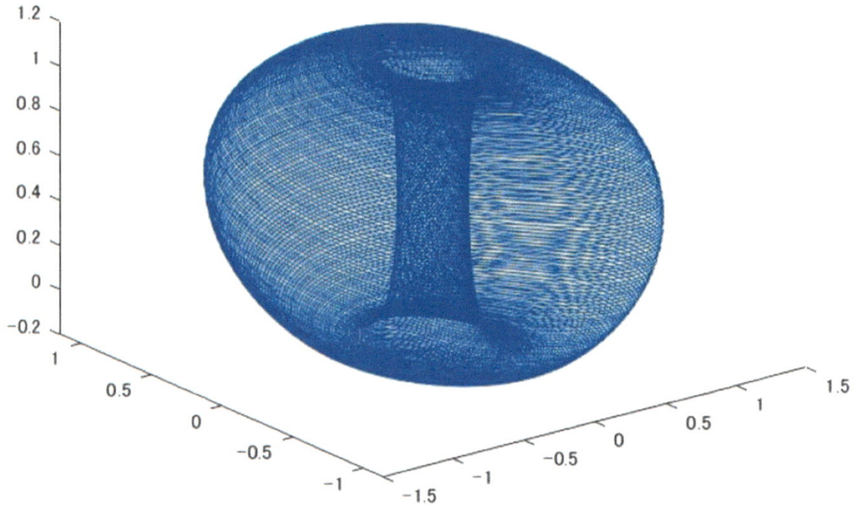

Fig. 12.20 The solution curve corresponding to (12.49) in the case when $\mu_1 = -0.50$

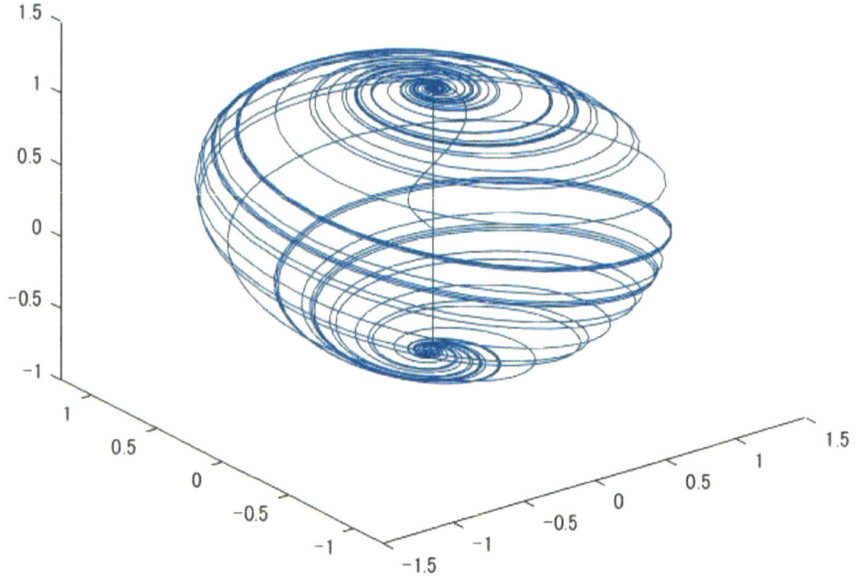

Fig. 12.21 The solution curve corresponding to (12.49) in the case when $\mu_1 = -0.40$

Figure 12.22 shows the graphs of x(t), y(t), and z(t) with respect to time t in (12.49).

Figure 12.23 shows the Lyapunov exponents for the three-dimensional solution curves by Sunday ChaosTimes. Estimated values of these Lyapunov exponents were

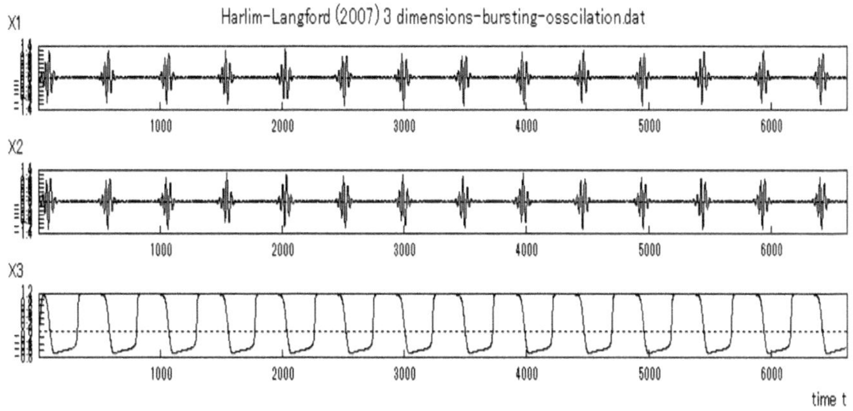

Fig. 12.22 The graphs of ×1, ×2, ×3 corresponding to x(t), y(t), and z(t), respectively, with respect to time t in (12.29)

0.0688, – 0.0227, and – 0.3865. Since the largest Lyapunov exponent of this system is positive, we may conclude that it is a chaos.

As a third route to chaos, we have *intermittency*, which was originated by Pomeau and Manneville [46]. According to Seydel ([51], p. 398), intermittency refers to oscillations that are periodic for certain time intervals, interrupted by bursts of aperiodic oscillations of finite duration. After the burst stops, it is usual that a new periodic phase starts. According to Pomeau and Manneville (op. cit., p.189), a number of investigators have observed in convective fluids an intermittent transition to turbulence. In discussing intermittency, they show an example of intermittency in the Lorenz model which has already been introduced in (12.48). In that study, they kept two of the three parameters of the model, σ, and$b$ equal to the original values, 10.0 and 8/3, respectively. On the other hand, they set the remaining parameter r around 166.06. They found that for r slightly larger than this value, these oscillations are interrupted by a burst.

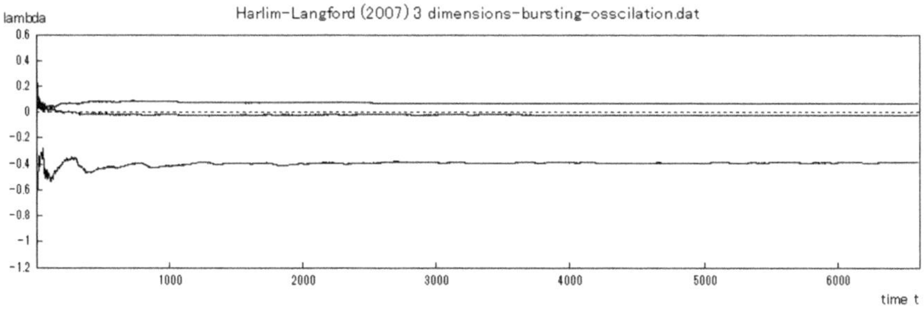

Fig. 12.23 The Lyapunov exponents for the three-dimensional solution curves by Sunday Chaos-Times

Considering their result, we computed three-dimensional solution curves to the Lorenz equation in which σ, r, and$b$ were set equal to 10.0, 166.08, and 8/3, respectively. Figure 12.24 shows this, and Fig. 12.25 shows the solution curve on the yz plane. The solution curve corresponding to the third axis denoted by *X3* in Fig. 12.24 after, say, 3000 iterations, is similar in shape to that obtained by b3 of Fig. 1a and b in Pomeau and Manneville (op. cit., p. 190). The Lyapunov exponents associated with the three-dimensional solution curves shown in Fig. 12.25 were 0.1139, – 0.1384, and – 0.8841. The Largest Lyapunov exponent indicates that this system is a chaos.

Remark Various examples of intermittency are shown in the literature other than Pomeau and Manneville (op. cit.). For example, according to Sprott ([53], pp. 139–140), it occurs to the left of the period-3 window in the logistic map, and in the tangent map $x_{n+1} = \tan x_n$ whose time series is shown in Fig. 6.9 on page 140. Addison [2] shows an example of the Lorenz equation. Manneville [37] reviews various types of intermittency in connection with the Rayleigh-Bénard convection.

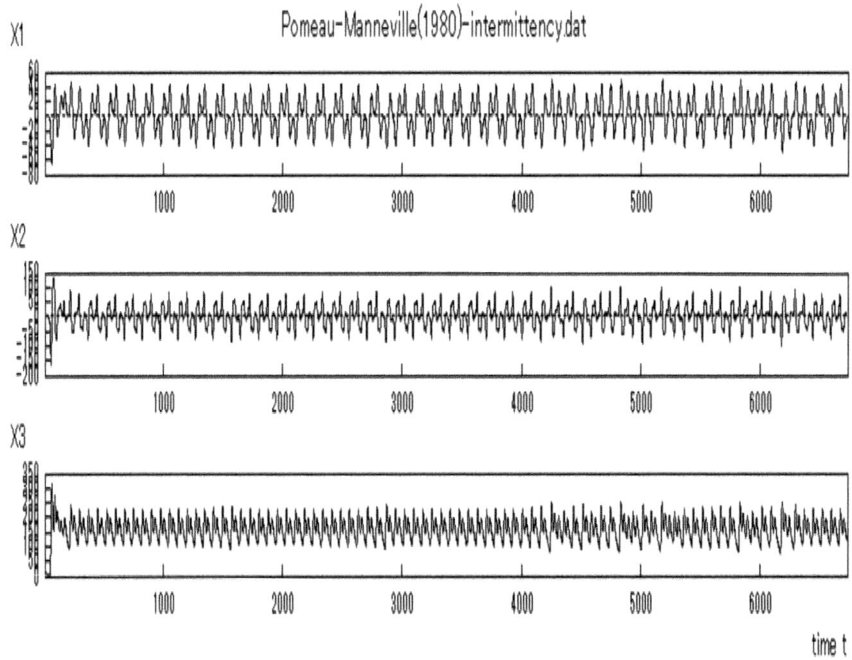

Fig. 12.24 Three-dimensional solution curves to the Lorenz equation in which σ, r, and$b$ were set equal to 10.0, 166.08, and 8/3, respectively

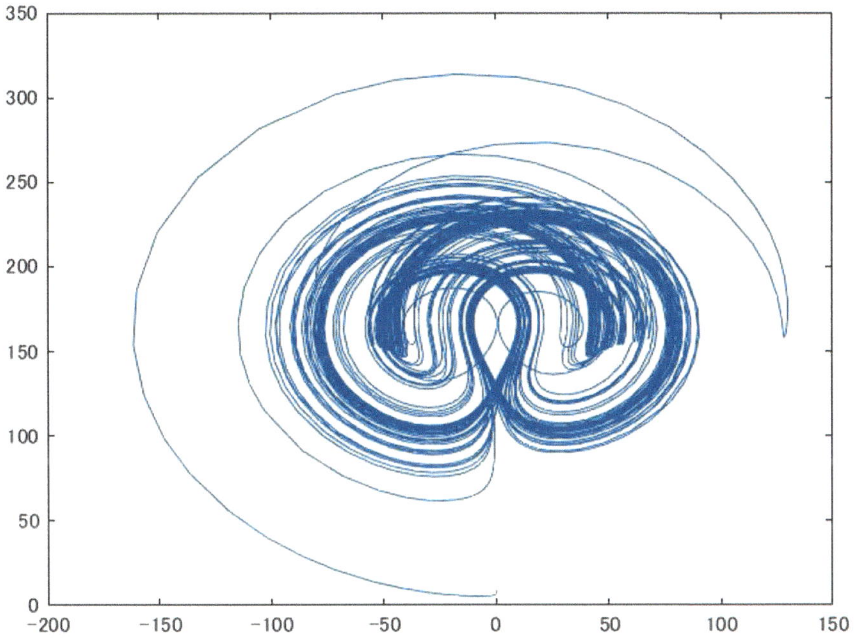

Fig. 12.25 The solution curve on the yz plane of the Lorenz equation

12.4 Basic Concepts in Difference Equation

In this section, we shall introduce basic concepts and theorems in difference equation. In contrast to differential equation discussed in 12.2, *time* is taken to be a discrete variable so that it is allowed to take only integer values. In this section we shall denote it by n. In general, a finite difference equation is expressed by

$$z_n = f(z_{n-1}, z_{n-2}, \ldots, z_{n-k}, n), \tag{12.50}$$

where f is a fixed *complex-valued function* and the integer k is called the order of the equation (e.g., Cull et al. [18]).

As for the above initial value problem, there is an important theorem on the existence and uniqueness of solution in the following.

Theorem 12.12 (Existence and uniqueness theorem) *Every initial value problem for a finite difference equation has a unique solution.*

For the proof of the above theorem, see, for example, Cull et al. (op. cit.).

Then, the initial value problem of (12.50) is the problem of finding a sequence z_n that satisfies (12.50) for a given function f and a fixed initial value

$$z_0 = (z_{k-1}, z_{k-2}, \ldots, z_1, z_0)^t. \tag{12.51}$$

In differential equations we have introduced the notion of the singular point of an autonomous system in Sect. 12.2.2. In contrast, we have the notion of the fixed point of a difference system which is a bit different from that of a differential equation.

Definition 12.23 (*fixed point*) Let a function f be analytic (i.e., differentiable). If a point z_f satisfies $f\left(z_f\right) = z_f$, then it is called a *fixed point* of the function.

Remark The fixed point is sometimes called a stationary point, equilibrium point, critical point, or singularity, in the context of difference equation. Elaydi ([21], p. 8) distinguishes between equilibrium point and fixed point. We shall introduce this distinction soon. Ebeling ([19], pp. 21, 140) uses the terms, fixed point, critical point, and singularity, but distinguishes the fixed point from the other two. However, we shall use the term, fixed point, in this book, unless otherwise stated.

In contrast to the above definition of a fixed point by Carleson and Gamelin ([11], p. 27), Cull et al. (op. cit., p. 218) define the fixed point as a special case of a periodic point as follows.

Definition 12.24 (*periodic point and fixed point*) For a function f defined on a set X, let $f^{(n)}$ denote the nth *iterate* of f,

$$f^{(n)} = \overbrace{f \circ f \circ \ldots f}^{ntimes}, \tag{12.52}$$

and the *orbit* of $x \in X$ is the sequence,

$$x, f(x), f^{(2)}(x), f^{(3)}(x), f^{(4)}(x), \ldots,$$

the sequence formed by starting with the value x and applying f again and again. If there is some $n > 0$ such that $f^{(n)}(x) = x$, the orbit of x is called a *periodic orbit*, x is called a *periodic point*, and its period is the least positive t such that $f^{(t)}(x) = x$. When t equals 1, x is called a *fixed point*.

As regards the periodic point defined above, Elaydi (op. cit., p. 29) defines it in more detail:

Definition 12.25 (*k-periodic and eventually k-periodic*) Let

$$x_{n+1} = f(x_n), \tag{12.53}$$

be a general difference equation, and let $\boldsymbol{p}$ be in the domain of f. Then

(1) $\boldsymbol{p}$ is called a *periodic point* of (12.53) if for some positive integer $k, f^{(k)}(\boldsymbol{p}) = \boldsymbol{p}$. Hence a point is *k-periodic* if it is a fixed point of $f^{(k)}$, that is, if it is an equilibrium point of the difference equation

$$x_{n+1} = g(x_n), \tag{12.54}$$

where $g = f^{(k)}$.

The periodic orbit of $\boldsymbol{p}$, $O(p) = \{p, f(p), f^{(2)}(p), \ldots, f^{(k-1)}(p)\}$, is often called a *k-cycle*.

(2) $\boldsymbol{p}$ is called *eventually k-periodic* if for some positive integer m, $f^m(\boldsymbol{p})$ is a *k-periodic* point. In other words, $\boldsymbol{p}$ is eventually k-periodic if

$$f^{(m+k)}(b) = f^m(b).$$

In the above definition, Elaydi distinguishes between equilibrium point and fixed point as follows:

Definition 12.26 A point $\boldsymbol{x}^*$ in the domain of f is said to be an *equilibrium point* of (12.53) if it is a *fixed point* of f, i.e., $f(\boldsymbol{x}^*) = \boldsymbol{x}^*$.

In Sect. 12.4.1, we shall introduce some of the real difference equations, which are special cases of (12.50), while in Sect. 12.5 we shall introduce some of the complex difference equations which are also special cases of (12.50).

12.4.1 Real Difference Equation and Its Fixed Point

In this section, we shall summarize several important notions and theorems on real difference equation. First, we shall introduce two simple difference equations (e.g., Elaydi [21]), and then introduce a system of linear difference equations:

A *linear homogeneous first-order difference equation* is defined by

$$x_{n+1} = a_n x_n, x_{n_0} = x_0, n \geq n_0 \geq 0, \tag{12.55}$$

and the associated nonhomogeneous difference equation is defined by

$$y_{n+1} = a_n y_n + g_n, y_{n_0} = y_0, n \geq n_0 \geq 0, \tag{12.56}$$

where it is assumed that $a_n \neq 0$, and a_n and g_n are real-valued functions defined for $n \geq n_0 \geq 0$.

Solutions of (12.55) and (12.56) are

$$x_n = \left[\prod_{i=n_0}^{n-1} a_i\right] x_0, \tag{12.57}$$

and

$$y_n = \left[\prod_{i=n_0}^{n-1} a_i\right] y_0 + \sum_{r=n_0}^{n-1} \left[\prod_{i=r+1}^{n-1} a_i\right] g_r, \tag{12.58}$$

respectively.

As another example, we shall introduce *a system of linear first-order difference equations* with initial value $\boldsymbol{x}_{n_0}$ for some $n_0 \geq 0$,

$$x_{n+1} = Ax_n, \quad x_{n_0} = x_0, \tag{12.59}$$

where

$$\boldsymbol{A} = \begin{bmatrix} a_{11} & a_{12} & \cdots & a_{1k} \\ a_{21} & a_{22} & \cdots & a_{11} \\ \vdots & \vdots & \vdots & \vdots \\ a_{k1} & a_{k1} & \cdots & a_{k1} \end{bmatrix} \tag{12.60}$$

Here, $\boldsymbol{A}$ is a $k \times k$ real nonsingular matrix.

The solution of (12.57) is obtained by simple iteration as

$$\boldsymbol{x}_{n,n_0,x_0} = \boldsymbol{A}^{n-n_0}\boldsymbol{x}_0, \tag{12.61}$$

where $\boldsymbol{A}^0 = \boldsymbol{I}$ is the $k \times k$identity matrix. If n_0 equals zero, then the $\boldsymbol{x}_{n,n_0,x_0}$ can be written as x_n. We can compute $\boldsymbol{A}^n$ utilizing the *Putzer algorithm* which is used to compute $e^{\boldsymbol{A}t}$ in differential equations (e.g.,Elaydi [21], pp. 106–112; Perko [43], p. 39), which are summarized below.

Let the *eigenvalues* of $\boldsymbol{A}$ be $\lambda_1, \lambda_2, \ldots, \lambda_k$. Since $\boldsymbol{A}$ is real, eigenvalues are real or complex. It should be remembered that $det(\boldsymbol{A} - \lambda\boldsymbol{I})$ is expressed as the *characteristic polynomial* which has already shown in Definition 11.28 in Chap. 11. Thus, we have the *characteristic equation* of $\boldsymbol{A}$, that is,

$$\lambda^k + a_1\lambda^{k-1} + a_2\lambda^{k-2} + \cdots + a_{k-1}\lambda + a_k = 0. \tag{12.62}$$

As regards the characteristic equation, it is well known that (12.62) holds if we replace λ with $\boldsymbol{A}$. This result is guaranteed by the Cayley-Hamilton theorem.

Theorem 12.13 (Cayley-Hamilton) *Every matrix satisfies its characteristic equation. That is,*

$$p(\boldsymbol{A}) = \prod_{j=1}^{k}(\boldsymbol{A} - \lambda_j\boldsymbol{I}) = 0, \tag{12.63}$$

or

$$\boldsymbol{A}^k + a_1\boldsymbol{A}^{k-1} + a_2\boldsymbol{A}^{k-2} + \cdots + a_{k-1}\boldsymbol{A} + a_k = 0. \tag{12.64}$$

Proof of this theorem is, for example, in Wilkinson ([57], pp. 38–39).

Let us now consider the following representation of $\boldsymbol{A}$:

$$\boldsymbol{A}^n = \sum_{j=1}^{s} u_j(n)\boldsymbol{M}(j-1), \tag{12.65}$$

where the $u_j(n)'$ s are scalar functions to be determined soon, and

$$\boldsymbol{M}(j) = \left(\boldsymbol{A} - \lambda_j I\right)\boldsymbol{M}(j-1), \ldots \boldsymbol{M}(0) = I. \tag{12.66}$$

By iteration, we have

$$\boldsymbol{M}(n) = \prod_{j=1}^{n}\left(\boldsymbol{A} - \lambda_j \boldsymbol{I}\right). \tag{12.67}$$

In this equation, $M(n) = O$ for all $n \geq k$ noticing the Cayley-Hamilton theorem. As a result we have

$$\boldsymbol{A}^n = \sum_{j=1}^{k} u_j(n)\boldsymbol{M}(j-1), \tag{12.68}$$

By examining (12.64) carefully, we finally obtain the following formula,

$$u_1(n) = \lambda_1^n, \ldots u_j(n) = \sum_{i=0}^{n-1} \lambda_j^{n-1-i} u_{j-1}(i), \ldots j = 2, 3, \ldots, k. \tag{12.69}$$

The Putzer algorithm computes the nth power of **A**, that is, $\boldsymbol{A}^n$, using (12.67) and (12.69). Then, we can compute the solution $\boldsymbol{x}_n$ using (12.61). For further details of the derivation of the Putzer algrithm, see Elaydi (op. cit., pp. 106–108).

12.4.2 Stability of Difference Equation

Next, we shall introduce the definitions of stability and semistability of difference equation, and then summarize the main stability results for some of the difference equations according to Elaydi (op. cit.).

Definition 12.27 Let us consider a general autonomous nonlinear difference equation described by (12.53), and its equilibrium point $\boldsymbol{x}^*$.

(a) The point $\boldsymbol{x}^*$ is *stable* if given $\varepsilon > 0$ there exists $\delta > 0$ such that $|\boldsymbol{x}_0 - \boldsymbol{x}^*| < \delta$ implies $\left|f^{(n)}(\boldsymbol{x}_0) - \boldsymbol{x}^*\right| < \delta$ for all $n > 0$. If $\boldsymbol{x}^*$ is not stable, then it is called *unstable*.
(b) The point $\boldsymbol{x}^*$ is attracting if there exists $\delta > 0$ such that $|\boldsymbol{x}_0 - \boldsymbol{x}^*| < \eta$ implies $\lim_{n\to\infty} x_0 = x^*$.

(c) The point $\boldsymbol{x}^*$ is *asymptotically stable* if it is stable and attracting. If $n = \infty$, $\boldsymbol{x}^*$ is said to be *globally asymptotically stable*.

Definition 12.28 Let us consider the difference equation described by (12.53), and its equilibrium point x^*. Then, the point $\boldsymbol{x}^*$ is *semistable* (from the right) if given $\varepsilon > 0$ there exists $\delta > 0$ such that if $x_0 > x^*$, $x_0 - x^* < \delta$, then \user2{x}_{n} - $\boldsymbol{x}^* < \epsilon$. Semistability from the left is defined similarly. If in addition, $\lim_{n\to\infty} \boldsymbol{x}_n = \boldsymbol{x}^*$ whenever $x_0 - x^* < \eta\{x^* - x_0 < \eta\}$, then $\boldsymbol{x}^*$ is said to be *semiasymptotically stable* from the right (or from the left, whatever the case may be).

The followings are main stability results for some of the difference equations.

First, we shall show the stability result for *a system of linear first-order difference equations with constant coefficients*, described by (12.59). Let $\rho(\boldsymbol{A})$ be the spectral radius of $\boldsymbol{A}$, which is defined in Chap. 11.

Theorem 12.14 *The zero solution of (12.59) is stable if and only if* $\rho(\boldsymbol{A}) \leq 1$ *and the eigenvalues of unit modulus are semisimple, while it is asymptotically stable if and only if* $\rho(\boldsymbol{A}) < 1$.

Proof of this theorem is, for example, in Elaydi ([21], p. 169).

Remark Semisimple means that an operator $\boldsymbol{A}$ on a (complex) vector space is diagonalizable (e.g., Hirsch and Smale [34], pp. 63–64).

There is another criterion for the stability of (12.59) in the special case when the order of $\boldsymbol{A}$ is two:

Theorem 12.15 *The zero solution of (12.59) when* $\boldsymbol{A}$ *is* 2×2 *is asymptotically stable if and only if*

$$|tr\,A| < 1 + \det A < 2. \tag{12.70}$$

Brief proof of this theorem is, for example, in Elaydi (op. cit., p. 170).

Example 12.8 In the special case when $k = 2$ in (12.60) and (12.59) becomes a set of difference equations with real constant coefficients:

$$\boldsymbol{x}_{n+1} = \boldsymbol{A}\boldsymbol{x}_n, \quad \boldsymbol{A} = \begin{bmatrix} a_{11} & a_{12} \\ a_{21} & a_{22} \end{bmatrix}. \tag{12.71}$$

Let the eigenvalues of $\boldsymbol{A}$ be $\lambda_1 and \lambda_2$. By a careful examination of the solution curves one can obtain the following patterns.

(1) *asymptotically stable node* when $\lambda_1 < \lambda_2 < 1$.
(2) *unstable node* when $\lambda_1 > \lambda_2$.
(3) *saddle* (unstable) when $0 < \lambda_1 \langle 1 and \lambda_2 \rangle 1$.
(4) *asymptotically stable node* when $0 < \lambda_1 = \lambda_2 < 1$..
(5) *degenerate node* when $\lambda_1 = 1$ and $\lambda_2 < \lambda_1$.
(6) *asymptotically stable node* when $\lambda_1 = \lambda_2 < 1$.

(7) *degenerate case* (unstable) when $\lambda_1 = \lambda_2 = 1$.
(8) *asymptotically stable focus* when $|\lambda| < 1$.

For further details, see, for example, Elaydi (op. cit., pp. 174–180).

Second, we shall show the stability result for a linear nonautonomous system described by

$$\boldsymbol{x}_{n+1} = \boldsymbol{A}_n \boldsymbol{x}_n, \ .\boldsymbol{n} \geq n_0 \geq 0, \tag{12.72}$$

where it is assumed that $\boldsymbol{A}_n$ is nonsingular for all $n \geq n_0$.

Before proceeding, let us define a *fundamental matrix* for (12.72) (e.g., Elaydi, op. cit., p. 113). Let $\boldsymbol{x}_{n1}, \boldsymbol{x}_{n2}, \ldots, \boldsymbol{x}_{nk}$ be the solutions of (12.72), and let $\boldsymbol{\Phi}_n$ be a $k \times k$ matrix whose columns are these solution vectors, that is,

$$\boldsymbol{\Phi}_n = [x_{n1}, x_{n2}, \ldots, x_{nk}]. \tag{12.73}$$

Then, according to Elaydi (op. cit., pp. 113–114), we have

$$\boldsymbol{\Phi}_{n+1} = \boldsymbol{A}_n \boldsymbol{\Phi}_n. \tag{12.74}$$

Moreover, if we write

$$\boldsymbol{\Phi}_{n,m} = \boldsymbol{\Phi}_n \boldsymbol{\Phi}_m^{-1}, \tag{12.75}$$

Equation (12.75) is a solution of the matrix difference equation

$$\boldsymbol{\Phi}_{n+1} = \boldsymbol{A}_n \boldsymbol{\Phi}_{n,m}. \tag{12.76}$$

Furthermore, if $\boldsymbol{A}_n = \boldsymbol{A}$, we have

$$\boldsymbol{\Phi}_{n,m} = \boldsymbol{\Phi}_{n-m} = \boldsymbol{A}^{n-m}. \tag{12.77}$$

Definition 12.29 If $\boldsymbol{\Phi}_n$ is a matrix that is nonsingular for all $n \geq n_0$ and satisfies (12.74), then it is said to be a *fundamental matrix* for system Eq. (12.72).

Theorem 12.16 *Consider system (12.72). Then its zero solution is*

(1) *stable if and only if there exists a positive constant M such that*

$$\Phi_n \leq M \ for \ n \geq n_0 \geq 0; \tag{12.78}$$

(2) *uniformly stable if and only if there exists a positive constant M such that*

$$\Phi_{n,m} \leq M \ for \ n_0 \leq m \leq n < \infty; \tag{12.79}$$

(3) *asymptotically stable if and only if*

$$\lim_{n\to\infty} \boldsymbol{\Phi}_n = 0,$$

(4) *uniformly asymptotically stable if and only if there exist positive constants M and* $\eta \in (0, 1)$ *such that*

$$\Phi_{n,m} \le M\eta^{n-m} for n_0 \le m \le n < \infty. \tag{12.80}$$

Proof of this theorem is, for example, in Elaydi (op. cit., pp. 117–168).

Third, we shall show the stability result for a periodic system

$$\boldsymbol{x}_{n+1} = \boldsymbol{A}_n \boldsymbol{x}_n, \quad \boldsymbol{A}_{n+N} = \boldsymbol{A}_n. \tag{12.81}$$

Before proceeding, we must define a *periodic matrix* and an exponent called *Floquet exponent*.

Theorem 12.17 *For every fundamental matrix* Φ_n *of the system (12.81), there exists a nonsingular periodic matrix* $\boldsymbol{P}_n$ *of period N such that*

$$\boldsymbol{\Phi}_n = \boldsymbol{P}_n \boldsymbol{B}^n, \tag{12.82}$$

Proof of this theorem is, for example, in Elaydi (op. cit., 137).

Remark If z_n is a solution of the system

$$z_{n+1} = \boldsymbol{B} z_n, \tag{12.83}$$

we have

$$\boldsymbol{x}_n = \boldsymbol{P}_n z_n. \tag{12.84}$$

For further details, see, for example, in Elaydi (op. cit., 138).

The *eigenvalues* λ of $\boldsymbol{B}$ in (12.83) are called the *Floquet exponents* of (12.81).

Theorem 12.18 *The zero solution of (12.81) is*

(1) *stable if and only if the Floquet exponents have modulus less than or equal to 1; those of modulus of 1 are semisimple.*
(2) *asymptotically stable if and only if all the Froquet exponents lie inside the unit disk.*

Proof of this theorem is, for example, in Elaydi (op. cit., p. 170).

Fourth, we shall show the stability result for a general nonlinear difference equation described by (12.53).

Theorem 12.19 *Let* $\boldsymbol{x}^*$ *be an equilibrium point of (12.53), and f is continuously differentiable at* $\boldsymbol{x}^*$. *The following statements then hold true*:

(1) *If* $\left|f'(x^*)\right| < 1$, *then* $\boldsymbol{x}^*$ *is an asymptotically stable point.*
(2) *If* $\left|f'(x^*)\right| > 1$, *then* $\boldsymbol{x}^*$ *is unstable.*

Part of the proof of this theorem is, for example, in Elaydi (op. cit., pp. 22–23).

What happens in the special case when $f'(\boldsymbol{x}^*) = 1$? We have the following theorem in this case.

Theorem 12.20 *Let* $f'(\boldsymbol{x}^*)$ *be 1 in (12.53). Then, we have the following results*:

(1) *If* $f''(x^*) \neq 0$, *then the equilibrium point* $\boldsymbol{x}^*$ *is unstable.*
(2) *If* $f''(\boldsymbol{x}^*) = 0$ *and* $f'''(x^*) > 0$, *then* $\boldsymbol{x}^*$ *is unstable.*
(3) *If* $f''(\boldsymbol{x}^*) = 0$ *and* $f'''(x^*) < 0$, *then* $\boldsymbol{x}^*$ *is asymptotically stable.*

Proof of this theorem is, for example, in Elaydi (op. cit., p. 24).

Finally, what happens in the special case when $f'(\boldsymbol{x}^*) = -1$? We have the following theorem in this case.

Theorem 12.21 *Let* $f'(\boldsymbol{x}^*)$ *be* -1 *in (12.53). Then, we have the following results*:

(1) *If* $Sf(\boldsymbol{x}^*) < 0$, *then* $\boldsymbol{x}^*$ *is asymptotically stable.*
(2) *If* $Sf(\boldsymbol{x}^*) > 0$, *then* $\boldsymbol{x}^*$ *is unstable,*

where $Sf(\boldsymbol{x})$ *is the Schwarzian derivative of the function* f:

$$Sf(\boldsymbol{x}) = \frac{f'''(\boldsymbol{x})}{f'(\boldsymbol{x})} - \frac{3}{2}\left[\frac{f''(\boldsymbol{x})}{f'(\boldsymbol{x})}\right]^2, \tag{12.85}$$

Proof of this theorem is, for example, in Elaydi (op. cit., pp. 26–27).

Example 12.9 In the previous section, we have introduced the logistic map described by (12.47), and have drawn a bifurcation diagram for this map. This map shows that its stability changes depending on the bifurcation parameter a in (12.47).

It is easy to compute the equilibrium points of this equation, according to Definition 12.23. Let us define a function,

$$F_a(x) = ax(1-x), \quad x \in [0,1], a > 0. \tag{12.86}$$

To find the equilibrium points (fixed points of F_a) of (12.47) we may solve the equation

$$F_a\left(x^*\right) = x^*. \tag{12.87}$$

As a result, we have two fixed points x^*,one being *0* and the other $(a-1)/a.(a-1)/a$.

Next, we shall compute the first derivative of F_a with respect to x in (12.85). We have

$$F_a'(x) = a(1 - 2x). \tag{12.88}$$

Applying Theorem 12.19 to (12.87), we conclude that,

(1) In the case when $x^* = 0$,

  (i) *0* is asymptotically stable if $0 < a < 1$ but unstable if $a > 1$.
  (ii) 0 is asymptotically stable if $a = 1$.

In this case, we may conclude from Definition 12.28 that 0 is unstable if we consider negative as well as positive initial points in the neighborhood of 0. However, since negative initial points are not in the domain of F_a defined by (12.86), we may discard them and consider only positive initial points. From Definition 12.28, 0 is semiasymptotically stable from the right, i.e., $x^* = 0$ is asymptotically stable in the domain [0, 1] (Elaydi, op. cit., p. 36).

(2) In the case when $x^* = (a - 1)/a$, $F_a'((a - 1)/a) = 2 - a$,

  (i) x^* is *asymptotically stable* if $1 < a \leq 3$ but *unstable* if $a > 3$.
  (ii) x^* is an unstable fixed point for $a > 3$.

For further details of the stability property of the Logistic map including 2-cycles and 2^2-cycles, see, for example, Elaydi (op. cit., pp. 36–40) and Zhang ([58], pp. 150–154).

Finally, we shall define the *Jacobian matrix* and the *Jacobian* of nonlinear difference system models. As regards the Jacobian matrix of a two-dimensional *real differential equation*, we have introduced this matrix in (8.4) in Chap. 8. The Jacobian matrix defined here is an extension of this matrix to a set of *complex difference equations*. First, we shall define a Jacobian matrix, according to Ebeling [20, p. 57] in the following.

Definition 12.30 (*Jacobian matrix*) Let $U \subset C^n$ be open. A map $f = (f_1, \ldots, f_m) : U \to C^m$ is called holomorphic if all the component functions f_μ are holomorphic in U.

Let $U \subset C^n$ be open, $f = (f_1, \ldots, f_m) : U \to C^m$ a holomorphic map. Then, the matrix

$$J_{f(w)} = \left(\frac{\vartheta f_\mu}{\vartheta z_j}\right)_{j=1,\ldots,n}^{\mu=1,\ldots,m}, \tag{12.89}$$

is called the holomorphic functional matrix or the Jacobian matrix of f at $w \in U$.

If this matrix is *square*, that is, $m = n$, *then* the determinant of this matrix is called *Jacobian*. Sometimes the Jacobian matrix and the Jacobian are referred to as the Jacobian. In real differential equation, we have already referred to the Jacobian matrix $Df(\boldsymbol{x}_s)$ in Theorem 12.9 in relation to the notion of structural stability. In that theorem, the Jacobian matrix plays an important role in *linearizing* the nonlinear differential equation at a fixed point (e.g., Guckenheimer and Holmes [28, pp. 12–13]. The Jacobian matrix defined in (12.89) is clearly an extension of the Jacobian matrix of a set of the real nonlinear differential equations to a set of the complex nonlinear differential equations.

Remark Ebeling calls the Jacobian matrix as *Jacobi matrix*.

12.5 Basic Concepts in Complex Difference Equation

In this section, we shall summarize several important notions and theorems on complex difference equation.

Definition 12.31 (*holomorphic function*) If $V \subset C$is an open set of complex numbers, a function $f : V \to C$ is called *holomorphic* (or "complex analytic") if the first derivative

$$z \Rightarrow f'(z) = \lim_{h \to \infty} (f(z+h) - f(z))/h, \tag{12.90}$$

is defined and continuous as a function from V to C, or equivalently if f has a power series expansion about any point $z_0 \in V$ which converges to f in some neighborhood of z_0. Such a function is *conformal* if the derivative $f'(z)$ never vanishes, and *univalent* if it is conformal and one-to-one.

Remark A complex-valued function is not necessarily holomorphic or complex differentiable. For example, $f(z) = \overline{z}$ (the conjugate of z) is not differentiable at any point z since.

$$(f(z+h) - f(z))/h = \overline{h}/h, \tag{12.91}$$

which equals $+1$ if h is real and -1 if h is purely imaginary.

As was pointed out in Chap. 9, earlier versions of DWD (Dynamical Weighted Digraph), which are nonlinear complex dynamical system models, assumed that its weight functions are not holomorphic (Chino [16]). This means that those earlier versions of DWD does not suit the needs of theoretical analysis. As a result, in later versions we have assumed that those weight functions are constants.

Definition 12.32 (*Riemann surface*) A *Riemann surface S* is a connected complex analytic manifold of complex dimension one.

According to Milnor [39, p. 2], we have the following theorem:

Theorem 12.22 (Uniformization theorem) *Any simply connected Riemann surface is conformally isomorphic either*

(1) *to the plane* C *consisting of all complex numbers* $z = x + iy$,
(2) *to the open disk* $D \subset C$ *consisting of all* z *with* $z^2 = x^2 + y^2 < 1$, *or*
(3) *to the Riemann sphere* $\hat{C}$ *consisting of* C *together with a point at infinity, using* $\zeta = 1/z$ *as local uniformizing parameter in a neighborhood of the point at infinity.*

Proof of this theorem is, for example, in Fisher et al. [25].

In complex dynamical system, there exist important concepts called the *Julia set and* the *Fatou set*. Although there are several definitions of these concepts in the literature, we shall adopt those by Beardon [7]. Before proceeding, it is necessary to define two notions, *extended complex (or Riemann) sphere* and *equicontinuous map* (e.g., Beardon, op. cit.).

Definition 12.33 (*extended complex plane*) The *extended complex* (or *Riemann*) *sphere* is the union,

$$C_\infty = C \cup \{\infty\}, \tag{12.92}$$

Remark There exist different notations for the Riemann sphere in the literature. For example, Beardon (op. cit., p. 27), Blanchard ([10], p. 87), and Milnor ([39], p. 2) denote it by C_∞, $\overline{C}$, and $\hat{C}$, respectively.

Definition 12.34 (*equicontinuous map*) A family of $\mathcal{F}$ of maps of (X, d)into (X_1, d_1) is equicontinuous at x_0 if and only if for every positive ε there exists a positive δ such that for all x in X, and for all f in $\mathcal{F}$,

$$d(x_0, x) < \delta \textit{ implies } d_1(f(x_0), f(x)) < \varepsilon. \tag{12.93}$$

The family $\mathcal{F}$ is equicontinuous on a subset X_0 of X if it is equicontinuous at each point x_0 of X.

Definition 12.35 (*Fatou set and Julia set*) Let R be a non-constant rational function. The *Fatou set* of R (denoted by $F(R)$) is the maximal open subset of C_∞ on which $\{R^n\}$ is equicontinuous, and the *Julia set* of R (denoted by $J(R)$) is its complement in C_∞.

Theorem 12.23 *For any non-constant rational map* R, *and any positive integer* p, $F(R^p) = F(R)$ *and* $J(R^p) = J(R)$.

Proof of this theorem is, for example, in Beardon (op. cit., p. 51).

12.5.1 Fixed Points and Periodic Points of Complex Difference Equation

In this section, we shall introduce important notions on fixed point and periodic points in complex difference equation.

Let S be a compact Riemann surface, let $f : S \to S$ be a non-constant holomorphic mapping, and let f^n be its n-fold iterate. A classical example is the case where S is the Riemann sphere C_∞. In general, any holomorphic map $f : C_\infty \to C_\infty$ on the Riemann sphere can be expressed as a rational function,

$$f(z) = p(z)/q(z), \tag{12.94}$$

of two polynomials. The degree d of $f = p/q$ is equal to the maximum of the degrees of p and q.

Now let us select a starting point z_0 in the complex plane C and apply f repeatedly constructing the points

$$z_0, z_1 = f(z_0), z_2 = f(z_1), \ldots, \tag{12.95}$$

Then, we may write this history as

$$z_n = f^n(z_0), \tag{12.96}$$

where $f^0 = I$, $\boldsymbol{I}$ being the identity matrix. We may also write it as

$$z_{n+1} = f(z_n). \tag{12.97}$$

Definition 12.36 **A** point ζ is a fixed point of f if f $(\zeta) = \zeta.(\zeta) = \zeta$.

We have the following classification of a fixed point (e.g., Beardon [7]):

Definition 12.37 Let ζ in C is a fixed point of an analytic function f. Then ζ is:

(1) *superattracting* if $f'(\zeta) = 0$,
(2) *attracting* if $0 < f'(\zeta) < 1$,
(3) *repelling* if $f'(\zeta) > 1$,
(4) *rationally indifferent* if $f'(\zeta)$ is a root of unity,
(5) *irrationally indifferent* if $f'(\zeta) = 1$, but $f'(\zeta)$ is not a root of unity.

Remark 1 $F'(\zeta)$ in the above definition is called the *multiplier* of f at ζ, which is the derivative of f with respect to ζ. Hereafter, we shall denote it by λ.

It should be noticed in the above definition that a superattracting fixed point is a critical point of f, while an attracting fixed point is not. Therefore, f has a local inverse near ζ if ζ is attracting, but not if it is superattracting (e.g., Beardon, op. cit., p. 99).

Remark 2 For example, Peitgen and Richter ([41], p. 27) and Blanchard [10] call the multilier λ the *eigenvalue* of ζ. Carleson and Gamelin ([11], p. 28) call "indifferent" in Definition 12.36 *neutral*.

The notions of periodic point and multiplier are natural extensions of those of fixed point and multiplier, respectively (e.g., Beardon, op. cit.; Milnor [39]).

Definition 12.38 A point is a *periodic point* (*cycle*) of a rational function f if it is a fixed point of some iterate f^m. For such a point ζ there is a positive integer n for which

$$\zeta, f(\zeta), f^2(\zeta), \ldots, f^{n-1}(\zeta), \tag{12.98}$$

are distinct but $f^n(\zeta) = \zeta$; the finite set of points in (12.98) is the *cycle* of ζ and the integer n is the period of ζ, which is sometimes called an n-cycle.

Definition 12.39 The *multiplier of the cycle* is defined as

$$\lambda = \prod_{k=0}^{n-1} f'(\zeta_k). \tag{12.99}$$

We have the following classification of a periodic point (e.g., Milnor, op. cit., p. 43):

Definition 12.40 A periodic point is either,attracting, repelling, or indifferent (=neutral) according as its multiplier satisfies $|\lambda| < 1$, $|\lambda| > 1$, or $|\lambda| = 1$.

The orbit is called

(1) *superattracting* if $\lambda = 0$,
(2) *geometrically attracting* if $0 < |\lambda| < 1$.

Blanchard ([10], p. 93) has a more detailed definition of the periodic orbit, which he denote by $O^+(z_0)$ where z_0 is a periodic point.

Definition 12.41 A periodic orbit $O^+(z_0)$ is:

(1) *superattracting* if $\lambda = 0$,
(2) *attracting* if $0 < |\lambda| < 1$,
(3) *repelling* if $|\lambda| > 1$, or
(4) *neutral* if $|\lambda| = 1$.

Example 12.10 Figure 12.26 shows an example of a Julia set, as in Blanchard (op. cit., p. 91, Fig. 2.1). This set is sometimes called *Douady's Rabbit*. This figure is the Julia set of the map $z = P(z) = z^2 + c$, where $c \approx -0.12256117 + 0.74486177$. Therefore, it is contained in the *Mandelbrot set* to be introduced soon. To be precise, Fig. 12.26 is the case in which $c \approx -0.1200138225 + 0.7400145i$, which was drawn by MATLAB.

Let us here examine some of the fixed points and the periodic points of this map. First, the fixed points of this map are the solutions of $P(z) = z^2 + c = z$, where $C \approx -0.12256117 + 0.74486177$. For simplicity, let $c = -0.12 + 0.74i$.

Then, the two roots of this equation are about $1.2737 - 0.4782i$ and $-0.2737 + 0.4782i$. Here it should be noted that $P'(z) = 2z$.Then, $|\lambda| = |P'(1.2737 - 0.4782i)| = |2.54747 - 0.9564i| > 1$, and the first fixed point is

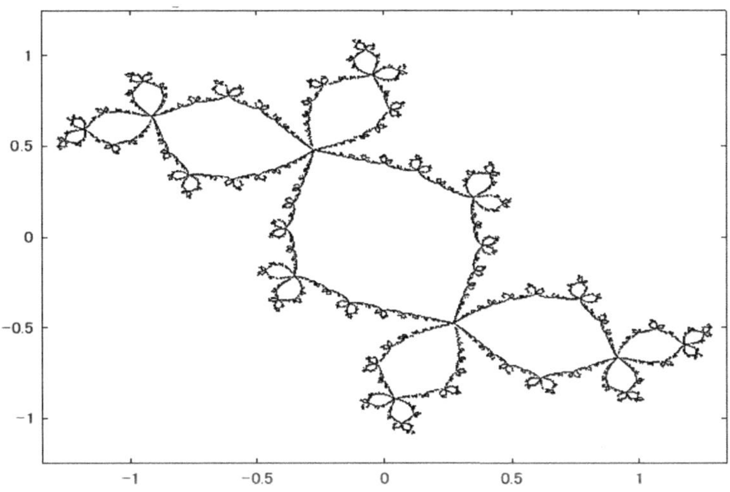

Fig. 12.26 The Julia set of the map $z = P(z) = z^2 + c$, where $C \approx -0.12256117 + 0.74486177$

repelling according to Definition 12.36. Similarly, $|\lambda| = \left|P'(-0.2737 + 0.4782i)\right| = |-0.5474 + 0.9564i| > 1$, and the second fixed point is also *repelling*. On the other hand, the two-cycles of this map are the solutions of $P^2(z) = \left(z^2 + c\right)^2 + c = z$, where $c = -0.12 + 0.74i$. That is, the 2-cycles are the solutions of the following polynomials of order 4:

$$z^4 + 2cz^2 - z + c(c + 1) = 0. \tag{12.100}$$

We can compute the four roots of this polynomial by the MATLAB function called *roots(x)*. The MATLAB code for computing these roots is shown in the Appendix. As a result, we have the four roots, two of them being the same values as those of the quadratic function $P(z)$ and the remaining two being about $-0.0866 - 0.8949i$ and $-0.9134 + 0.8949i$. From the values of the multiplier corresponding the remaining two, we can conclude that they constitute a repelling 2-cycle. Similarly we can compute the eight roots of the polynomial. As a result we have the eight roots, four of them being the same values as before and the three of the remaining 6 roots being a repelling 3-cycle whose values are about 0.9911-0.9087i, $0.0367 - 1.0612i$, and $-1.2449 + 0.6622i$, and the remaining three being an attracting 3-cycle whose values are about $-0.6531+0.5625i, -0.1199+0.7399i$, and $-0.0099+0.0053i. -0.0099 + 0.0053i$.

Example 12.11 Figure 12.27 shows two 3-cycles and a repelling 2-cycle on the Julia set shown in Fig. 12.26. The purple orbit and yellow orbit show these 3-cycles and a blue orbit shows a repelling 2-cycle. We conclude from Lemma 12.2 which will be introduced later that the repelling periodic points are on the Julia set, while the attracting periodic points are not.

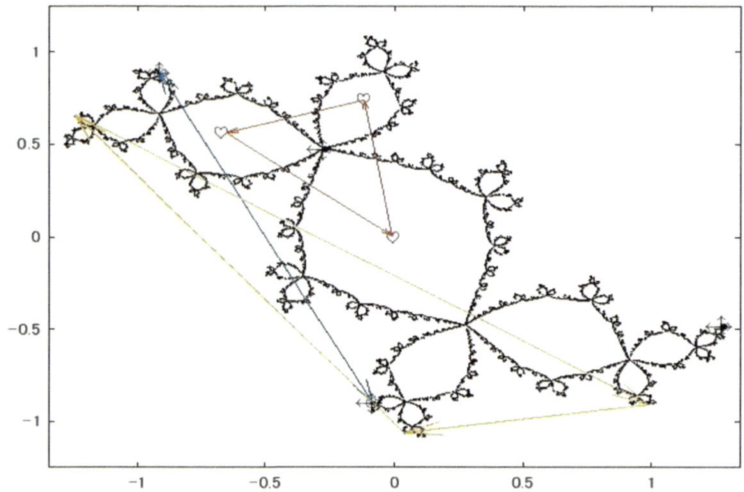

Fig. 12.27 Example 12.10 in which two threefoldperiodic points and a repelling periodic point are added

12.5.2 *Mandelbrot Set, Julia Set, and Fatou Set*

So far, we have introduced the complex dynamical systems whose maps are a rational function described by (12.94), and a special quadratic polynomial function

$$P_c(z) = z^2 + c. \tag{12.101}$$

Of course, this function is a special case of a more general polynomial function of order k, that is,

$$P(z) = z^k + c_1 z^{k-1} + \cdots + c_{k-1} z + c_k. \tag{12.102}$$

Moreover, Eq. (12.99) is also a special case of a general polynomial function of order 2, that is,

$$R(z) = az^2 + 2bz + d. \tag{12.103}$$

However, if we choose c as $c = ad + b - b^2$ and if we suppose that $M(z) = az + b$, we have

$$M^{-1} \circ P_c \circ M(z) = R(z). \tag{12.104}$$

This means that we need only study the class of quadratic polynomials of the form (12.101) to understand the dynamics of all quadratic polynomials of order 2 as, for example, Blanchard ([10], p. 89) points out.

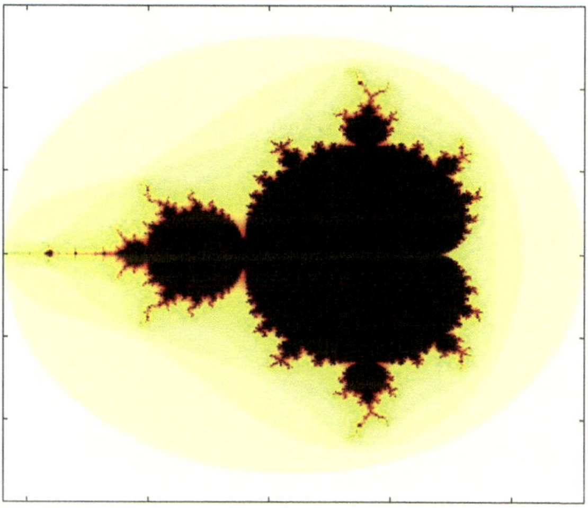

Fig. 12.28 The Mandelbrot set $\mathcal{M}$

Let P_c^n be the nth iterate of the map described by (12.101). Then, we have

Theorem 12.24 (the Mandelbrot set) *The Mandelbrot set* $\mathcal{M}$ *is a closed simply connected subset of the disk* $\{|c| < 2\}$, *which meets the real axis in the interval* $[-2, 1/4]$. *Moreover,* $\mathcal{M}$ *consists of precisely those* c *such that* $\left|P_c^n\right| \leq 2$ *for all* $n \geq 1$.

Proof of the above theorem is, for example, in Carleson and Gamelin ([11], p. 124).

Figure 12.28 shows the Mandelbrot set. The c-value of Fig. 12.26 in Example 12.10 before is located at the center of the large bulb at the top of its main part of $\mathcal{M}$(Peitgen and Richter [41], p. 12).

Example 12.12 Figure 12.29 is another example of the Julia set, which is sometimes called "cauliflower" (e.g., Milnor [39], p. 112). Although Milnor as well as Blanchard ([10], p. 101) define this map as $P(z) = z^2 + z$, it is equivalent to the map $P_c(z) = z^2 + c$, where $c = 0.25$, if we remember the relation between $R(z)$ in (12.103) and $P_c(z)$ in (12.101) discussed previously. It is interesting to note that the cauliflower includes a *rationally indifferent (or neutral) fixed point* which is defined in Definition 12.37. The rationally indifferent (or neutral) point is sometimes called the *parabolic point*.

Example 12.13 Figure 12.30 shows an orbit after 54 iterations, whose initial point is almost on the Julia set in Fig. 12.28, that is, $0.700607338144023705 - 0.1i$ with the parabolic fixed point which is surrounded by the three tiny arrows to mark its place. We define the initial point which is almost on the Julia set as the one just before the trajectory falls inside the Julia set visually. It was computed by MATLAB. One can see that the orbit moves irregularly on the Julia set.

Example 12.14 Figure 12.31 shows the orbits starting from the four initial points, three being located inside the Julia set and one being located a bit to the right of

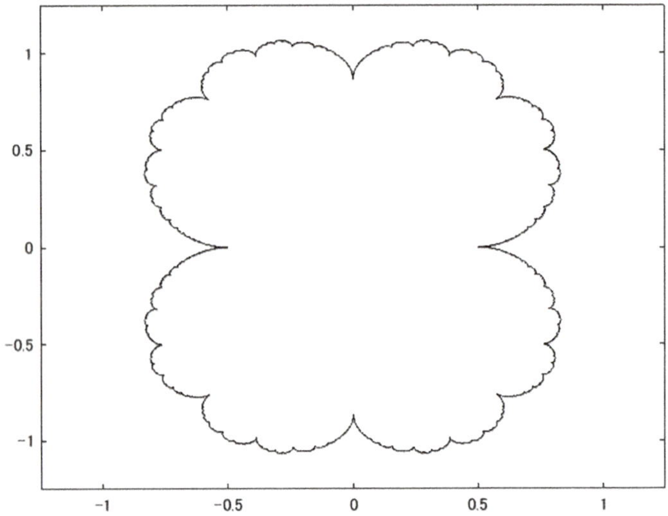

Fig. 12.29 A Julia set called "cauliflower"

Fig. 12.30 An orbit whose initial point is almost on the Julia set in Fig. 12.28

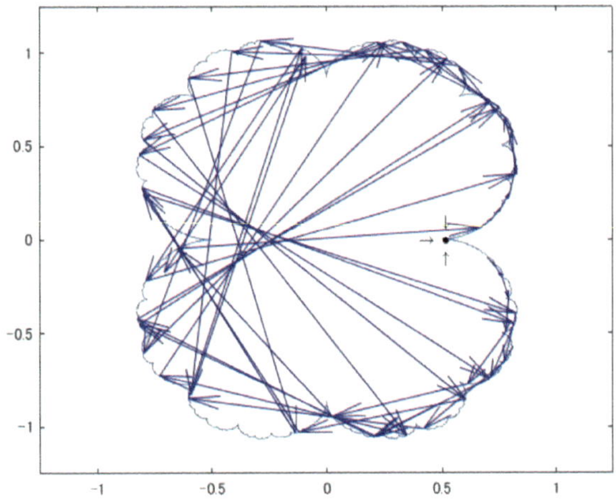

the parabolic fixed point on the Julia set. It is apparent that all the orbits inside the Julia set converge to the parabolic fixed point as time proceeds. To be precise, these orbits are tangent to thereal axis. On the other hand, the initial point which is located a bit to the right of the parabolic fixed point diverges to infinity on the positive real axis. This is the very feature of the parabolic fixed point which is well known (e.g., Carleson and Gamelin [11], p. 37).

Example 12.15 Figure 12.32 shows the nearby orbits which go in or out of the parabolic fixed point $z_0^{(5)}$ in Fig. 12.30. These orbits exhibit the features of the parabolic attracting region at 0 in Fig. 1 of Carleson and Gamelin (op. cit., p. 37).

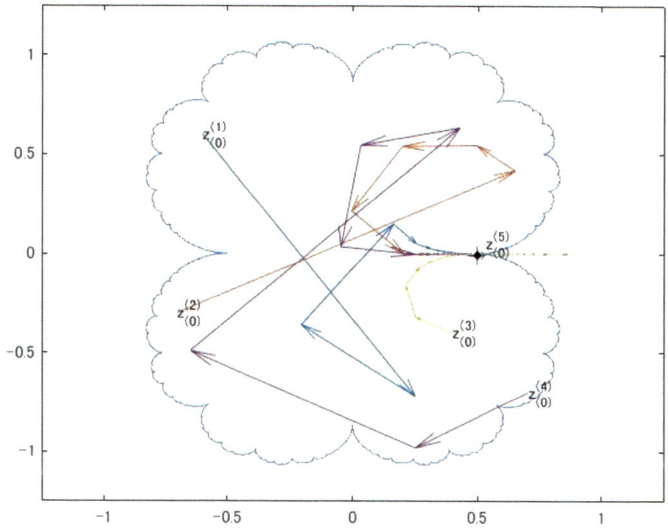

Fig. 12.31 Three orbits within the Julia set

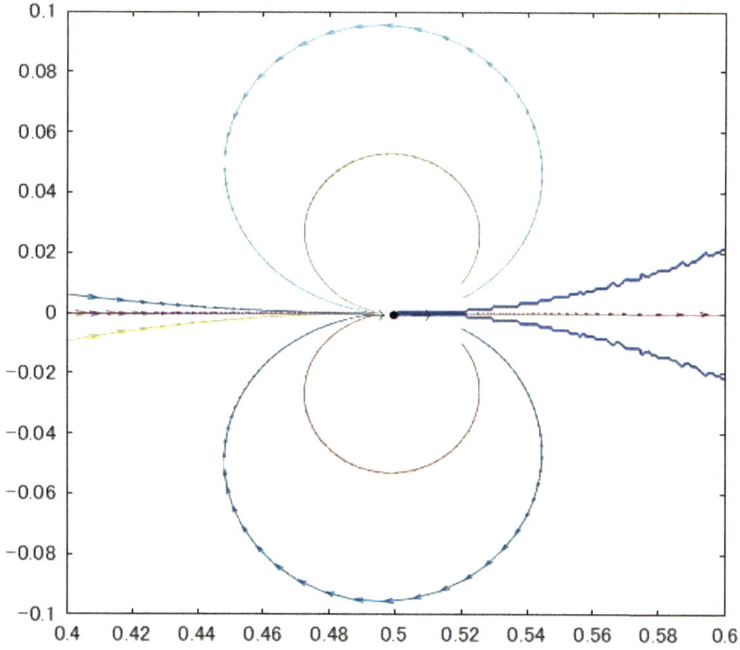

Fig. 12.32 Nearby orbits which go in or out of the parabolic fixed point in Fig. 12.30

There are several lemmas concerning the k-fold iterate and a cycle which are related to the Julia set and Fatou set (e.g., Blanchard, op. cit.; Carleson and Gamelin, op. cit.; Milnor [39]). Further details, see, for example, these papers and textbooks.

Lemma 12.1 For any $k > 0$, the Julia set $J(f^k)$ of the k-fold iterate coincides with the Julia set $J(f)$.

The outline of the proof of the above lemma is, for example, in Milnor (op. cit., p. 41).

Definition 12.42 (*basin of attraction*) If $\mathcal{O}$ is an attracting periodic orbit (cycle) of period m, the *basin of attraction* is an open set $\mathcal{A} \subset S$ consisting of all points $z \in S$ for which the successive iterates $f^m(z)$, $f^{2m}(z)$,... converge towards some point of $\mathcal{O}$.

Lemma 12.2 (basins and repelling points) *Every attracting periodic orbit is contained in the Fatou set of f. The entire basin of attraction $\mathcal{A}$ for an attracting periodic orbit is contained in the Fatou set, while every repelling periodic orbit is contained in the Julia set.*

The proof of the above lemma is, for example, in Milnor (op. cit., p. 43).

Definition 12.43 (*parabolic periodic point*) A periodic point $z_0 = f^n(z_0)$ is called *parabolic* if the multiplier λ at z_0 is equal to $+1$, yet f^n is not the identity map, or more generally if λ is a root of unity, yet no iterate of f is the identity map.

Lemma 12.3 (parabolic points) *Every parabolic periodic point belongs to the Julia set.*

Proof of the above lemma is, for example, in Milnor (op. cit., p. 44).

Lemma 12.4 (J is not empty) *If f is rational of degree two or more, then the Julia set $J(f)$ is non-vacuous.*

Proof of the above lemma is, for example, in Milnor (op. cit., p. 44).

Theorem 12.25 *Suppose that U is a periodic component of the Fatou set $\mathcal{F}$. Then exactly one of the following holds:*

(1) *U contains an attracting periodic point,*
(2) *U is parabolic,*
(3) *U is a Siegel disk,*
(4) *U is a Herman ring.*

Proof of this theorem is, for example, in Carleson and Gamelin ([11], pp. 74–79). For the details of Siegel disk and Herman ring, we shall, for example, refer to Milnor (op. cit.).

So far, we have introduced rational functions (rational maps) and polynomial functions (polynomial maps) in examining dynamical properties of complex difference equations. In general, these functions belong to a family of functions called the *entire function* which is analytic at each point in the "entire" finite plane (e.g., Ablowitz and Fokas [1]). Specifically, an entire function that is not a polynomial is called *transcendental* (e.g., Gelbaum and Olmsted [26]). Although there have been a body of literature which deal with the dynamical properties of complex difference equations (e.g., Baker [6]; Eremenko [22]; Eremenko and Lyubich [23]), we shall not introduce them here, since these studies are beyond the scope of this book.

Appendix

In this appendix, we shall show a few examples of the MATLAB codes for depicting the figures shown in this chapter.

Example A.1 (Codes for Fig. 12.14)

The following MATLAB commands are written to draw Fig. 12.14 in Sect. 12.3 which is a bifurcation diagram for the logistic map in the range $3.825 \leq a \leq 3.86$.

```
% Logistics Map
%  Classic chaos example. Plots semi-stable values of
%     x(n+1) = r*x(n)*(1-x(n)) as r increases to 4.
%
%  Originally written by Michael Hanchak, Dayton OH, USA, 2011, and also
%  Hanchak, M. (2023), logistic map, MATLAB Central File Exchange,
%     January 21, 2023.
%  It was revised by N. Chino a bit, on January 22, 2023
%
clear
scale = 10000; % determines the level of rounding
maxpoints = 200; % determines maximum values to plot
N = 3000; % number of "r" values to simulate
% a = 2.0; % starting value of "r"
% a=2.80;
% a=3.40;
a=3.825;
%
% b=3.70; final value of "r"... anything higher diverges.
% b=4.0;
% b=3.86;
b=3.83;
rs = linspace(a,b,N); % vector of "r" values
M = 500; % number of iterations of logistics equation
```

```
    % Loop through the "r" values
    for j = 1:length(rs)

      r=rs(j); % get current "r"
      x=zeros(M,1); % allocate memory
      x(1) = 0.5; % initial condition (can be anything from 0 to 1)

      for i = 2:M, % iterate
      x(i) = r*x(i-1)*(1-x(i-1));
    end
    % only save those unique, semi-stable values
    out{j} = unique(round(scale*x(end-maxpoints:end)));
end
% Rearrange cell array into a large n-by-2 vector for plotting
data = [];
for k = 1:length(rs)
    n = length(out{k});
    data = [data;  rs(k)*ones(n,1),out{k}];
end

          % Plot the data
          figure(97);clf
          h=plot(data(:,1),data(:,2)/scale,'k.');
          set(h,'markersize',1)
          axis tight
          %set(gca,'units','normalized','position',[0 0 1 1])
          %set(gcf,'color','white')
          %axis off
```

Example A.2 (Codes for Fig. 12.21)

The following MATLAB commands are written to draw Fig. 12.21 in Sect. 12.3 which is the solution curve corresponding to (12.49) in the case when $\mu_1 = -0.40$. In this code, BD, CD, AD, DD, and ED correspond, respectively, to μ_1, μ_2, μ_3, ω, and k in (12.49). Out of the five figures depicted by this code, Fig. 5 corresponds to Fig. 12.21. The directory path in the print statements in the following code should be revised according to the user's directory structure of the personal computer used.

```
% program name: ODE_autonomous_Harlim_langford_Springer.m
%
%                        written by N. Chino, January 28, 2023
%
% Case 11 BD=-0.4 bursting oscillation
%
%   Langford equation (Harlim & Langford, W. F. (2007), Fig.11(a)
%
tspn=[0,500]; y0=[0.1;0.1;0.1];

% Case 11
AD=1; BD=-0.4; CD=0.4; DD=3.5; ED=-1.0;
%
ode45(@(t,y) langford2007(t,y,AD,BD,CD,DD,ED),tspn,y0);
[t,y]=ode45(@(t,y) langford2007(t,y,AD,BD,CD,DD,ED),tspn,y0);
pause;
%
figure(2);
plot(t,y(:,1),t,y(:,2),t,y(:,3))
pause;
%
xt=y(:,1);
figure(3);
plot(xt);
ylabel('x1');
xlabel('time t');
%
% Case 11
print -depsc eps-files/Harlim-Langford-Case11-signals-for-x-variable.eps
pause;
%
maxt=size(xt);
maxtm1=maxt-2;
tau=1;
for n=1:tau:maxtm1
   yt(n)=xt(n+tau);
   zt(n)=xt(n+2*tau);
   v=[xt(n),yt(n),zt(n)]; disp(v)
end;
```

```
%
for i=1:maxtm1
xtt(i)=xt(i);
ytt(i)=y(i);
end
%
figure(4);
plot3(xtt,ytt,zt);
xlabel('x-value');
ylabel('y-value');
zlabel('z-value');
%
% Case 11
print -depsc eps-files/chino/Harlim-Langford-Case11-first-var-3dim.eps
pause;
%
figure(5);
plot3(y(:,1),y(:,2),y(:,3))
%
% Case 11
print -depsc eps-files/Harlim-Langford-limit-cycle-Case11.eps
%
yt=y(:,1);
%
% Case 11
save('data/chino/HarlimL-Langford-limit-cycle-Case11.dat', 'yt', '-ASCII');
save('data/chino/Harlim-Langford (2007) 3dimensions-bursting-osscilation.dat',
'y', '-ASCII');
%
[coeff,score,latent]=pca(y);
latent
coeff
```

```
function dy=langford2007(t,y,AD,BD,CD,DD,ED)
dy=zeros(3,1);   % a column vector
dy(1)=(y(3)+BD)*y(1)-DD*y(2);
dy(2)=DD*y(1)+(y(3)+BD)*y(2);
dy(3)=CD+AD*y(3)-y(3)^3+ED*(y(1)^2+y(2)^2);
```

Example A.3 (Codes for Fig. 12.21)

The following MATLAB commands are written to compute the fixed points and the periodic points of the complex map $P_c(z) = z^2 + c$ in (12.101). Moreover, these commands compute not only the multipliers of the fixed points but also those of the periodic points (or cycles) up to 3-cycle. The value of c must be specified by the user. In this commands, this value is set equal to

$$c = -0.390540870218400 - 0.586787907346969i.$$

The user may classify the fixed points as well as periodic points, according to the output values of the multipliers of the fixed points shown in Definition 12.36 as well as those of the multipliers of the periodic points shown in Definition 12.39.

```
%--------------------------------------------------------------------------------
%
% file name: roots_of_quadratic_equations.m
%           version 1.1  written by N. Chino on June 26, 2017
%           version 1.3  revised by N. Chino on August 212.2017.
%
%  This program computes up to the periodic points with cycle 4
% of a special quadratic difference equation, z_{n+1}=z_{n}^2 + c.
%
%--------------------------------------------------------------------------------
%
format long;
% c=-0.12+0.74*i; % Peitgen & Richter, 1986, Fig.4.
%c=-0.1200138225+0.7400145i; % almost the same value as P-R (1986), Fig.4.
%
c=-0.390540870218400-0.586787907346969i; % P-R (1986), XII, Fig.9.
%
% c=-1;        % Milnor, 2000, p89, Fig. 9.
% c=0.25;       % Chino-Shiraiwa, 2014, dyadic relation, Case 1.
% c=0.25-0.00005i; % Chino-Shiraiwa, 2014, dyadic relation, Case 2, 3.
%
disp('1. Input value of c of the quadratic difference equation');
disp(c);
%
```

```
disp('2-1. Fixed points of the quadratic difference equation');
coeff1=[1 -1 c];
f=roots(coeff1);
disp(f);
%
disp('2-2. Multipliers of the fixed points');
mul=2*f;
disp(mul);
%
disp('2-3. Absolute values of the multipliers of the fixed points');
amul=abs(mul);
disp(amul);
%
disp('3-1. Periodic points with cycle 2');
coeff2=[1 0 2*c -1 c*(c+1)];
p2=roots(coeff2);
disp(p2);
%
disp('12.2. Multipliers of the periodic points with cycle 2');
mul=4.*p2.*(p2.*p2+c);
disp(mul);
%
disp('3-3. Absolute values of the multipliers of the 2-cycle');
amul=abs(mul);
disp(amul);
%
disp('4-1. Periodic points with cycle 3');
coeff3=[1 0 4*c 0 2*c*(3*c+1) 0 4*c*c*(c+1) -1 c*(c^3+2*c^2+c+1)];
p3=roots(coeff3);
```

```
disp(p3);
%
disp('4-2. Multipliers of the periodic points with cycle 3');
mul=8.*p3.^7+24.*c.*p3.^5+8.*c.*(3.*c+1).*p3.^3+8.*c.^2.*(c+1).*p3;
disp(mul);
%
disp('4-3. Absolute values of the multipliers of the 3-cycle');
amul=abs(mul);
disp(amul);
%
disp('4-1. Periodic points with cycle 4');
% Concatination is done of the following sentence using ',...'.
coeff4=[1 0 8*c 0 4*c*(7*c+1) 0 8*c^2*(7*c+3)
 0 2*c*(35*c^3+30*c^2+3*c+1)
 0 8*c^2*(7*c^3+10*c^2+3*c+1),...
 0 4*c^2*(7*c^4+15*c^3+9*c^2+4*c+1) 0 8*c^3*(c+1)*(c^3+2*c^2+c+1) -1
 c*(c*(c^3+2*c^2+c+1)^2+1)];
p4=roots(coeff4);
disp(p4);
%
disp('4-2. Multipliers of the periodic points with cycle 4');
mul1=16.*p4.^15+112.*c.*p4.^13+48.*c.*(7.*c+1).*p4.^11;
mul2=80.*c.^2.*(7.*c+3).*p4.^9+16.*c.*(35.*c.^3+30.*c.^2+3.*c+1).*p4.^7;
mul3=48.*c.^2.*(7.*c.^3+10.*c.^2+3.*c+1).*p4.^5;
mul4=16.*c.^2.*(7.*c.^4+15.*c.^3+9.*c.^2+4.*c+1).*p4.^3;
mul5=16.*c.^3.*(c+1).*(c.^3+2.*c.^2+c+1).*p4;
mul=mul1+mul2+mul3+mul4+mul5;
disp(mul);
%
disp('4-3. Absolute values of the multipliers of the 4-cycle');
amul=abs(mul);
disp(amul);
```

References

1. Ablowitz, M. J., & Fokas, A. S. (1997). *Complex variables: Introduction and applications.* Cambridge University Press.
2. Addison, P. S. (2001). *Fractals and chaos—An illustrated course.* Institute of Physics Publishing.
3. Aihara, K. (ed.) (2005). *Kaos Jikeiretsu Kaiseki no Kiso to Ouyo* [*Fundamentals and Applications of Chaotic Time Series Analysis*]. Sangyo-Tosho.
4. Anderson, T. W. (1971). *The statistical analysis of time series.* Wiley.
5. Arnord, V. I. (1983). *Geometrical methods in the theory of ordinary differential equations.* Springer.

6. Baker, I. N. (1984). Wandering domains in the iteration of entire functions. In *Proceedings of the London Mathematical Society* (pp. 563–576).
7. Beardon, A. F. (1991). *Iteration of rational functions—Complex analytic dynamical systems*. Springer.
8. Berry, M. V. (1979). Diffractals. *Journal of Physics A: Mathematical and General, 12*, 781–797.
9. Birkhoff, C., & Rota, G.-C. (1978). *Ordinary differential equations* (3rd ed.). Wiley.
10. Blanchard, P. (1984). Complex analytic dynamics of the Riemann sphere. *Bulletin (New Series) of The American Mathematical Society, 11*, 85–141.
11. Carleson, L., & Gamelin, T. W. (1993). *Complex dynamics*. Springer.
12. Chino, N. (1981). Some trials for a unified geometrical interpretation of several methods for the analysis of asymmetry and related techniques. In *A letter from N. Chino to Dr. T. M. Newcomb*.
13. Chino, N. (1984). Spatio-temporal structure of Newcomb's data. *Bulletin of The Faculty of Humanities of Aichigakuin University, 14*, 84–100. (in Japanese).
14. Chino, N. (1984b). Toward a theory of dynamical system in group dynamics. In *Working paper of the specific researches in education and psychology by the aid of The Ministry of Education in Japan*, Nagoya, Nagoya University.
15. Chino, N. (1988). A bifurcation model of changes in interdependence structure among objects. *Bulletin of The Faculty of Humanities of Aichigakuin University, 17*, 85–109.
16. Chino, N. (2014). A general non-Newtonian n-body problem and dynamical scenarios of solutions. In *Proceedings of the 42th Annual Meeting of the Behaviormtric Society of Japan* (pp. 48–51). Sendai, Japan.
17. Chino, N., & Nakagawa, M. (1990). A bifurcation model of change in group structure. *The Japanese Journal of Experimental Social Psychology, 29*, 25–38.
18. Cull, P., Flahive, M., & Robson, R. (2005). *Difference equations—From Rabbits to Chaos*. Springer.
19. Ebeling, W. (2007). *Functions of several complex variables and their singularities*. American Mathematical Society.
20. Eckmann, J.-P. (1981). Roads to turbulence in dissipative dynamical systems. *Reviews of Modern Physics, 53*(No. 4, Part I), 643–654.
21. Elaydi, S. N. (1999). *An introduction to difference equations* (2nd ed.). Springer.
22. Eremenko, A. E. (1989). On the iteration of entire functions. In *Dynamical systems and Ergodic Theory Branch Center Publications* (Vol. 23, pp. 339–345). Polish Scientific Publishers.
23. Eremenko, A. E., & Lyubich, M. Y. (1992). Dynamical properties of some classes of entire functions. *Annales de l'institut Fourier, 42*, 989–1020.
24. Falconer, K. (2003). *Fractal geometry-mathematical foundations and applications* (2nd ed.). Wiley.
25. Fisher, Y., Hubbard, J. H., & Wittner, B. S. (1988). A proof of the uniformization theorem for arbitrary plane domains. *Proceedings of The American Mathematical Society, 104*, 413–418.
26. Gelbaum, B. R., & Olmsted, J. M. H. (1994). *Counterexamples in analysis*. Dover Publications Inc.
27. Grobman, D. M. (1959). Homeomorphisms of systems of differential equations. *Doklady Akademii Nauk SSSR, 128*, 880.
28. Guckenheimer, J., & Holmes, P. (1983). *Nonlinear oscillations, dynamical systems, and bifurcations of vector fields*. Springer.
29. Hanchak, M. (2011). *Logistics Map*. MATLAB Central File Exchange. https://www.mathworks.com/matlabcentral/fileexchange/12.2424-logistics-map
30. Harlim, J., & Langford, W. F. (2007). The cusp-Hopf bifurcation. *International Journal of Bifurcation and Chaos, 17*, 2547–2570.
31. Hartman, P. (1960). A lemma in the theory of structural stability of differential equations. *Proceedings of the American Mathematical Society, 11*, 610–620.
32. Hartman, P. (1964, 1982). *Ordinary differential equations*. Birkhäuser.
33. Higuchi, T. (1988). Approach to an irregular time series on the basis of the fractal theory. *Physica D: Nonlinear Phenomena, 31*, 277–283.

34. Hirsch, M. W., & Smale, S. (1974). *Differential equations, dynamical systems, and linear algebra*. Academic Press.
35. Lorenz, E. N. (1963). Deterministic nonperiodic flow. *Journal of Atmospheric Science, 20*, 130–141.
36. Mandelbrot, B. B. (1977). *The fractal geometry of nature*. W. H. Freeman and Company.
37. Manneville, P. (2006). Rayleigh-Bénard convection, thirty years of experimental, theoretical, and modeling work. In I. Mutabazi, E. Guyon, & J. E. Wesfreid (Eds.), *Dynamics of spatio-temporal cellular structures—Henri Bénard Centenary review* (pp. 1–24). Springer.
38. May, R. M. (1976). Simple mathematical models with very complicated dynamics. *Nature, 261*, 459–467.
39. Milnor, J. (2000). *Dynamics in one complex variable-introductory lectures* (2nd ed.). Vieweg & Sohn.
40. Palmer, K. J. (1973). A generalization of Hartman's linearization theorem. *Journal of Mathematical Analysis and Applications, 41*, 753–758.
41. Peitgen, H.-O., & Richter, P. H. (1986). *The beauty of fractals-images of complex dynamical systems*. Springer.
42. Peixoto, M. M. (1962). Structural stability on two-dimensional manifolds. *Topology, 1*(12), 1120.
43. Perko, L. (2001). *Differential equations and dynamical systems* (3rd ed.). Springer.
44. Pinto, M., & Robledo, G. (2018). A Grobman-Hartman theorem for a differential equation with piecewise constant arguments of mixed type. *Zeitschrift für Analysis and ihre Anwendungen, 37*(12), 1126.
45. Poincaré, H. (1892). *Les Méthodes Nouvelles de la Méchanique Celeste*. Gauthier Villars.
46. Pomeau, Y., & Manneville, P. (1980). Intermittent transition to turbulence in dissipative dynamical systems. *Communications in Mathematical Physics, 74*, 189–197.
47. R&D Team. (2002). *Sunday Chaos times*. AIHARA Electrical Engineering Co., Ltd.
48. Rosen, R. (1970). *Dynamical system theory in biology*. Wiley-Interscience.
49. Schuster, H. G. (1988). *Deterministic Chaos—An introduction* (2nd ed.). VCH.
50. Seydel, R. (1988). *From Equilibrium to Chaos—Practical bifurcation and stability analysis*. Elsevier Science.
51. Seydel, R. (2010). *Practical bifurcation and stability analysis* (3rd ed.). Springer.
52. Shimada, I., & Nagashima, T. (1979). A numerical approach to ergodic problem of dissipative dynamical system. *Progress of Theoretical Physics, 61*, 1605–1616.
53. Sprott, J. C. (2003). *Chaos and time-series analysis*. Oxford University Press.
54. Stevenson, A. (Ed.). (2010). *Oxford dictionary of English* (3rd ed.). Oxford University Press.
55. Thiel, M., Kurths, J., Moura, A., & Károlyi, G. (Eds.). (2010). *Nonlinear dynamics and chaos: Advances and perspectives*. Springer.
56. Wiggins, S. (1990). *Introduction to applied nonlinear dynamical systems and chaos*. Springer.
57. Wilkinson, J. H. (1965). *The Algebraic eigenvalue problem*. Clarendon Press.
58. Zhang, W.-B. (2006). *Discrete dynamical systems, bifurcations and chaos in economics*. Elsevier.
59. Zhang, W. M., Liu, P., & Lei, X. (2021). On H lder dependence of the parametrized Hartman-Grobman theorem. *Acta Mathematica Sinica, English Series, 38*, 137–147.

Index

N. Chino, *Structure and Dynamics of Asymmetric Interactions*,
https://doi.org/10.1007/978-981-97-8269-7

T

Y

Z

GPSR Compliance

The European Union's (EU) General Product Safety Regulation (GPSR) is a set of rules that requires consumer products to be safe and our obligations to ensure this.

If you have any concerns about our products, you can contact us on ProductSafety@springernature.com

In case Publisher is established outside the EU, the EU authorized representative is:

Springer Nature Customer Service Center GmbH
Europaplatz 3
69115 Heidelberg, Germany

Batch number: 10732076

Printed by Printforce, the Netherlands